Linear Algebra

Linear Algebra

Michael L. O'Leary

Registered Office
John Wiley & Sons, Inc., 111 River Street, Hoboken, NJ 07030, USA

Editorial Office
111 River Street, Hoboken, NJ 07030, USA

For details of our global editorial offices, customer services, and more information about Wiley products visit us at www.wiley.com.

Wiley also publishes its books in a variety of electronic formats and by print-on-demand. Some content that appears in standard print versions of this book may not be available in other formats.

Library of Congress Cataloging-in-Publication Data

Names: O'Leary, Michael L., author.
Title: Linear algebra / Michael L. O'Leary.
Description: Hoboken, NJ : Wiley, 2021. | Includes bibliographical
   references and index.
Identifiers: LCCN 2021008522 (print) | LCCN 2021008523 (ebook) | ISBN
   9781119437444 (hardback) | ISBN 9781119437475 (adobe pdf) | ISBN
   9781119437604 (epub) | ISBN 9781119437437 (obook)
Subjects: LCSH: Algebras, Linear.
Classification: LCC QA184 .O425 2021 (print) | LCC QA184 (ebook) | DDC
   512/.5–dc23
LC record available at https://lccn.loc.gov/2021008522
LC ebook record available at https://lccn.loc.gov/2021008523

Cover design by Wiley
Cover image: © Hank Erdmann/Shutterstock

Set in 10/12pt, STIXTwoText by SPi Global, Chennai, India

For my niece,
Lindsay

Contents

Preface

This book is an introduction to linear algebra. Its goal is to develop the standard first topics of the subject. Although there are many computations in the sections, which is expected, the focus is on proving the results and learning how to do this. For this reason, the book starts with a chapter dedicated to basic logic, set theory, and proof-writing. Although linear algebra has many important applications ranging from electrical circuitry and quantum mechanics to cryptography and computer gaming, these topics will need to wait for another day. The goal here is to master the mathematics so that one is ready for a second course in the subject, either abstract or applied. This may go against current trends in mathematics education, but if any mathematical subject can stand on its own and be learned for its own sake, it is the amazing and beautiful linear algebra.

In addition to the focus on proofs, linear transformations play a central role. For this reason, functions are introduced early, and once the important sets of $\mathbb{R}^n$ are defined in the second chapter, linear transformations are described in the third chapter and motivate the introduction of matrices and their operations. From there, invertible linear transformations and invertible matrices are encountered in the fourth chapter followed by a complete generalization of all previous topics in the fifth with the definition of abstract vector spaces. Geometries are added to the abstractions in the sixth chapter, and the book concludes with nice matrix representations. Therefore, the book's structure is as follows.

Logic and Set Theory Statements and truth tables are introduced. This includes logical equivalence so that the reader becomes familiar with the logic of statements. This is particularly important when dealing with implications and reasoning that involves De Morgan's laws. Sets and their operations follow with an introduction to quantification including how to negate both universal and existential sentences. Proof methods are next, including direct and indirect proof, and these are applied to proofs involving subsets. Mathematical induction is also presented. The chapter closes with an introduction to functions, including the concepts of one-to-one, onto, and binary operation.

Euclidean Space The definition of $\mathbb{R}^n$ is the focus of the second chapter with the main interpretation being that of arrows originating at the origin. Euclidean distance and length are defined, and these are followed by the dot and cross products. Applications include planes and lines, areas and volumes, and the orthogonal projection.

Transformations and Matrices Now that functions have been defined and interesting sets to serve as their domains and codomains have been given, linear transformations are introduced. After some basic properties, it is shown that these functions have nice representations as matrices. The matrix operations come next, their definitions being motivated by the definitions of the function operations. Linear operators on $\mathbb{R}^2$ and $\mathbb{R}^3$ serve as important examples of linear transformations. These include the reflections, rotations, contractions, dilations, and shears. The introduction of the kernel and the range is next. Issues with finding these sets motivate the need for easier techniques. Thus, Gauss–Jordan elimination and Gaussian elimination finally make their appearance.

Invertibility The fourth chapter introduces the idea of an invertible matrix and ties it to the invertible linear operator. The standard procedure of how to find an inverse is given using elementary matrices, and inverses are then used to solve certain systems of linear equations. The determinant with its basic properties is next. How the elementary row operations affect the determinant is explained and carefully proved using mathematical induction. The next section combines the inverse and the determinant, and important results concerning both are proved. The chapter concludes with some mathematical applications including orthogonal matrices, Cramer's Rule, and how the determinant can be used to compute the area or volume of the image of a polygon or a solid under a linear transformation.

Abstract Vectors Now that the concrete work has been done, it is time to generalize. Vector spaces lead the way as the generalization of $\mathbb{R}^n$, and these are quickly followed by linear transformations between these abstract vector spaces. The important topics of subspace, linear dependence and linear independence, and basis and dimension soon follow. The proof that every vector space has a basis is given for the sake of completion, but, other than for the result, the techniques are not pursued very far because this book is, after all, an introduction to the subject. Rank and nullity are defined, both in terms of linear transformations and in terms of matrices. The chapter then concludes with probably the most important topic of the book, isomorphism. Along with isomorphism, coordinates, coordinate maps, and change of basis matrices are presented. The section and chapter concludes with the discovery of the standard matrix of a linear transformation. Although there is more to come, a standing ovation for the standard matrix and its diagram would not be inappropriate.

Inner Product Spaces Although $\mathbb{R}^n$ is usually viewed as Cartesian space, it is technically just a set of $n \times 1$ matrices. Any geometry that it has was given to it in the second chapter, even though its geometry is a copy of the geometry of Cartesian space. A close examination reveals that the geometry of $\mathbb{R}^n$ is based on the dot product. Mimicking this, an abstract vector space is given its geometry with an inner product, which is a function defined so that it has the same basic properties as the dot product. The vector space then becomes an inner product space so that distances, lengths, and angles can be found using objects like matrices, polynomials, and functions. Other topics related to the inner product include a generalization of the orthogonal projection, orthonormal bases, direct sums, and the Gram–Schmidt process.

Matrix Theory The book concludes with an introduction to the powerful concepts of eigenvalues and eigenvectors. Both the characteristic polynomial and the minimal polynomial are defined and used throughout the chapter. Generalized eigenvectors are presented and used to write $\mathbb{R}^n$ as a direct sum of subspaces. The concept of similar matrices is given, and if a matrix does not have enough eigenvectors, it is proved that such matrices are similar to matrices with a nice form. This is where Schur's Lemma makes its appearance. However, if a matrix does have enough eigenvectors, the matrix is similar to a very nice diagonal matrix. This is the last section of the book, which includes orthogonal diagonalization, simultaneous diagonalization, and a quick introduction to quadratic forms and how to use eigenvalues to find an equation for a conic section without a middle term.

As with any textbook, where the course is taught influences how the book is used. Many universities and colleges have an introduction to proof course. Because such courses serve as a prerequisite for any proof-intensive mathematics course, the first chapter of this book can be passed over at these institutions and used only as a reference. If there is no such prerequisite, the first chapter serves as a detailed introduction to proof-writing that is short enough not to infringe too much on the time spent on purely linear algebra topics. Wherever the book finds itself, the course outline can easily be adjusted with any excluded topics serving as bonus reading for the eager student.

Now for some technical comments. Theorems, definitions, and examples are numbered sequentially as a group in the now common chapter.section.number format. Although some proofs find their way into the text, most start with **Proof**, end with ■, and are indented. Examples, on the other hand, are simply indented. Some equations are numbered as (chapter.number) and are referred to simply using (chapter.number). Most if not all of the mathematical notation should be clear. It was decided to represent vectors as columns. This leads to some interesting typesetting, but the clarity and consistency probably more than makes up for any formatting issues. Vectors are boldface, such as $\mathbf{u}$ and $\mathbf{v}$, and scalars are not. Most sums are written like $\mathbf{u}_1 + \mathbf{u}_2 + \cdots + \mathbf{u}_k$. There is a similar notation for products. However, there are times when summation and product notation must be used. Therefore, if $\mathbf{u}_1, \mathbf{u}_2, \ldots, \mathbf{u}_k$ are vectors,

$$\sum_{i=1}^{k} \mathbf{u}_i = \mathbf{u}_1 + \mathbf{u}_2 + \cdots + \mathbf{u}_k \quad \text{and} \quad \sum_{i \neq 2} \mathbf{u}_i = \mathbf{u}_1 + \mathbf{u}_3 + \cdots + \mathbf{u}_k,$$

and if $r_1, r_2, \ldots, r_k$ are real numbers,

$$\prod_{i=1}^{k} r_i = r_1 r_2 \cdots r_k \quad \text{and} \quad \prod_{i \neq 2} r_i = r_1 r_3 \cdots r_k.$$

Each section ends with a list of exercises. Some are computations, some are verifications where the job is to make a computation that illustrates a theorem from the section, and some involve proving results where remembering one's logic and set theory and how to prove sentences will go a long way.

Solution manuals, one for students and one for instructors, are available. See the book's page at *wiley.com*.

Lastly, this book was typeset using LATEX from the free software distribution of TEX Live running in Arch Linux with the KDE Plasma desktop. The diagrams were created using LibreOffice Draw.

MICHAEL L. O'LEARY

Glen Ellyn, Illinois
September, 2020

Acknowledgments

Thanks are due to Kalli Schultea, Senior Editor at Wiley, for her support of this project. Thanks are also due to Kimberly Monroe-Hill and Gayathree Sekar, also at Wiley, for their work in producing this book.

I would like to thank some of my colleagues at College of DuPage who helped at various stages of this project. They are James Adduci, Christopher Bailey, Patrick Bradley, Jennifer-Anne Hill, Rita Patel, and Matthew Wechter. I also thank my linear algebra classes of 2017, 2019, and 2020 who were good sports as I experimented on them with various drafts of this book. They found many errors and provided needed corrections.

On a personal note, I would like to express my gratitude to my parents for their continued caring and support; to my brother and his wife, who will make sure my niece learns her math; to my dissertation advisor, Paul Eklof, who taught me both set theory and algebra; to Robert Meyer, who introduced me to linear algebra; to David Elfman, who taught me about logic through programming on an Apple II; and to my wife, Barb, whose love and patience supported me as I finished this book.

About the Companion Website

This book is accompanied by a companion website:

www.wiley.com/go/o'leary/linearalgebra

The website includes the solutions manual and will be live in the fall of 2021.

CHAPTER 1

Logic and Set Theory

1.1 Statements

A sentence that is true or false but not both is called a **statement**. Here are some sentences, some of which are statements.

- Please read the linear algebra book.
  — This is not a statement because it is a request. It is neither true nor false.
- All quadrilaterals have 4 sides.
  — This is a true statement.
- Some triangles have 5 sides.
  — This is a false statement.
- $x + y = y + x$.
  — This is not a statement since the variables have not been assigned values.
- $x + y = y + x$ for all integers x and y.
  — This is a true statement.
- $x + y = 10$ for all real numbers y.
  — This is not a statement because x has not been assigned a value.

Linear Algebra, First Edition. Michael L. O'Leary.
© 2021 John Wiley & Sons, Inc. Published 2021 by John Wiley & Sons, Inc.
Companion Website: www.wiley.com/go/o'leary/linearalgebra

Connectives

Let p and q represent sentences. These variables combined with the **(logical) connectives**, which are $\sim$ (**not**), $\wedge$ (**and**), $\vee$ (**or**), $\rightarrow$ (**if...then**), $\leftrightarrow$ (**if and only if**), can be used to represent **compound sentences**. To illustrate the meanings of the connectives, let p be the statement, *lines intersect in a point*, and q be the sentence, *planes intersect in a line*. These sentences can be combined using the connectives:

$\sim p$	Lines do not intersect in a point.
$p \wedge q$	Lines intersect in a point, and planes intersect in a line.
$p \vee q$	Lines intersect in a point, or planes intersect in a line.
$p \rightarrow q$	If lines intersect in a point, then planes intersect in a line.
$p \leftrightarrow q$	Lines intersect in a point if and only if planes intersect in a line.

Let p and q be statements. This means that the **truth value** of p is either true (T) or false (F). The same can be said of q. Joining p and q with a connective yields a statement. The truth value of the resulting statement depends on the truth values of p and q. A **truth table** is used to identify all of the statement's possible truth values.

■ Definition 1.1.1

The sentence $\sim p$ is the **negation** of p. If p is a statement, the truth table of $\sim p$ is:

p	$\sim p$
T	F
F	T

■ Definition 1.1.2

The sentence $p \wedge q$ is the **conjunction** of p and q, and the sentence $p \vee q$ is the **disjunction** of p and q. If p and q are statements, the truth tables of $p \wedge q$ and $p \vee q$ are:

p	q	$p \wedge q$
T	T	T
T	F	F
F	T	F
F	F	F

p	q	$p \vee q$
T	T	T
T	F	T
F	T	T
F	F	F

■ Definition 1.1.3

The sentence $p \rightarrow q$ is an **implication** or **conditional sentence**, where p is called the **antecedent** and q is called the **consequent**. The sentence $p \leftrightarrow q$ is a **biconditional**. If p and q are statements, the truth tables of $p \rightarrow q$ and $p \leftrightarrow q$ are:

p	q	$p \to q$
T	T	T
T	F	F
F	T	T
F	F	T

p	q	$p \leftrightarrow q$
T	T	T
T	F	F
F	T	F
F	F	T

■ **Example 1.1.4**

The sentence $p \leftrightarrow (q \vee \sim p)$ is read by assuming that parentheses work as grouping symbols like in algebra and by attaching any $\sim$ to the first sentence to its immediate right. This implies that $p \leftrightarrow (q \vee \sim p)$ is interpreted by examining its sentences using an order starting with the variables:

$$p, q, \sim p, q \vee \sim p, p \leftrightarrow (q \vee \sim p),$$

which, if $p \leftrightarrow (q \vee \sim p)$ is a statement, produces the truth table:

p	q	$\sim p$	$q \vee \sim p$	$p \leftrightarrow (q \vee \sim p)$
T	T	F	T	T
T	F	F	F	F
F	T	T	T	F
F	F	T	T	F

Each column of the truth table on the right-hand side requires values from columns to its left. For example, evaluating $q \vee \sim p$ requires the truth values in the second and third columns.

Logical Equivalence

Two statements p and q are **(logically) equivalent** (written $p \equiv q$) means that they always have the same truth values. This is often proved using a truth table.

■ **Example 1.1.5**

The sentence $p \leftrightarrow q$ is the conjunction of two implications. Specifically, $p \leftrightarrow q$ means p *if* q, *and* p *only if* q. The first implication is $q \to p$, and the second implication is $p \to q$. Therefore, if p and q are statements,

$$p \leftrightarrow q \equiv (p \to q) \wedge (q \to p).$$

This is confirmed with the truth table:

p	q	$p \leftrightarrow q$	$p \to q$	$q \to p$	$(p \to q) \wedge (q \to p)$
T	T	**T**	T	T	**T**
T	F	**F**	F	T	**F**
F	T	**F**	T	F	**F**
F	F	**T**	T	T	**T**

■ **Example 1.1.6**

Let p and q be statements.

- The **converse** of $p \to q$ is $q \to p$.
- The **contrapositive** of $p \to q$ is $\sim q \to \sim p$.

For example, given the implication

if lines intersect in a point, planes intersect in a line,

its converse is

if planes intersect in a line, lines intersect in a point,

and its contrapositive is

if planes do not intersect in a line, lines do not intersect in a point.

An implication and its converse are not logically equivalent, but an implication and its contrapositive are logically equivalent. The third and sixth columns (in boldface) of the truth table show $p \to q \equiv \sim q \to \sim p$, while the third and last columns show $p \to q \not\equiv q \to p$.

p	q	$p \to q$	$\sim q$	$\sim p$	$\sim q \to \sim p$	$q \to p$
T	T	**T**	F	F	**T**	T
T	F	**F**	T	F	**F**	T
F	T	**T**	F	T	**T**	F
F	F	**T**	T	T	**T**	T

■ **Example 1.1.7**

There are some equivalences that are quite famous. Observe that

$$p \to q \equiv \sim p \vee q. \tag{1.1}$$

This is seen by the truth table:

p	q	$p \to q$	$\sim p$	$\sim p \vee q$
T	T	**T**	F	**T**
T	F	**F**	F	**F**
F	T	**T**	T	**T**
F	F	**T**	T	**T**

There are also **De Morgan's Laws**,

$$\sim(p \wedge q) \equiv \sim p \vee \sim q, \tag{1.2}$$
$$\sim(p \vee q) \equiv \sim p \wedge \sim q,$$

where (1.2) is confirmed by the truth table:

p	q	$p \wedge q$	$\sim(p \wedge q)$	$\sim p$	$\sim q$	$\sim p \vee \sim q$
T	T	T	**F**	F	F	**F**
T	F	F	**T**	F	T	**T**
F	T	F	**T**	T	F	**T**
F	F	F	**T**	T	T	**T**

To illustrate how De Morgan's Law (1.2) works, use $\equiv$ to assign

$$p \equiv 2 + 2 = 4,$$
$$q \equiv 3 + 5 = 10.$$

Then, (1.2) can be written as

it is false that both $2 + 2 = 4$ and $3 + 5 = 10$,

which is logically equivalent to

$2 + 2 \neq 4$ or $3 + 5 \neq 10$.

Another example is the **Double Negation Rule**,

$$\sim\sim p \equiv p, \tag{1.3}$$

which is proved by the simple truth table:

p	$\sim p$	$\sim\sim p$
T	F	**T**
T	F	**T**
F	T	**F**
F	T	**F**

■ **Example 1.1.8**

To understand the meaning of $p \rightarrow q$, notice that in addition to (1.1),

$$p \rightarrow q \equiv \sim(p \wedge \sim q). \tag{1.4}$$

This is proved by the truth table:

p	q	$p \rightarrow q$	$\sim q$	$p \wedge \sim q$	$\sim(p \wedge \sim q)$
T	T	**T**	F	F	**T**
T	F	**F**	T	T	**F**
F	T	**T**	F	F	**T**
F	F	**T**	T	F	**T**

The statement $\sim(p \wedge \sim q)$ claims that it is not the case that p is true but q is false. This is the exact meaning of $p \rightarrow q$. An alternate proof of (1.4) involves applying De Morgan's Law (1.2) and Double Negation (1.3) to obtain

$$\sim(p \wedge \sim q) \equiv \sim p \vee \sim\sim q \equiv \sim p \vee q. \tag{1.5}$$

Then, (1.5) combined with (1.1) gives (1.4). Also, Double Negation (1.3) with (1.4) implies that

$$\sim(p \rightarrow q) \equiv p \wedge \sim q, \tag{1.6}$$

so to show that an implication is false, it must be demonstrated that the antecedent can be true at the same time that the consequent is false.

Exercises

1. Determine the truth value of each sentence that is a statement.

    (a) $42 + 13 = 55$

    (b) For all real numbers x, $x < 5$.

    (c) Study linear algebra.

    (d) There are seven prime integers.

    (e) $x = 7$

    (f) This statement is false.

2. Define: $p \equiv$ The sum of two odd integers is an odd integer.
    $q \equiv$ The angle sum of a rectangle is 2π.
    $r \equiv$ The tangent function is differentiable everywhere.

    For each of the given statements, write the statement using p, q, or r and the appropriate logical connectives and find its truth value.

    (a) The tangent function is differentiable everywhere, and the angle sum of a rectangle is 2π.

    (b) The sum of two odd integers is an odd integer, or the sum of odd integers is an odd integer.

    (c) If the angle sum of a rectangle is 2π, the tangent function is not differentiable everywhere.

    (d) The sum of two odd integers is an odd integer if and only if it is true that the tangent function is differentiable everywhere.

    (e) The tangent function is differentiable everywhere if and only if the angle sum of a rectangle is 2π, and the sum of two odd integers is an even integer.

    (f) It is not the case that the angle sum of a rectangle is not 2π.

3. Write each sentence in the form *if p then q* and determine its truth value. Some words may need to be changed so that the answer is grammatically correct.

    (a) If a rectangle has adjacent congruent sides, a square has adjacent congruent sides.

    (b) Polynomials have at most two roots if quadratic polynomials have two complex roots.

    (c) Trigonometric functions are periodic only if polynomials are periodic.

    (d) The derivative of a constant function is zero.

    (e) A necessary condition for the opposite angles of a parallelogram to be congruent is that parallel lines intersect.

    (f) A sufficient condition for all systems of linear equations to have a solution is that some systems of linear equations have a solution.

4. Write the converse and contrapositive for the implications in Exercise 3.

5. Without using a truth table, explain the meaning of De Morgan's Laws found in Example 1.1.7.

6. Let p and q be true statements but r and s be false statements. Find the truth values.

   (a) $(p \wedge q) \vee r$

   (b) $q \leftrightarrow (r \vee \sim q)$

   (c) $p \rightarrow (q \rightarrow [r \rightarrow s])$

   (d) $(\sim p \wedge q) \vee ([p \rightarrow q] \wedge \sim s)$

   (e) $([p \wedge q] \rightarrow q) \wedge (p \rightarrow q)$

   (f) $\sim \sim \sim p \leftrightarrow (q \wedge r)$

   (g) $([p \rightarrow q] \vee [q \rightarrow r]) \vee s$

   (h) $(p \rightarrow q) \vee ([q \rightarrow r] \vee s)$

   (i) $(p \vee q) \wedge (q \vee r)$

   (j) $([p \vee q] \wedge q) \vee ([p \vee q] \wedge r)$

7. Write the truth table.

   (a) $\sim p \rightarrow p$

   (b) $p \rightarrow \sim q$

   (c) $(p \vee q) \wedge \sim (p \wedge q)$

   (d) $(p \rightarrow q) \vee (q \leftrightarrow p)$

   (e) $p \rightarrow (q \wedge \sim p)$

   (f) $(p \rightarrow q) \wedge \sim p$

   (g) $(\sim p \vee q) \wedge ([p \rightarrow q] \vee \sim p)$

   (h) $(p \wedge q) \vee r$

   (i) $p \wedge (q \vee r)$

   (j) $(p \vee q) \rightarrow r$

   (k) $p \vee (q \rightarrow r)$

   (l) $(p \rightarrow q) \wedge \sim (r \vee p)$

   (m) $(p \rightarrow q) \leftrightarrow (r \rightarrow s)$

   (n) $p \vee ([\sim q \leftrightarrow r] \wedge q)$

8. Prove the given famous logical equivalences.

   (a) Associative Laws: $\begin{aligned} (p \wedge q) \wedge r &\equiv p \wedge (q \wedge r) \\ (p \vee q) \vee r &\equiv p \vee (q \vee r) \end{aligned}$

   (b) Commutative Laws: $\begin{aligned} p \wedge q &\equiv q \wedge p \\ p \vee q &\equiv q \vee p \end{aligned}$

   (c) Distributive Laws: $\begin{aligned} p \wedge (q \vee r) &\equiv (p \wedge q) \vee (p \wedge r) \\ p \vee (q \wedge r) &\equiv (p \vee q) \wedge (p \vee r) \end{aligned}$

9. Prove using truth tables or by using logical equivalences as in Example 1.1.8.

   (a) $p \vee \sim p \equiv p \rightarrow p$

   (b) $p \wedge q \equiv (p \leftrightarrow q) \wedge (p \vee q)$

   (c) $p \vee \sim p \equiv (p \vee q) \vee \sim (p \wedge q)$

   (d) $p \rightarrow (q \wedge r) \equiv (p \rightarrow q) \wedge (p \rightarrow r)$

   (e) $r \wedge (p \rightarrow q) \equiv r \wedge (\sim q \rightarrow \sim p)$

   (f) $(p \wedge q) \rightarrow r \equiv (p \wedge \sim r) \rightarrow \sim q$

   (g) $(p \vee q) \vee r \equiv (q \vee p) \vee r$

   (h) $\sim p \wedge \sim (q \vee r) \equiv \sim (p \vee [q \vee r])$

   (i) $\sim (p \vee q) \vee r \equiv (p \vee q) \rightarrow r$

   (j) $(p \wedge q) \wedge r \equiv r \wedge (p \wedge q)$

1.2 Sets and Quantification

A **set** is a collection of things called **elements**. Anything can be an element, but in linear algebra, elements are typically numbers, matrices, functions, or vectors. If a is an element of the set A, write $a \in A$. If both a and b are elements of A, write $a, b \in A$. If c is not an element of A, write $c \notin A$. If B is a set that has exactly the same elements as A, write $A = B$, which means that A is **equal** to B. If $A \neq B$, there is an element that is in one set but not the other. If A contains no elements, write $A = \varnothing$, where $\varnothing$ is the **empty set**, the set with no elements.

Some examples of famous sets, written in **roster form**, are the following:

- $\{1, 2, 3, 4, 5, 6, 7, 8, 9, 10\}$ = the set of integers from 1 to 10.

- $\mathbb{N} = \{0, 1, 2, 3, \dots\}$ = the set of **natural numbers**.

- $\mathbb{Z} = \{\dots, -2, -1, 0, 1, 2, \dots\}$ = the set of **integers**.

- $\mathbb{Z}^+ = \{1, 2, 3, \dots\}$ = the set of **positive integers**.

This approach to writing sets has the elements of each set found within braces and uses **ellipses** (...) to represent a repeating pattern. In general, writing

$$A = \{a_1, a_2, \dots, a_n\}$$

means that A has n distinct elements, $a_1, a_2, \dots, a_n$, and writing

$$A = \{a_1, a_2, a_3, \dots\}$$

means that A has infinitely many elements, $a_1, a_2, a_3, \dots$, where $a_i \neq a_j$ if $i \neq j$.

The problem with roster form is that it is not good for describing most sets, like

$$\mathbb{R} = \text{ the set of real numbers.}$$

What is needed is the ability to write a condition that describes exactly when an element is in a set.

Universal Quantifiers

The statement

$$\text{for all } x \in \mathbb{R}, x + 42 = 42 + x, \tag{1.7}$$

or, equivalently, $x + 42 = 42 + x$ for every $x \in \mathbb{R}$, claims that $x + 42 = 42 + x$ is true for every substitution of a real number for x. Use the function notation $p(x)$ to represent $x + 42 = 42 + x$. Substitutions work and are denoted as expected. For example, $p(7) \equiv 7 + 42 = 42 + 7$. Letting $\forall$ represent "for all," write (1.7) as $(\forall x \in \mathbb{R})p(x)$. If x is assumed to be a real number, write (1.7) as $(\forall x)p(x)$, or denote the inclusion of x in $\mathbb{R}$ by writing $(\forall x)[x \in \mathbb{R} \rightarrow p(x)]$.

■ Definition 1.2.1

If A is a set, then $(\forall x \in A)p(x) \equiv (\forall x)[x \in A \rightarrow p(x)]$. Both sentences claim that $p(a)$ is true for every $a \in A$. The symbol $\forall$ is called the **universal quantifier**.

■ **Example 1.2.2**

The statement $(\forall x \in \varnothing)p(x)$ is true because $x \in \varnothing$ is false for all substitutions of x, from which follows that $x \in \varnothing \to p(x)$ is true for all substitutions of x by Definition 1.1.3.

■ **Example 1.2.3**

$(\forall x \in \mathbb{R})x + 10 = 5$ is false because $1 + 10 \neq 5$ and $1 \in \mathbb{R}$. The statement

$$(\forall x \in \mathbb{R})(\forall y \in \mathbb{R})x + y = y + x$$

is true because

$$(\forall y \in \mathbb{R})a + y = y + a \tag{1.8}$$

is true for every $a \in \mathbb{R}$, and the reason that (1.8) is true is that the statement $a + b = b + a$ is true for every $b \in \mathbb{R}$.

Example 1.2.3 suggests a method for proving true a statement with a universal quantifier. Consider

$$(\forall x \in \mathbb{R})x + 0 = x. \tag{1.9}$$

To prove it, let $a \in \mathbb{R}$. Because the only property that a is assigned is that it is a real number, it is considered an arbitrary or randomly chosen real number. It is known that 0 has the property that $a + 0 = a$. Thus, because a is arbitrary, (1.9) is true.

Existential Quantifiers

The statement

$$\text{there exists } x \in \mathbb{R} \text{ such that } x + 27 = 42 \tag{1.10}$$

or, equivalently, $x + 27 = 42$ for some $x \in \mathbb{R}$, claims that $x + 27 = 42$ is true for at least one substitution of a real number for x. Letting $\exists$ represent "there exists" and $p(x)$ represent $x + 27 = 42$, (1.10) can be written as $(\exists x \in \mathbb{R})p(x)$. If x is assumed to be a real number, write (1.10) as $(\exists x)p(x)$, or denote the inclusion of x in $\mathbb{R}$ by writing $(\exists x)[x \in \mathbb{R} \wedge p(x)]$.

■ **Definition 1.2.4**

If A is a set, then $(\exists x \in A)p(x) \equiv (\exists x)[x \in A \wedge p(x)]$. Both statements claim that $p(a)$ is true for at least one $a \in A$. The symbol $\exists$ is called the **existential quantifier**.

■ **Example 1.2.5**

For any sentence $p(x)$, the statement $(\exists x \in \varnothing)p(x)$ is false. This is because $x \in \varnothing$ is false for all substitutions of x, from which follows $x \in \varnothing \wedge p(x)$ is false for all substitutions of x by Definition 1.1.2.

■ **Example 1.2.6**

$(\exists x \in \mathbb{R})x + 0 = x + 1$ is false because there is no real number a such that $a + 0 = a + 1$. The statement

$$(\exists x \in \mathbb{R})(\exists y \in \mathbb{R})x + y = 13$$

is true because

$$(\exists y \in \mathbb{R})5 + y = 13 \qquad\qquad (1.11)$$

is true, and (1.11) is true because $5 + 8 = 13$ is true.

Example 1.2.6 suggests a method for proving true a statement with an existential quantifier. Consider $(\exists x \in \mathbb{R})x + 3 = 9$. To prove it, it is enough to find a real number a such that $a + 3 = 9$ is true. Taking $a = 6$ does it.

Negating Quantifiers

As observed in Example 1.2.3, the statement

$$(\forall x \in \mathbb{R})x + 10 = 5$$

is false because $1 + 10 \neq 5$ and $1 \in \mathbb{R}$. This means that

$$(\exists x \in \mathbb{R})x + 10 \neq 5$$

is true. Generalizing, $(\forall x \in A)p(x)$ is false if there exists $a \in A$ such that $p(a)$ is false, which means that $(\exists x \in A)\sim p(x)$ is true. The element a that demonstrates that a statement starting with a universal quantifier is false is a **counterexample**.

Likewise, as noted in Example 1.2.6,

$$(\exists x \in \mathbb{R})x + 0 = x + 1$$

is false because $a + 0 \neq a + 1$ for every real number a. This means that

$$(\forall x \in \mathbb{R})x + 0 \neq x + 1$$

is true. Generalizing, $(\exists x \in A)p(x)$ is false if there is no $a \in A$ such that $p(a)$ is true, which means that $(\forall x \in A)\sim p(x)$ is true. These two results are summarized in the next theorem.

■ **Theorem 1.2.7**

Let $p(x)$ be a sentence and let A be a set.

    (a) $\sim(\forall x \in A)p(x) \equiv (\exists x \in A)\sim p(x)$.

    (b) $\sim(\exists x \in A)p(x) \equiv (\forall x \in A)\sim p(x)$.

■ **Example 1.2.8**

De Morgan's Law (1.2) and (1.6) implies

$$\sim(\forall x \in A)[p(x) \rightarrow q(x)] \equiv (\exists x \in A)\sim[p(x) \rightarrow q(x)]$$
$$\equiv (\exists x \in A)[p(x) \wedge \sim q(x)],$$

and also by De Morgan's Law,

$$\sim(\exists x \in A)[p(x) \wedge q(x)] \equiv (\forall x \in A)\sim[p(x) \wedge q(x)]$$
$$\equiv (\forall x \in A)[\sim p(x) \vee \sim q(x)].$$

A statement can have multiple quantifiers. Consider $2x - 7y = 1$. This line can be graphed by writing an x-y table. Values for y-coordinates are calculated based on the values chosen for x resulting in the points to be plotted. This process demonstrates that

$$(\forall x \in \mathbb{R})(\exists y \in \mathbb{R})2x - 7y = 1. \tag{1.12}$$

The statement (1.12) has a universal quantifier followed by the sentence

$$(\exists y \in \mathbb{R})2x - 7y = 1,$$

and since there is no quantifier on x, substitutions can be made for x. Conclude that (1.12) is true because whenever x is replaced with an arbitrary real number, a real number y can be found to satisfy $2x - 7y = 1$. Specifically, that real number is $y = (2x - 1)/7$.

■ **Example 1.2.9**

Find the negation of $(\exists x \in A)(\forall y \in A)[p(x) \rightarrow q(y)]$.

$$\sim(\exists x \in A)(\forall y \in A)[p(x) \rightarrow q(y)] \equiv (\forall x \in A)\sim(\forall y \in A)[p(x) \rightarrow q(y)]$$
$$\equiv (\forall x \in A)(\exists y \in A)\sim[p(x) \rightarrow q(y)]$$
$$\equiv (\forall x \in A)(\exists y \in A)[p(x) \wedge \sim q(y)].$$

■ **Example 1.2.10**

Prove or show false.

- $(\forall x \in \mathbb{R})(\forall y \in \mathbb{R})x + y = 3$
  — Because $7 + 10 \neq 3$, the numbers 7 and 10 are counterexamples, so the given statement is false.
- $(\forall x \in \mathbb{R})(\exists y \in \mathbb{R})x + y = 3$
  — Let $a \in \mathbb{R}$. Then, $(\exists y \in \mathbb{R})a + y = 3$ states that there exists a real number y such that $a + y = 3$, which is true because $a + (3 - a) = 3$. This means that the given statement is true.
- $(\exists x \in \mathbb{R})(\forall y \in \mathbb{R})x + y = 3$
  — The statement is false. To see this, take $a \in \mathbb{R}$. Then, $a + (2 - a) \neq 3$, proving $(\forall x \in \mathbb{R})(\exists y \in \mathbb{R})x + y \neq 3$ is true.
- $(\exists x \in \mathbb{R})(\exists y \in \mathbb{R})x + y = 3$
  — This is true since $3 + 0 = 3$.

Set-Builder Notation

Let A be a set. Consider a sentence $p(x)$ such that for every a,

$$a \in A \text{ if and only if } p(a) \text{ is true.}$$

Such a sentence serves as a condition that an element must satisfy in order to be in A. For example, let $E = \{\dots, -4, -2, 0, 2, 4, \dots\}$ and $p(x)$ be the sentence

$$(\exists n \in \mathbb{Z})x = 2n. \tag{1.13}$$

Notice that $p(x)$ completely describes E because the even integers are exactly those elements a such that $p(a)$ is true. In particular, $p(0)$ and $p(-10)$ are true but $p(3)$ is false. Thus, 0 and -10 are elements of E, but 3 is not. Sentences like $p(x)$ can be used to define sets.

■ **Definition 1.2.11**

Let A be a set. If $p(x)$ is a sentence such that $a \in A$ if and only if $p(a)$, write

$$A = \{x : p(x)\}.$$

This is called **set-builder notation**. Read $\{x : p(x)\}$ as "the set of all x such that p of x."

Using Definition 1.2.11, write E with (1.13) using set-builder notation as

$$E = \{x : (\exists n \in \mathbb{Z})x = 2n\} = \{2n : n \in \mathbb{Z}\}.$$

■ **Example 1.2.12**

$A = \{-4, 4\}$ is the set of roots of the polynomial $x^2 - 16$. Using set-builder notation, A can be written as

$$A = \{x : x^2 - 16 = 0 \text{ and } x \in \mathbb{R}\} = \{x \in \mathbb{R} : (x + 4)(x - 4) = 0\}.$$

■ **Example 1.2.13**

- $\varnothing = \{x \in \mathbb{R} : x \neq x\}$.
- $\left\{\dots, -\dfrac{3}{5}, -\dfrac{2}{5}, -\dfrac{1}{5}, \dfrac{0}{5}, \dfrac{1}{5}, \dfrac{2}{5}, \dfrac{3}{5}, \dots\right\} = \left\{\dfrac{n}{5} : n \in \mathbb{Z}\right\}$.
- $\{\dots, x - 4, x - 2, x, x + 2, x + 4, \dots\} = \{x + 2n : n \in \mathbb{Z}\}$.

■ **Example 1.2.14**

- $\mathbb{Q} = \left\{\dfrac{a}{b} : a, b \in \mathbb{Z} \text{ and } b \neq 0\right\} = $ the set of **rational numbers**.
- $\mathbb{R} = \left\{a : \lim\limits_{n \to \infty} a_n = a \wedge (\forall i \in \mathbb{Z}^+)\, a_i \in \mathbb{Q}\right\} = $ the set of **real numbers**.
- $\mathbb{R}^+ = \{x : x \in \mathbb{R} \wedge x > 0\} = $ the set of **positive real numbers**.
- $\mathbb{C} = \{a + bi : a, b \in \mathbb{R}\} = $ the set of **complex numbers**, where $i^2 = -1$.

Set Operations

As the logical connectives enabled new statements to be written using given statements, the **set operations** allow new sets to be defined from given sets.

■ **Definition 1.2.15**

Let A and B be sets.

(a) $A \setminus B = \{x : x \in A \wedge x \notin B\}$ = the **set difference** of B from A.

(b) $A \cap B = \{x : x \in A \wedge x \in B\}$ = the **intersection** of A and B.

(c) $A \cup B = \{x : x \in A \vee x \in B\}$ = the **union** of A and B.

For example, if $A = \{1, 2, 3, 4, 5\}$ and $B = \{3, 4, 5, 6, 7\}$, then

$$A \setminus B = \{1, 2\}, \quad B \setminus A = \{6, 7\}, \quad A \cap B = \{3, 4, 5\}, \quad A \cup B = \{1, 2, 3, 4, 5, 6, 7\}.$$

■ **Example 1.2.16**

As with the logical connectives, parentheses behave as grouping symbols.

- $\{0, 1\} \cup (\{2\} \setminus \{1, 3\}) = \{0, 1\} \cup \{2\} = \{0, 1, 2\}$.
- $(\{0, 1\} \cup \{2\}) \setminus \{1, 3\} = \{0, 1, 2\} \setminus \{1, 3\} = \{0, 2\}$.

If only one operation is used when writing a set and the operation is $\cap$ or $\cup$, the order in which they are used is not relevant.

- $\{1\} \cap \{2\} \cap \{3\} \cap \{3, 4\} = \varnothing$.
- $\{1\} \cup \{2\} \cup \{3\} \cup \{3, 4\} = \{1, 2, 3, 4\}$.

Another set operation should be introduced at this time. It is named after René Descartes.

■ **Definition 1.2.17**

Let $n \in \mathbb{Z}$ such that $n > 1$. The **Cartesian product** of $A_1, A_2, \ldots, A_n$ is

$$A_1 \times A_2 \times \cdots \times A_n = \{(a_1, a_2, \ldots, a_n) : a_i \in A_i \text{ for } i = 1, 2, \ldots, n\}.$$

The elements of $A_1 \times A_2 \times \cdots \times A_n$ are called **ordered n-tuples**.

■ **Example 1.2.18**

- $\{2, 4, 6\} \times \{3, 5\} = \{(2, 3), (2, 5), (4, 3), (4, 5), (6, 3), (6, 5)\}$.
- $\{7, 9\} \times \{\varnothing, \{0\}\} = \{(7, \varnothing), (7, \{0\}), (9, \varnothing), (9, \{0\})\}$.
- $\varnothing \times A = A \times \varnothing = \varnothing$ for every set A.

Note that $\{2, 4, 6\} \times \{3, 5\}$ has six elements. This is because ordered n-tuples are equal if and only if their corresponding coordinates are equal.

Using this notation, $\mathbb{R} \times \mathbb{R}$ is the **Cartesian plane**, $\mathbb{R} \times \mathbb{R} \times \mathbb{R}$ is **Cartesian space**, and **Cartesian n-space** is

$$\underbrace{\mathbb{R} \times \mathbb{R} \times \cdots \times \mathbb{R}}_{n \text{ times}}.$$

Families of Sets

Sets can contain other sets as elements such as in Example 1.2.18. Consider,

$$\mathscr{F} = \{\{1, 2, \ldots, n\} : n \in \mathbb{Z}^+\}.$$

As a roster,

$$\mathscr{F} = \{\{1\}, \{1, 2\}, \{1, 2, 3\}, \ldots\}.$$

Call $\mathscr{F}$ a **family of sets**. It is often important to identify the elements of such a set. For example, the elements of $\mathscr{F}$ can be named as $A_n = \{1, 2, \ldots, n\}$, where n is a positive integer. This means that

$$\mathscr{F} = \{A_n : n \in \mathbb{Z}^+\}. \tag{1.14}$$

In (1.14), the set $\mathbb{Z}^+$ is called an **index set**, a set that is used to label the elements of the family. Each element of an index set is called an **index**. For example, the element of $\mathscr{F}$ indexed by 7 is $A_7 = \{1, 2, 3, 4, 5, 6, 7\}$.

The set operations of intersection and union (Definition 1.2.15) can be generalized to families of sets.

■ **Definition 1.2.19**

Let $\mathscr{F} = \{A_i : i \in I\}$ be a family of sets.

(a) $\bigcap \mathscr{F} = \bigcap_{i \in I} A_i = \{x : (\forall i \in I) x \in A_i\} = $ the **intersection** of $\mathscr{F}$.

(b) $\bigcup \mathscr{F} = \bigcup_{i \in I} A_i = \{x : (\exists i \in I) x \in A_i\} = $ the **union** of $\mathscr{F}$.

Consider $\mathscr{E} = \{[n, n+1] : n \in \mathbb{Z}\}$. Then,

$$\bigcup \mathscr{E} = \bigcup_{n \in \mathbb{Z}} [n, n+1] = \mathbb{R},$$

and

$$\bigcap \mathscr{E} = \bigcap_{n \in \mathbb{Z}} [n, n+1] = \emptyset. \tag{1.15}$$

The equation in (1.15) leads to the next definition.

■ **Definition 1.2.20**

Let $\mathscr{F}$ be a family of sets.

(a) If $\bigcap \mathscr{F} = \emptyset$, then $\mathscr{F}$ is **disjoint**.

(b) If $A \cap B = \emptyset$ for all $A, B \in \mathscr{F}$ such that $A \neq B$, then $\mathscr{F}$ is **pairwise disjoint**.

For example, $\{\{2, 4, 6\}, \{1, 3, 5\}, \{7, 8\}\}$ is both disjoint and pairwise disjoint, but the set $\{\{1\}, \{2\}, \{1, 2\}\}$ is disjoint but not pairwise disjoint.

■ **Example 1.2.21**

Let m be a positive integer. For all $n \in \mathbb{Z}$, define $A_n = \{km + n : k \in \mathbb{Z}\}$. Collect these sets into the family, $\mathscr{P} = \{A_n : n \in \mathbb{Z}\}$. Observe that $\mathscr{P}$ is a pairwise disjoint family of sets such that

$$\bigcup \mathscr{P} = \mathbb{Z}.$$

Such a family is called a **partition**.

Exercises

1. Write the given sets in roster form.

   (a) The set of all integers from 1 to 100.

   (b) The set of all nonpositive integers.

   (c) The set of all odd integers.

   (d) The set of integers in the open interval $(-5, 6)$.

   (e) The set of rational numbers in the closed interval $[0, 1]$ that can be represented with exactly two decimal places.

   (f) The set of linear polynomials with even integer coefficients.

   (g) $\mathbb{Q}$.

2. True or false.

   (a) $1/2 \in \mathbb{Z} \cap \mathbb{Q}$ (d) $-4 \in \mathbb{Z} \setminus \mathbb{Q}$ (g) $\{1, 2, 3\} = \{3, 2, 1\}$

   (b) $1/2 \in \mathbb{Z} \cup \mathbb{Q}$ (e) $4 + \pi \in \mathbb{R} \setminus \mathbb{C}$ (h) $0 \in \emptyset$

   (c) $-4 \in \mathbb{Q} \setminus \mathbb{Z}$ (f) $4.34534 \in \mathbb{C} \setminus \mathbb{R}$ (i) $\emptyset \in \emptyset$

3. True or false.

   (a) Every rectangle is a square.

   (b) No rectangle is a square.

   (c) There is a rectangle that is a square.

   (d) There is a square that is a rectangle.

   (e) For all $x \in \mathbb{R}$, there exists $y \in \mathbb{R}$ such that $2x + y = 100$.

   (f) $2x + y = 100$ for all $x, y \in \mathbb{R}$.

   (g) $2x + y = 100$ for some $x, y \in \mathbb{R}$.

   (h) There exists $x \in \mathbb{R}$ such that $2x + y = 100$ for all $y \in \mathbb{R}$.

4. Negate.

   (a) $(\exists x)[q(x) \rightarrow r(x)]$

   (b) $(\forall x)[p(x) \vee q(x)]$

   (c) $(\forall x)(\exists y)[p(x) \wedge q(y)]$

   (d) $(\exists x)(\exists y)[p(x) \vee q(y)]$

   (e) $(\forall x)(\forall y)[p(x) \vee (\exists z)q(z)]$

    (f) $(\forall x)(\exists y)[p(x) \leftrightarrow q(y)]$

    (g) $(\exists x){\sim}r(x) \vee (\forall x)[q(x) \leftrightarrow {\sim}p(x)]$

    (h) $(\forall x)(\forall y)(\exists z)(p(x) \rightarrow [q(y) \wedge r(z)]) \rightarrow (\exists x){\sim}q(x)$

    (i) $(\forall x)(\exists y)(\forall z)([p(x) \vee r(z)] \rightarrow {\sim}[q(y) \vee s(z)])$

5. Write the negation of the given statements.

    (a) For all $x \in \mathbb{R}$, there exists $y \in \mathbb{R}$ such that $xy = 5$.

    (b) There exists a real number x such that $x + y = 1$ for all real numbers y.

    (c) Every multiple of 4 is a multiple of 2.

    (d) For every integer x, there exists an integer y such that $xy = 1$.

    (e) No interval contains an irrational number.

    (f) There is an interval that contains a rational number.

6. Provide counterexamples for each of the following false statements.

    (a) Every integer is a solution to $x + 10 = 2$.

    (b) For every integer x, there exists an integer y such that $xy = 1$.

    (c) The sum of any two odd integers is odd.

    (d) For every integer n, if n is even, then n^2 is a multiple of 8.

7. Find real numbers a, b, and c that provide a counterexample for

$$(\forall a \in \mathbb{R})(\forall b \in \mathbb{R})(\forall c \in \mathbb{R})(\exists x \in \mathbb{R})(ax^2 + bx + c = 0).$$

8. Determine whether the given statements are true or false.

    (a) $1 \in \{x : p(x)\}$ when ${\sim}p(1)$ is true.

    (b) $7 \in \{x \in \mathbb{R} : x^2 - 5x = 14\}$

    (c) $x^2 - 5x \in \{a_2 x^2 + a_1 x + a_0 : a_i \in \mathbb{Q}\}$

    (d) $xy \in \{2k + 1 : k \in \mathbb{Z}\}$ if x is an even integer and y is an odd integer.

    (e) $\cos\theta \in \{a\cos\theta + b\sin\theta : a, b \in \mathbb{R}\}$

    (f) $\{1, -5\} = \{x : (x - 1)(x + 5) = 0\}$

    (g) $\{1, -5\} = \{x : (x - 1)^2(x + 5)^2 = 0\}$

9. Write the given sets in roster form.

    (a) $\{4n + 1 : n \in \mathbb{Z}\}$ (d) $\{n\tan\theta : n \in \mathbb{Z}\}$

    (b) $\{-5n : n \in \mathbb{Z}\}$ (e) $\{ax^2 + ax + a : a \in \mathbb{Z}^+\}$

    (c) $\{0 \cdot n : n \in \mathbb{Q}\}$ (f) $\{a\cos\theta + b\sin\theta : a, b \in \mathbb{R}\}$

10. Write each set using set-builder notation.

    (a) All positive rational numbers.

    (b) All functions f such that $f(0) = 1$.

    (c) All odd integers.

    (d) All integer multiples of 7.

    (e) All integers that have a remainder of 1 when divided by 3.

(f) All **ordered pairs** (2-tuples) of real numbers in which the x-coordinate is positive and the y-coordinate is negative.

(g) All complex numbers with real part equal to 0.

(h) All polynomials of degree at most 3 with real coefficients.

11. Let $A = \{0, 2, 4, 6\}$, $B = \{3, 4, 5, 6\}$, and $C = \{0, 1, 2\}$. Write the given sets in roster notation.

(a) $A \cup B$

(b) $A \cap B$

(c) $A \setminus B$

(d) $B \setminus A$

(e) $A \times B$

(f) $(A \cap B) \times (C \setminus A)$

(g) $(A \cup B) \cap C$

(h) $A \cup (B \cap C)$

(i) $A \cap B \cup C$

(j) $(A \setminus B) \setminus C$

(k) $A \setminus (B \setminus C)$

(l) $(A \cup B) \setminus (A \cap C)$

12. Write the given Cartesian products in roster form.

(a) $\{1, 2, 3\} \times \{a, b, c\}$

(b) $\{1, 2\} \times \{3, 4\} \times \{1, 2\}$

(c) $\mathbb{Z} \times \{0\} \times \{1, 2\}$

(d) $\mathbb{Q} \times \emptyset \times \mathbb{Z}$

13. Find the union and intersection of the given families of sets.

(a) $\{\{1, 2, 3\}, \{2, 3, 4\}, \{3, 4, 5\}\}$

(b) $\{\{0, 1, 2, \dots, n\} : n \in \mathbb{Z}^+\}$

(c) $\{[0, 2n) : n \in \mathbb{Z}^+\}$

(d) $\{(n, n+1) : n \in \mathbb{Z}\}$

(e) $\{[m, m+1] \times \{n\} : m \in \mathbb{Z}, n \in \mathbb{R}\}$

(f) $\{\{x\} : x \in \mathbb{R}\}$

14. Indicate whether the given family of sets is a partition of the given set.

(a) $\{\{1, 2\}, \{3, 4\}, \{5\}\}$ of $\{1, 2, 3, 4, 5\}$

(b) $\{\{5n + k : n \in \mathbb{Z}\} : k \in \mathbb{Z}\}$ of $\mathbb{Z}$

(c) $\{(n, n+1) : n \in \mathbb{Z}\}$ of $\mathbb{R}$

(d) $\{[n, n+1) : n \in \mathbb{Z}\}$ of $\mathbb{R}$

(e) $\{[n, n+1] : n \in \mathbb{Z}\}$ of $\mathbb{R}$

(f) $\{\{x\} : x \in \mathbb{R}\}$ of $\mathbb{R}$

15. Prove. See Example 1.2.10.

(a) There exists $x \in \mathbb{Z}$ such that $x^2 + 2x - 3 = 0$.

(b) $x^2 z + 2xyz + y^2 z = z(x + y)^2$ for all $x, y, z \in \mathbb{R}$.

(c) There exist real numbers u and v such that $2u + 5v = -29$.

(d) For all integers x, there exists an integer y so that $x - y = 10$.

(e) There exists an integer x such that for all integers y, $yx = x$.

(f) For all $a, b, c \in \mathbb{R}$, there exists $x \in \mathbb{C}$ such that $ax^2 + bx + c = 0$.

(g) $(x + 1)^3 = x^3 + 3x^2 + 3x + 1$ for all real numbers x.

(h) For all integers x, there exists an integer y such that $x + y = 0$.

(i) There exist $a, b \in \mathbb{R}$ so that $(a + 2b)x + (2a - b) = 2x - 6$ for every $x \in \mathbb{R}$.

1.3   Sets and Proofs

Consider the statement,

$$\text{for all } x \in \mathbb{Z}, \text{ if } x \text{ is odd, then } 3x \text{ is odd.} \tag{1.16}$$

How could this statement be proved? First, the mathematical meaning of all of the words of the statement must be known. In this case (1.16) requires the following two definitions. The second definition is directly used in sentence (1.16), but the second definition requires the first.

■ **Definition 1.3.1**

An integer m **divides** an integer n means that $m \neq 0$ and there exists $k \in \mathbb{Z}$ such that $n = mk$.

■ **Definition 1.3.2**

An integer n is **even** if 2 divides n, and n is **odd** if 2 divides $n + 1$.

As examples, 4 divides 20 because $20 = 4(5)$, but 8 does not divide 20 because $20 \neq 8k$ for all $k \in \mathbb{Z}$. In addition, 14 is even because 2 divides 14, and 11 is odd because 2 divides 12.

Since all of the other words and symbols in (1.16) are understood, the next goal is to understand the logic required to prove the sentence. The statement starts with a universal quantifier, and these have already been encountered (Example 1.2.10). This means that the proof should start with

$$\text{let } a \in \mathbb{Z}.$$

The statement that must now be shown is

$$\text{if } a \text{ is odd, then } 3a \text{ is odd.} \tag{1.17}$$

Definition 1.1.3 states that (1.17) is true for every case other than when a is odd and $3a$ is even, so this needs to be shown to be impossible. This is done by proving that if the antecedent is true, it must be the case that the consequent is also true. To do this, assume that

$$a \text{ is odd} \tag{1.18}$$

and show that

$$3a \text{ is odd.} \tag{1.19}$$

To accomplish this, notice that (1.18) implies that 2 divides $a+1$ by Definition 1.3.2, which means that $a + 1 = 2k$ for some integer k by Definition 1.3.1. Hence,

$$3a + 1 = 3(2k - 1) + 1 = 6k - 2 = 2(3k - 1),$$

so 2 divides $3a + 1$ because $3k - 1$ is an integer. This means that $3a$ is odd, and the proof is done.

The proof in the previous paragraph can be summarized like this:

1. Let $a \in \mathbb{Z}$.   Assumption
2. Assume that a is odd.   Assumption
3. 2 divides $a + 1$.   Definition
4. $a + 1 = 2k$ for some $k \in \mathbb{Z}$.   Definition
5. $3a + 1 = 2(3k - 1)$.   Theorem/Axiom
6. 2 divides $3a + 1$.   Definition
7. $3a$ is odd.   Definition
8. If a is odd, then $3a$ is odd.   Logic
9. For all $x \in \mathbb{Z}$, if x is odd, then $3x$ is odd.   Logic

Each of the statements of the proof are listed on the left, and the reason for each line is listed on the right. A word of explanation is needed for the reasons.

- Assumption

  An **assumption** is made based on the requirements of a particular rule of logic. For example, a variable is introduced to prove a statement with a universal quantifier, or an antecedent is assumed to be true in order to prove an implication.

  Line (1) begins the proof of (1.16) because the statement begins with a universal quantifier, and (2) is the start of the proof of the implication. Both of these assumptions are there for logical reasons.

- Logic

  The statement follows due to a **rule of logic**. This rule may refer back to an assumption or might simply be a deduction from a previous line.

  The assumption of (2) and the conclusion in (7) proves that (2) implies (7), and the assumption of (1) followed by (8) allows the deduction of the universally quantified statement of (9).

- Definition

  A **definition** gives meaning to a word or a symbol. If a statement is included in a proof because of a definition, it is because the translation of a particular word or symbol is needed to proceed with the proof.

  Lines (3) and (7) are based on the definition of an odd integer, and (4) and (6) are based on the definition of an integer dividing an integer.

- Theorem/Axiom

  A **theorem** is a previously proved statement. A statement can be entered into a proof if it follows when a theorem is applied to previous statements. Some theorems play specific roles, such as a **lemma**, which is a technical result mainly used to help prove one particular theorem, or a **corollary**, which is a theorem that quickly follows from another theorem. Some statements are used like theorems, but they are assumed true instead of proven. Such a statement is called an **axiom** or a **postulate**.

Only line (5) follows by the application of an axiom. Specifically, line (5) requires substitution and the distributive law from basic algebra.

A careful examination of the proof of (1.16) reveals that it has all of the expected pieces of a proof, so it can be generalized to the next definition.

■ Definition 1.3.3

Let $p_1, p_2, \dots, p_n$ and q be statements. A **proof** of the **conclusion** q from the **premises** $p_1, p_2, \dots, p_n$ is a sequence of statements starting with the premises and ending with the conclusion such that each statement p in the proof that is not a premise follows from a prior statement using one of these reasons:

(a) p is an assumption introduced for a rule of logic.

(b) p follows by a rule of logic.

(c) p follows by using a definition.

(d) p follows by applying a theorem or an axiom.

The premises $p_1, p_2, \dots, p_n$ **prove** the conclusion q ($p_1, p_2, \dots, p_n \Rightarrow q$) if there is a proof starting with $p_1, p_2, \dots, p_n$ and ending with q.

Direct Proof

Lines (2)–(7) of the proof of (1.16) show that

$$a \text{ is odd } \Rightarrow 3a \text{ is odd},$$

from which it is immediately concluded in (8) that when a is an integer,

$$\text{if } a \text{ is odd, then } 3a \text{ is odd}.$$

Proving implications like this is very common in mathematics, so this strategy receives a name.

■ Theorem 1.3.4 [Direct Proof]

Let p and q be statements. If $p \Rightarrow q$, then $p \to q$.

The reason that Direct Proof (1.3.4) works is that $p \to q$ is false only when p is true but q is false (Definition 1.1.3). Therefore, if it can be shown that q is true whenever p is true, $p \to q$ follows.

■ Example 1.3.5

Prove that an integer n is even if and only if n^2 is even. As noted in Example 1.1.5, a biconditional statement is the conjunction of two implications. In this case, both of the implications will be proved with Direct Proof. Let $n \in \mathbb{Z}$.

- Suppose that n is even. This means that $n = 2k$ for some $k \in \mathbb{Z}$. Thus, $n^2 = 4k^2$, which is even because it is divisible by 2. Therefore,

$$\text{if } n \text{ is even, } n^2 \text{ is even.} \tag{1.20}$$

- The converse of (1.20) is difficult to prove, so try to prove the contrapositive of its converse instead (Example 1.1.6). Assume that the integer n is odd. This means that $n + 1 = 2k$ for some $k \in \mathbb{Z}$. Then,

$$n^2 + 1 = (2k - 1)^2 + 1 = 4k^2 - 4k + 2 = 2(2k^2 - 2k + 1),$$

so n^2 is odd because $2k^2 - 2k + 1 \in \mathbb{Z}$. Hence, if n^2 is even, n is even.

■ **Example 1.3.6**

Prove that the following are equivalent for all integers n:

(a) n is even.

(b) n^2 is even.

(c) $n + 1$ is odd.

The standard procedure is to show that (a) implies (b), (b) implies (c), and (c) implies (a). There are other paths that can be followed as long as all sentences are proven to imply each other. Example 1.3.5 has already shown that (a) is equivalent to (b), so it remains to show that (c) can be added to the list. One method is to show that (a) is equivalent to (c).

- Let $n = 2k$ for some $k \in \mathbb{Z}$. Then, $(n + 1) + 1 = 2(k + 1)$, so $n + 1$ is odd.
- Let $n + 2 = 2l$ for some integer l. Then, $n = 2(l - 1)$, so n is even.

■ **Example 1.3.7**

Suppose that it is required to prove that

$$4x - 7 = 5 \text{ has a unique real solution.} \tag{1.21}$$

The statement in (1.21) claims the existence of an element of $\mathbb{R}$ that satisfies an equation and that there is only one such element, so a demonstration of (1.21) requires the proof of two statements.

- **Existence:** $4x - 7 = 5$ for some $x \in \mathbb{R}$.
- **Uniqueness:** For all $x, y \in \mathbb{R}$, if $4x - 7 = 5$ and $4y - 7 = 5$, then $x = y$.

The proof is as follows.

- $x = 3$ is a solution because $4(3) - 7 = 5$.
- Take $a, b \in \mathbb{R}$ and assume that $4a - 7 = 5$ and $4b - 7 = 5$. Then,

$$4a - 7 = 4b - 7,$$
$$4a = 4b,$$
$$a = b.$$

■ **Example 1.3.8**

Prove for all $x, y \in \mathbb{Z}$, if $xy = 0$, then $x = 0$ or $y = 0$. Let a and b be integers and suppose that $ab = 0$. Because $p \vee q \equiv {\sim}p \rightarrow q$, also suppose that $a \neq 0$ to begin the Direct Proof. This implies that a^{-1} exists, so $a^{-1}ab = a^{-1} \cdot 0$. Hence, $b = 0$, and it follows that $a = 0$ or $b = 0$.

Subsets

Let $C = \{1, 2, 3, 4\}$ and $D = \{0, 1, 2, 3, 4, 5, 6\}$. Observe that every element of C is an element of D. This important relation between some sets is defined using an implication and usually proved by Direct Proof.

■ **Definition 1.3.9**

The set A is a **subset** of the set B $(A \subseteq B)$ means

$$(\forall x)(x \in A \rightarrow x \in B).$$

If $A \subseteq B$ but $A \neq B$, A is a **proper subset** of B $(A \subset B)$. If A is not a subset of B $(A \nsubseteq B)$,

$$(\exists x)(x \in A \wedge x \notin B).$$

To show that $A \nsubseteq B$, find an element in A that is not an element of B. For example, if $A = \{1, 2, 3, 4\}$ and $B = \{3, 4, 5\}$, then $A \nsubseteq B$ because $1 \in A$ but $1 \notin B$.

■ **Example 1.3.10**

Let $A = \{9x + 8y : x, y \in \mathbb{Z}\}$ and $B = \{3x + 2y : x, y \in \mathbb{Z}\}$. Take $a \in A$. This means that $a = 9m + 8n = 3(3m) + 4(2n)$ for some $m, n \in \mathbb{Z}$. Because $3m$ and $2n$ are integers, $a \in B$. Hence, $A \subseteq B$.

■ **Example 1.3.11**

To prove if $A \subseteq B$, then $\{1\} \times A \subseteq \{1\} \times B$, assume that A is a subset of B. Let $(1, a) \in \{1\} \times A$. Since $a \in A$, it follows that $a \in B$. Therefore, $(1, a) \in \{1\} \times B$.

The next theorem enumerates some basic subset results. All are true because each is closely linked to certain rules of logic.

■ **Theorem 1.3.12**

Let A, B, and C be sets.

   (a) $\varnothing \subseteq A$
   (b) $A \subseteq A$
   (c) If $A \subseteq B$ and $B \subseteq C$, then $A \subseteq C$.
   (d) If $A \subseteq B$ and $x \in A$, then $x \in B$.
   (e) If $A \subseteq B$ and $x \notin B$, then $x \notin A$.

Proof

(a) is true because $a \in \varnothing$ is false for every element a. This implies that $a \in \varnothing \to a \in A$ is true. To prove (d), assume that $A \subseteq B$ and take $a \in A$. Because $a \in A \to a \in B$, it follows that $a \in B$. The remaining parts are left to Exercise 13. ∎

■ **Example 1.3.13**

The proof of Theorem 1.3.12(d) used a rule of logic. It states that $p \to q$ and p prove q for all statements p and q. That is, $p \to q, p \Rightarrow q$. This reasoning is justified by examining the truth table of

$$[(p \to q) \wedge p] \to q, \tag{1.22}$$

which is as follows:

p	q	$p \to q$	$(p \to q) \wedge p$	$[(p \to q) \wedge p] \to q$
T	T	T	T	T
T	F	F	F	T
F	T	T	F	T
F	F	T	F	T

The last column of the truth table displays that (1.22) is always true. It is true when $p \to q$ or p are false. More notably, it is true when both $p \to q$ and p are true. This makes sense because if p gives q and p is true, q should follow.

There are other rules of logic similar to (1.22) that can also be proved using truth tables. Each rule starts with one or two premises and then has an immediate conclusion. In fact, each can be considered proofs, but very short ones. For this reason, the notation of Definition 1.3.3 is used. For example, the rule of logic used to prove Theorem 1.3.12(d) can be written as $p \to q, p \Rightarrow q$. A related logic rule, which is used to prove Theorem 1.3.12(c), is $p \to q, q \to r \Rightarrow p \to r$.

Set Equality

Two sets are equal when they contain exactly the same elements. This can be defined using the concept of a subset.

■ **Definition 1.3.14**

The sets A and B are **equal** ($A = B$) means $A \subseteq B$ and $B \subseteq A$.

■ **Example 1.3.15**

Let $a, b, n \in \mathbb{Z}$. Prove $S = T$, where

$$S = \{x \in \mathbb{Z} : x \text{ divides } a + nb \text{ and } x \text{ divides } b\}$$

and

$$T = \{x \in \mathbb{Z} : x \text{ divides } a \text{ and } x \text{ divides } b\}.$$

First, prove $S \subseteq T$. Let $u \in S$. Then, $a + nb = ul$ and $b = uk$ for some integers l and k. Substituting, $a + nuk = ul$. Thus, $u \in T$ because $l - nk \in \mathbb{Z}$ and $a = ul - nuk = u(l - nk)$.

Second, prove $T \subseteq S$. Let $u \in T$. This means that u divides both a and b, so there are integers l and k such that $a = ul$ and $b = uk$. Hence, $u \in S$ because $a + nb = ul + nuk = u(l + nk)$ and $l + nk$ is an integer.

■ **Example 1.3.16**

To prove that the set A is nonempty, find an element of A. For instance, $A = \{x \in \mathbb{R} : (x + 4)(x - 9) = 0\} \neq \varnothing$ because $9 \in A$.

■ **Example 1.3.17**

Let $A = \{(a, b) \in \mathbb{R} \times \mathbb{R} : a + b = 0\}$ and $B = \{(0, b) : b \in \mathbb{R}\}$. The family $\{A, B\}$ is not disjoint since $(0, 0)$ is an element of both A and B. However, $A \neq B$ because $(1, -1) \in A$ but $(1, -1) \notin B$, so $\{A, B\}$ contains 2 elements.

Indirect Proof

Suppose that it is required to show that $p_1, p_2, \dots, p_n \Rightarrow q$, but it is very difficult to show that q follows from the premises. What might work is to assume $p_1, p_2, \dots, p_n, \sim q$ and show that what follows is a **contradiction**, a statement that is always false. Since all contradictions are logically equivalent, represent the contradiction by $r \wedge \sim r$. If the contradiction can be reached,

$$p_1, p_2, \dots, p_n, \sim q \Rightarrow r \wedge \sim r.$$

By Direct Proof (Theorem 1.3.4),

$$p_1, p_2, \dots, p_n \Rightarrow \sim q \to r \wedge \sim r.$$

However, since $r \wedge \sim r$ must be false, in order for the conclusion to follow from the premises, $\sim q$ must be false, which implies that q must be true given $p_1, p_2, \dots, p_n$. This argument justifies the next proof technique, which is also known as **proof by contradiction** and **reductio ad absurdum**.

■ **Theorem 1.3.18 [Indirect Proof]**

Let $p_1, p_2, \dots, p_n, q, r$ be propositions. If $p_1, p_2, \dots, p_n, \sim q \Rightarrow r \wedge \sim r$, then $p_1, p_2, \dots, p_n \Rightarrow q$.

■ **Example 1.3.19**

Let $A \subseteq B$. To see that $A \setminus B = \varnothing$, towards contradiction, take $a \in A \setminus B$. This means that $a \in A$ but $a \notin B$, so $a \notin A$ by Theorem 1.3.12(e), which is a contradiction when paired with $a \in A$. Therefore, $A \setminus B$ is empty.

■ **Example 1.3.20**

Prove for all integers n, if n^2 is even, then n is even. In Example 1.3.5, this was proved by showing the contrapositive. The approach here is to try an indirect proof. Take an integer n and let n^2 be even. In order to obtain a contradiction, assume that $n + 1 = 2k$ for some $k \in \mathbb{Z}$. Substituting yields

$$n^2 = (2k - 1)^2 = 4k^2 - 4k + 1 = 2(2k^2 - 2k - 1) + 1,$$

showing that n^2 is odd. This is impossible, which implies that n is even.

■ **Example 1.3.21**

Show that $\sqrt{2}$ is an irrational number. Suppose instead that $\sqrt{2} = a/b$, where $a, b \in \mathbb{Z}$ with $b \neq 0$ and a/b has been reduced. Then, $2 = a^2/b^2$. Therefore, $a^2 = 2b^2$, so $a = 2k$ for some $k \in \mathbb{Z}$ (Example 1.3.5). It follows that $b^2 = 4k^2$, and this implies that b is also even, contradicting the fact that the fraction was reduced.

Mathematical Induction

Consider the equation,

$$\sum_{i=1}^{k} i^2 = \frac{k(k + 1)(2k + 1)}{6}. \tag{1.23}$$

One way to prove that (1.23) is true for all positive integers k is to show that

$$S = \left\{ k \in \mathbb{Z}^+ : \sum_{i=1}^{k} i^2 = \frac{k(k + 1)(2k + 1)}{6} \right\} = \mathbb{Z}^+.$$

Since $S \subseteq \mathbb{Z}^+$, it is enough to prove that $\mathbb{Z}^+ \subseteq S$. This can be accomplished in two steps.

- First, show $1 \in S$. This is true because

$$\frac{1(1 + 1)(2 \cdot 1 + 1)}{6} = \frac{1(2)(3)}{6} = 1.$$

- Second, use Direct Proof (Theorem 1.3.4) to demonstrate that if $n \in S$, then $n + 1 \in S$. To accomplish this, suppose that $n \in S$, so

$$\sum_{i=1}^{n} i^2 = \frac{n(n + 1)(2n + 1)}{6}.$$

Then, $n + 1 \in S$ because

$$\sum_{i=1}^{n+1} i^2 = \sum_{i=1}^{n} i^2 + (n+1)^2$$

$$= \frac{n(n+1)(2n+1)}{6} + (n+1)^2$$

$$= \frac{(n+1)(2n^2+7n+6)}{6}$$

$$= \frac{(n+1)(n+2)(2n+3)}{6}$$

$$= \frac{(n+1)([n+1]+1)(2[n+1]+1)}{6}.$$

In the proof of (1.23), it was proved that both $1 \in S$ and for all $n \in \mathbb{Z}^+$,

$$\text{if } n \in S, \text{ then } n+1 \in S. \tag{1.24}$$

That is, (1.24) gives an infinite sequence of implications,

$$\text{if } 1 \in S, \text{ then } 2 \in S, \quad \text{if } 2 \in S, \text{ then } 3 \in S, \quad \text{if } 3 \in S, \text{ then } 4 \in S, \quad \ldots$$

Because it was proved that $1 \in S$, conclude that $2 \in S$, which in turn implies that $3 \in S$, and then $4 \in S$, and so on forever to discover that each positive integer is an element of S. However, although this process appears reasonable, one cannot actually perform an infinite number of steps. For this reason, that this works is taken as an assumption.

■ Axiom 1.3.22 [Mathematical Induction]

Let S be a set of positive integers. Then, $S = \mathbb{Z}^+$ if

(a) $1 \in S$,

(b) for all $n \in \mathbb{Z}^+$, if $n \in S$, then $n+1 \in S$.

Part (a) of Axiom 1.3.22 is called the **basis case**. The proof of part (b) of Axiom 1.3.22 is called the **induction step**. This is usually proved using Direct Proof with the assumption of the proof being called the **induction hypothesis**.

Since the main work of a proof using mathematical induction focuses on the sentence that describes the set S, the proof often skips the explicit definition of the set and simply works with the sentence. This is the purpose of the next theorem.

■ Theorem 1.3.23

If $p(x)$ is a sentence, $p(n)$ is true for all $n \in \mathbb{Z}^+$ provided that

(a) $p(1)$,

(b) for all $n \in \mathbb{Z}^+$, if $p(n)$, then $p(n+1)$.

Proof

Let $p(1)$ and $p(n) \to p(n + 1)$ for all $n \in \mathbb{Z}^+$. Define $S = \{n \in \mathbb{Z}^+ : p(n)\}$. First, $1 \in S$ because $p(1)$ is true. Second, suppose that $n \in S$. By definition of S, $p(n)$ holds, so $p(n + 1)$ is true. This implies that that $n + 1 \in S$. By Axiom 1.3.22 it follows that $S = \mathbb{Z}^+$. Hence, $p(n)$ is true for all $n \in \mathbb{Z}^+$. ■

■ **Example 1.3.24**

Let $x \geq 0$. Prove by mathematical induction that for any positive integer k,

$$(x + 1)^k \geq x^k + 1.$$

For the basis case, $(x + 1)^1 = x^1 + 1$. Next, let $n \in \mathbb{Z}^+$. Then, **by induction**, that is, by assuming $(x + 1)^n \geq x^n + 1$,

$$(x + 1)^{n+1} = (x + 1)^n(x + 1)$$
$$\geq (x^n + 1)(x + 1)$$
$$= x^{n+1} + x^n + x + 1$$
$$\geq x^{n+1} + 1.$$

Exercises

1. Let p, q, and r be statements. As in Example 1.3.13, use truth tables to confirm the given rules of logic.

   (a) $p \to q, \sim q \Rightarrow \sim p$ (d) $p, q \Rightarrow p \wedge q$

   (b) $p \vee q, \sim q \Rightarrow p$ (e) $p \wedge q \Rightarrow p$

   (c) $p \to q, q \to r \Rightarrow p \to r$ (f) $p \Rightarrow p \vee q$

2. Let $a, b, c, d \in \mathbb{Z}$. Prove using Direct Proof and identify the parts of the proof as *assumption*, *logic*, *definition*, or *theorem/axiom*.

   (a) If a divides b, then a divides bd.

   (b) If a divides b and a divides d, then a^2 divides bd.

   (c) If a divides b and c divides d, then ac divides bd.

   (d) If a divides b and b divides c, then a divides c.

3. Let $a, b \in \mathbb{Z}$. Prove.

   (a) If a and b are even integers, $a^4 + b^4 + 32$ is divisible by 8.

   (b) If a and b are odd integers, 4 divides $a^4 + b^4 + 6$.

4. Prove and identify the parts of the proof as *assumption*, *logic*, *definition*, or *theorem/axiom*.

   (a) The square of every even integer is even.

   (b) The square of some integer is odd.

   (c) There exists an integer that divides every integer.

   (d) The sum of two odd integers is even.

(e) The sum of an even integer and an odd integer is odd.

(f) The product of two odd integers is odd.

(g) The product of an even integer and an odd integer is even.

5. Prove the results of Exercise 4 using Indirect Proof.

6. Prove the following by using Direct Proof to prove the contrapositive.

(a) For every $n \in \mathbb{Z}$, if n^4 is even, then n is even.

(b) For all $n \in \mathbb{Z}$, if $n^3 + n^2$ is odd, then n is odd.

(c) For all $a, b \in \mathbb{Z}$, if ab is even, then a is even or b is even.

7. Prove the results of Exercise 6 using Indirect Proof.

8. Prove that n is odd if and only if $n^3 + n^2 + n$ is odd for all positive integers n.

9. Let $a, b, c, d \in \mathbb{Z}$. Prove and identify the parts of the proof as *assumption, logic, definition,* or *theorem/axiom*.

(a) a is even if and only if a^2 is even.

(b) a is odd if and only if $a + 1$ is even.

(c) a is even if and only if $a + 2$ is even.

(d) If $c \neq 0$, then a divides b if and only if ac divides bc.

10. Prove that the following are equivalent for all $a, b \in \mathbb{Z}^+$. (See Example 1.3.6.)

- a divides b.
- a divides $-b$.
- $-a$ divides b.
- $-a$ divides $-b$.

11. Let $a \in \mathbb{Z}$. Prove that the following are equivalent.

- a is divisible by 3.
- $3a$ is divisible by 9.
- $a + 3$ is divisible by 3.

12. Answer true or false.

(a) $\varnothing \subseteq \varnothing$

(b) $\varnothing \subseteq \{1\}$

(c) $1 \subseteq \mathbb{Z}$

(d) $\{1\} \subseteq \varnothing$

(e) $0 \subseteq \varnothing$

(f) $\{1\} \subseteq \mathbb{Z}$

(g) $\varnothing \subseteq \varnothing \setminus \varnothing$

(h) $\varnothing \in \mathbb{R}$

13. Prove the remaining parts of Theorem 1.3.12.

14. Prove or show false: $A \cup B \subseteq A$ for all sets A and B.

15. Prove.

(a) $\{x \in \mathbb{R} : x^2 - 3x + 2 = 0\} \subseteq \mathbb{N}$

(b) $(0, 1) \subseteq [0, 1]$

(c) $[0, 1] \not\subseteq (0, 1)$

(d) $\mathbb{Z} \times \mathbb{Z} \not\subseteq \mathbb{Z} \times \mathbb{Z}^+$

(e) $(0, 1) \cap \mathbb{Q} \not\subseteq [0, 1] \cap \mathbb{Z}$

(f) $\mathbb{R} \subseteq \mathbb{C}$

(g) $\{5 + bi : b \in \mathbb{R}\} \subseteq \mathbb{C}$

(h) $\mathbb{C} \not\subseteq \mathbb{R}$

(i) $A \cap B \subseteq A$

(j) $A \subseteq A \cup B$

(k) $A \setminus B \subseteq A$

(l) If $A \subseteq B$, then $A \cup C \subseteq B \cup C$.

(m) If $A \subseteq B$, then $A \cap C \subseteq B \cap C$.

(n) If $A \subseteq B$, then $C \setminus B \subseteq C \setminus A$.

16. Let $A \subseteq C$ and $B \subseteq D$. Prove that $A \times B \subseteq C \times D$.

17. Show that $A \subseteq B$ does not imply that $B \subseteq A$.

18. Prove: $\{na : n \in \mathbb{Z}\} \subseteq \{na + mb : n, m \in \mathbb{Z}\}$ for all $a, b \in \mathbb{Z}$.

19. Take $a, b, c \in \mathbb{Z}^+$ such that a divides b and b divides c. Prove that $A \subseteq C$, where $A = \{n \in \mathbb{Z}^+ : n \text{ divides } a\}$ and $C = \{n \in \mathbb{Z}^+ : n \text{ divides } c\}$.

20. Let A, B, and C be sets. Prove.

    (a) $A \cup A = A$

    (b) $A \cap B = B \cap C$.

    (c) $A \cup (B \cup C) = (A \cup B) \cup C$.

    (d) $A \cap (B \cup C) = (A \cap B) \cup (A \cap C)$

    (e) $C \setminus (A \cup B) = (C \setminus A) \cap (C \setminus B)$

21. For each equality of Exercise 20, prove or show false that exchanging union and intersection results in another equality.

22. Prove that for all sets A, B, and C, if $A = B$ and $B = C$, then $A = C$.

23. Let $a, c, m \in \mathbb{Z}$ and $A = \{a + mk : k \in \mathbb{Z}\}$ and $B = \{a + m(c + k) : k \in \mathbb{Z}\}$. Prove that $A = B$.

24. Prove.

    (a) The equation $x - 10 = 23$ has a unique solution.

    (b) The equation $\sqrt{2x - 5} = 2$ has a unique solution.

    (c) For every real number y, the equation $2x + 5y = 10$ has a unique solution.

    (d) The equation $x^2 + 5x + 6 = 0$ has at most two integer solutions.

25. Prove by using Direct Proof but do not use the contrapositive: for all integers a and b, if ab is even, then a is even or b is even.

26. Use mathematical induction to prove that the given equations hold for all positive integers n.

    (a) $1 + 2 + 3 + \cdots + n = \dfrac{n(n + 1)}{2}$

    (b) $1 + 3 + 5 + \cdots + (2n - 1) = n^2$

    (c) $1^2 + 3^2 + 5^2 + \cdots + (2n - 1)^2 = \dfrac{n(2n - 1)(2n + 1)}{3}$

    (d) $1^3 + 2^3 + 3^3 + \cdots + n^3 = \dfrac{n^2(n + 1)^2}{4}$

    (e) $1 + r + r^2 + \cdots + r^n = \dfrac{1 - r^{n+1}}{1 - r} \ (r \neq 1)$

    (f) $1 \cdot 1! + 2 \cdot 2! + \cdots + n \cdot n! = (n + 1)! - 1$

    (g) $\dfrac{1}{2!} + \dfrac{2}{3!} + \cdots + \dfrac{n}{(n + 1)!} = 1 - \dfrac{1}{(n + 1)!}$

    (h) $2 \cdot 6 \cdot 10 \cdot 14 \cdot \cdots \cdot (4n - 2) = \dfrac{(2n)!}{n!}$

27. Prove the given inequalities for all positive integers n.

    (a) $n < 2^n$

    (b) $n! < n^n$

    (c) $\displaystyle\sum_{i=1}^{n} \frac{1}{i^2} \leq 2 - \frac{1}{n}$

28. Let $S = \{n \in \mathbb{Z} : p(n)\}$ and $k \in \mathbb{Z}$. Prove that if $k \in S$ and $n \in S$ implies that $n + 1 \in S$ for all $n = k, k + 1, k + 2, \dots$, then $S = \{k, k + 1, k + 2, \dots\}$.

29. Let n be an integer. Use Exercise 28 to prove the given inequalities.

    (a) $n^2 < 2^n$ for all $n \geq 5$.

    (b) $n^3 < 2^n$ for all $n \geq 10$.

    (c) $2^n < n!$ for all $n \geq 4$.

1.4 Functions

Functions are often viewed as rules that given an input value returns a unique output value. These rules generate ordered pairs of inputs and outputs that are often graphed in the Cartesian plane. This means that functions are actually sets of ordered pairs.

■ **Definition 1.4.1**

Let A and B be sets. A **function** or a **map** f is a subset of $A \times B$ such that

$$\text{for all } (x_1, y_1), (x_2, y_2) \in f, \text{if } x_1 = x_2, \text{then } y_1 = y_2. \qquad (1.25)$$

Since the second coordinate depends only on the first coordinate, define $f(x)$ to be the unique second coordinate such that $(x, f(x)) \in f$. Call $f(x)$ the **image** of x and say that f **maps** x to $f(x)$. Call x a **pre-image** of $f(x)$. Furthermore, with every function there are two important sets. The **domain** of f is the subset of A defined by

$$\text{dom } f = \{x \in A : (x, y) \in f \text{ for some } y \in B\},$$

and the **range** or **image** of f is the subset of B defined by

$$\text{ran } f = \{y \in B : (x, y) \in f \text{ for some } x \in A\} = \{f(x) : x \in \text{dom } f\}.$$

If D is the domain of the function f, then f is a function $D \to B$ or, simply,

$$f : D \to B.$$

The set B is called a **codomain** of f. All of this is illustrated in Figure 1.1.

■ **Example 1.4.2**

The set $f = \{(1, 2), (3, 7), (5, 7)\}$ is a function $\{1, 3, 5\} \to \mathbb{R}$. The domain of f is $\{1, 3, 5\}$ and its range is $\{2, 7\}$. Observe that $1 \in \text{dom } f$ is because $(1, 2) \in f$, and $7 \in \text{ran } f$ because $f(3) = 7$.

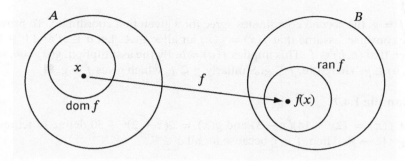

Figure 1.1   A function $f \subseteq A \times B$.

■ **Example 1.4.3**

Define $f = \{(x, 3 + 7\sin \pi x) : x \in \mathbb{R}\}$. Since sine is a function, f is also a function because taking $a_1, a_2 \in \mathbb{R}$,

$$a_1 = a_2,$$
$$\pi a_1 = \pi a_2,$$
$$\sin \pi a_1 = \sin \pi a_2,$$
$$7\sin \pi a_1 = 7\sin \pi a_2,$$
$$3 + 7\sin \pi a_1 = 3 + 7\sin \pi a_2.$$

■ **Example 1.4.4**

For any set A, the **identity map** on A is the function $i_A : A \to A$ defined by
$$i_A = \{(x, x) : x \in A\}.$$
In particular, $i_{\mathbb{R}}(42) = 42$, and $\operatorname{ran} i_{\mathbb{Z}} = \mathbb{Z}$.

A set of ordered-pairs is not a function when there exists $(x, y_1), (x, y_2) \in f$ such that $y_1 \neq y_2$. This is simply the negation of (1.25).

■ **Example 1.4.5**

The set $A = \{(1, 2), (1, 7), (5, 7)\}$ is not a function because $(1, 2), (1, 7) \in A$. Also, $B = \{(|y|, y) : y \in \mathbb{R}\}$ is not a function since $(1, 1) \in B$ and $(1, -1) \in B$.

Because functions are sets, they are equal when they are subsets of each other (Definition 1.3.14). This implies that if the domains of two functions are not equal, the functions cannot be equal. However, when the domains are the same, the next theorem can be used.

■ **Theorem 1.4.6**

The functions $f, g : A \to B$ are equal if and only if $f(x) = g(x)$ for all $x \in A$.

Proof

If $f = g$, the second coordinates agree for a given first coordinate. To prove the converse, assume that $f(x) = g(x)$ for all $x \in A$. Let $a \in A$ and $b \in B$ such that $(a, b) \in f$. This implies $f(a) = b$. By the assumption, $g(a) = b$, so $(a, b) \in g$. Therefore, $f \subseteq g$. Similarly, $g \subseteq f$, which gives $f = g$. ∎

■ **Example 1.4.7**

Let $f(x) = (2x + 14)(x - 3)$ and $g(x) = 2(x + 2)^2 - 50$ define functions $f, g : \mathbb{R} \to \mathbb{R}$. Then, $f = g$ because for all $a \in \mathbb{R}$,

$$f(a) = (2a + 14)(a - 3) = 2a^2 + 8a - 42 = 2(a + 2)^2 - 50 = g(a).$$

■ **Example 1.4.8**

By Theorem 1.4.6, functions f and g with a common domain are not equal when there exists x such that $f(x) \neq g(x)$. For instance, sin and cos are not equal because although their domains are equal, $\sin \pi = 0$, but $\cos \pi = -1$.

■ **Definition 1.4.9**

Let $g : A \to B$ and $f : C \to D$ be functions with ran $g \subseteq C$. The **composition** of f and g is

$$f \circ g = \{(x, f(g(x))) : x \in A\}.$$

Notice that the composition as defined in Definition 1.4.9 is a function. To prove this, take $(a_1, f(g(a_1))), (a_2, f(g(a_2))) \in f \circ g$. Let $a_1 = a_2$. Since g is a function, $g(a_1) = g(a_2)$, and because f is a function, $f(g(a_1)) = f(g(a_2))$. The function $f \circ g$ is illustrated in Figure 1.2, where the set B is not included to simplify the diagram.

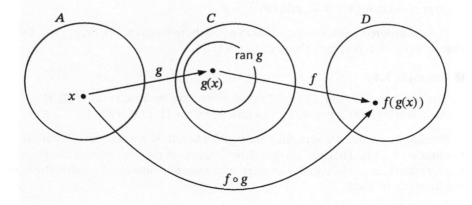

Figure 1.2   The composition of f and g.

■ **Example 1.4.10**

Let $f(x) = 4x - 3$ and $g(x) = x^2 + 3$. Since ran $f \subseteq$ dom g and ran $g \subseteq$ dom f, both compositions can be defined. So,

$$(f \circ g)(x) = f(g(x)) = 4(x^2 + 3) - 3 = 4x^2 + 9,$$

and

$$(g \circ f)(x) = g(f(x)) = (4x - 3)^2 + 3 = 16x^2 - 24x + 12.$$

Therefore, $(f \circ g)(0) = 9$ and $(g \circ g)(0) = 12$. This shows that $f \circ g \neq g \circ f$.

Injections

Let f be a function. Suppose that f also satisfies the converse of (1.25) in Definition 1.4.1. This means that

$$\text{for all } (x_1, y_1), (x_2, y_2) \in f, \text{ if } y_1 = y_2, \text{ then } x_1 = x_2. \tag{1.26}$$

What does this mean about the function? Consider

$$g = \{(1, 2), (2, 4), (3, 4)\} \quad \text{and} \quad h = \{(1, 2), (2, 4), (3, 5)\}. \tag{1.27}$$

Both functions can be viewed as $\{1, 2, 3\} \rightarrow \{1, 2, 3, 4, 5\}$. The only difference between g and h is the second coordinate in the third ordered pair. This difference causes g to have the property that it has a second coordinate that is paired with 2 first coordinates and causes h to have the property that every second coordinate is paired with exactly 1 first coordinate. Essentially, this means that when h is reversed, the result is a function. This cannot be said about g.

■ **Definition 1.4.11**

The function $f : A \rightarrow B$ is **one-to-one** or an **injection** means that for all $x_1, x_2 \in A$, if $f(x_1) = f(x_2)$, then $x_1 = x_2$.

Using the same functions from (1.27), g is not one-to-one (Figure 1.3), but h is one-to-one (Figure 1.4). Observe that the arrows in Figure 1.4 can be reversed to yield a function with ran h as its domain and A as a codomain.

■ **Example 1.4.12**

To show that $f : \mathbb{R} \rightarrow \mathbb{R}$ defined by $f(x) = 7x - 3$ is one-to-one, let $a_1, a_2 \in \mathbb{R}$ and suppose $f(a_1) = f(a_2)$. Then,

$$7a_1 - 3 = 7a_2 - 3,$$
$$7a_1 = 7a_2,$$
$$a_1 = a_2.$$

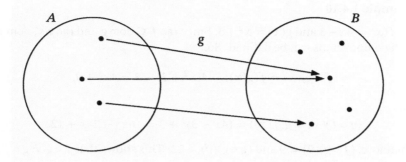

Figure 1.3 g is not a one-to-one function.

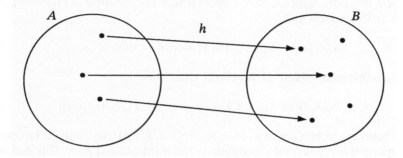

Figure 1.4 h is a one-to-one function.

■ **Example 1.4.13**

Let $g : \mathbb{R} \times \mathbb{R} \to \mathbb{R} \times \mathbb{R} \times \mathbb{R}$ be the function $g(x, y) = (x, y, 0)$. To prove that g is one-to-one, take $(a_1, b_1), (a_2, b_2) \in \mathbb{R} \times \mathbb{R}$ and let $g(a_1, b_1) = g(a_2, b_2)$. This means that $(a_1, b_1, 0) = (a_2, b_2, 0)$. Hence, $a_1 = a_2$ and $b_1 = b_2$, so $(a_1, b_1) = (a_2, b_2)$.

The negation of a function f being one-to-one (Definition 1.4.11) is

there exists $(x_1, y_1), (x_2, y_2) \in f$ such that $y_1 = y_2$ and $x_1 \neq x_2$.

Hence, to show that a function is not one-to-one, it is necessary to find an element of the range that has at least two pre-images (Figure 1.3). An example of such a function is $f : \mathbb{R} \to \mathbb{R}$ defined by $f(x) = |x|$. This function is not an injection because $f(1) = 1$ and $f(-1) = 1$.

Both $f(x) = 5x$ and $g(x) = \sqrt{x}$ are one-to-one functions. Their compositions,

$$(f \circ g)(x) = 5\sqrt{x} \quad \text{and} \quad (g \circ f)(x) = \sqrt{5x},$$

are also one-to-one. This is true for all compositions of one-to-one functions.

■ **Theorem 1.4.14**

Let $g : A \to B$ and $f : C \to D$ be functions such that ran $g \subseteq C$. If f and g are one-to-one, $f \circ g$ is one-to-one.

Proof

Let f and g be injections. Take $a_1, a_2 \in A$ and let $(f \circ g)(a_1) = (f \circ g)(a_2)$. Because f is one-to-one, $g(a_1) = g(a_2)$, and since g is one-to-one, $a_1 = a_2$. ■

Surjections

The range and codomain of a function need not be equal, but there is a name for functions in which they are.

■ **Definition 1.4.15**

The function $f : A \to B$ is **onto** or a **surjection** means for every $y \in B$, there exists $x \in A$ such that $f(x) = y$.

Figures 1.3 and 1.4 are diagrams of one-to-one functions that are not onto, and the diagram of Figure 1.5 is that of a function that is onto but is not one-to-one. The purpose of the figures is to represent how to deal with the idea of an onto function. Figure 1.5 illustrates that to show that a function is onto, it must be demonstrated that every element of its codomain has a pre-image.

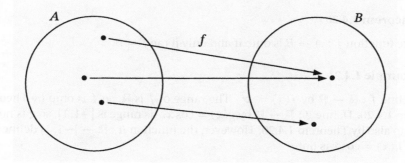

Figure 1.5 f is an onto function.

■ **Example 1.4.16**

To show that $f : \mathbb{R} \to \mathbb{R}$ defined by $f(x) = 7x - 3$ is onto, let $b \in \mathbb{R}$. Then, $(b + 3)/7 \in \mathbb{R}$ and

$$f\left(\frac{b + 3}{7}\right) = 7\left(\frac{b + 3}{7}\right) - 3 = b.$$

■ **Example 1.4.17**

Define $f : \mathbb{R} \times \mathbb{R} \times \mathbb{R} \to \mathbb{R} \times \mathbb{R}$ by $f(x, y, z) = (x, y)$, which is not one-to-one because $f(0, 1, 2) = f(0, 1, 3) = (0, 1)$. However, f is onto because for all $(a, b) \in \mathbb{R}^2$, it is the case that $f(a, b, 0) = (a, b)$.

Figures 1.3 and 1.4 illustrate that to show that a function is not onto, an element of the codomain that does not have a pre-image must be found.

■ **Example 1.4.18**

Define $h : \mathbb{C} \to \mathbb{C}$ by $h(a + bi) = a$. Because i has no pre-image, h is not onto. Notice that if h was redefined as a function $\mathbb{C} \to \mathbb{R}$, then h would be a surjection.

■ **Example 1.4.19**

The function $f : \mathbb{R} \times \mathbb{R} \to \mathbb{R} \times \mathbb{R} \times \mathbb{R}$ defined by $f(a, b) = (a, b, 0)$ is not onto because $(1, 1, 1)$ does not have a pre-image in $\mathbb{R}^2$. Indeed, every element of the range of f has a 0 in its third coordinate.

Let $f : A \to B$ be an onto function. Since ran $f \subseteq B$, to show equality, take $b \in B$. By Definition 1.4.15, there exists $a \in A$ such that $f(a) = b$. This implies that $b \in$ ran f, so if f is onto, its range equals its codomain. To prove the converse, let ran $f = B$. Take $b \in B =$ ran f. Hence, there exists $a \in A$ such that $f(a) = b$, so f is onto. This proves the next result.

■ **Theorem 1.4.20**

The function $f : A \to B$ is onto if and only if ran $f = B$.

■ **Example 1.4.21**

Define $f : \mathbb{R} \to \mathbb{R}$ by $f(x) = x^3$. The range of f is $\mathbb{R}$, so f is onto by Theorem 1.4.20. Define $g : \mathbb{R} \to \mathbb{R}$ by $g(x) = \cos x$. Its range is $[-1, 1]$, so g is not onto also by Theorem 1.4.20. However, the function $h : \mathbb{R} \to [-1, 1]$ defined by $h(x) = \cos x$ is onto.

Consider the functions $f : \mathbb{R} \to [0, \infty)$ defined by $f(x) = x^2$ and $g : [0, \infty) \to \mathbb{R}$ defined by

$$g(x) = \begin{cases} \ln x & \text{if } x > 0, \\ 0 & \text{if } x = 0. \end{cases}$$

Both are surjections. Their compositions,

$$(f \circ g)(x) = \begin{cases} (\ln x)^2 & \text{if } x > 0, \\ 0 & \text{if } x = 0, \end{cases}$$

which is a function $[0, \infty) \to [0, \infty)$, and

$$(g \circ f)(x) = \begin{cases} \ln x^2 & \text{if } x \neq 0, \\ 0 & \text{if } x = 0, \end{cases}$$

which is a function $\mathbb{R} \to \mathbb{R}$, are also surjections. This is true for all compositions of onto functions under the right conditions.

■ **Theorem 1.4.22**

Let $g : A \to B$ and $f : C \to D$ be functions such that ran $g = C$. If f is onto, $f \circ g$ is onto.

Proof

Assume that f and g are onto. Let $d \in D$. Since f is onto, there exists $c \in C$ such that $f(c) = d$. Because c is also an element of the range of g, there exists $a \in A$ such that $g(a) = c$. Thus, $(f \circ g)(a) = f(g(a)) = f(c) = d$. ■

Bijections and Inverses

Examples 1.4.12 and 1.4.16 suggest that a linear function that is not a horizontal line is both one-to-one and onto (Exercise 17). There are many functions that share both of these properties. They are named using the following terminology.

■ **Definition 1.4.23**

A **bijection** or a **one-to-one correspondence** is a function that is both an injection and a surjection.

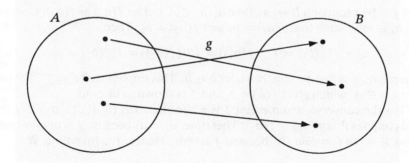

Figure 1.6 g is a bijection.

A bijection can be represented as in Figure 1.6. Make this concrete by setting

$$g = \{(1, 5), (2, 4), (3, 6)\}. \tag{1.28}$$

This set is a bijection $\{1, 2, 3\} \to \{4, 5, 6\}$. This means that every first coordinate of an ordered pair of g is paired with exactly one second coordinate (function), every second coordinate from g is paired with exactly one first coordinate (one-to-one), and every element of the codomain is an element of the range (onto). Notice that all of this implies that if the ordered pairs were reversed, the result would be the bijection, $\{(5, 1), (4, 2), (6, 3)\}$. Of course, this does not work all of the time. For example, if $h = \{(1, 2), (4, 7), (6, 7)\}$, then switching the coordinates results in the set $\{(2, 1), (7, 4), (7, 6)\}$, which is not a function.

■ Definition 1.4.24

Given a function $f : A \rightarrow B$, the **inverse** of f is defined as

$$f^{-1} = \{(y, x) : (x, y) \in f\},$$

and f is **invertible** means that f^{-1} is a function $B \rightarrow A$.

This means that $g^{-1} = \{(5, 1), (4, 2), (6, 3)\}$, where g is defined in (1.28), and because g^{-1} is a function $\{4, 5, 6\} \rightarrow \{1, 2, 3\}$, the function g is invertible. Also, recall that $f(x) = 7x - 3$ is a bijection $\mathbb{R} \rightarrow \mathbb{R}$. Using the standard algebra technique, f^{-1} is a function $\mathbb{R} \rightarrow \mathbb{R}$ given by

$$f^{-1}(x) = \frac{x + 3}{7}. \tag{1.29}$$

These examples suggest that there is a connection between a function being a bijection and a function being invertible.

■ Theorem 1.4.25

$f : A \rightarrow B$ is invertible if and only if f is a bijection.

Proof

Let f^{-1} be a function $B \rightarrow A$. Take $a_1, a_2 \in A$ and let $f(a_1) = f(a_2)$. Choose $b_1, b_2 \in \operatorname{ran} f$ such that $f(a_1) = b_1$ and $f(a_2) = b_2$. Then,

$$a_1 = f^{-1}(b_1) = f^{-1}(f(a_1)) = f^{-1}(f(a_2)) = f^{-1}(b_2) = a_2.$$

Therefore, f is one-to-one. Next, let $b \in B$. This implies that $a = f^{-1}(b)$ for some $a \in A$, which gives $f(a) = b$, and f is shown to be onto.

For the converse, assume that f is a bijection. Let $(b, a_1), (b, a_2) \in f^{-1}$. This implies that $f(a_1) = f(a_2)$. Therefore, $a_1 = a_2$ because f is one-to-one. Also, $B = \operatorname{ran} f = \operatorname{dom} f^{-1}$ because f is onto. Hence, f is invertible. ■

■ Example 1.4.26

The function $f(x) = 4x - 3$ is invertible. To see this, examine its inverse,

$$f^{-1} = \{(4x - 3, x) : x \in \mathbb{R}\}.$$

Suppose that b_1 and b_2 are real numbers and let $4b_1 - 3 = 4b_2 - 3$. Some simple algebra leads to $b_1 = b_2$. Therefore, f^{-1} is a function, so f is invertible and, by Theorem 1.4.25, f is also a bijection.

■ **Example 1.4.27**

The function $g(x) = x^2 + 3$ is not invertible because $g(1) = g(-1)$. However, by restricting the domain of g, a function can be defined that is invertible. Let $h(x) = x^2 + 3$ with $x \geq 0$. Then, h is a bijection $\mathbb{R} \to [3, \infty)$, so h is invertible.

Composition of functions provides a test for whether two functions are inverses of each other.

■ **Theorem 1.4.28**

Let $f : A \to B$ and $g : B \to A$ be functions. The following are equivalent.

(a) $f^{-1} = g$.

(b) $(f \circ g)(x) = x$ for all $x \in B$ and $(g \circ f)(y) = y$ for all $y \in A$.

Proof

First, assume (a). Let $a \in A$. Then, there exists $b \in B$ such that $f(a) = b$, which implies that $g(b) = a$ because $f^{-1} = g$. Hence,

$$(g \circ f)(a) = g(f(a)) = g(b) = a.$$

The first statement of (b) follows similarly.

Second, assume (b). Since the domains of f^{-1} and g might be different, Theorem 1.4.6 does not apply. For this reason, take $(b, a) \in f^{-1}$, so $f(a) = b$. Then, $(b, a) \in g$ because $g(f(a)) = a$. Thus, $f^{-1} \subseteq g$. Similar reasoning shows that $g \subseteq f^{-1}$. ■

■ **Example 1.4.29**

Take $f(x) = 4x - 3$ from Example 1.4.26. To see that $g(x) = (x + 3)/4$ is the inverse of f, use Theorem 1.4.28 and confirm that

$$(f \circ g)(x) = 4\left(\frac{x+3}{4}\right) - 3 = x \quad \text{and} \quad (g \circ f)(x) = \frac{(4x-3)+3}{4} = x.$$

The next two results follow from Theorems 1.4.14, 1.4.22, and 1.4.25.

■ **Theorem 1.4.30**

If $g : A \to B$ and $f : B \to C$ are bijections, f^{-1} and $f \circ g$ are bijections.

■ **Corollary 1.4.31**

If $g : A \to B$ and $f : B \to C$ are invertible, f^{-1} and $f \circ g$ are invertible.

To illustrate Theorem 1.4.30, the function f^{-1} given in (1.29) is a bijection with domain and codomain equal to $\mathbb{R}$. Also, $\tan^{-1} : \mathbb{R} \to (-\pi/2, \pi/2)$ is a bijection. Thus, $\tan^{-1} \circ f^{-1} : \mathbb{R} \to (-\pi/2, \pi/2)$ is a bijection.

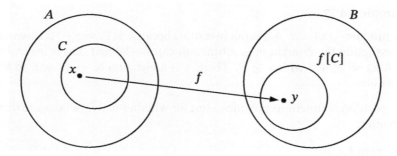

Figure 1.7   The image of C under f.

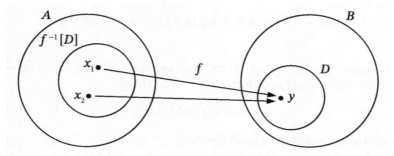

Figure 1.8   The inverse image of D under f.

Images and Inverse Images

There are times when it is important to know the images of all of the elements in a subset of the domain of a function. Suppose, for example, that

$$f = \{(1, 2), (3, 4), (5, 6), (7, 8)\}. \tag{1.30}$$

The set $\{1, 3\}$ is a subset of the domain of f, and the set of images of this subset is the set $\{2, 4\}$. Occasionally it is required to know the pre-images of a subset of the range of a function. The set $\{6, 8\}$ is a subset of the range of f, and the set of pre-images of this subset is the set $\{5, 7\}$. This leads to the next two definitions, which are illustrated in Figures 1.7 and 1.8.

■ **Definition 1.4.32**

Let $f : A \to B$ be a function and C be a set.

(a) $f[C] = \{f(x) : x \in C\} =$ the **image** of C under f.

(b) $f^{-1}[C] = \{x \in A : f(x) \in C\} =$ the **inverse image** of C under f.

Definition 1.4.32 implies that $f[A] = \operatorname{ran} f$ and $f^{-1}[B] = \operatorname{dom} f$ for any function $f : A \to B$. Also, $f[\varnothing] = \varnothing$, and $f^{-1}[\varnothing] = \varnothing$.

■ **Example 1.4.33**

Define $f : \mathbb{R} \to \mathbb{R}$ by $f(x) = x^2 - 1$.

- $f[(1, 2)] = (0, 3)$.
- $f[(-2, -1)] = (0, 3)$.
- $f^{-1}[(0, 3)] = (-2, -1) \cup (1, 2)$.

- $f^{-1}[(-2, -1]] = \{0\}$.
- $f^{-1}[(-2, -1)] = \varnothing$.

Example 1.4.33 (proofs left to Exercise 31) shows that $f^{-1}[D] = C$ does not follow from $f[C] = D$. To do so requires an additional hypotheses on the function as noted in Exercise 32.

Operations

Addition and multiplication of real numbers share the property that given two possibly equal real numbers, the sum of those numbers will always be the same real number and the product of those numbers will always be the same real number. This means that both addition and multiplication are functions $\mathbb{R} \times \mathbb{R} \to \mathbb{R}$.

■ **Definition 1.4.34**

For any set A, let $*$ be a function $A \times A \to A$. This function can be considered a **binary operation** on A when $*(x_1, x_2)$ is written as $x_1 * x_2$ for all $x_1, x_2 \in A$.

For example, addition of real numbers is a binary operation denoted by the $+$ symbol. Because addition is a function $\mathbb{R} \times \mathbb{R} \to \mathbb{R}$, write $+(4, 2) = 6$, which is usually written as $4 + 2 = 6$. Also, given $a, b \in \mathbb{R}$, it follows that $a + b \in \mathbb{R}$ since the codomain of addition is $\mathbb{R}$. This can be generalized to any binary operation.

■ **Definition 1.4.35**

The set A is **closed** under the binary operation $*$ means $x * y \in A$ for all $x, y \in A$.

■ **Example 1.4.36**

Define $*$ so that $x * y = 4x - 2y$ for all $x, y \in \mathbb{R}$. To prove that $*$ is a binary operation on $\mathbb{R}$, show that $*$ is a function with domain $\mathbb{R} \times \mathbb{R}$ and that $\mathbb{R}$ is closed under $*$.

- Let $(a_1, b_1), (a_2, b_2) \in \mathbb{R} \times \mathbb{R}$ such that $(a_1, b_1) = (a_2, b_2)$. Then,

$$a_1 * b_1 = 4a_1 - 2b_1 = 4a_2 - 2b_2 = a_2 * b_2.$$

- Take $a, b \in \mathbb{R}$. It is known that $\mathbb{R}$ is closed under multiplication and subtraction, so $a * b \in \mathbb{R}$ because $a * b = 4a + 2b$.

■ **Example 1.4.37**

For all sets A and B, define

$$^AB = \{f \ : \ f \text{ is a function } A \to B\}. \tag{1.31}$$

To illustrate, if $h(x) = 2x + 1$ with dom $h = \mathbb{R}$, then $h \in {}^{\mathbb{R}}\mathbb{R}$. Now, fix a set A. Define the binary operations $+$ (**function addition**) and $-$ (**function subtraction**) on $^A\mathbb{R}$ by taking functions $f, g : A \to \mathbb{R}$ and setting

$$f + g = \{(x, f(x) + g(x)) \ : \ x \in A\},$$
$$f - g = \{(x, f(x) - g(x)) \ : \ x \in A\}.$$

Similar definitions can be made for multiplication and division (Exercise 43). To confirm that $f + g$ is a binary operation, proceed as follows:

- To prove that $f+g$ is a function, let $(x_1, y_1), (x_1, y_2) \in f+g$ with $x_1 = x_2$. Since f and g are functions,

$$y_1 = f(x_1) + g(x_1) = f(x_2) + g(x_2) = y_2.$$

- Since $f + g$ is a function with domain A and codomain $\mathbb{R}$, it follows that $^A\mathbb{R}$ is closed under function addition.

That $f - g$ is a binary operation is proved similarly.

Many binary operations share many properties with the operations of addition and multiplication of real numbers. The next definition gives four of these properties.

■ **Definition 1.4.38**

Let $*$ be a binary operation on A. It is assumed that multiple instances of $*$ are read from left to right unless grouped by parentheses or brackets, which have precedence. If $*$ is called **addition**, it is usually denoted by $+$. If $*$ is called **multiplication**, it is usually denoted by $\cdot$ or excluded altogether as in $a \cdot b = ab$. Multiplication will always have precedence over addition unless overridden by parentheses or brackets.

(a) $*$ is **associative** means that for all $x, y, z \in A$,

$$x * y * z = x * (y * z).$$

(b) $*$ is **commutative** means that for all $x, y \in A$.

$$x * y = y * x.$$

(c) $e \in A$ is an **identity** with respect to $*$ when for all $x \in A$,

$$e * x = x * e = x.$$

If $*$ is called addition, e is an **additive identity** and usually denoted by 0, and if $*$ is called multiplication, e is a **multiplicative identity** and is usually denoted by 1.

(d) Let $e \in A$ be an identity respect to $*$. For all $x \in A$, the element $x' \in A$ is an **inverse** of x with respect to $*$ if

$$x * x' = x' * x = e.$$

If $*$ is called addition, x' is an **additive inverse** and usually denoted by $-x$, and if $*$ is called multiplication, x' is a **multiplicative inverse** and is usually denoted by x^{-1}.

It is taken for granted that addition and multiplication are both associative and commutative on $\mathbb{R}$, that the real number 0 is an additive identity and the real number 1 is a multiplicative identity, and that every real number has an additive inverse and every nonzero real number has a multiplicative inverse. It is also known that these values are unique.

■ Theorem 1.4.39

Let $*$ be a binary operation on A.

(a) If A has an identity with respect to $*$, it is unique.

(b) If $*$ is associative and $a \in A$ has an inverse with respect to $*$, the inverse is unique.

Proof

Let e_1 and e_2 be identities of A. Then, $e_1 = e_1 * e_2 = e_2$. Also, let $*$ be associative. Take $a \in A$ and suppose that a' and a'' are inverses of a. Then,

$$a' = a' * e = a' * (a * a'') = a' * a * a'' = e * a'' = a''. ■$$

■ Example 1.4.40

- Addition and multiplication of functions is associative and commutative (Exercise 38).

- Define the **zero function** on the set A by

$$0_A = \{(x, 0) : x \in A\}.$$

This is a function by Exercise 39. To prove that 0_A is the additive identity of $^A\mathbb{R}$, let f be a function $A \to \mathbb{R}$. Then, for all $x \in A$,

$$(0_A + f)(x) = 0_A(x) + f(x) = 0 + f(x) = f(x).$$

Since function addition is commutative, $f + 0_A = 0_A + f = f$ by Theorem 1.4.6.

- Let $f \in {}^A\mathbb{R}$. The **negative** of f is

$$-f = \{(x, -f(x)) : x \in A\},$$

and is a function by Exercise 40. To see that $-f$ is the additive inverse of f, evaluate for all $x \in A$,

$$[f + (-f)](x) = f(x) + (-f)(x) = f(x) + [-f(x)] = 0.$$

Therefore, $f + (-f) = -f + f = 0_A$.

There is a second type of operation besides the binary operation given in Definition 1.4.34. Its main difference is that its left-hand input is always a real number.

■ **Definition 1.4.41**

For any set A, let $\cdot$ be a function $\mathbb{R} \times A \to A$. This function can be considered a **scalar operation** on A when $\cdot(r, x)$ is written as $r \cdot x$ or rx for all $x \in A$ and $r \in \mathbb{R}$. The value rx is called the **scalar multiple** of x by r.

■ **Definition 1.4.42**

The set A is **closed** under the scalar operation $\cdot$ means $r \cdot x = rx \in A$ for all $x \in A$ and $r \in \mathbb{R}$.

■ **Example 1.4.43**

Given $f : A \to \mathbb{R}$ and $r, s \in \mathbb{R}$, the scalar multiple of f by r is

$$rf = \{(x, rf(x)) : x \in A\}. \tag{1.32}$$

As an example, if $f(x) = \tan x$, then $7f(x) = 7 \tan x$. That rf is always a function (Exercise 41) proves that (1.32) is a scalar operation $\mathbb{R} \times {}^{\mathbb{R}}\mathbb{R} \to {}^{\mathbb{R}}\mathbb{R}$. Furthermore, as seen in Exercise 42,

$$
\begin{aligned}
rsf &= r(sf), & r(f + g) &= rf + rg, \\
(r + s)f &= rf + sf, & 1f &= f.
\end{aligned}
$$

Exercises

1. Indicate whether each given set is a function. If a set is not a function, find a first coordinate that is paired with 2 second coordinates.

    (a) $\{(1, 2), (2, 3), (3, 4), (4, 5), (5, 1)\}$ (d) $\{(x, \pm\sqrt{|x|}) : x \in \mathbb{R}\}$

    (b) $\{(1, 1), (1, 2), (1, 3), (1, 4), (1, 5)\}$ (e) $\{(x, x^2) : x \in \mathbb{R}\}$

    (c) $\{(x, \sqrt{|x|}) : x \in \mathbb{R}\}$

2. Prove that the given sets are functions. Identify the range of each.

    (a) $\{(x, 1/x) : x \in \mathbb{R} \setminus \{0\}\}$ (c) $\{(x, |x|) : x \in \mathbb{R}\}$

    (b) $\{(x, x + 1) : x \in \mathbb{Z}\}$ (d) $\{(x, \sqrt{x}) : x \in [0, \infty)\}$

3. Let $f = \{(x, y) \in \mathbb{R}^2 : 2x + y = 1\}$. Show that f is a function with domain and codomain equal to the set of real numbers.

4. Let $f, g : \mathbb{R} \to \mathbb{R}$ be functions. Prove that $h(x, y) = (f(x), g(y))$ is a function with domain and codomain equal to $\mathbb{R} \times \mathbb{R}$.

5. Explain why each of the given equations does not describe a function.

    (a) $y = 5 \pm x$ (c) $x = 4y^2 - 1$

    (b) $x^2 + y^2 = 1$ (d) $y^2 - x^2 = 9$

6. Let f and g be functions. Prove the following.

    (a) If f and g are functions, $f \cap g$ is a function.

    (b) $f \cup g$ is a function if and only if $f(x) = g(x)$ for all $x \in \operatorname{dom} f \cap \operatorname{dom} g$.

7. Show that $\emptyset$ is a function and find its domain and range.

8. Prove that the given pairs of functions are equal.

    (a) $f(x) = (x-1)(x-2)(x+3)$ and $g(x) = x^3 - 7x + 6$ where $f, g : \mathbb{R} \to \mathbb{R}$.

    (b) $f(a, b) = a + b$ and $g(a, b) = b + a$ where $f, g : \mathbb{Z} \times \mathbb{Z} \to \mathbb{Z}$.

9. Show that the given pairs of functions are not equal.

    (a) $f(x) = x$ and $g(x) = 2x$ where $f, g : \mathbb{R} \to \mathbb{R}$.

    (b) $f(x) = x - 3$ and $g(x) = x + 3$ where $f, g : \mathbb{R} \to \mathbb{R}$.

10. Prove that each function f is one-to-one.

    (a) $f : \mathbb{R} \to \mathbb{R}, \quad f(x) = 2x + 1$

    (b) $f : \mathbb{R} \times \mathbb{R} \to \mathbb{R} \times \mathbb{R}, \quad f(x, y) = (3y, 2x)$

    (c) $f : \mathbb{R} \setminus \{9\} \to \mathbb{R} \setminus \{0\}, \quad f(x) = 1/(x - 9)$

    (d) $f : \mathbb{Z} \times \mathbb{R} \to \mathbb{Z} \times \mathbb{R}, \quad f(n, x) = (3n, e^x)$

11. Prove that each function f is not one-to-one.

    (a) $f : \mathbb{R} \to \mathbb{R}, \quad f(x) = x^4 + 3$

    (b) $f : \mathbb{R} \to \mathbb{R}, \quad f(x) = |x - 2| + 4$

12. Prove that each function f is onto.

    (a) $f : \mathbb{R} \to \mathbb{R}, \quad f(x) = 2x + 1$

    (b) $f : \mathbb{R} \to \mathbb{R}^+, \quad f(x) = e^x$

    (c) $f : \mathbb{R} \setminus \{0\} \to \mathbb{R} \setminus \{0\}, \quad f(x) = 1/x$

    (d) $f : \mathbb{Z} \times \mathbb{Z} \to \mathbb{Z}, \quad f(a, b) = a + b$

13. Prove that each function f is not onto.

    (a) $f : \mathbb{R} \to \mathbb{R}, \quad f(x) = e^x$

    (b) $f : \mathbb{R} \to \mathbb{R}, \quad f(x) = |x|$

    (c) $f : \mathbb{Z} \times \mathbb{Z} \to \mathbb{Z} \times \mathbb{Z}, \quad f(a, b) = (3a, b^2)$

14. A function $f : \mathbb{R} \to \mathbb{R}$ is **increasing** means that for all $x_1, x_2 \in \mathbb{R}$, if $x_1 < x_2$, then $f(x_1) < f(x_2)$. Prove that if $f : \mathbb{R} \to \mathbb{R}$ is increasing, then f is one-to-one.

15. Let $f : \mathbb{R} \to \mathbb{R}$ be a function. As in Exercise 14, define what it means for f to be **decreasing** and prove that a decreasing function is one-to-one.

16. Find an increasing function $\mathbb{R} \to \mathbb{R}$ that is not onto.

17. Prove that a linear function that is neither a horizontal nor a vertical line is both one-to-one and onto.

18. Define $f : \mathbb{Q} \times \mathbb{Z} \to \mathbb{Z} \times \mathbb{Q}$ by $f(x, y) = (y, x)$. Show that f is a bijection.

19. Show that the function $g : A \times B \to C \times D$ defined by $g(a, b) = (f_1(a), f_2(b))$ is a bijection if both $f_1 : A \to C$ and $f_2 : B \to D$ are bijections. Is the converse true?

20. Define $f : A \times (B \times C) \to (A \times B) \times C$ by $f(a, (b, c)) = ((a, b), c)$. Prove f is a bijection.

21. Prove that the empty set is a bijection with domain and range equal to $\varnothing$.

22. Let $g : A \to B$ and $f : B \to C$ be functions. Prove or show false.

    (a) If $f \circ g$ is one-to-one, f is one-to-one.
    (b) If $f \circ g$ is one-to-one, g is one-to-one.
    (c) If $f \circ g$ is onto, f is onto.
    (d) If $f \circ g$ is onto, g is onto.

23. Let $f : A \to B$ and $g : B \to C$ be functions. Find examples if possible.

    (a) f is one-to-one, but $g \circ f$ is not one-to-one.
    (b) g is one-to-one, but $g \circ f$ is not one-to-one.
    (c) f is onto, but $g \circ f$ is not onto.
    (d) g is onto, but $g \circ f$ is not onto.

24. Given its definition, domain, and codomain, find the inverse of f.

    (a) $f(x) = 4x + 8$, $\mathbb{R} \to \mathbb{R}$ (c) $f(x) = \sin x$, $\mathbb{R} \to \mathbb{R}$
    (b) $f(x) = |x|$, $\mathbb{R} \to [0, \infty)$ (d) $f(x) = \ln x$, $(0, \infty) \to \mathbb{R}$

25. Given its definition, domain, and codomain, prove whether f is invertible.

    (a) $f(x) = 2^x + 1$, $\mathbb{R} \to \mathbb{R}$
    (b) $f(x) = |x|$, $(-\infty, 0] \to [0, \infty)$
    (c) $f(x) = \tan x$, $(-\pi/2, \pi/2) \to \mathbb{R}$
    (d) $f(x, y) = (2x + y, x - y)$, $\mathbb{R} \times \mathbb{R} \to \mathbb{R} \times \mathbb{R}$

26. Use Theorem 1.4.28 to prove that the inverse of $f(x) = \dfrac{x + 3}{x + 5}$ is $g(x) = \dfrac{3 - 5x}{x - 1}$.

27. Find $f[A]$ given the definitions of f and A. Assume $A \subseteq \operatorname{dom} f$.

    (a) $f(x) = 5x - 2$, $A = [2, 5]$
    (b) $f(x) = e^x$, $A = \mathbb{R}$
    (c) $f(x) = \csc x$, $A = (0, \pi)$
    (d) $f(x, y) = (3x, y)$, $A = [-2, 2] \times [-2, 2]$

28. Find $f^{-1}[A]$ given the definitions of f and A. Assume $A \subseteq \operatorname{ran} f$.

    (a) $f(x) = x + 1$, $A = [-3, 5]$
    (b) $f(x) = x^4$, $A = [1, 16]$
    (c) $f(x) = \arctan x$, $A = [0, \pi/4]$
    (d) $f(x, y) = (3x, y)$, $A = [-2, 2] \times [-2, 2]$

29. Given $f : A \rightarrow B$, let $C, D \subseteq A$ and $E, F \subseteq B$. Prove the following.

    (a) $f[C \cup D] = f[C] \cup f[D]$ (c) $f^{-1}[E \cup F] = f^{-1}[E] \cup f^{-1}[F]$

    (b) $f[C \cap D] \subseteq f[C] \cap f[D]$ (d) $f^{-1}[E \cap F] = f^{-1}[E] \cap f^{-1}[F]$

30. Prove or show false: for all $f : A \rightarrow B$, if $C \subseteq A$, then $f^{-1}[f[C]] \subseteq C$.

31. Prove the equations of Example 1.4.33.

32. Let f be a function $A \rightarrow B$. Prove.

    (a) f is one-to-one if and only if $f^{-1}[f[C]] = C$ for all subsets C of A.

    (b) f is onto if and only if $f[f^{-1}[D]] = D$ for all subsets D of B.

33. Prove whether each set is closed under the given binary operation.

    (a) $\mathbb{Z}$, division

    (b) $\mathbb{N}$, subtraction

    (c) $\{2n + 1 : n \in \mathbb{Z}\}$, addition

    (d) $\mathbb{C} \times \mathbb{Z}$, $(z_1, n_1) * (z_2, n_2) = (n_1 z_1 z_2, 3n_1 + n_2)$

34. Determine if $*$ is a binary operation on A and prove the result.

    (a) $x * y = y - x$, $A = \mathbb{Z}$

    (b) $x * y = e^{xy}$, $A = \mathbb{R}$

    (c) $x * y = 3(x + y)$, $A = \{2n + 1 : n \in \mathbb{Z}\}$

    (d) $(x_1, y_1) * (x_2, y_2) = (x_1 + y_1, x_2 y_2)$, $A = \mathbb{Z} \times \mathbb{Z}$

35. Determine which binary operations from Exercise 34 are associative and which are commutative.

36. Which binary operations from Exercise 34 have an identity?

37. For each binary operation in Exercise 34, find which elements have an inverse and find that inverse.

38. Prove that addition of functions is associative and commutative. Can the same be said of function subtraction? Explain.

39. Given a set A, prove that 0_A as defined in Example 1.4.40 is a function.

40. Given a set A, prove that $-f$ as defined in Example 1.4.40 is a function.

41. Let f be a function $\mathbb{R} \rightarrow \mathbb{R}$ and let $r \in \mathbb{R}$. Prove that rf is a function. See Example 1.4.43.

42. Let f and g be functions $\mathbb{R} \rightarrow \mathbb{R}$ and let $r, s \in \mathbb{R}$. Prove that $rsf = r(sf)$, $r(f + g) = rf + rg$, $(r + s)f = rf + sf$, and $1f = f$.

43. Following Example 1.4.37, define the operations of function multiplication and function division. Are these operations associative and commutative? Do these operations have an identity and inverses? Prove the results.

CHAPTER 2

Euclidean Space

2.1 Vectors

Basic algebra is the study of $\mathbb{R}$ with various operations, such as addition and multiplication. This is extended to $\mathbb{R} \times \mathbb{R}$ with equations for lines, distances between points, and angle measures. In calculus, eventually $\mathbb{R} \times \mathbb{R} \times \mathbb{R}$ is studied. At this point, vectors are introduced, which are visualized as arrows either in a plane or in space. Although there is a close connection between a point and a vector, points and vectors are typically represented by different notation. This is also done here but using a notation different from that usually found in a calculus course. The reason for this will become apparent later.

■ **Definition 2.1.1**

Let n be a positive integer and define

$$\mathbb{R}^n = \left\{ \begin{bmatrix} x_1 \\ x_2 \\ \vdots \\ x_n \end{bmatrix} : x_1, x_2, \ldots, x_n \in \mathbb{R} \right\}.$$

Each element of $\mathbb{R}^n$ is called a **vector** or sometimes a **column vector** or a **Cartesian vector**, and each real number a_i in a vector is called a **coordinate** of the vector. The integer n is the **dimension** of $\mathbb{R}^n$.

The definition of $\mathbb{R}^n$ is **formal** in that the given form or pattern is what is used to define the elements of the set. Since the definition is abstract like this, it is convenient to find more concrete interpretations to help understand $\mathbb{R}^n$. It is possible to interpret $\mathbb{R}^n$ as a set of points. Specifically, $\mathbb{R}^2$ can be considered as the Cartesian plane, $\mathbb{R}^3$ can be viewed as Cartesian space, and $\mathbb{R}^n$ can be viewed as a set of points existing in n dimensions.

However, the elements of $\mathbb{R}^n$ are called *vectors*, and the term *vector* is typically associated with arrows as when vectors are studied in calculus or physics. Each arrow has an **initial point**, the start of the arrow, and a **terminal point**, the end of the arrow where the arrowhead is found. These arrows are used to interpret what a vector is in one of two ways.

- First, the arrow may be restricted to having the origin as its initial point. The advantage to this interpretation is that there is one arrow for every vector. For example, in $\mathbb{R}^2$ the arrow that interprets $\begin{bmatrix} a \\ b \end{bmatrix}$ is the one that originates at the origin and points to (a, b). Another advantage is that the arrow and point interpretations of a vector can be easily exchanged depending on the context.

- Second, it is common to represent a single vector by infinitely many arrows as with vector fields in calculus. Any two points (a_1, b_1) and (a_2, b_2) such that $a = a_2 - a_1$ and $b = b_2 - b_1$ can be the initial and terminal points, respectively, of an arrow that represents $\begin{bmatrix} a \\ b \end{bmatrix}$. To draw this arrow, simply start at (a_1, b_1) and move a units horizontally and b units vertically. Although this can be done with the first interpretation as well, the subtraction of coordinates in the second interpretation is very reminiscent of the idea of slope.

Figure 2.1 illustrates both of these interpretations.

■ Example 2.1.2

The arrow drawn from the origin to the point $(5, -3)$ can represent the vector $\begin{bmatrix} 5 \\ -3 \end{bmatrix}$, which illustrates the first interpretation. The arrow drawn from $(3, 2)$ to $(8, -1)$ can also represent the vector $\begin{bmatrix} 5 \\ -3 \end{bmatrix}$ because both $8 - 3 = 5$ and $-1 - 2 = -3$. This is the second interpretation. In this case, both arrows can be viewed as the same vector, and such arrows are called **equivalent**.

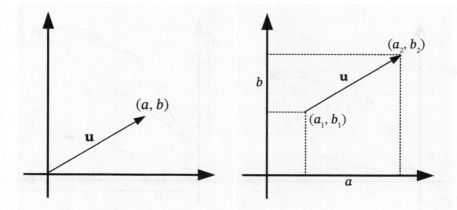

Figure 2.1   Two interpretations of *vector* as an arrow.

Vector Operations

Before any operations can be performed on vectors, what it means for vectors to be equal must be defined.

■ Definition 2.1.3

If $\mathbf{u}, \mathbf{v} \in \mathbb{R}^n$ have coordinates $u_1, u_2, \dots, u_n$ and $v_1, v_2, \dots, v_n$, respectively, then $\mathbf{u}$ and $\mathbf{v}$ are **equal**, written as $\mathbf{u} = \mathbf{v}$, means that $u_i = v_i$ for every $i = 1, 2, \dots, n$. Otherwise, the vectors are not equal, written as $\mathbf{u} \neq \mathbf{v}$.

There are two standard operations on arrows. The first is **arrow addition**. How this works depends on where the arrows originate. If the initial point of each arrow is fixed at the origin, the sum of the arrows is the diagonal of the parallelogram formed by the two arrows. Consider arrows in $\mathbb{R} \times \mathbb{R}$. Let (a, b) be the terminal point of arrow $\mathbf{u}$ and let (c, d) be the terminal point of arrow $\mathbf{v}$. An investigation of Figure 2.2a reveals that the terminal point of $\mathbf{u} + \mathbf{v}$ is $(a + c, b + d)$ because the shaded triangles are congruent (Exercise 4). If an arrow can instead begin at any point in the plane, the sum of the arrows can be defined using the parallelogram as in the first case, but it is usually done by taking the first arrow $\mathbf{u}$ and joining an arrow $\mathbf{v}'$ equivalent to $\mathbf{v}$ so that the terminal point of $\mathbf{u}$ equals the initial point of $\mathbf{v}'$. Figure 2.2b reveals that $\mathbf{u} + \mathbf{v}$ has horizontal change $a + c$ and vertical change $b + d$. In either case, the sum of the arrows is seen to represent the vector $\begin{bmatrix} a + c \\ b + d \end{bmatrix}$.

The second operation is **arrow scalar multiplication**. This operation is the same whether all of the arrows originate at the origin or not, so for this discussion suppose all arrows have the origin as their initial point. When an arrow $\mathbf{u}$ is multiplied by $r \in \mathbb{R}$, the arrow is increased by a factor of r, so, as in Figure 2.3, if the terminal point of $\mathbf{u}$ is (a, b), the terminal point of $r\mathbf{u}$ is (ra, rb), and the scalar

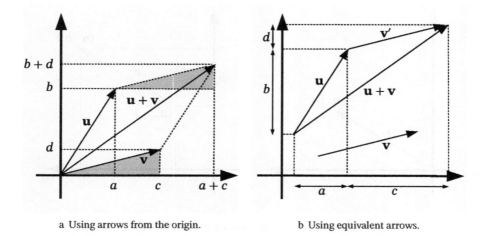

a Using arrows from the origin.   b Using equivalent arrows.

Figure 2.2   Addition of arrows.

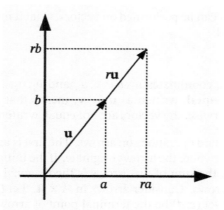

Figure 2.3   Scaling of arrows.

multiple of the arrow is seen to represent the vector $\begin{bmatrix} ra \\ rb \end{bmatrix}$. If $|r| > 1$, the length of $r\mathbf{u}$ is greater than the length of $\mathbf{u}$. If $0 < |r| < 1$, then the length of $r\mathbf{u}$ is less than the length of $\mathbf{u}$. Also, if $r = 0$, then $r\mathbf{u}$ is simply a point. If $r > 0$, the arrow $r\mathbf{u}$ is in the direction of $\mathbf{u}$. Lastly, if $r < 0$, then $r\mathbf{u}$ is in the opposite direction of $\mathbf{u}$.

Since arrows are one of the main interpretations of *vector*, it is natural to define the vector operations to coincide with the arrow operations. The first operation will be a binary operation (Definition 1.4.34), and the second will be a scalar operation (Definition 1.4.41).

■ Definition 2.1.4

Let $\mathbf{u}, \mathbf{v} \in \mathbb{R}^n$ and write $\mathbf{u} = \begin{bmatrix} u_1 \\ u_2 \\ \vdots \\ u_n \end{bmatrix}$ and $\mathbf{v} = \begin{bmatrix} v_1 \\ v_2 \\ \vdots \\ v_n \end{bmatrix}$. Let $r \in \mathbb{R}$. Define the

binary operation $+$ on $\mathbb{R}^n$ by

$$\mathbf{u} + \mathbf{v} = \begin{bmatrix} u_1 + v_1 \\ u_2 + v_2 \\ \vdots \\ u_n + v_n \end{bmatrix}$$

and define the scalar operation $\cdot$ on $\mathbb{R}^n$ by

$$r \cdot \mathbf{u} = \begin{bmatrix} ru_1 \\ ru_1 \\ \vdots \\ ru_n \end{bmatrix},$$

where $r \cdot \mathbf{u}$ is usually written as $r\mathbf{u}$. The first operation is called **vector addition**, and the second operation is called **scalar multiplication**.

For example, working with vectors from $\mathbb{R}^4$,

$$\begin{bmatrix} 3 \\ 5 \\ -2 \\ 0 \end{bmatrix} + \begin{bmatrix} -4 \\ 0 \\ 6 \\ -3 \end{bmatrix} = \begin{bmatrix} -1 \\ 5 \\ 4 \\ -3 \end{bmatrix} \quad \text{and} \quad 5 \begin{bmatrix} 3 \\ 5 \\ -2 \\ 0 \end{bmatrix} = \begin{bmatrix} 15 \\ 25 \\ -10 \\ 0 \end{bmatrix}.$$

Definition 2.1.4 is used to define a system of algebra on $\mathbb{R}^n$, one that involves vectors instead of numbers. The additive identity [Definition 1.4.38(c)] of this system is the **zero vector**, which is the vector in $\mathbb{R}^n$ with all of its coordinates equal to 0 and is denoted by

$$\mathbf{0}_n = \begin{bmatrix} 0 \\ 0 \\ \vdots \\ 0 \end{bmatrix}.$$

To prove this, take $\begin{bmatrix} u_1 \\ u_2 \\ \vdots \\ u_n \end{bmatrix} \in \mathbb{R}^n$ and compute

$$\begin{bmatrix} 0 \\ 0 \\ \vdots \\ 0 \end{bmatrix} + \begin{bmatrix} u_1 \\ u_2 \\ \vdots \\ u_n \end{bmatrix} = \begin{bmatrix} 0 + u_1 \\ 0 + u_2 \\ \vdots \\ 0 + u_n \end{bmatrix} = \begin{bmatrix} u_1 \\ u_2 \\ \vdots \\ u_n \end{bmatrix} = \begin{bmatrix} u_1 + 0 \\ u_2 + 0 \\ \vdots \\ u_n + 0 \end{bmatrix} = \begin{bmatrix} u_1 \\ u_2 \\ \vdots \\ u_n \end{bmatrix} + \begin{bmatrix} 0 \\ 0 \\ \vdots \\ 0 \end{bmatrix}.$$

The system has a type of multiplicative identity in the scalar 1 because

$$1 \begin{bmatrix} u_1 \\ u_2 \\ \vdots \\ u_n \end{bmatrix} = \begin{bmatrix} 1u_1 \\ 1u_2 \\ \vdots \\ 1u_n \end{bmatrix} = \begin{bmatrix} u_1 \\ u_2 \\ \vdots \\ u_n \end{bmatrix}.$$

The algebraic system on $\mathbb{R}^n$ also has the property that every vector has an additive inverse [Definition 1.4.38(d)] because

$$\begin{bmatrix} u_1 \\ u_2 \\ \vdots \\ u_n \end{bmatrix} + \begin{bmatrix} -u_1 \\ -u_2 \\ \vdots \\ -u_n \end{bmatrix} = \begin{bmatrix} u_1 - u_1 \\ u_2 - u_2 \\ \vdots \\ u_n - u_n \end{bmatrix} = \begin{bmatrix} 0 \\ 0 \\ \vdots \\ 0 \end{bmatrix} = \begin{bmatrix} -u_1 + u_1 \\ -u_2 + u_2 \\ \vdots \\ -u_n + u_n \end{bmatrix} = \begin{bmatrix} -u_1 \\ -u_2 \\ \vdots \\ -u_n \end{bmatrix} + \begin{bmatrix} u_1 \\ u_2 \\ \vdots \\ u_n \end{bmatrix}.$$

Since,

$$\begin{bmatrix} -u_1 \\ -u_2 \\ \vdots \\ -u_n \end{bmatrix} = (-1) \begin{bmatrix} u_1 \\ u_2 \\ \vdots \\ u_n \end{bmatrix},$$

it follows that

$$- \begin{bmatrix} u_1 \\ u_2 \\ \vdots \\ u_n \end{bmatrix} = (-1) \begin{bmatrix} u_1 \\ u_2 \\ \vdots \\ u_n \end{bmatrix}.$$

The algebraic system on $\mathbb{R}^n$ has these and other properties that are summarized in the following theorem. The operations are read from left to right unless overridden by grouping symbols. Also, the context of the theorem makes it clear that the zero vector is from $\mathbb{R}^n$, so $\mathbf{0}$ is written instead of $\mathbf{0}_n$.

■ **Theorem 2.1.5**

Let $\mathbf{u}, \mathbf{v}, \mathbf{w} \in \mathbb{R}^n$ and $r, s \in \mathbb{R}$.

(a) $\mathbf{u} + \mathbf{v} = \mathbf{v} + \mathbf{u}$ (**Commutative Law**).

(b) $\mathbf{u} + \mathbf{v} + \mathbf{w} = \mathbf{u} + (\mathbf{v} + \mathbf{w})$ (**Associative Law**).

(c) $\mathbf{u} + \mathbf{0} = \mathbf{0} + \mathbf{u} = \mathbf{u}$ (**Additive Identity Law**).

(d) $\mathbf{u} + (-1)\mathbf{u} = (-1)\mathbf{u} + \mathbf{u} = \mathbf{0}$ (**Additive Inverse Law**).

(e) $rs\mathbf{u} = r(s\mathbf{u})$ (**Associative Law**).

(f) $r(\mathbf{u} + \mathbf{v}) = r\mathbf{u} + r\mathbf{v}$ (**Distributive Law**).

(g) $(r + s)\mathbf{u} = r\mathbf{u} + s\mathbf{u}$ (**Distributive Law**).

(h) $1\mathbf{u} = \mathbf{u}$ (**Multiplicative Identity**).

Proof

Write $\mathbf{u} = \begin{bmatrix} u_1 \\ u_2 \\ \vdots \\ u_n \end{bmatrix}$, $\mathbf{v} = \begin{bmatrix} v_1 \\ v_2 \\ \vdots \\ v_n \end{bmatrix}$, and $\mathbf{w} = \begin{bmatrix} w_1 \\ w_2 \\ \vdots \\ w_n \end{bmatrix}$. To prove (b), first compute

$$\mathbf{u} + \mathbf{v} + \mathbf{w} = \begin{bmatrix} u_1 + v_1 \\ u_2 + v_2 \\ \vdots \\ u_n + v_n \end{bmatrix} + \begin{bmatrix} w_1 \\ w_2 \\ \vdots \\ w_n \end{bmatrix} = \begin{bmatrix} u_1 + v_1 + w_1 \\ u_2 + v_2 + w_2 \\ \vdots \\ u_n + v_n + w_n \end{bmatrix}$$

and

$$\mathbf{u} + (\mathbf{v} + \mathbf{w}) = \begin{bmatrix} u_1 \\ u_2 \\ \vdots \\ u_n \end{bmatrix} + \begin{bmatrix} v_1 + w_1 \\ v_2 + w_2 \\ \vdots \\ v_n + w_n \end{bmatrix} = \begin{bmatrix} u_1 + (v_1 + w_1) \\ u_2 + (v_2 + w_2) \\ \vdots \\ u_n + (v_n + w_n) \end{bmatrix}.$$

Then, (b) follows because for all $i = 1, 2, \ldots, n$, the associative law of real numbers implies that

$$u_i + v_i + w_i = u_i + (v_i + w_i).$$

To prove (g), use the distributive law of real numbers to derive

$$(r + s)\mathbf{u} = \begin{bmatrix} (r+s)u_1 \\ (r+s)u_2 \\ \vdots \\ (r+s)u_n \end{bmatrix} = \begin{bmatrix} ru_1 + su_1 \\ ru_2 + su_2 \\ \vdots \\ ru_n + su_n \end{bmatrix} = \begin{bmatrix} ru_1 \\ ru_2 \\ \vdots \\ ru_n \end{bmatrix} + \begin{bmatrix} su_1 \\ su_2 \\ \vdots \\ su_n \end{bmatrix} = r\mathbf{u} + s\mathbf{u}.$$

The remaining parts are left to Exercise 5. ■

Using Theorem 2.1.5(d), define **vector subtraction** by

$$\mathbf{u} - \mathbf{v} = \mathbf{u} + (-\mathbf{v}) = \mathbf{u} + (-1)\mathbf{v}.$$

For example, $\begin{bmatrix} 7 \\ 3 \end{bmatrix} - \begin{bmatrix} 2 \\ 9 \end{bmatrix} = \begin{bmatrix} 5 \\ -6 \end{bmatrix}$.

■ **Example 2.1.6**

If there is an algebra on $\mathbb{R}^n$, there should be equations to solve. Consider the vector equation in $\mathbb{R}^2$,

$$4\left(\begin{bmatrix} 1 \\ 4 \end{bmatrix} + \begin{bmatrix} x \\ y \end{bmatrix} \right) = \begin{bmatrix} 8 \\ 0 \end{bmatrix}. \tag{2.1}$$

Solve the equation as usual but write all of the steps:

$$\frac{1}{4}\left[4\left(\begin{bmatrix} 1 \\ 4 \end{bmatrix} + \begin{bmatrix} x \\ y \end{bmatrix} \right) \right] = \frac{1}{4}\begin{bmatrix} 8 \\ 0 \end{bmatrix}, \tag{2.2}$$

$$\left(\frac{1}{4} \cdot 4\right)\left(\begin{bmatrix} 1 \\ 4 \end{bmatrix} + \begin{bmatrix} x \\ y \end{bmatrix}\right) = \frac{1}{4}\begin{bmatrix} 8 \\ 0 \end{bmatrix}, \tag{2.3}$$

$$1\left(\begin{bmatrix} 1 \\ 4 \end{bmatrix} + \begin{bmatrix} x \\ y \end{bmatrix}\right) = \frac{1}{4}\begin{bmatrix} 8 \\ 0 \end{bmatrix}, \tag{2.4}$$

$$1\left(\begin{bmatrix} 1 \\ 4 \end{bmatrix} + \begin{bmatrix} x \\ y \end{bmatrix}\right) = \begin{bmatrix} 2 \\ 0 \end{bmatrix}, \tag{2.5}$$

$$\begin{bmatrix} 1 \\ 4 \end{bmatrix} + \begin{bmatrix} x \\ y \end{bmatrix} = \begin{bmatrix} 2 \\ 0 \end{bmatrix}, \tag{2.6}$$

$$\begin{bmatrix} -1 \\ -4 \end{bmatrix} + \left(\begin{bmatrix} 1 \\ 4 \end{bmatrix} + \begin{bmatrix} x \\ y \end{bmatrix}\right) = \begin{bmatrix} -1 \\ -4 \end{bmatrix} + \begin{bmatrix} 2 \\ 0 \end{bmatrix}, \tag{2.7}$$

$$\left(\begin{bmatrix} -1 \\ -4 \end{bmatrix} + \begin{bmatrix} 1 \\ 4 \end{bmatrix}\right) + \begin{bmatrix} x \\ y \end{bmatrix} = \begin{bmatrix} -1 \\ -4 \end{bmatrix} + \begin{bmatrix} 2 \\ 0 \end{bmatrix}, \tag{2.8}$$

$$\begin{bmatrix} 0 \\ 0 \end{bmatrix} + \begin{bmatrix} x \\ y \end{bmatrix} = \begin{bmatrix} -1 \\ -4 \end{bmatrix} + \begin{bmatrix} 2 \\ 0 \end{bmatrix}, \tag{2.9}$$

$$\begin{bmatrix} 0 \\ 0 \end{bmatrix} + \begin{bmatrix} x \\ y \end{bmatrix} = \begin{bmatrix} 1 \\ -4 \end{bmatrix}, \tag{2.10}$$

$$\begin{bmatrix} x \\ y \end{bmatrix} = \begin{bmatrix} 1 \\ -4 \end{bmatrix}. \tag{2.11}$$

Each step can be justified (Exercise 6) by either using the definition of the operation, appealing to Theorem 2.1.5, or using basic arithmetic. For example, the reason that step (2.2) follows from (2.1) is that scalar multiplication is a function that takes the scalar as its left input (Definition 1.4.41), and step (2.11) follows from (2.10) because of Theorem 2.1.5(c).

To advance the next theorem, recall that for all $x, y \in \mathbb{R}$, $0 \cdot x = x \cdot 0 = 0$ and $xy = 0$ implies that $x = 0$ or $y = 0$. The zero vector possesses the same properties.

■ **Theorem 2.1.7**

Let $\mathbf{u} \in \mathbb{R}^n$ and $r \in \mathbb{R}$.

(a) $0\mathbf{u} = \mathbf{0}$.

(b) $r\mathbf{0} = \mathbf{0}$.

(c) If $r\mathbf{u} = \mathbf{0}$, then $r = 0$ or $\mathbf{u} = \mathbf{0}$.

Proof

Let $\mathbf{u} \in \mathbb{R}^n$ with coordinates $u_1, u_2, \ldots, u_n$ and take $r \in \mathbb{R}$. Then,

$$0\mathbf{u} = \begin{bmatrix} 0 \cdot u_1 \\ 0 \cdot u_2 \\ \vdots \\ 0 \cdot u_n \end{bmatrix} = \begin{bmatrix} 0 \\ 0 \\ \vdots \\ 0 \end{bmatrix}, \quad \text{and} \quad r\mathbf{0} = \begin{bmatrix} r \cdot 0 \\ r \cdot 0 \\ \vdots \\ r \cdot 0 \end{bmatrix} = \begin{bmatrix} 0 \\ 0 \\ \vdots \\ 0 \end{bmatrix},$$

proving (a) and (b). To prove (c), suppose that $r\mathbf{u} = \mathbf{0}$ and $r \neq 0$. This implies that $ru_i = 0$ for $i = 1, 2, \ldots, n$. Since r is nonzero, $u_i = 0$ for $i = 1, 2, \ldots, n$, so $\mathbf{u}$ is the zero vector. ∎

The cancellation laws are a direct result of Theorem 2.1.7(c). The proof is left to Exercise 10.

■ Corollary 2.1.8 [Cancellation]

Let $\mathbf{u}, \mathbf{v} \in \mathbb{R}^n$ and $r, s \in \mathbb{R}$.

   (a) If $r\mathbf{u} = s\mathbf{u}$ and $\mathbf{u} \neq \mathbf{0}$, then $r = s$.

   (b) If $r\mathbf{u} = r\mathbf{v}$ and $r \neq 0$, then $\mathbf{u} = \mathbf{v}$.

Distance and Length

The interpretations of vector as either a point or an arrow means that sets of vectors can be naturally viewed geometrically. This means, for example, that in $\mathbb{R}^n$ distances, lengths, and angle measures should be able to be computed and lines and planes should be able to be found. Start with distance. Given (u_1, u_2) and (v_1, v_2) in $\mathbb{R} \times \mathbb{R}$, the distance between these two points is

$$\sqrt{(u_1 - v_1)^2 + (u_2 - v_2)^2}. \tag{2.12}$$

Distances in $\mathbb{R} \times \mathbb{R} \times \mathbb{R}$ are found similarly. Let $A = (u_1, u_2, u_3)$, $B = (v_1, v_2, v_3)$, $C = (u_1, u_2, v_3)$, and $D = (u_1, v_2, v_3)$ in Cartesian space as in Figure 2.4. Because $\triangle CDB$ lies in the plane $z = v_3$,

$$DB = |u_1 - v_1|, \quad CD = |u_2 - v_2|.$$

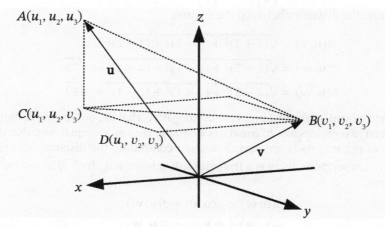

Figure 2.4 Finding the distance between vectors.

Since $\angle CDB$ is a right angle, an application of the Pythagorean Theorem gives

$$CB = \sqrt{(u_1 - v_1)^2 + (u_2 - v_2)^2}.$$

Next, because AC is parallel to the z-axis,

$$AC = |u_3 - v_3|,$$

and then because $\angle ACB$ is a right angle, another application of the Pythagorean Theorem gives

$$AB = \sqrt{CB^2 + (u_3 - v_3)^2} = \sqrt{(u_1 - v_1)^2 + (u_2 - v_2)^2 + (u_3 - v_3)^2}.$$

If $\mathbf{u}, \mathbf{v} \in \mathbb{R}^3$ are interpreted as arrows in $\mathbb{R} \times \mathbb{R} \times \mathbb{R}$ as in Figure 2.4, it appears appropriate to generalize this argument to n dimensions to define the distance between two vectors.

■ **Definition 2.1.9**

Writing $\mathbf{u} = \begin{bmatrix} u_1 \\ u_2 \\ \vdots \\ u_n \end{bmatrix}$ and $\mathbf{v} = \begin{bmatrix} v_1 \\ v_2 \\ \vdots \\ v_n \end{bmatrix}$ so that $\mathbf{u}, \mathbf{v} \in \mathbb{R}^n$, define the **(Euclidean)**
distance between $\mathbf{u}$ and $\mathbf{v}$ to be

$$d(\mathbf{u}, \mathbf{v}) = \sqrt{(u_1 - v_1)^2 + (u_2 - v_2)^2 + \cdots + (u_n - v_n)^2}.$$

■ **Example 2.1.10**

Given the vectors $\mathbf{u} = \begin{bmatrix} 1 \\ 0 \\ 4 \end{bmatrix}$, $\mathbf{v} = \begin{bmatrix} 2 \\ 2 \\ 2 \end{bmatrix}$, and $\mathbf{w} = \begin{bmatrix} 5 \\ -1 \\ 1 \end{bmatrix}$ in $\mathbb{R}^3$, it is easy to
compute the distances between the vectors:

$$d(\mathbf{u}, \mathbf{v}) = \sqrt{(1 - 2)^2 + (0 - 2)^2 + (4 - 2)^2} = 3,$$
$$d(\mathbf{u}, \mathbf{w}) = \sqrt{(1 - 5)^2 + (0 + 1)^2 + (4 - 1)^2} = \sqrt{26},$$
$$d(\mathbf{v}, \mathbf{w}) = \sqrt{(2 - 5)^2 + (2 + 1)^2 + (2 - 1)^2} = \sqrt{19}.$$

Observe that all of the distances are nonnegative, that the only way a distance between two vectors would equal 0 is when the vectors are equal, and that the order of the vectors is unimportant when computing the distance between them. Furthermore, there is a triangle with sides of lengths 3, $\sqrt{26}$, and $\sqrt{19}$ because

$$d(\mathbf{u}, \mathbf{w}) < d(\mathbf{u}, \mathbf{v}) + d(\mathbf{v}, \mathbf{w}),$$
$$d(\mathbf{v}, \mathbf{w}) < d(\mathbf{v}, \mathbf{u}) + d(\mathbf{u}, \mathbf{w}), \tag{2.13}$$
$$d(\mathbf{u}, \mathbf{v}) < d(\mathbf{u}, \mathbf{w}) + d(\mathbf{w}, \mathbf{v}).$$

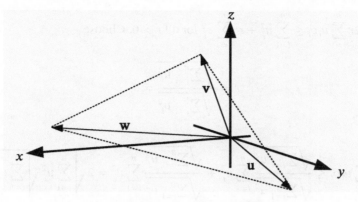

Figure 2.5 A triangle in $\mathbb{R}^3$.

This triangle formed by the vectors can be visualized as in Figure 2.5. If any of the inequalities of (2.13) were an equality, the vectors would not form a triangle. Instead, the vectors would be **collinear**, meaning that they are scalar multiples of each other (Exercise 18).

In general, the distance function of Definition 2.1.9 has all of the properties seen in Example 2.1.10. The proofs are straightforward, except for that of the inequalities of (2.13). It requires a technical but famous inequality attributed to Augustin-Louis Cauchy and Karl Hermann Amandus Schwarz.

■ **Theorem 2.1.11 [Cauchy–Schwartz Inequality]**

For all $u_i, v_i \in \mathbb{R}$ with $i = 1, 2, \dots, n$,

$$\left| \sum_{i=1}^{n} u_i v_i \right| \leq \sqrt{\sum_{i=1}^{n} u_i^2} \sqrt{\sum_{i=1}^{n} v_i^2}.$$

Proof

Suppose that

$$\sum_{i=1}^{n} u_i^2 \neq 0 \text{ and } \sum_{i=1}^{n} v_i^2 \neq 0,$$

because if not, $u_i = 0$ for all i or $v_i = 0$ for all i, and the inequality holds. Let r be a positive real number. Because

$$0 \leq \sum_{i=1}^{n} (u_i - r v_i)^2,$$

by expanding the sum it follows that

$$0 \leq \sum_{i=1}^{n} u_i^2 - 2r \sum_{i=1}^{n} u_i v_i + r^2 \sum_{i=1}^{n} v_i^2.$$

Then, $2r \sum_{i=1}^{n} u_i v_i \leq \sum_{i=1}^{n} u_i^2 + r^2 \sum_{i=1}^{n} v_i^2$ for all $r > 0$. Choose

$$r = \frac{\sqrt{\sum_{i=1}^{n} u_i^2}}{\sqrt{\sum_{i=1}^{n} v_i^2}} \neq 0,$$

so

$$2 \sum_{i=1}^{n} u_i v_i \leq \frac{\sqrt{\sum_{i=1}^{n} v_i^2}}{\sqrt{\sum_{i=1}^{n} u_i^2}} \sum_{i=1}^{n} u_i^2 + \frac{\sqrt{\sum_{i=1}^{n} u_i^2}}{\sqrt{\sum_{i=1}^{n} v_i^2}} \sum_{i=1}^{n} v_i^2 = 2 \sqrt{\sum_{i=1}^{n} u_i^2} \sqrt{\sum_{i=1}^{n} v_i^2}.$$

Therefore, the inequality follows because

$$-\sqrt{\sum_{i=1}^{n} u_i^2} \sqrt{\sum_{i=1}^{n} v_i^2} \leq \sum_{i=1}^{n} u_i v_i \leq \sqrt{\sum_{i=1}^{n} u_i^2} \sqrt{\sum_{i=1}^{n} v_i^2}. \blacksquare$$

■ **Theorem 2.1.12**

Let $\mathbf{u}, \mathbf{v}, \mathbf{w} \in \mathbb{R}^n$.

   (a) $d(\mathbf{u}, \mathbf{v}) \geq 0$.
   (b) $d(\mathbf{u}, \mathbf{v}) = 0$ if and only if $\mathbf{u} = \mathbf{v}$.
   (c) $d(\mathbf{u}, \mathbf{v}) = d(\mathbf{v}, \mathbf{u})$.
   (d) $d(\mathbf{u}, \mathbf{v}) \leq d(\mathbf{u}, \mathbf{w}) + d(\mathbf{w}, \mathbf{v})$ (**Triangle Inequality**).

Proof

The proofs of the first three parts are left to Exercise 14. For the last part, let $u_i, v_i, w_i \in \mathbb{R}$ for all $i = 1, 2, \ldots, n$. By the Cauchy–Schwartz Inequality,

$$\sum_{i=1}^{n}(u_i - v_i)^2$$

$$= \sum_{i=1}^{n}[(u_i - w_i) + (w_i - v_i)]^2$$

$$= \sum_{i=1}^{n}(u_i - w_i)^2 + 2 \sum_{i=1}^{n}(u_i - w_i)(w_i - v_i) + \sum_{i=1}^{n}(w_i - v_i)^2$$

$$\leq \sum_{i=1}^{n}(u_i - w_i)^2 + 2 \sqrt{\sum_{i=1}^{n}(u_i - w_i)^2} \sqrt{\sum_{i=1}^{n}(w_i - v_i)^2} + \sum_{i=1}^{n}(w_i - v_i)^2$$

$$= \left(\sqrt{\sum_{i=1}^{n}(u_i - w_i)^2} + \sqrt{\sum_{i=1}^{n}(w_i - v_i)^2} \right)^2.$$

By taking square roots,

$$\sqrt{\sum_{i=1}^{n}(u_i - v_i)^2} \leq \sqrt{\sum_{i=1}^{n}(u_i - w_i)^2} + \sqrt{\sum_{i=1}^{n}(w_i - v_i)^2}. \blacksquare$$

To measure the length of an arrow, simply find the distance between its initial and terminal points. For example, imagining an arrow in Cartesian n-space with initial point $(a_1, a_2, \ldots, a_n)$ and terminal point $(b_1, b_2, \ldots, b_n)$, its length is

$$\sqrt{(b_1 - a_1)^2 + (b_2 - a_2)^2 \cdots + (b_n - a_n)^2}.$$

If the arrow's initial point is the origin, its length is the distance from its terminal point to $(0, 0, \ldots, 0)$,

$$\sqrt{b_1^2 + b_2^2 + \cdots + b_n^2}.$$

For this reason, the length of the vector $\mathbf{u}$ can be calculated as

$$\|\mathbf{u}\| = d(\mathbf{u}, \mathbf{0}), \tag{2.14}$$

which is the reason behind the next definition.

■ **Definition 2.1.13**

If $\mathbf{u} = \begin{bmatrix} u_1 \\ u_2 \\ \vdots \\ u_n \end{bmatrix} \in \mathbb{R}^n$, the **length** or **magnitude** or **norm** of $\mathbf{u}$ is

$$\|\mathbf{u}\| = \sqrt{u_1^2 + u_2^2 + \cdots + u_n^2}.$$

Let $\mathbf{u} = \begin{bmatrix} u_1 \\ u_2 \\ \vdots \\ u_n \end{bmatrix}$ and $\mathbf{v} = \begin{bmatrix} v_1 \\ v_2 \\ \vdots \\ v_n \end{bmatrix}$. Then, as in Definition 2.1.9,

$$d(\mathbf{u}, \mathbf{v}) = \sqrt{(u_1 - v_1)^2 + (u_2 - v_2)^2 + \cdots + (u_n - v_n)^2}.$$

Because

$$\mathbf{u} - \mathbf{v} = \begin{bmatrix} u_1 - v_1 \\ u_2 - v_2 \\ \vdots \\ u_n - v_n \end{bmatrix},$$

it follows that $d(\mathbf{u}, \mathbf{v}) = \|\mathbf{u} - \mathbf{v}\|$, which is illustrated in Figure 2.6. This proves the next theorem.

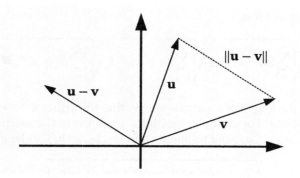

Figure 2.6   The distance between $\mathbf{u}$ and $\mathbf{v}$ is $\|\mathbf{u} - \mathbf{v}\|$.

■ Theorem 2.1.14

If $\mathbf{u}, \mathbf{v} \in \mathbb{R}^n$, then $d(\mathbf{u}, \mathbf{v}) = \|\mathbf{u} - \mathbf{v}\|$.

In many respects, due to (2.14), the computation of $\|\mathbf{u}\|$ for any $\mathbf{u} \in \mathbb{R}^n$ shares many properties with the absolute value function. For example, for the following vectors in $\mathbb{R}^3$, if $\mathbf{u} = \begin{bmatrix} 1 \\ 2 \\ 3 \end{bmatrix}$, then $\|\mathbf{u}\| = \sqrt{14}$. Also, $\|\mathbf{0}\| = 0$. This suggests that the length of a vector is also nonnegative and that the only vector that has a length of 0 is the zero vector. Moreover, as $|ab| = |a| \cdot |b|$ and $|a + b| \leq |a| + |b|$ for all $a, b \in \mathbb{R}$, working with the lengths of some sample vectors, like

$$\left\| 2 \begin{bmatrix} 3 \\ 4 \\ 5 \end{bmatrix} \right\| = \left\| \begin{bmatrix} 6 \\ 8 \\ 10 \end{bmatrix} \right\| = \sqrt{200} = 2\sqrt{50} = 2 \left\| \begin{bmatrix} 3 \\ 4 \\ 5 \end{bmatrix} \right\|$$

and

$$\left\| \begin{bmatrix} 3 \\ 4 \\ 5 \end{bmatrix} + \begin{bmatrix} 1 \\ 2 \\ 3 \end{bmatrix} \right\| = \left\| \begin{bmatrix} 4 \\ 6 \\ 8 \end{bmatrix} \right\| = \sqrt{116} \leq \sqrt{50} + \sqrt{14} = \left\| \begin{bmatrix} 3 \\ 4 \\ 5 \end{bmatrix} \right\| + \left\| \begin{bmatrix} 1 \\ 2 \\ 3 \end{bmatrix} \right\|,$$

reveal similar results that are proved in the next theorem.

■ Theorem 2.1.15

Let $\mathbf{u}, \mathbf{v} \in \mathbb{R}^n$ and $r \in \mathbb{R}$.

    (a) $\|\mathbf{u}\| \geq 0$.

    (b) $\|\mathbf{u}\| = 0$ if and only if $\mathbf{u} = \mathbf{0}$.

    (c) $\|r\mathbf{u}\| = |r| \cdot \|\mathbf{u}\|$.

    (d) $\|\mathbf{u} + \mathbf{v}\| \leq \|\mathbf{u}\| + \|\mathbf{v}\|$ (**Triangle Inequality**).

Proof

To prove (c), let the coordinates of $\mathbf{u} \in \mathbb{R}^n$ be $u_1, u_2, \ldots, u_n$. Then,

$$\|r\mathbf{u}\| = \sqrt{(ru_1)^2 + (ru_2)^2 + \cdots + (ru_n)^2}$$
$$= \sqrt{r^2(u_1^2 + u_2^2 + \cdots + u_n^2)}$$
$$= |r|\sqrt{u_1^2 + u_2^2 + \cdots + u_n^2} = |r| \cdot \|\mathbf{u}\|.$$

To prove (d), apply Theorem 2.1.12(c) and (d), part (c), and (2.14),

$$\|\mathbf{u} + \mathbf{v}\| = \|\mathbf{u} - (-\mathbf{v})\| = d(\mathbf{u}, -\mathbf{v}) \leq d(\mathbf{u}, \mathbf{0}) + d(-\mathbf{v}, \mathbf{0}) = \|\mathbf{u}\| + \|\mathbf{v}\|.$$

The proof of the other parts is left to Exercise 19. ■

Theorem 2.1.15(c) states that for every $\mathbf{u} \in \mathbb{R}^n$ and $r \in \mathbb{R}$, when $\mathbf{u}$ is multiplied by the scalar r, the length of the resulting vector is $|r|$ times the length of $\mathbf{u}$. There will be times when it will be convenient for $\|\mathbf{u}\| = 1$ because in this case the length of $r\mathbf{u}$ will simply be $|r|$.

■ **Definition 2.1.16**

A vector of length 1 is called a **unit vector**.

Every nonzero vector $\mathbf{u} \in \mathbb{R}^n$ can be **normalized**, which means there is a scalar r so that $r\mathbf{u}$ is a unit vector. The resulting unit vector is called the **normalization** of $\mathbf{u}$ and can be considered the **direction** of the vector. For example,

$$\left\| \frac{1}{\sqrt{14}} \begin{bmatrix} 1 \\ 2 \\ 3 \end{bmatrix} \right\| = \sqrt{\frac{1}{14} + \frac{4}{14} + \frac{9}{14}} = 1,$$

so view $\begin{bmatrix} 1/\sqrt{14} \\ 2/\sqrt{14} \\ 3/\sqrt{14} \end{bmatrix}$ as the direction of $\begin{bmatrix} 1 \\ 2 \\ 3 \end{bmatrix}$. The needed scalar to accomplish this is simply the reciprocal of the length of the given vector.

■ **Theorem 2.1.17**

$\frac{1}{\|\mathbf{u}\|}\mathbf{u}$ is a unit vector for all $\mathbf{u} \in \mathbb{R}^n$.

Proof

Take $\mathbf{u} \in \mathbb{R}^n$ and compute its magnitude using Theorem 2.1.15(c),

$$\left\| \frac{1}{\|\mathbf{u}\|}\mathbf{u} \right\| = \frac{1}{\|\mathbf{u}\|}\|\mathbf{u}\| = 1. ■$$

Lines and Planes

Consider the following computations from $\mathbb{R}^3$.

$$\begin{bmatrix} 4 \\ 6 \\ 10 \end{bmatrix} = 4 \begin{bmatrix} 1 \\ 0 \\ 0 \end{bmatrix} + 6 \begin{bmatrix} 0 \\ 1 \\ 0 \end{bmatrix} + 10 \begin{bmatrix} 0 \\ 0 \\ 1 \end{bmatrix}, \tag{2.15}$$

$$\begin{bmatrix} 4 \\ 6 \\ 10 \end{bmatrix} = \frac{21}{16} \begin{bmatrix} 6 \\ 0 \\ -1 \end{bmatrix} + \frac{31}{32} \begin{bmatrix} -4 \\ -3 \\ 27 \end{bmatrix} + \frac{95}{32} \begin{bmatrix} 0 \\ 3 \\ -5 \end{bmatrix}, \tag{2.16}$$

$$\begin{bmatrix} 4 \\ 6 \\ 10 \end{bmatrix} = 4 \begin{bmatrix} 1 \\ 1 \\ 0 \end{bmatrix} + 2 \begin{bmatrix} 0 \\ 1 \\ 1 \end{bmatrix} + 2 \begin{bmatrix} 0 \\ 0 \\ 4 \end{bmatrix} + 0 \begin{bmatrix} 4 \\ 9 \\ 3 \end{bmatrix}. \tag{2.17}$$

This leads to an important definition.

■ **Definition 2.1.18**

If $\mathbf{u}, \mathbf{m}_1, \mathbf{m}_2, \ldots, \mathbf{m}_k \in \mathbb{R}^n$ such that there exists real numbers $r_1, r_2, \ldots, r_k$ so that $\mathbf{u} = r_1 \mathbf{m}_1 + r_2 \mathbf{m}_2 + \cdots + r_k \mathbf{m}_k$, then $\mathbf{u}$ is a **linear combination** of $\{\mathbf{m}_1, \mathbf{m}_2, \ldots, \mathbf{m}_k\}$.

(2.15) shows that $\mathbf{u} = \begin{bmatrix} 4 \\ 6 \\ 10 \end{bmatrix}$ is a linear combination of

$$\left\{ \begin{bmatrix} 1 \\ 0 \\ 0 \end{bmatrix}, \begin{bmatrix} 0 \\ 1 \\ 0 \end{bmatrix}, \begin{bmatrix} 0 \\ 0 \\ 1 \end{bmatrix} \right\}. \tag{2.18}$$

(2.16) and (2.17) show that $\mathbf{u}$ is also a linear combination of

$$\left\{ \begin{bmatrix} 6 \\ 0 \\ -1 \end{bmatrix}, \begin{bmatrix} -4 \\ -3 \\ 27 \end{bmatrix}, \begin{bmatrix} 0 \\ 3 \\ -5 \end{bmatrix} \right\} \text{ and } \left\{ \begin{bmatrix} 1 \\ 1 \\ 0 \end{bmatrix}, \begin{bmatrix} 0 \\ 1 \\ 1 \end{bmatrix}, \begin{bmatrix} 0 \\ 0 \\ 4 \end{bmatrix}, \begin{bmatrix} 4 \\ 9 \\ 3 \end{bmatrix} \right\}.$$

The unit vectors of (2.18) stand out because every vector of $\mathbb{R}^3$ can be written as a linear combination of those vectors in a very natural way using unique scalars. Therefore, they receive a name and are generalized in the following.

■ **Definition 2.1.19**

For all $i = 1, 2, \ldots, n$, define the ith **standard vector** of $\mathbb{R}^n$ to be the vector $\mathbf{e}_i$ that has all coordinates equal to 0 except the ith coordinate, which is equal to 1. That is, $\mathbf{e}_1, \mathbf{e}_2, \ldots, \mathbf{e}_n$ are the standard vectors of $\mathbb{R}^n$ when

$$\mathbf{e}_1 = \begin{bmatrix} 1 \\ 0 \\ \vdots \\ 0 \end{bmatrix}, \quad \mathbf{e}_2 = \begin{bmatrix} 0 \\ 1 \\ \vdots \\ 0 \end{bmatrix}, \quad \ldots, \quad \mathbf{e}_n = \begin{bmatrix} 0 \\ 0 \\ \vdots \\ 1 \end{bmatrix}.$$

The main purpose of the standard vectors is to generate $\mathbb{R}^n$. For example,

$$\mathbb{R}^2 = \left\{ r \begin{bmatrix} 1 \\ 0 \end{bmatrix} + s \begin{bmatrix} 0 \\ 1 \end{bmatrix} : r, s \in \mathbb{R} \right\}$$

and

$$\mathbb{R}^3 = \left\{ r \begin{bmatrix} 1 \\ 0 \\ 0 \end{bmatrix} + s \begin{bmatrix} 0 \\ 1 \\ 0 \end{bmatrix} + t \begin{bmatrix} 0 \\ 0 \\ 1 \end{bmatrix} : r, s, t \in \mathbb{R} \right\}.$$

The standard vectors also generate subsets of $\mathbb{R}^n$. For example,

$$M = \left\{ r \begin{bmatrix} 1 \\ 0 \\ 0 \end{bmatrix} + s \begin{bmatrix} 0 \\ 1 \\ 0 \end{bmatrix} : r, s \in \mathbb{R} \right\} \tag{2.19}$$

is considered the xy-plane, and

$$N = \left\{ \begin{bmatrix} 0 \\ 0 \\ 5 \end{bmatrix} + r \begin{bmatrix} 1 \\ 0 \\ 0 \end{bmatrix} + s \begin{bmatrix} 0 \\ 1 \\ 0 \end{bmatrix} : r, s \in \mathbb{R} \right\} \tag{2.20}$$

is considered the plane parallel to the xy-plane but five units above it. Although the standard vectors serve as natural building blocks for $\mathbb{R}^n$ and some of its subsets, other vectors can be chosen to generate these and other sets.

■ **Definition 2.1.20**

Let $\mathbf{u}, \mathbf{m}_1, \mathbf{m}_2, \dots, \mathbf{m}_k \in \mathbb{R}^n$.

(a) The set

$$S(\mathbf{m}_1, \mathbf{m}_2, \dots, \mathbf{m}_k) = \left\{ \sum_{i=1}^{k} r_i \mathbf{m}_i : r_1, r_2, \dots, r_k \in \mathbb{R} \right\}$$

is the set **spanned** by $\mathbf{m}_1, \mathbf{m}_2, \dots, \mathbf{m}_k$. If $\mathbb{R}^n = S(\mathbf{m}_1, \mathbf{m}_2, \dots, \mathbf{m}_k)$, the set $\{\mathbf{m}_1, \mathbf{m}_2, \dots, \mathbf{m}_k\}$ is said to **span** $\mathbb{R}^n$ and is called a **spanning set** of $\mathbb{R}^n$.

(b) The set

$$\mathbf{u} + S(\mathbf{m}_1, \mathbf{m}_2, \dots, \mathbf{m}_k) = \left\{ \mathbf{u} + \sum_{i=1}^{k} r_i \mathbf{m}_i : r_1, r_2, \dots, r_k \in \mathbb{R} \right\}$$

is called a **coset**. The vector $\mathbf{u}$ is called the **position vector** and each $\mathbf{m}_i$ ($i = 1, 2, \dots, k$) is called a **direction vector**. Also, every vector in $S(\mathbf{m}_1, \mathbf{m}_2, \dots, \mathbf{m}_k)$ is **parallel** to the coset.

Using Definition 2.1.20, the set M of (2.19) can be written as

$$M = S\left(\begin{bmatrix} 1 \\ 0 \\ 0 \end{bmatrix}, \begin{bmatrix} 0 \\ 1 \\ 0 \end{bmatrix}\right) = \begin{bmatrix} 0 \\ 0 \\ 0 \end{bmatrix} + S\left(\begin{bmatrix} 1 \\ 0 \\ 0 \end{bmatrix}, \begin{bmatrix} 0 \\ 1 \\ 0 \end{bmatrix}\right), \tag{2.21}$$

so $\{\mathbf{e}_1, \mathbf{e}_2\}$ is a spanning set of M. The set N of (2.20) has no spanning set, but it can be written as

$$N = \begin{bmatrix} 0 \\ 0 \\ 5 \end{bmatrix} + S\left(\begin{bmatrix} 1 \\ 0 \\ 0 \end{bmatrix}, \begin{bmatrix} 0 \\ 1 \\ 0 \end{bmatrix}\right). \tag{2.22}$$

An examination of Definition 2.1.20 is in order. Recall that to determine a line in the Cartesian plane it is sufficient to provide a point and a slope. The point identifies the location of the line, and the slope gives its direction. This is generalized to $\mathbb{R}^n$ using vectors.

■ **Definition 2.1.21**

Let $n \geq 1$ and $\mathbf{u}, \mathbf{m} \in \mathbb{R}^n$ with $\mathbf{m} \neq \mathbf{0}$. The coset $\mathbf{u} + S(\mathbf{m})$ is the **line** containing $\mathbf{u}$ in the direction of $\mathbf{m}$. Two lines with different position vectors are **parallel** if their direction vectors are scalar multiples of each other.

■ **Example 2.1.22**

The set of standard vectors of $\mathbb{R}^n$ (Definition 2.1.19) spans $\mathbb{R}^n$. For instance, $S(\mathbf{e}_1, \mathbf{e}_2) = \mathbb{R}^2$ and $S(\mathbf{e}_1, \mathbf{e}_2, \mathbf{e}_3) = \mathbb{R}^3$. Furthermore, in $\mathbb{R}^2$ define

$$S(\mathbf{e}_1) = x\text{-axis} \quad \text{and} \quad S(\mathbf{e}_2) = y\text{-axis},$$

and in $\mathbb{R}^3$ define

$$S(\mathbf{e}_1) = x\text{-axis}, \quad S(\mathbf{e}_2) = y\text{-axis}, \quad S(\mathbf{e}_3) = z\text{-axis}.$$

This implies that an axis is nothing more than a representation of all multiples of a standard vector.

The interpretation of vectors as arrows originating at the origin in the Cartesian plane will be used to visualize the line of Definition 2.1.21. Suppose that $\mathbf{m} = \begin{bmatrix} m_1 \\ m_2 \end{bmatrix}$ and $\mathbf{u} = \begin{bmatrix} u_1 \\ u_2 \end{bmatrix}$. These define the line

$$L = \begin{bmatrix} u_1 \\ u_2 \end{bmatrix} + S\left(\begin{bmatrix} m_1 \\ m_2 \end{bmatrix}\right) = \left\{ \begin{bmatrix} u_1 + tm_1 \\ u_2 + tm_2 \end{bmatrix} : t \in \mathbb{R} \right\}.$$

Every vector $\mathbf{x}$ in L can be interpreted as as an arrow pointing to the points of the considered line. Each of the vectors are found by drawing the parallelogram formed

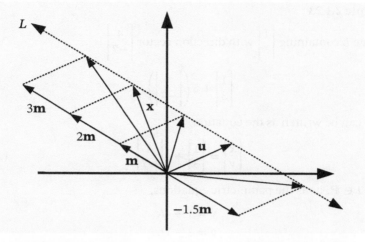

Figure 2.7   The line L containing $\mathbf{u}$ with direction vector $\mathbf{m}$.

by $\mathbf{u}$ and a multiple of $\mathbf{m}$. As seen in Figure 2.7, the fixed vector $\mathbf{u}$ determines the location of the line, and the multiples of $\mathbf{m}$ stretch or contract depending on the scalar creating the vectors in L when added to $\mathbf{u}$. It follows that for all $\mathbf{x} \in L$, there exists $t \in \mathbb{R}$ such that

$$\mathbf{x} = \mathbf{u} + t\mathbf{m}. \tag{2.23}$$

Writing $\mathbf{x} = \begin{bmatrix} x \\ y \end{bmatrix}$, (2.23) implies that the vectors in L satisfy the equation

$$\begin{bmatrix} x \\ y \end{bmatrix} = \begin{bmatrix} u_1 \\ u_2 \end{bmatrix} + t \begin{bmatrix} m_1 \\ m_2 \end{bmatrix},$$

which in turn can be written as the **parametric equations**,

$$x = u_1 + tm_1,$$
$$y = u_2 + tm_2,$$

where $t \in \mathbb{R}$ is the **parameter**. Eliminating the parameter t yields the equation

$$y = \frac{m_2}{m_1}x + \frac{m_1 u_2 - m_2 u_1}{m_1},$$

which is the slope-intercept form of the line. Notice that the slope is m_2/m_1. This is expected because $\mathbf{m}$ gives the direction of the line. Furthermore, since the slopes of parallel lines are equal, the direction vectors of parallel lines will be multiples of each other.

■ **Example 2.1.23**

The line L containing $\begin{bmatrix} 1 \\ 3 \end{bmatrix}$ with direction vector $\begin{bmatrix} 4 \\ -7 \end{bmatrix}$ is

$$\begin{bmatrix} 1 \\ 3 \end{bmatrix} + s\left(\begin{bmatrix} 4 \\ -7 \end{bmatrix}\right),$$

which can be written as the equation,

$$\begin{bmatrix} x \\ y \end{bmatrix} = \begin{bmatrix} 1 \\ 3 \end{bmatrix} + t\begin{bmatrix} 4 \\ -7 \end{bmatrix}, \tag{2.24}$$

where $t \in \mathbb{R}$, or as the parametric equations,

$$x = 1 + 4t,$$
$$y = 3 - 7t.$$

■ **Example 2.1.24**

Continuing Example 2.1.23, notice that $\mathbf{u} = \begin{bmatrix} 5 \\ 4 \end{bmatrix}$ is not in the line L. To find the **perpendicular distance** from $\mathbf{u}$ to L, first find the line L' that contains $\mathbf{u}$ but is perpendicular to L. Since the direction vector of L is $\begin{bmatrix} 4 \\ -7 \end{bmatrix}$, the slope of L is $-7/4$. Hence, the slope of a line perpendicular to L is $4/7$, which means its direction vector can be taken to be $\begin{bmatrix} 7 \\ 4 \end{bmatrix}$. This implies that a set of parametric equations for L' are

$$x = 5 + 7t,$$
$$y = 4 + 4t. \tag{2.25}$$

Next, find the vector $\mathbf{v}$ in $L \cap L'$. To do this, take $a, b \in \mathbb{R}$ such that

$$1 + 4a = 5 + 7b,$$
$$3 - 7a = 4 + 4b.$$

This system of linear equations is equivalent to

$$4a - 7b = 4,$$
$$7a + 4b = -1,$$

which has $a = 9/65$ and $b = -32/65$ as a solution. Hence, by (2.25),

$$\mathbf{v} = \begin{bmatrix} 101/65 \\ 132/65 \end{bmatrix}.$$

Finally, to find the distance from $\mathbf{u}$ to L, compute $d(\mathbf{u}, \mathbf{v}) \approx 3.97$.

The definition of a line (Definition 2.1.21) required only one direction vector because lines are one-dimensional. Because planes have two dimensions, their definition will require two vectors that are not scalar multiples of each other.

■ **Definition 2.1.25**

Let $n \geq 2$ and $\mathbf{u}, \mathbf{m}, \mathbf{n} \in \mathbb{R}^n$ such that $\mathbf{m}$ and $\mathbf{n}$ are not collinear. The coset

$$\mathbf{u} + \mathsf{S}(\mathbf{m}, \mathbf{n})$$

is called the **plane** containing $\mathbf{u}$ in the direction of $\mathbf{m}$ and $\mathbf{n}$. Two planes with different position vectors are **parallel** if their direction vectors form corresponding pairs that are scalar multiples of each other.

■ **Example 2.1.26**

Individually, the three standard vectors of $\mathbb{R}^3$ span the three axes. Taken in pairs, the standard vectors span the three main planes of $\mathbb{R}^3$. Specifically, define

$$\begin{aligned}
\mathsf{S}(\mathbf{e}_1, \mathbf{e}_2) &= xy\text{-plane}, \\
\mathsf{S}(\mathbf{e}_1, \mathbf{e}_3) &= xz\text{-plane}, \\
\mathsf{S}(\mathbf{e}_2, \mathbf{e}_3) &= yz\text{-plane}.
\end{aligned}$$

Like the interpretation of a line in Figure 2.7, visualize a plane using arrows originating at the origin as in Figure 2.8. The vector $\mathbf{u}$ is in the plane and determines its position. When linear combinations of $\mathbf{m}$ and $\mathbf{n}$ are added to $\mathbf{u}$, a vector $\mathbf{x}$ is obtained in the plane. The plane P is the set of all such vectors $\mathbf{x}$.

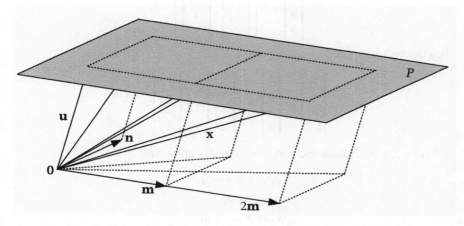

Figure 2.8 The plane P containing $\mathbf{u}$ with direction vectors $\mathbf{m}$ and $\mathbf{n}$.

■ **Example 2.1.27**

The coset of (2.21) is a plane containing **0**. It has vector equation,

$$
\begin{bmatrix} x \\ y \\ z \end{bmatrix} = x \begin{bmatrix} 1 \\ 0 \\ 0 \end{bmatrix} + y \begin{bmatrix} 0 \\ 1 \\ 0 \end{bmatrix},
$$

and parametric equations,

$$
\begin{aligned}
x &= r, \\
y &= s, \\
z &= 0,
\end{aligned}
$$

with parameters $r, s \in \mathbb{R}$. The coset N of (2.22) is a plane parallel to M. It has vector equation,

$$
\begin{bmatrix} x \\ y \\ z \end{bmatrix} = \begin{bmatrix} 0 \\ 0 \\ 5 \end{bmatrix} + x \begin{bmatrix} 1 \\ 0 \\ 0 \end{bmatrix} + y \begin{bmatrix} 0 \\ 1 \\ 0 \end{bmatrix},
$$

and parametric equations,

$$
\begin{aligned}
x &= r, \\
y &= s, \\
z &= 5.
\end{aligned}
$$

■ **Example 2.1.28**

The plane in $\mathbb{R}^4$,

$$
\begin{bmatrix} 1 \\ 2 \\ 3 \\ 4 \end{bmatrix} + S \left(\begin{bmatrix} 1 \\ 0 \\ -1 \\ 5 \end{bmatrix}, \begin{bmatrix} -3 \\ -1 \\ 4 \\ 9 \end{bmatrix} \right),
$$

has the vector equation,

$$
\begin{bmatrix} x_1 \\ x_2 \\ x_3 \\ x_4 \end{bmatrix} = \begin{bmatrix} 1 \\ 2 \\ 3 \\ 4 \end{bmatrix} + s \begin{bmatrix} 1 \\ 0 \\ -1 \\ 5 \end{bmatrix} + t \begin{bmatrix} -3 \\ -1 \\ 4 \\ 9 \end{bmatrix},
$$

where $s, t \in \mathbb{R}$. Parametric equations for this plane are

$$
\begin{aligned}
x_1 &= 1 + s - 3t, \\
x_2 &= 2 - t, \\
x_3 &= 3 - s + 4t, \\
x_4 &= 4 + 5s + 9t.
\end{aligned}
$$

Exercises

1. Let $\mathbf{u} = \begin{bmatrix} 3 \\ 5 \end{bmatrix}$, $\mathbf{v} = \begin{bmatrix} 2 \\ -1 \end{bmatrix}$, and $\mathbf{w} = \begin{bmatrix} 0 \\ 6 \end{bmatrix}$. Compute and interpret using arrows originating at the origin.

    (a) $\mathbf{u} + \mathbf{v}$ (d) $\mathbf{u} + \mathbf{v} + \mathbf{w}$

    (b) $\mathbf{w} - \mathbf{v}$ (e) $2\mathbf{u} + 5\mathbf{v} - \mathbf{w}$

    (c) $3\mathbf{u} + 4\mathbf{w}$ (f) $4(\mathbf{v} + \mathbf{w})$

2. Confirm the results of Theorem 2.1.5 by computing using the vectors defined in Exercise 1.

    (a) $\mathbf{u} + \mathbf{v}$ and $\mathbf{v} + \mathbf{u}$ (e) $(3 \cdot 5)\mathbf{u}$ and $3(5\mathbf{u})$

    (b) $\mathbf{u} + \mathbf{v} + \mathbf{w}$ and $\mathbf{u} + (\mathbf{v} + \mathbf{w})$ (f) $6(\mathbf{u} + \mathbf{v})$ and $6\mathbf{u} + 6\mathbf{v}$

    (c) $\mathbf{u} + \mathbf{0}$ (g) $(7 + 4)\mathbf{u}$ and $7\mathbf{u} + 4\mathbf{u}$

    (d) $\mathbf{u} + (-1)\mathbf{u}$ (h) $1\mathbf{u}$

3. Repeat Exercises 1 and 2 using $\mathbf{u} = \begin{bmatrix} 4 \\ -2 \\ 0 \end{bmatrix}$, $\mathbf{v} = \begin{bmatrix} 1 \\ 2 \\ 3 \end{bmatrix}$, and $\mathbf{w} = \begin{bmatrix} 1 \\ 0 \\ -3 \end{bmatrix}$.

4. Prove that the shaded triangles of Figure 2.2a are congruent.

5. Prove the remaining parts of Theorem 2.1.5.

6. Justify the remaining steps in Example 2.1.6.

7. Solve the given equations, justifying each step as in Example 2.1.6.

    (a) $\begin{bmatrix} 3 \\ 7 \end{bmatrix} - \begin{bmatrix} x \\ y \end{bmatrix} = \begin{bmatrix} 10 \\ -3 \end{bmatrix}$

    (b) $4\left(\begin{bmatrix} 2 \\ 0 \end{bmatrix} + \begin{bmatrix} x \\ y \end{bmatrix} \right) = -7 \left(\begin{bmatrix} x \\ y \end{bmatrix} + \begin{bmatrix} 8 \\ 5 \end{bmatrix} \right)$

    (c) $\begin{bmatrix} 1 \\ 0 \\ 3 \end{bmatrix} + \left(\begin{bmatrix} 4 \\ 2 \\ -3 \end{bmatrix} + \begin{bmatrix} x \\ y \\ z \end{bmatrix} \right) = \begin{bmatrix} 9 \\ -5 \\ 4 \end{bmatrix}$

    (d) $\begin{bmatrix} 1 \\ 0 \\ 3 \end{bmatrix} + 6 \left(\begin{bmatrix} 1 \\ 2 \\ 3 \end{bmatrix} - \begin{bmatrix} x \\ y \\ z \end{bmatrix} \right) = 10 \left(\begin{bmatrix} x \\ y \\ z \end{bmatrix} + \begin{bmatrix} 4 \\ 6 \\ 0 \end{bmatrix} \right)$

8. Redo the proof of parts (a) and (b) of Theorem 2.1.7 without using coordinates.

9. Prove Theorem 2.1.7(c) by assuming $\mathbf{u} \neq \mathbf{0}$ and showing $r = 0$.

10. Prove Corollary 2.1.8.

11. Find all $\begin{bmatrix} x \\ y \\ z \end{bmatrix} \in \mathbb{R}^3$ that solve the equation, $(x - 3) \left(\begin{bmatrix} x \\ y \\ z \end{bmatrix} - \begin{bmatrix} 1 \\ 3 \\ -7 \end{bmatrix} \right) = \begin{bmatrix} 0 \\ 0 \\ 0 \end{bmatrix}$.

12. Let $\mathbf{u} = \begin{bmatrix} 4 \\ -3 \end{bmatrix}$ and $\mathbf{v} = \begin{bmatrix} -6 \\ 1 \end{bmatrix}$.

    (a) Write the Cauchy–Schwartz Inequality (Theorem 2.1.11) for $n = 2$.
    (b) Use $\mathbf{u}$ and $\mathbf{v}$ to confirm the Cauchy–Schwartz Inequality.

13. Let $\mathbf{u}, \mathbf{v}$ be nonzero vectors in $\mathbb{R}^n$ with coordinates $u_i, v_i \in \mathbb{R}$ for $i = 1, 2, \dots, n$. Prove that $\mathbf{u}$ is a scalar multiple of $\mathbf{v}$ if and only if

$$\left| \sum_{i=1}^{n} u_i v_i \right| = \sqrt{\sum_{i=1}^{n} u_i^2} \sqrt{\sum_{i=1}^{n} v_i^2}.$$

    What happens if one of the vectors is the zero vector?

14. Prove the first three parts of Theorem 2.1.12.

15. Find $d(\mathbf{u}, \mathbf{v})$.

    (a) $\mathbf{u} = \begin{bmatrix} 3 \\ -5 \end{bmatrix}$ and $\mathbf{v} = \begin{bmatrix} -3 \\ 14 \end{bmatrix}$ (c) $\mathbf{u} = \begin{bmatrix} 7 \\ 3 \end{bmatrix}$ and $\mathbf{v} = \begin{bmatrix} 0 \\ 0 \end{bmatrix}$

    (b) $\mathbf{u} = \begin{bmatrix} 1 \\ -2 \\ 3 \end{bmatrix}$ and $\mathbf{v} = \begin{bmatrix} -5 \\ 6 \\ -2 \end{bmatrix}$ (d) $\mathbf{u} = 4\begin{bmatrix} 2 \\ 4 \\ -6 \end{bmatrix}$ and $\mathbf{v} = 7\begin{bmatrix} -7 \\ 5 \\ 3 \end{bmatrix}$

16. Find the lengths of each of the vectors $\mathbf{v}$ in Exercise 15.

17. Find the unit vectors in the direction of each of the vectors $\mathbf{u}$ in Exercise 15.

18. Let $\mathbf{u}, \mathbf{v}, \mathbf{w} \in \mathbb{R}^2$ such that $d(\mathbf{u}, \mathbf{v}) + d(\mathbf{v}, \mathbf{w}) = d(\mathbf{u}, \mathbf{w})$. Prove that $\mathbf{u}, \mathbf{v}, \mathbf{w}$ are scalar multiples of each other.

19. Prove the remaining parts of Theorem 2.1.15.

20. Let $\mathbf{u}, \mathbf{v} \in \mathbb{R}^n$. Prove that $\big| \|\mathbf{u}\| - \|\mathbf{v}\| \big| \leq \|\mathbf{u} - \mathbf{v}\|$.

21. Prove the **Parallelogram Law**: For all $\mathbf{u}, \mathbf{v} \in \mathbb{R}^n$,

$$\|\mathbf{u} + \mathbf{v}\|^2 + \|\mathbf{u} - \mathbf{v}\|^2 = 2\|\mathbf{u}\|^2 + 2\|\mathbf{v}\|^2.$$

22. Find a spanning set of $\mathbb{R}^2$ that contains only unit vectors but does not contain any standard vectors. Prove your result.

23. Find a spanning set of $\mathbb{R}^3$ that does not contain any standard vectors.

24. Find parametric equations for the given lines.

    (a) $\mathrm{S}\left(\begin{bmatrix} 3 \\ -2 \end{bmatrix} \right)$

    (b) $\begin{bmatrix} 3 \\ -8 \end{bmatrix} + \mathrm{S}\left(\begin{bmatrix} 5 \\ 6 \end{bmatrix} \right)$

    (c) The line that contains $\begin{bmatrix} 1 \\ 0 \\ -4 \\ 5 \end{bmatrix}$ and $\begin{bmatrix} 3 \\ 1 \\ 0 \\ -3 \end{bmatrix}$

25. Write the line given in Exercise 24(b) in slope-intercept form.

26. Find parametric equations for the given planes.

(a) $\begin{bmatrix} 7 \\ 2 \\ 0 \end{bmatrix} + S\left(\begin{bmatrix} 1 \\ 2 \\ 3 \end{bmatrix}, \begin{bmatrix} 0 \\ 4 \\ -1 \end{bmatrix}\right)$

(b) The plane that contains $\begin{bmatrix} 1 \\ 0 \\ -4 \\ 5 \end{bmatrix}, \begin{bmatrix} 3 \\ 1 \\ 0 \\ -3 \end{bmatrix}$, and $\begin{bmatrix} 1 \\ 1 \\ 1 \\ 1 \end{bmatrix}$

27. Find the intersection of the lines

$$\begin{bmatrix} 1 \\ 0 \end{bmatrix} + S\left(\begin{bmatrix} 3 \\ -2 \end{bmatrix}\right) \quad \text{and} \quad \begin{bmatrix} 2 \\ 4 \end{bmatrix} + S\left(\begin{bmatrix} 2 \\ 3 \end{bmatrix}\right).$$

28. Find the intersection of

$$S\left(\begin{bmatrix} 1 \\ 2 \\ 3 \end{bmatrix}\right) \quad \text{and} \quad \begin{bmatrix} 1 \\ 0 \\ -1 \end{bmatrix} + S\left(\begin{bmatrix} 2 \\ -1 \\ 4 \end{bmatrix}, \begin{bmatrix} 1 \\ 1 \\ 1 \end{bmatrix}\right).$$

29. Writing the answer using both the notation of a coset and as parametric equations, find the intersection of the planes

$$\begin{bmatrix} 3 \\ -3 \\ 0 \end{bmatrix} + S\left(\begin{bmatrix} 2 \\ 3 \\ 0 \end{bmatrix}, \begin{bmatrix} 5 \\ 0 \\ -15 \end{bmatrix}\right) \quad \text{and} \quad \begin{bmatrix} 0 \\ 3 \\ 1 \end{bmatrix} + S\left(\begin{bmatrix} 1 \\ 0 \\ 1 \end{bmatrix}, \begin{bmatrix} 0 \\ 1 \\ 4 \end{bmatrix}\right).$$

30. Prove that the intersection in $\mathbb{R}^3$ of the xy-plane and the xz-plane is the x-axis.

31. Find the perpendicular distance from $\begin{bmatrix} 1 \\ 3 \end{bmatrix}$ to the line $\begin{bmatrix} 3 \\ -1 \end{bmatrix} + S\left(\begin{bmatrix} 1 \\ 5 \end{bmatrix}\right).$

32. Show that $\begin{bmatrix} 19 \\ 8 \end{bmatrix}$ is in the line $\begin{bmatrix} 7 \\ 4 \end{bmatrix} + S\left(\begin{bmatrix} 6 \\ 2 \end{bmatrix}\right).$

33. Show that $\begin{bmatrix} 15 \\ 3 \\ 6 \end{bmatrix}$ is in the plane $\begin{bmatrix} 1 \\ 7 \\ 4 \end{bmatrix} + S\left(\begin{bmatrix} 6 \\ 0 \\ 2 \end{bmatrix}, \begin{bmatrix} 1 \\ 1 \\ 1 \end{bmatrix}\right).$

34. Show that $\left\{\begin{bmatrix} 3 \\ 4 \end{bmatrix}, \begin{bmatrix} -1 \\ 7 \end{bmatrix}\right\}$ spans $\mathbb{R}^2$ but $\mathscr{B} = \left\{\begin{bmatrix} 1 \\ 2 \end{bmatrix}, \begin{bmatrix} 4 \\ 8 \end{bmatrix}\right\}$ does not span $\mathbb{R}^2$. Why does $\mathscr{B}$ not span $\mathbb{R}^2$?

35. Show that $\left\{\begin{bmatrix} 1 \\ 2 \\ 0 \end{bmatrix}, \begin{bmatrix} 1 \\ 0 \\ 1 \end{bmatrix}, \begin{bmatrix} 0 \\ 2 \\ -3 \end{bmatrix}\right\}$ spans $\mathbb{R}^3$ but $\mathscr{C} = \left\{\begin{bmatrix} 1 \\ 2 \\ 0 \end{bmatrix}, \begin{bmatrix} 1 \\ 0 \\ 1 \end{bmatrix}\right\}$ does not span $\mathbb{R}^3$. Why does $\mathscr{C}$ not span $\mathbb{R}^3$?

36. Let $\mathbf{m}_1, \mathbf{m}_2, \ldots, \mathbf{m}_l \in \mathbb{R}^n$ and $1 \leq k \leq l$. Prove that $S(\mathbf{m}_1, \mathbf{m}_2, \ldots, \mathbf{m}_k)$ is a subset of $S(\mathbf{m}_1, \mathbf{m}_2, \ldots, \mathbf{m}_l)$.

37. Let $\mathbf{m}_1, \mathbf{m}_2 \in \mathbb{R}^n$ and $r \in \mathbb{R}$. Prove.
    (a) $S(\mathbf{m}_1) = S(r\mathbf{m}_1)$
    (b) $S(\mathbf{m}_1, \mathbf{m}_2) = S(\mathbf{m}_1, r\mathbf{m}_1 + \mathbf{m}_2)$

38. Suppose that each of the vectors $\mathbf{m}_1, \mathbf{m}_2, \ldots, \mathbf{m}_k \in \mathbb{R}^n$ are linear combinations of the vectors $\mathbf{n}_1, \mathbf{n}_1, \ldots, \mathbf{n}_l \in \mathbb{R}^n$ and that each of the vectors $\mathbf{n}_1, \mathbf{n}_1, \ldots, \mathbf{n}_l$ are linear combinations of $\mathbf{m}_1, \mathbf{m}_2, \ldots, \mathbf{m}_k$. Prove that

$$S(\mathbf{m}_1, \mathbf{m}_2, \ldots, \mathbf{m}_k) = S(\mathbf{n}_1, \mathbf{n}_2, \ldots, \mathbf{n}_l).$$

39. Let $\mathbf{u}, \mathbf{v} \in \mathbb{R}^n$. Explain why $\mathbf{x} = \mathbf{u} + t(\mathbf{u} - \mathbf{v})$ is a vector equation for the line containing $\mathbf{u}$ and $\mathbf{v}$.

40. Find $L_1 \cap L_2$, where $L_1 = \left\{ \begin{bmatrix} r \\ 2r + 3 \end{bmatrix} : r \in \mathbb{R} \right\}$ and $L_2 = \left\{ \begin{bmatrix} r \\ -5r + 6 \end{bmatrix} : r \in \mathbb{R} \right\}$.

41. Find parametric equations for the line segment joining $\begin{bmatrix} a \\ b \end{bmatrix}$ and $\begin{bmatrix} c \\ d \end{bmatrix}$. Also, what vector should be considered the midpoint of this segment? Explain.

2.2 Dot Product

The distance function introduced the geometry of $\mathbb{R}^n$ in Section 2.1. The other necessary aspect of the geometry of $\mathbb{R}^n$ can now be defined. This is angle measure. Take two nonzero vectors $\mathbf{u}$ and $\mathbf{v}$. View the vectors $\mathbf{u}$ and $\mathbf{v}$ as arrows with initial points at the origin. Because

$$\|\mathbf{u} - \mathbf{v}\| = \|\mathbf{u} + (-\mathbf{v})\| \leq \|\mathbf{u}\| + \|\mathbf{v}\|,$$

there is a (possibly degenerate) triangle with sides of lengths $\|\mathbf{u}\|$, $\|\mathbf{v}\|$, and $\|\mathbf{u} - \mathbf{v}\|$ as illustrated in Figure 2.6. Let the interior angle formed by $\mathbf{u}$ and $\mathbf{v}$ have measure θ. By the Law of Cosines,

$$\|\mathbf{u} - \mathbf{v}\|^2 = \|\mathbf{u}\|^2 + \|\mathbf{v}\|^2 - 2\|\mathbf{u}\|\,\|\mathbf{v}\|\cos\theta.$$

Writing $\mathbf{u} = \begin{bmatrix} u_1 \\ u_2 \\ \vdots \\ u_n \end{bmatrix}$ and $\mathbf{v} = \begin{bmatrix} v_1 \\ v_2 \\ \vdots \\ v_n \end{bmatrix}$ yields

$$\sum_{i=1}^{n}(u_i - v_i)^2 = \sum_{i=1}^{n}u_i^2 + \sum_{i=1}^{n}v_i^2 - 2\|\mathbf{u}\|\,\|\mathbf{v}\|\cos\theta,$$

so
$$-2u_1v_1 - 2u_2v_2 - \cdots - 2u_nv_n = -2\|\mathbf{u}\|\,\|\mathbf{v}\|\cos\theta,$$
and this gives
$$\cos\theta = \frac{u_1v_1 + u_2v_2 + \cdots + u_nv_n}{\|\mathbf{u}\|\,\|\mathbf{v}\|}, \tag{2.26}$$
which can be used to find θ. To simplify the definition of this angle, notation is introduced to identify the numerator in (2.26).

■ **Definition 2.2.1**

For all $\mathbf{u}, \mathbf{v} \in \mathbb{R}^n$ with coordinates $u_1, u_2, \ldots, u_n$ and $v_1, v_2, \ldots, v_n$, respectively, the **dot product** (or **Euclidean inner product**) of $\mathbf{u}$ and $\mathbf{v}$ is
$$\mathbf{u} \bullet \mathbf{v} = u_1v_1 + u_2v_2 + \cdots + u_nv_n.$$

■ **Definition 2.2.2**

For all nonzero $\mathbf{u}, \mathbf{v} \in \mathbb{R}^n$, the **angle between u and v** has measure θ if $0 \le \theta \le \pi$ and
$$\cos\theta = \frac{\mathbf{u} \bullet \mathbf{v}}{\|\mathbf{u}\|\,\|\mathbf{v}\|}.$$

■ **Example 2.2.3**

Let $\mathbf{u} = \begin{bmatrix} 3 \\ 0 \\ -5 \end{bmatrix}$ and $\mathbf{v} = \begin{bmatrix} 6 \\ 1 \\ 4 \end{bmatrix}$. Then, $\mathbf{u} \bullet \mathbf{v} = -2$. In addition, $\|\mathbf{u}\| = \sqrt{34}$ and $\|\mathbf{v}\| = \sqrt{53}$. Hence, by Definition 2.2.2,
$$\cos\theta = \frac{-2}{\sqrt{1802}},$$
and restricting θ to $0 \le \theta \le \pi$, the measure θ is 1.62 radians (about 93°).

■ **Example 2.2.4**

In $\mathbb{R}^2$, letting $\mathbf{u} = \begin{bmatrix} 2 \\ 2\sqrt{3} \end{bmatrix}$ and $\mathbf{v} = \begin{bmatrix} 5\sqrt{2} \\ -5\sqrt{2} \end{bmatrix}$,
$$\mathbf{u} \bullet \mathbf{v} = 10\left(\sqrt{2} - \sqrt{6}\right).$$
Some trigonometry will reveal that the angle formed by $\mathbf{u}$ and $\mathbf{v}$ has measure $7\pi/12$, so, alternately, because
$$\mathbf{u} \bullet \mathbf{v} = \|\mathbf{u}\|\,\|\mathbf{v}\|\cos\theta, \tag{2.27}$$
the dot product can be rewritten as
$$\mathbf{u} \bullet \mathbf{v} = \left\|\begin{bmatrix} 2 \\ 2\sqrt{3} \end{bmatrix}\right\| \left\|\begin{bmatrix} 5\sqrt{2} \\ -5\sqrt{2} \end{bmatrix}\right\| \cos\frac{7\pi}{12} = -40\frac{\sqrt{3}-1}{2\sqrt{2}} = 10\left(\sqrt{2} - \sqrt{6}\right).$$

■ Example 2.2.5

Compute

$$\begin{bmatrix} 2 \\ 3 \\ 7 \end{bmatrix} \cdot \begin{bmatrix} 4 \\ -1 \\ 5 \end{bmatrix} = 2 \cdot 4 + 3 \cdot -1 + 7 \cdot 5 = 40.$$

Also, the dot products of $\begin{bmatrix} 2 \\ 3 \\ 7 \end{bmatrix}$ with the standard vectors from $\mathbb{R}^3$ yield

$$\begin{bmatrix} 2 \\ 3 \\ 7 \end{bmatrix} \cdot \begin{bmatrix} 1 \\ 0 \\ 0 \end{bmatrix} = 2, \quad \begin{bmatrix} 2 \\ 3 \\ 7 \end{bmatrix} \cdot \begin{bmatrix} 0 \\ 1 \\ 0 \end{bmatrix} = 3, \quad \begin{bmatrix} 2 \\ 3 \\ 7 \end{bmatrix} \cdot \begin{bmatrix} 0 \\ 0 \\ 1 \end{bmatrix} = 7.$$

In general, if $\mathbf{e}_i$ is the ith standard vector from $\mathbb{R}^n$ and $\mathbf{u} \in \mathbb{R}^n$ has coordinates $u_1, u_2, \ldots, u_n$,

$$\mathbf{u} \cdot \mathbf{e}_i = \begin{bmatrix} u_1 \\ \vdots \\ u_{i-1} \\ u_i \\ u_{i+1} \\ \vdots \\ u_n \end{bmatrix} \cdot \begin{bmatrix} 0 \\ \vdots \\ 0 \\ 1 \\ 0 \\ \vdots \\ 0 \end{bmatrix} = u_i. \tag{2.28}$$

This means that computing the dot product of a given vector with the ith standard vector is a means by which to obtain the ith coordinate of a vector.

Although the dot product is a common computation in linear algebra, it is not a binary operation because its output is a real number despite its inputs being vectors. Instead, the dot product is a function $\mathbb{R}^n \times \mathbb{R}^n \to \mathbb{R}$. It has properties reminiscent of multiplication including the assumption that it has precedence over vector addition. Scalar multiplication, however, has precedence over the dot product, and this is seen in the following property list.

■ Theorem 2.2.6

Let $\mathbf{u}, \mathbf{v}, \mathbf{w} \in \mathbb{R}^n$ and $r \in \mathbb{R}$.

(a) $\mathbf{u} \cdot \mathbf{v} = \mathbf{v} \cdot \mathbf{u}$.

(b) $\mathbf{u} \cdot (\mathbf{v} + \mathbf{w}) = \mathbf{u} \cdot \mathbf{v} + \mathbf{u} \cdot \mathbf{w}$.

(c) $r\mathbf{u} \cdot \mathbf{v} = \mathbf{u} \cdot r\mathbf{v} = r(\mathbf{u} \cdot \mathbf{v})$.

(d) $\mathbf{u} \cdot \mathbf{u} \geq 0$.

(e) $\mathbf{u} \cdot \mathbf{u} = 0$ if and only if $\mathbf{u} = \mathbf{0}$.

Proof

Let **u** and **v** have coordinates $u_1, u_2, \ldots, u_n$ and $v_1, v_2, \ldots, v_n$, respectively. To prove (c),

$$r\mathbf{u} \cdot \mathbf{v} = ru_1v_1 + ru_2v_1 + \cdots + ru_nv_n$$
$$= u_1(rv_1) + u_2(rv_2) + \cdots + u_n(rv_n) = \mathbf{u} \cdot r\mathbf{v}$$
$$= r(u_1v_1 + u_2v_2 + \cdots + u_nv_n) = r(\mathbf{u} \cdot \mathbf{v}).$$

To prove (d), observe that $\mathbf{u} \cdot \mathbf{u} = u_1^2 + u_2^2 + \cdots + u_n^2$, which cannot be negative since $u_i \in \mathbb{R}$ for $i = 1, 2, \ldots, n$. The other parts are left to Exercise 4. ■

Observe that the slope of the line $\mathsf{S}\left(\begin{bmatrix} 4 \\ 7 \end{bmatrix}\right)$ is $7/4$ and the slope of $\mathsf{S}\left(\begin{bmatrix} -14 \\ 8 \end{bmatrix}\right)$ is $-4/7$, so these lines are perpendicular. Because $7/4$ and $-4/7$ are negative reciprocals of each other, it follows that

$$\begin{bmatrix} 4 \\ 7 \end{bmatrix} \cdot \begin{bmatrix} -14 \\ 8 \end{bmatrix} = 0.$$

The idea of perpendicular lines is generalized to vectors with the next definition.

■ **Definition 2.2.7**

Let $\mathbf{u}, \mathbf{v} \in \mathbb{R}^n$ and θ be the measure of the angle between **u** and **v**.

(a) **u** is **perpendicular** to **v** if $\mathbf{u} \neq \mathbf{0}$, $\mathbf{v} \neq \mathbf{0}$, and $\theta = \pi/2$.

(b) **u** and **v** are **orthogonal** if $\mathbf{u} \cdot \mathbf{v} = 0$.

Notice that if **u** is perpendicular to **v**, the vectors are also orthogonal, but the converse does not follow. For example, $\begin{bmatrix} 0 \\ 0 \end{bmatrix}$ is orthogonal to $\begin{bmatrix} 3 \\ 2 \end{bmatrix}$, but $\begin{bmatrix} 0 \\ 0 \end{bmatrix}$ is not perpendicular to $\begin{bmatrix} 3 \\ 2 \end{bmatrix}$ because an angle requires two sides of nonzero length. When the zero vector is excluded from consideration, the next result follows.

■ **Theorem 2.2.8**

The nonzero vectors **u** and **v** in $\mathbb{R}^n$ are perpendicular if and only if $\mathbf{u} \cdot \mathbf{v} = 0$.

Proof

Let $\mathbf{u}, \mathbf{v} \in \mathbb{R}^n$ such that $\mathbf{u} \neq \mathbf{0}$ and $\mathbf{v} \neq \mathbf{0}$. Let θ be the angle between these vectors. Assume that $\theta = \pi/2$. Then, $\mathbf{u} \cdot \mathbf{v} = 0$ by the equation of Definition 2.2.2. To prove the converse, let $\mathbf{u} \cdot \mathbf{v} = 0$. Hence, $\cos\theta = 0$ by (2.27), which implies that $\theta = \pi/2$ since $0 \leq \theta \leq \pi$. ■

Lines and Planes

The dot product can be used to describe lines and planes in $\mathbb{R}^n$. To do this, a definition is required.

■ **Definition 2.2.9**

Let $\mathbf{n}, \mathbf{u}, \mathbf{m}_1, \mathbf{m}_2, \dots, \mathbf{m}_k \in \mathbb{R}^n$. Define $\mathbf{n}$ to be a **normal vector** (or $\mathbf{n}$ is **normal**) to the coset $\mathbf{u} + S(\mathbf{m}_1, \mathbf{m}_2, \dots, \mathbf{m}_k)$ if $\mathbf{n}$ is orthogonal to $\mathbf{w}$ for all $\mathbf{w} \in S(\mathbf{m}_1, \mathbf{m}_2, \dots, \mathbf{m}_k)$.

To understand the significance of Definition 2.2.9, consider the line L in $\mathbb{R}^2$ containing $\mathbf{0}$ in Figure 2.9. Its direction can be determined by its normal vector $\mathbf{n}$ because every vector in L is orthogonal to $\mathbf{n}$. Furthermore, every line parallel to L will also have $\mathbf{n}$ as a normal vector. An example in Figure 2.9 is L', which is L shifted from the origin using $\mathbf{u}$. In the figure, both $\mathbf{u}$ and $\mathbf{v}$ are vectors in L'. The figure also suggests that although $\mathbf{n}$ is not orthogonal to vectors in L', the normal $\mathbf{n}$ is orthogonal to $\mathbf{u} - \mathbf{v}$. To confirm this, let $\mathbf{m}$ be a direction vector such that $L = S(\mathbf{m})$ and $L' = \mathbf{u} + S(\mathbf{m})$. Since $\mathbf{v} \in L'$, it follows that $\mathbf{v} = \mathbf{u} + r\mathbf{m}$ for some $r \in \mathbb{R}$. Therefore, because $\mathbf{n}$ is normal to L and L', Theorem 2.2.6(c) implies

$$(\mathbf{u} - \mathbf{v}) \bullet \mathbf{n} = (\mathbf{u} - [\mathbf{u} + r\mathbf{m}]) \bullet \mathbf{n} = -r\mathbf{m} \bullet \mathbf{n} = -r(\mathbf{m} \bullet \mathbf{n}) = 0. \qquad (2.29)$$

(2.29) can be generalized to the next theorem (Exercise 14).

■ **Theorem 2.2.10**

If $C \subseteq \mathbb{R}^n$ is a coset with normal vector $\mathbf{n}$, then $\mathbf{n}$ is orthogonal to $\mathbf{u} - \mathbf{v}$ for all $\mathbf{u}, \mathbf{v} \in C$.

■ **Example 2.2.11**

Using Figure 2.9, let $\mathbf{n} = \begin{bmatrix} 3 \\ 5 \end{bmatrix}$ and $\mathbf{m} = \begin{bmatrix} 10 \\ -6 \end{bmatrix}$. The vector $\mathbf{n}$ is normal to the

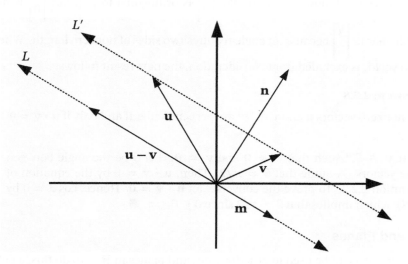

Figure 2.9 Lines L and L' with normal $\mathbf{n}$.

line $L = \mathsf{S}\left(\begin{bmatrix} 10 \\ -6 \end{bmatrix}\right)$ in $\mathbb{R}^2$ because for all $r \in \mathbb{R}$,

$$\begin{bmatrix} 3 \\ 5 \end{bmatrix} \cdot \begin{bmatrix} 10r \\ -6r \end{bmatrix} = 30r - 30r = 0.$$

Thus, $\begin{bmatrix} 3 \\ 5 \end{bmatrix}$ is also normal to $\mathbf{u} + \mathsf{S}\left(\begin{bmatrix} 10 \\ -6 \end{bmatrix}\right)$ for all $\mathbf{u} \in \mathbb{R}^2$. In particular, let

$$L' = \begin{bmatrix} 3 \\ -1 \end{bmatrix} + \mathsf{S}\left(\begin{bmatrix} 10 \\ -6 \end{bmatrix}\right).$$

Then, $\begin{bmatrix} 3 \\ -1 \end{bmatrix}, \begin{bmatrix} 23 \\ -13 \end{bmatrix} \in L'$. Notice that

$$\begin{bmatrix} 3 \\ 5 \end{bmatrix} \cdot \begin{bmatrix} 3 \\ -1 \end{bmatrix} = 4 \quad \text{and} \quad \begin{bmatrix} 3 \\ 5 \end{bmatrix} \cdot \begin{bmatrix} 23 \\ -13 \end{bmatrix} = 4,$$

but

$$\begin{bmatrix} 3 \\ 5 \end{bmatrix} \cdot \left(\begin{bmatrix} 3 \\ -1 \end{bmatrix} - \begin{bmatrix} 23 \\ -13 \end{bmatrix}\right) = \begin{bmatrix} 3 \\ 5 \end{bmatrix} \cdot \begin{bmatrix} -20 \\ 12 \end{bmatrix} = 0.$$

■ **Example 2.2.12**

$\begin{bmatrix} 2 \\ -3 \\ 5 \end{bmatrix}$ is normal to the plane $\begin{bmatrix} -5 \\ 0 \\ 3 \end{bmatrix} + \mathsf{S}\left(\begin{bmatrix} 3 \\ 2 \\ 0 \end{bmatrix}, \begin{bmatrix} 1 \\ 4 \\ 2 \end{bmatrix}\right)$ because

$$\begin{bmatrix} 2 \\ -3 \\ 5 \end{bmatrix} \cdot \begin{bmatrix} 3 \\ 2 \\ 0 \end{bmatrix} = 0 \quad \text{and} \quad \begin{bmatrix} 2 \\ -3 \\ 5 \end{bmatrix} \cdot \begin{bmatrix} 1 \\ 4 \\ 2 \end{bmatrix} = 0,$$

which implies that for all $r, s \in \mathbb{R}$,

$$\begin{bmatrix} 2 \\ -3 \\ 5 \end{bmatrix} \cdot \left(r \begin{bmatrix} 3 \\ 2 \\ 0 \end{bmatrix} + s \begin{bmatrix} 1 \\ 4 \\ 2 \end{bmatrix}\right) = 0.$$

Example 2.2.12 suggests the next theorem.

■ **Theorem 2.2.13**

Let $\mathbf{u}, \mathbf{m}_1, \mathbf{m}_2, \dots, \mathbf{m}_k$ be vectors in $\mathbb{R}^n$. The vector $\mathbf{n} \in \mathbb{R}^n$ is a normal vector to the coset $\mathbf{u} + \mathsf{S}(\mathbf{m}_1, \mathbf{m}_2, \dots, \mathbf{m}_k)$ if and only if $\mathbf{n}$ is orthogonal to $\mathbf{m}_i$ for all $i = 1, 2, \dots, k$.

Proof

Suppose that $\mathbf{n} \bullet \mathbf{m}_i = 0$ for $i = 1, 2, \ldots, k$. Take $\mathbf{w} \in S(\mathbf{m}_1, \mathbf{m}_2, \ldots, \mathbf{m}_k)$ and write $\mathbf{w} = r_1\mathbf{m}_1 + r_2\mathbf{m}_2 + \cdots + r_k\mathbf{m}_k$ for some $r_1, r_2, \ldots, r_k \in \mathbb{R}$. Then, by Theorem 2.2.6,

$$\begin{aligned}
\mathbf{n} \bullet \mathbf{w} &= \mathbf{n} \bullet (r_1\mathbf{m}_1 + r_2\mathbf{m}_2 + \cdots + r_k\mathbf{m}_k) \\
&= \mathbf{n} \bullet r_1\mathbf{m}_1 + \mathbf{n} \bullet r_2\mathbf{m}_2 + \cdots + \mathbf{n} \bullet r_k\mathbf{m}_k \\
&= r_1(\mathbf{n} \bullet \mathbf{m}_1) + r_2(\mathbf{n} \bullet \mathbf{m}_2) + \cdots + r_k(\mathbf{n} \bullet \mathbf{m}_k) = 0.
\end{aligned}$$

The converse is an immediate consequence of Definition 2.2.9. ∎

Normal vectors can be used to identify the direction vectors of a coset. Consider the line in $\mathbb{R}^2$ with position vector $\mathbf{u} = \begin{bmatrix} u_1 \\ u_2 \end{bmatrix}$ and normal vector $\mathbf{n} = \begin{bmatrix} a \\ b \end{bmatrix}$. Suppose that $\mathbf{x} = \begin{bmatrix} x \\ y \end{bmatrix}$ is in the line. Theorem 2.2.10 gives

$$\begin{bmatrix} a \\ b \end{bmatrix} \bullet \begin{bmatrix} x - u_1 \\ y - u_2 \end{bmatrix} = 0. \tag{2.30}$$

From (2.30) a **vector-normal equation** (or **point-normal equation**) of the line is derived,

$$a(x - u_1) + b(y - u_2) = 0, \tag{2.31}$$

from which follows an equation for the line in **standard form**,

$$ax + by = au_1 + bu_2. \tag{2.32}$$

Observe that (2.32) can be written as

$$\mathbf{n} \bullet \mathbf{x} = \mathbf{n} \bullet \mathbf{u}, \tag{2.33}$$

remembering that $\mathbf{n}$ is normal to the line and $\mathbf{u}$ is a fixed vector in the line.

■ **Example 2.2.14**

Let $\begin{bmatrix} 1 \\ 3 \end{bmatrix}$ be normal to a line L in $\mathbb{R}^2$ containing $\begin{bmatrix} 4 \\ 5 \end{bmatrix}$. Because $\begin{bmatrix} 1 \\ 3 \end{bmatrix} \bullet \begin{bmatrix} x - 4 \\ y - 5 \end{bmatrix} = 0$, a vector-normal equation (2.31) for L is

$$(x - 4) + 3(y - 5) = 0,$$

and because $\begin{bmatrix} 1 \\ 3 \end{bmatrix} \bullet \begin{bmatrix} 4 \\ 5 \end{bmatrix} = 19$, an equation in standard form for L is

$$x + 3y = 19.$$

Lastly, a set of parametric equations for L is

$$x = -3r + 19,$$
$$y = r.$$

The same can be done for planes in $\mathbb{R}^3$. Let $\mathbf{u} = \begin{bmatrix} u_1 \\ u_2 \\ u_3 \end{bmatrix}$ be a position vector for a

plane with normal vector $\mathbf{n} = \begin{bmatrix} a \\ b \\ c \end{bmatrix}$. Suppose that $\mathbf{x} = \begin{bmatrix} x \\ y \\ z \end{bmatrix}$ is in the plane. Then, by Theorem 2.2.10,

$$\begin{bmatrix} a \\ b \\ c \end{bmatrix} \cdot \begin{bmatrix} x - u_1 \\ y - u_2 \\ z - u_3 \end{bmatrix} = 0,$$

from which a **vector-normal equation** of the plane can be written,

$$a(x - u_1) + b(y - u_2) + c(z - u_3) = 0. \tag{2.34}$$

Therefore, an equation in **standard form** for the plane is

$$ax + by + cz = au_1 + bu_2 + cu_3. \tag{2.35}$$

Observe that (2.35) can also be written in the form of (2.33).

■ **Example 2.2.15**

Let $\mathbf{n} = \begin{bmatrix} 3 \\ 5 \\ -9 \end{bmatrix}$ be a normal vector for a plane P in $\mathbb{R}^3$ containing $\begin{bmatrix} 1 \\ 2 \\ 3 \end{bmatrix}$. A vector-normal equation for P is

$$3(x - 1) + 5(y - 2) - 9(z - 3) = 0, \tag{2.36}$$

and an equation in standard form for P is

$$3x + 5y - 9z = -14. \tag{2.37}$$

The normal $\mathbf{n}$ is easily seen in (2.36) and (2.37). Also, parametric equations for P can be found, such as

$$x = -5r + 3s - 14/3,$$
$$y = 3r,$$
$$z = s.$$

Notice that as in Example 2.2.14, the number of parameters indicates the dimension of the coset.

A line in $\mathbb{R}^2$ and a plane in $\mathbb{R}^3$ both have one fewer dimensions than the surrounding space. This is generalized as follows.

■ **Definition 2.2.16**

Let $a_1, a_2, \ldots, a_n, c \in \mathbb{R}$. A **hyperplane** is the set of solutions in $\mathbb{R}^n$ to

$$a_1 x_1 + a_2 x_2 + \cdots + a_n x_n = c.$$

Orthogonal Projection

In Example 2.1.24, the perpendicular distance from a line to a vector not in the line was found. A similar problem is to take a plane and a vector not in the plane and find the perpendicular distance from the vector to the plane. For example, what is the distance from $\begin{bmatrix} 3 \\ 6 \\ 1 \end{bmatrix}$ to the plane given by $3x + 7y - 2z = 5$? Modifying the method of Example 2.1.24 to three dimensions, the distance can be determined by finding the equation of the line containing $\begin{bmatrix} 3 \\ 6 \\ 1 \end{bmatrix}$ that is perpendicular to the given plane. This can be done because a normal vector to the plane is easily found. Then, find the point of intersection of the line with the given plane and use Definition 2.1.9 to find the distance (Exercises 25 and 26). It should be noted that the dot product can help compute that distance.

■ **Theorem 2.2.17**

Let $\mathbf{u}, \mathbf{v} \in \mathbb{R}^n$.

(a) $\|\mathbf{u}\| = \sqrt{\mathbf{u} \cdot \mathbf{u}}$.

(b) $d(\mathbf{u}, \mathbf{v}) = \sqrt{(\mathbf{u} - \mathbf{v}) \cdot (\mathbf{u} - \mathbf{v})}$.

Proof

Take $\mathbf{u}, \mathbf{v} \in \mathbb{R}^n$ and write

$$\mathbf{u} = \begin{bmatrix} u_1 \\ u_2 \\ \vdots \\ u_n \end{bmatrix} \quad \text{and} \quad \mathbf{v} = \begin{bmatrix} v_1 \\ v_2 \\ \vdots \\ v_n \end{bmatrix}.$$

First, because $\mathbf{u} \cdot \mathbf{u} = u_1^2 + u_2^2 + \cdots + u_n^2$, Definition 2.1.13 gives (a). Second, (b) follows from (a) and Theorem 2.1.14. ■

There is, however, a simpler way to find the distance from a vector to a plane, but it requires some initial work. Let $\mathbf{u}, \mathbf{v} \in \mathbb{R}^n$ and $\mathbf{v} \neq \mathbf{0}$. The first problem to be solved is to find a scalar r and vector $\mathbf{w}$ such that $\mathbf{w}$ is orthogonal to $\mathbf{v}$ and $\mathbf{u} = r\mathbf{v} + \mathbf{w}$. This is illustrated in Figure 2.10. Because $\mathbf{w} = \mathbf{u} - r\mathbf{v}$, it is enough to find r. To do this, suppose that

$$(\mathbf{u} - r\mathbf{v}) \cdot \mathbf{v} = 0.$$

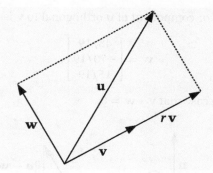

Figure 2.10 $\mathbf{u} = r\mathbf{v} + \mathbf{w}$ with $\mathbf{w}$ orthogonal to $\mathbf{v}$.

Then,

$$\mathbf{u} \cdot \mathbf{v} - r\mathbf{v} \cdot \mathbf{v} = 0$$

from which follows

$$r = \frac{\mathbf{u} \cdot \mathbf{v}}{\mathbf{v} \cdot \mathbf{v}},$$

so the desired vector is

$$r\mathbf{v} = \frac{\mathbf{u} \cdot \mathbf{v}}{\mathbf{v} \cdot \mathbf{v}}\mathbf{v}. \tag{2.38}$$

Since $\mathbf{v} \cdot \mathbf{v} = \|\mathbf{v}\|^2$ by Theorem 2.2.17(a), the vector of (2.38) is named with the next definition.

■ **Definition 2.2.18**

Let $\mathbf{u}, \mathbf{v} \in \mathbb{R}^n$ with $\mathbf{v} \neq \mathbf{0}$. The **orthogonal projection** (or simply the **projection**) of $\mathbf{u}$ in the direction of $\mathbf{v}$ is

$$\text{proj}_\mathbf{v}\mathbf{u} = \frac{\mathbf{u} \cdot \mathbf{v}}{\|\mathbf{v}\|^2}\mathbf{v},$$

and $\mathbf{u} - \text{proj}_\mathbf{v}\mathbf{u}$ is called the **vector component of u orthogonal to v**.

The projection received its name because it can be viewed as the shadow the vector $\mathbf{u}$ casts on the line with direction vector $\mathbf{v}$ assuming that a light source is casting rays that are perpendicular to $\mathbf{v}$.

■ **Example 2.2.19**

Let $\mathbf{u} = \begin{bmatrix} 2 \\ -4 \\ 1 \end{bmatrix}$ and $\mathbf{v} = \begin{bmatrix} 5 \\ 3 \\ -2 \end{bmatrix}$. Then, $\mathbf{u} \cdot \mathbf{v} = -4$ and $\|\mathbf{v}\|^2 = 38$. Hence,

$$\text{proj}_\mathbf{v}\mathbf{u} = \begin{bmatrix} -10/19 \\ -6/19 \\ 4/19 \end{bmatrix}.$$

Notice that the vector component of **u** orthogonal to **v** is

$$\mathbf{w} = \begin{bmatrix} 48/19 \\ -70/19 \\ 15/19 \end{bmatrix},$$

and it is indeed the case that $\mathbf{v} \cdot \mathbf{w} = 0$.

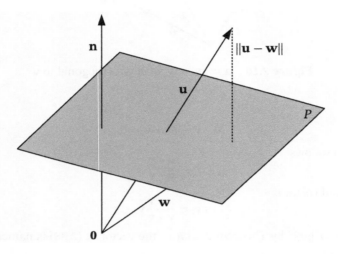

Figure 2.11   The distance from the vector **u** to the plane P.

Now to solve the distance to the plane problem. Let P be a plane in $\mathbb{R}^3$ with normal vector **n**. The perpendicular distance from $\mathbf{u} \in \mathbb{R}^3$ to P is $\|\mathbf{u} - \mathbf{w}\|$, where $\mathbf{w} \in P$ such that $\mathbf{u} - \mathbf{w}$ is collinear with **n** (Figure 2.11). Since **w** is probably difficult to find, attempt another method. To simplify the work, view the plane P from the side so that P can be considered a line. This is the vantage point of Figure 2.12, where the vectors are interpreted as arrows originating from the origin O. Let $\mathbf{v} \in P$. Using the labels in Figure 2.12, **v** and $\mathbf{u} - \mathbf{v}$ form parallelogram $ODAC$. Let E be in $\mathsf{S}\,(\mathbf{n})$ so that DE is orthogonal to $\mathsf{S}\,(\mathbf{n})$, and let B be in P so that BA is orthogonal to P and is the sought-after distance. Using basic geometry, it can be shown that $\triangle EOD$ is congruent to $\triangle BAC$ (Exercise 28). Therefore, since $\mathrm{proj}_{\mathbf{n}}(\mathbf{u} - \mathbf{v})$ is segment OE,

$$\|\mathrm{proj}_{\mathbf{n}}(\mathbf{u} - \mathbf{v})\| = BA. \tag{2.39}$$

The calculation of (2.39) can be simplified with the next theorem.

■ **Theorem 2.2.20**

For all $\mathbf{u}, \mathbf{v} \in \mathbb{R}^n$, if $\mathbf{v} \neq 0$, then $\|\mathrm{proj}_{\mathbf{v}}\mathbf{u}\| = \dfrac{|\mathbf{u} \cdot \mathbf{v}|}{\|\mathbf{v}\|}$.

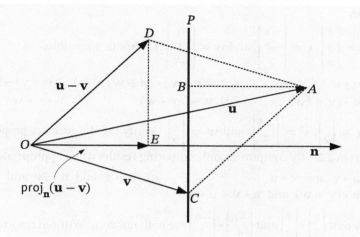

Figure 2.12 The distance from the vector **u** to the plane P, side view.

Proof

Let $\mathbf{v} \neq \mathbf{0}$. Then,

$$\|\text{proj}_{\mathbf{v}}\mathbf{u}\| = \left\|\frac{\mathbf{u} \cdot \mathbf{v}}{\|\mathbf{v}\|^2}\mathbf{v}\right\| = \frac{|\mathbf{u} \cdot \mathbf{v}|}{\|\mathbf{v}\|}\left\|\frac{1}{\|\mathbf{v}\|}\mathbf{v}\right\| = \frac{|\mathbf{u} \cdot \mathbf{v}|}{\|\mathbf{v}\|}. \blacksquare$$

■ Example 2.2.21

Suppose $\mathbf{u} = \begin{bmatrix} 3 \\ 6 \\ 1 \end{bmatrix}$ and P is the plane given by the equation $3x + 7y + 2z = 6$.

The vector **v** can be taken to be any vector in P, so let $\mathbf{v} = \begin{bmatrix} 0 \\ 0 \\ 3 \end{bmatrix}$, which implies

that

$$\mathbf{u} - \mathbf{v} = \begin{bmatrix} 3 \\ 6 \\ -2 \end{bmatrix}.$$

Since a normal vector to P is $\mathbf{n} = \begin{bmatrix} 3 \\ 7 \\ 2 \end{bmatrix}$, the perpendicular distance of **u** to the

plane is

$$\|\text{proj}_{\mathbf{n}}(\mathbf{u} - \mathbf{v})\| = \frac{47\sqrt{62}}{62}.$$

A similar strategy yields the distance from a point to a line (Exercise 31).

Exercises

1. Let $\mathbf{u} = \begin{bmatrix} 4 \\ 5 \\ 0 \end{bmatrix}$, $\mathbf{v} = \begin{bmatrix} -1 \\ -4 \\ 2 \end{bmatrix}$, and $\mathbf{w} = \begin{bmatrix} 0 \\ 3 \\ -7 \end{bmatrix}$. Compute if possible.

    (a) $\mathbf{u} \cdot \mathbf{v}$ (c) $2(\mathbf{u} + \mathbf{v}) \cdot (\mathbf{u} + \mathbf{w})$ (e) $3(\mathbf{w} \cdot \mathbf{v}) - 9(\mathbf{v} \cdot \mathbf{u})$

    (b) $\mathbf{u} \cdot (\mathbf{v} + 5\mathbf{w})$ (d) $\mathbf{u} + (6\mathbf{v} \cdot \mathbf{w})$ (f) $\mathbf{u} \cdot \mathbf{v} \cdot \mathbf{w}$

2. Let $\mathbf{u} = \begin{bmatrix} 7 \\ 2 \end{bmatrix}$, $\mathbf{v} = \begin{bmatrix} -6 \\ 4 \end{bmatrix}$, and $\mathbf{w} = \begin{bmatrix} 1 \\ 5 \end{bmatrix}$. Verify the dot product properties of
    Theorem 2.2.6 by computing and comparing results when appropriate.

    (a) $\mathbf{u} \cdot \mathbf{v}$ and $\mathbf{v} \cdot \mathbf{u}$ (c) $7\mathbf{u} \cdot \mathbf{v}$ and $\mathbf{u} \cdot 7\mathbf{v}$ and $7(\mathbf{u} \cdot \mathbf{v})$

    (b) $\mathbf{u} \cdot (\mathbf{v} + \mathbf{w})$ and $\mathbf{u} \cdot \mathbf{v} + \mathbf{u} \cdot \mathbf{w}$

3. Verify that $\begin{bmatrix} 7 \\ 2 \end{bmatrix} \cdot \begin{bmatrix} 7 \\ 2 \end{bmatrix}$ and $\begin{bmatrix} -6 \\ 4 \end{bmatrix} \cdot \begin{bmatrix} -6 \\ 4 \end{bmatrix}$ are both nonzero. Without resorting to an
    official proof, explain why the zero vector is the only vector with the property
    that the result is 0 when it is dotted with itself.

4. Finish the proof of Theorem 2.2.6.

5. Let $\mathbf{e}_1$, $\mathbf{e}_2$, and $\mathbf{e}_3$ be the standard vectors of $\mathbb{R}^3$. Take $\mathbf{u} \in \mathbb{R}^3$ so that $\mathbf{u} \cdot \mathbf{e}_1 = 6$,
    $\mathbf{u} \cdot \mathbf{e}_2 = 7$, and $\mathbf{u} \cdot \mathbf{e}_3 = -3$. Find $\mathbf{u}$.

6. Let $\mathbf{u}, \mathbf{v}, \mathbf{w} \in \mathbb{R}^n$. If $\mathbf{u}$ is orthogonal to $\mathbf{v}$ and $\mathbf{v}$ is orthogonal to $\mathbf{w}$, must it be
    the case that $\mathbf{u}$ is orthogonal to $\mathbf{w}$? Explain.

7. Let $\mathbf{u}, \mathbf{v}_1, \mathbf{v}_2, \dots, \mathbf{v}_k \in \mathbb{R}^n$. Prove that if $\mathbf{u}$ is orthogonal to $\mathbf{v}_i$ for all $i = 1, 2, \dots, k$,
    then $\mathbf{u}$ is orthogonal to $\mathbf{w}$ for all $\mathbf{w} \in S(\mathbf{v}_1, \mathbf{v}_2, \dots, \mathbf{v}_k)$.

8. Are $\begin{bmatrix} 4 \\ 0 \end{bmatrix}$ and $\begin{bmatrix} 0 \\ 7 \end{bmatrix}$ perpendicular? Explain.

9. Are $\begin{bmatrix} 0 \\ 0 \end{bmatrix}$ and $\begin{bmatrix} 2 \\ 9 \end{bmatrix}$ perpendicular? Explain.

10. Are $\begin{bmatrix} 0 \\ 0 \end{bmatrix}$ and $\begin{bmatrix} 2 \\ 9 \end{bmatrix}$ orthogonal? Explain.

11. Find the angles formed by the given pairs of vectors. Identify which pairs are
    orthogonal.

    (a) $\begin{bmatrix} 1 \\ 2 \\ 3 \\ 4 \end{bmatrix}$ and $\begin{bmatrix} 1 \\ 0 \\ -1 \\ 0 \end{bmatrix}$.

    (b) $\begin{bmatrix} 1 \\ 4 \\ 3 \end{bmatrix}$ and any vector from $S\left(\begin{bmatrix} 3 \\ 0 \\ -1 \end{bmatrix}, \begin{bmatrix} 1 \\ 2 \\ -3 \end{bmatrix} \right)$.

    (c) The normal vectors of $2x + 5y - z = 0$ and $x + 10z = 0$.

12. Let **u** and **v** be nonzero vectors in $\mathbb{R}^2$. Find a vector that bisects the angle formed by **u** and **v** and prove the result.

13. Let $\mathbf{u} = \begin{bmatrix} 3 \\ 6 \end{bmatrix}$ and $\mathbf{v} = \begin{bmatrix} -3 \\ 8 \end{bmatrix}$. Use Theorem 2.2.17 to compute.

    (a) $d(\mathbf{u}, \mathbf{v})$

    (b) $\|\mathbf{u}\|$

    (c) $d(3\mathbf{u} + \mathbf{v}, \mathbf{u} - 7\mathbf{v})$

    (d) $\|4\mathbf{v}\|$

14. Prove Theorem 2.2.10.

15. For every $\mathbf{u}, \mathbf{v} \in \mathbb{R}^n$, prove that $(\mathbf{u} + \mathbf{v}) \cdot (\mathbf{u} - \mathbf{v}) = \|\mathbf{u}\|^2 - \|\mathbf{v}\|^2$.

16. Prove the Pythagorean Theorem: For all $\mathbf{u}, \mathbf{v} \in \mathbb{R}^n$, the vector **u** is orthogonal to the vector **v** if and only if $\|\mathbf{u} + \mathbf{v}\|^2 = \|\mathbf{u}\|^2 + \|\mathbf{v}\|^2$.

17. Use Exercise 16 to prove that $\begin{bmatrix} 1 \\ -2 \\ 3 \end{bmatrix}$ and $\begin{bmatrix} 5 \\ 4 \\ 1 \end{bmatrix}$ are orthogonal.

18. Let $\mathbf{u}, \mathbf{v} \in \mathbb{R}^n$. Prove that if $\mathbf{u} \cdot \mathbf{v} = 0$, then $\|\mathbf{u} + \mathbf{v}\| = \|\mathbf{u} - \mathbf{v}\|$. Is the converse true?

19. Prove that $\mathbf{u} \cdot \mathbf{v} = \frac{1}{2} (\|\mathbf{u}\|^2 + \|\mathbf{v}\|^2 - \|\mathbf{u} - \mathbf{v}\|^2)$ for all $\mathbf{u}, \mathbf{v} \in \mathbb{R}^n$.

20. Let **n** be normal to a line L that does not contain **0**. Is it possible to find a vector $\mathbf{u} \in L$ such that **n** is orthogonal to **u**? Explain.

21. Find the parametric equations, vector-normal form, and standard form for the given lines in $\mathbb{R}^2$.

    (a) Contains the vectors $\begin{bmatrix} 7 \\ 10 \end{bmatrix}$ and $\begin{bmatrix} -5 \\ 22 \end{bmatrix}$.

    (b) Horizontal line containing $\begin{bmatrix} 4 \\ -7 \end{bmatrix}$.

    (c) $\begin{bmatrix} 4 \\ 2 \end{bmatrix} + s\left(\begin{bmatrix} 1 \\ 7 \end{bmatrix} \right)$.

22. The three lines of Exercise 21 can be combined to form three pairs of lines. Find the intersection of each of those pairs of lines.

23. Find parametric equations and an equation in standard form for the given planes in $\mathbb{R}^3$.

    (a) Contains $\begin{bmatrix} 2 \\ 4 \\ 0 \end{bmatrix}$ and has normal $\begin{bmatrix} 3 \\ -6 \\ 7 \end{bmatrix}$.

    (b) Containing $\begin{bmatrix} 3 \\ 2 \\ 1 \end{bmatrix}, \begin{bmatrix} 3 \\ 0 \\ 2 \end{bmatrix}$ and $\begin{bmatrix} 4 \\ 3 \\ 0 \end{bmatrix}$.

    (c) The xy-plane.

24. Find parametric equations for the line containing $\begin{bmatrix} 1 \\ 5 \\ -2 \end{bmatrix}$ and orthogonal to the plane $3x - 4y + 8z = 10$.

25. Find parametric equations for the line containing $\begin{bmatrix} 3 \\ 6 \\ 10 \end{bmatrix}$ and orthogonal to the plane $3x + 7y - 2z = 5$.

26. Find the vector **u** in the intersection of the line found in Exercise 25 and the plane $3x + 7y - 2z = 5$. Also, find the distance between **u** and $\begin{bmatrix} 3 \\ 6 \\ 10 \end{bmatrix}$.

27. Find the plane containing $\begin{bmatrix} 3 \\ 6 \\ 10 \end{bmatrix}$ which is parallel to the plane $3x + 7y - 2z = 5$.

28. Prove that $\triangle EOD$ is congruent to $\triangle BAC$ in Figure 2.12.

29. Find $\text{proj}_v\mathbf{u}$ using the given vectors.

    (a) $\mathbf{u} = \begin{bmatrix} 3 \\ 7 \end{bmatrix}$ and $\mathbf{v} = \begin{bmatrix} 8 \\ 2 \end{bmatrix}$ (c) $\mathbf{u} = \begin{bmatrix} 6 \\ 3 \end{bmatrix}$ and $\mathbf{v} = \begin{bmatrix} 6 \\ 3 \end{bmatrix}$

    (b) $\mathbf{u} = \begin{bmatrix} 2 \\ 0 \\ 1 \end{bmatrix}$ and $\mathbf{v} = \begin{bmatrix} 3 \\ -3 \\ 2 \end{bmatrix}$ (d) $\mathbf{u} = \begin{bmatrix} 1 \\ 2 \\ 3 \end{bmatrix}$ and $\mathbf{v} = \begin{bmatrix} 0 \\ 0 \\ 0 \end{bmatrix}$

30. For each **u** and **v** in Exercise 29, find both the vector component of **u** orthogonal to **v** and the length of $\text{proj}_v\mathbf{u}$.

31. Find the distance from $\begin{bmatrix} 4 \\ -7 \end{bmatrix}$ to the line $8x + 5y = 1$.

32. Find the distance from $\begin{bmatrix} 3 \\ 0 \\ 5 \end{bmatrix}$ to the plane $3x + y - 2z = 0$.

33. Find the distance between the lines $3x + 5y = 4$ and
$$x = -10t + 1,$$
$$y = 6t - 1.$$

34. Find the distance between the planes $4x + 3y + z = 6$ and $4x + 3y + z = -5$.

2.3   Cross Product

Two vectors are enough to determine a plane (Definition 2.1.25). However, it is known from basic geometry that three noncollinear points are enough to determine a plane. These ideas can be combined to find an equation for a given plane.

Let $\mathbf{u}$, $\mathbf{v}$, and $\mathbf{w}$ be three noncollinear vectors in a plane P existing in $\mathbb{R}^3$. To use these to find a normal vector $\mathbf{n} = \begin{bmatrix} n_1 \\ n_2 \\ n_3 \end{bmatrix}$ to P, write

$$\mathbf{u} - \mathbf{v} = \begin{bmatrix} a \\ b \\ c \end{bmatrix} \quad \text{and} \quad \mathbf{u} - \mathbf{w} = \begin{bmatrix} d \\ e \\ f \end{bmatrix}.$$

These vectors are orthogonal to $\mathbf{n}$ by Theorem 2.2.10, so

$$\begin{aligned} a n_1 + b n_2 + c n_3 &= 0, \\ d n_1 + e n_2 + f n_3 &= 0. \end{aligned} \tag{2.40}$$

To solve this system of linear equations, assume that $a \neq 0$ and $ae - bd \neq 0$, leaving the cases when $a = 0$ or $ae - bd = 0$ to Exercise 9. To solve this system, multiply the first equation by $1/a$ to obtain

$$\begin{aligned} n_1 + (b/a)n_2 + (c/a)n_3 &= 0, \\ d n_1 + e n_2 + f n_3 &= 0, \end{aligned}$$

and then eliminate n_1 from the second equation by multiplying the first equation by d and adding to the second to give

$$\begin{aligned} n_1 + (b/a)n_2 + (c/a)n_3 &= 0, \\ [(ae - db)/a]n_2 + [(af - cd)/a]n_3 &= 0. \end{aligned}$$

Multiply the second equation by $a/(ae - db)$. The result is

$$\begin{aligned} n_1 + (b/a)n_2 + (c/a)n_3 &= 0, \\ n_2 + [(af - dc)/(ae - db)]n_3 &= 0. \end{aligned}$$

Lastly, eliminate n_2 from the first equation by multiplying the second equation by $-b/a$ and adding to the first. This gives

$$\begin{aligned} n_1 + [(ce - bf)/(ae - db)]n_3 &= 0, \\ n_2 + [(af - dc)/(ae - db)]n_3 &= 0. \end{aligned}$$

For any $r \in \mathbb{R}$, a solution to this system in parametric form is

$$n_1 = -\frac{r(ce - bf)}{ae - bd},$$

$$n_2 = -\frac{r(af - dc)}{ae - db},$$

$$n_3 = r,$$

so taking $r = ae - bd$, it is seen that

$$\mathbf{n} = \begin{bmatrix} bf - ce \\ dc - af \\ ae - db \end{bmatrix}$$

is a solution to (2.40) and, hence, a normal for P. Therefore, any two given non-collinear vectors parallel to a plane in $\mathbb{R}^3$ can be used to compute a normal to the plane. This computation is given a name.

■ **Definition 2.3.1**

For all $\mathbf{u} = \begin{bmatrix} a \\ b \\ c \end{bmatrix}$ and $\mathbf{v} = \begin{bmatrix} d \\ e \\ f \end{bmatrix}$ in $\mathbb{R}^3$, the **cross product** of $\mathbf{u}$ and $\mathbf{v}$ is

$$\mathbf{u} \times \mathbf{v} = \begin{bmatrix} bf - ec \\ dc - af \\ ae - db \end{bmatrix}.$$

How to compute the cross product can be remembered by using Figure 2.13. To start, ignore the first column, cross multiply finding bf (paired by a solid line) and ec (paired with a dashed line), and subtract the products giving $bf - ec$. Next, ignore the second column and follow the same procedure to find $dc - af$, noting the switch of the solid and dashed lines. Lastly, compute $ae - db$. These three values are the coordinates of the cross product.

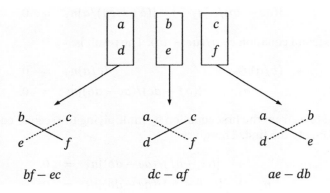

Figure 2.13   Computing the cross product.

The proof that the cross product of two given vectors is orthogonal to those vectors is demonstrated by (2.40) and stated here for future reference.

■ **Theorem 2.3.2**

If $\mathbf{u}, \mathbf{v} \in \mathbb{R}^3$, then $\mathbf{u} \times \mathbf{v}$ is orthogonal to both $\mathbf{u}$ and $\mathbf{v}$.

■ **Example 2.3.3**

Take the vectors $\begin{bmatrix} 1 \\ 2 \\ 3 \end{bmatrix}$, $\begin{bmatrix} 4 \\ 0 \\ -1 \end{bmatrix}$, and $\begin{bmatrix} 0 \\ 1 \\ 1 \end{bmatrix}$. These determine a plane in $\mathbb{R}^3$. The

vectors $\begin{bmatrix} -3 \\ 2 \\ 4 \end{bmatrix}$ and $\begin{bmatrix} 1 \\ 1 \\ 2 \end{bmatrix}$ are direction vectors for the plane. The cross product

of these vectors,

$$\begin{bmatrix} -3 \\ 2 \\ 4 \end{bmatrix} \times \begin{bmatrix} 1 \\ 1 \\ 2 \end{bmatrix} = \begin{bmatrix} 0 \\ 10 \\ -5 \end{bmatrix}, \tag{2.41}$$

gives a normal to the plane. Notice that $\begin{bmatrix} 0 \\ 2 \\ -1 \end{bmatrix}$ and $\begin{bmatrix} 0 \\ -20 \\ 10 \end{bmatrix}$ are also normals

to the plane. Further notice that $\begin{bmatrix} 0 \\ 10 \\ -5 \end{bmatrix}$ is indeed orthogonal to every point of

the plane because

$$\begin{bmatrix} 0 \\ 10 \\ -5 \end{bmatrix} \cdot \begin{bmatrix} -3 \\ 2 \\ 4 \end{bmatrix} = 0 \text{ and } \begin{bmatrix} 0 \\ 10 \\ -5 \end{bmatrix} \cdot \begin{bmatrix} 1 \\ 1 \\ 2 \end{bmatrix} = 0.$$

Hence, if $\begin{bmatrix} x \\ y \\ z \end{bmatrix}$ is any vector in the plane, $\begin{bmatrix} x-1 \\ y-2 \\ z-3 \end{bmatrix}$ is parallel to the plane and

$$\begin{bmatrix} 0 \\ 10 \\ -5 \end{bmatrix} \cdot \begin{bmatrix} x-1 \\ y-2 \\ z-3 \end{bmatrix} = 0,$$

which implies that $10(y-2) - 5(z-3) = 0$, and this gives an equation in standard form for the plane, $10y - 5z = 5$.

Properties

Unlike standard multiplication of real numbers and unlike the dot product, the cross product is not commutative. For example,

$$\begin{bmatrix} 1 \\ 1 \\ 2 \end{bmatrix} \times \begin{bmatrix} -3 \\ 2 \\ 4 \end{bmatrix} = \begin{bmatrix} 0 \\ -10 \\ 5 \end{bmatrix},$$

so by (2.41),

$$\begin{bmatrix} 1 \\ 1 \\ 2 \end{bmatrix} \times \begin{bmatrix} -3 \\ 2 \\ 4 \end{bmatrix} \neq \begin{bmatrix} -3 \\ 2 \\ 4 \end{bmatrix} \times \begin{bmatrix} 1 \\ 1 \\ 2 \end{bmatrix}.$$

The cross product does have some expected properties, however, and these are listed in the next result. It is assumed that the cross product has precedence over vector addition but scalar multiplication has precedence over the cross product.

■ **Theorem 2.3.4**

Let $\mathbf{u}, \mathbf{v}, \mathbf{w} \in \mathbb{R}^3$ and $r \in \mathbb{R}$.

(a) $\mathbf{u} \times \mathbf{v} = -(\mathbf{v} \times \mathbf{u})$.

(b) $\mathbf{u} \times (\mathbf{v} + \mathbf{w}) = \mathbf{u} \times \mathbf{v} + \mathbf{u} \times \mathbf{w}$.

(c) $(\mathbf{u} + \mathbf{v}) \times \mathbf{w} = \mathbf{u} \times \mathbf{w} + \mathbf{v} \times \mathbf{w}$.

(d) $r(\mathbf{u} \times \mathbf{v}) = r\mathbf{u} \times \mathbf{v} = \mathbf{u} \times r\mathbf{v}$.

(e) $\mathbf{u} \times \mathbf{0} = \mathbf{0} \times \mathbf{u} = \mathbf{0}$.

(f) $\mathbf{u} \times \mathbf{u} = \mathbf{0}$.

Proof

Write $\mathbf{u} = \begin{bmatrix} u_1 \\ u_2 \\ u_3 \end{bmatrix}$, $\mathbf{v} = \begin{bmatrix} v_1 \\ v_2 \\ v_3 \end{bmatrix}$, and $\mathbf{w} = \begin{bmatrix} w_1 \\ w_2 \\ w_3 \end{bmatrix}$. First, (a) follows because

$$\mathbf{u} \times \mathbf{v} = \begin{bmatrix} u_2v_3 - v_2u_3 \\ v_1u_3 - u_3v_1 \\ u_1v_2 - v_1u_2 \end{bmatrix} = -\begin{bmatrix} v_2u_3 - u_2v_3 \\ u_1v_3 - v_3u_1 \\ v_1u_2 - u_1v_2 \end{bmatrix} = -(\mathbf{v} \times \mathbf{u}).$$

Second, (b) follows because

$$\mathbf{u} \times (\mathbf{v} + \mathbf{w}) = \begin{bmatrix} u_2(v_3 + w_3) - (v_2 + w_2)u_3 \\ (v_1 + w_1)u_3 - u_3(v_1 + w_1) \\ u_1(v_2 + w_2) - (v_1 + w_1)u_2 \end{bmatrix}$$

$$= \begin{bmatrix} u_2v_3 - v_2u_3 \\ v_1u_3 - u_3v_1 \\ u_1v_2 - v_1u_2 \end{bmatrix} + \begin{bmatrix} u_2w_3 - w_2u_3 \\ w_1u_3 - u_3w_1 \\ u_1w_2 - w_1u_2 \end{bmatrix} = \mathbf{u} \times \mathbf{v} + \mathbf{u} \times \mathbf{w}.$$

Third, (c) can be proven using (a) and (b) because

$$(\mathbf{u} + \mathbf{v}) \times \mathbf{w} = -[\mathbf{w} \times (\mathbf{u} + \mathbf{v})]$$
$$= -(\mathbf{w} \times \mathbf{u} + \mathbf{w} \times \mathbf{v})$$
$$= -(\mathbf{w} \times \mathbf{u}) - (\mathbf{w} \times \mathbf{v}) = \mathbf{u} \times \mathbf{w} + \mathbf{v} \times \mathbf{w}.$$

The remainder of the proof is left to Exercise 10. ■

The next result connects the dot product with the cross product. It will also help in computing an area. It is credited to Joseph-Louis Lagrange.

■ **Theorem 2.3.5 [Lagrange's Identity]**

If $\mathbf{u}, \mathbf{v} \in \mathbb{R}^3$, then $\|\mathbf{u} \times \mathbf{v}\|^2 = \|\mathbf{u}\|^2 \|\mathbf{v}\|^2 - (\mathbf{u} \cdot \mathbf{v})^2$.

Proof

Assume that the coordinates of $\mathbf{u}$ are $u_1, u_2, u_3 \in \mathbb{R}$ and the coordinates of $\mathbf{v}$ are $v_1, v_2, v_3 \in \mathbb{R}$. Notice that

$$\|\mathbf{u}\|^2 \|\mathbf{v}\|^2 = \left(u_1^2 + u_2^2 + u_3^2\right)\left(v_1^2 + v_2^2 + v_3^2\right) = \sum_{i=1}^{3}\sum_{j=1}^{3} u_i^2 v_j^2$$

and

$$(\mathbf{u} \cdot \mathbf{v})^2 = (u_1 v_1 + u_2 v_2 + u_3 v_3)^2 = \sum_{i=1}^{3}\sum_{j=1}^{3} u_i u_j v_i v_j$$

Therefore,

$$\|\mathbf{u} \times \mathbf{v}\|^2 = (u_2 v_3 - u_3 v_2)^2 + (u_1 v_3 - u_3 v_1)^2 + (u_1 v_2 - u_2 v_1)^2$$

$$= \sum_{\substack{i=1 \\ }}^{3}\sum_{\substack{j=1 \\ j\neq i}}^{3} u_i^2 v_j^2 - 2\sum_{\substack{i=1 \\ }}^{3}\sum_{\substack{j=1 \\ j\neq i}}^{3} u_i u_j v_i v_j$$

$$= \sum_{i=1}^{3}\sum_{j=1}^{3} u_i^2 v_j^2 - \sum_{i=1}^{3}\sum_{j=1}^{3} u_i u_j v_i v_j = \|\mathbf{u}\|^2 \|\mathbf{v}\|^2 - (\mathbf{u} \cdot \mathbf{v})^2. \blacksquare$$

Areas and Volumes

As the dot product can be computed using cosine as in (2.27), the cross product can be computed using sine.

■ Theorem 2.3.6

For all nonzero vectors $\mathbf{u}$ and $\mathbf{v}$ in $\mathbb{R}^3$, if θ is the measure of the angle between $\mathbf{u}$ and $\mathbf{v}$, then $\|\mathbf{u} \times \mathbf{v}\| = \|\mathbf{u}\| \|\mathbf{v}\| \, |\sin \theta|$.

Proof

Starting with Lagrange's Identity (Theorem 2.3.5),

$$\|\mathbf{u} \times \mathbf{v}\|^2 = \|\mathbf{u}\|^2 \|\mathbf{v}\|^2 - (\mathbf{u} \cdot \mathbf{v})^2$$
$$= \|\mathbf{u}\|^2 \|\mathbf{v}\|^2 - \|\mathbf{u}\|^2 \|\mathbf{v}\|^2 \cos^2 \theta$$
$$= \|\mathbf{u}\|^2 \|\mathbf{v}\|^2 (1 - \cos^2 \theta) = \|\mathbf{u}\|^2 \|\mathbf{v}\|^2 \sin^2 \theta.$$

Taking square roots gives $\|\mathbf{u} \times \mathbf{v}\| = \|\mathbf{u}\| \|\mathbf{v}\| \, |\sin \theta|$. $\blacksquare$

Consider the parallelogram described by the noncollinear and nonzero vectors $\mathbf{u}$ and $\mathbf{v}$ in $\mathbb{R}^3$ (Figure 2.14). If the base has length $\|\mathbf{u}\|$, its height is $\|\mathbf{v}\| \sin \theta$. This implies the following result using Theorem 2.3.6.

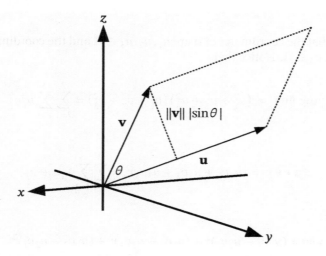

Figure 2.14   The parallelogram described by **u** and **v**.

■ **Corollary 2.3.7**

Let **u**, **v** be nonzero vectors in $\mathbb{R}^3$. The area of the parallelogram formed by **u** and **v** is $\|\mathbf{u} \times \mathbf{v}\|$.

■ **Example 2.3.8**

The vectors $\begin{bmatrix} 1 \\ 3 \\ -2 \end{bmatrix}$ and $\begin{bmatrix} 0 \\ 4 \\ -1 \end{bmatrix}$ are not multiples of each other, so they form a parallelogram. The area of this parallelogram is

$$\left\| \begin{bmatrix} 1 \\ 3 \\ -2 \end{bmatrix} \times \begin{bmatrix} 0 \\ 4 \\ -1 \end{bmatrix} \right\| = \left\| \begin{bmatrix} 5 \\ 1 \\ 4 \end{bmatrix} \right\| = \sqrt{42}.$$

■ **Example 2.3.9**

Take $\mathbf{u}, \mathbf{v}, \mathbf{w} \in \mathbb{R}^3$ such that **u** and **v** are not collinear and **w** is not in the plane formed by **u** and **v**. These three vectors describe a parallelepiped in $\mathbb{R}^3$ (Figure 2.15). Consider the base of the parallelepiped to be the parallelogram formed by **u** and **v**. By Corollary 2.3.7, the area of this base is $\|\mathbf{u} \times \mathbf{v}\|$. The height of the parallelepiped is given by the distance between the base and the face opposite the base. To find this distance, project **w** onto $\mathbf{u} \times \mathbf{v}$, which is normal to the plane containing the base. By Theorem 2.2.20, the volume of this parallelepiped is

$$\|\mathbf{u} \times \mathbf{v}\| \, \|\mathrm{proj}_{\mathbf{u} \times \mathbf{v}} \mathbf{w}\| = \|\mathbf{u} \times \mathbf{v}\| \frac{|\mathbf{w} \cdot (\mathbf{u} \times \mathbf{v})|}{\|\mathbf{u} \times \mathbf{v}\|} = |\mathbf{w} \cdot (\mathbf{u} \times \mathbf{v})|.$$

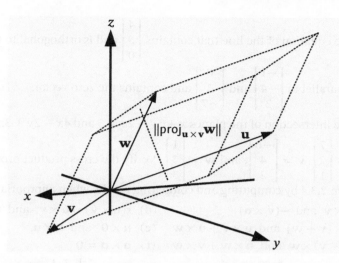

Figure 2.15 The parallelepiped described by **u**, **v**, and **w**.

The value $\mathbf{w} \cdot (\mathbf{u} \times \mathbf{v})$ is called the **triple scalar product**. For example, letting

$$\mathbf{u} = \begin{bmatrix} 1 \\ 3 \\ -2 \end{bmatrix}, \quad \mathbf{v} = \begin{bmatrix} 0 \\ 4 \\ -1 \end{bmatrix}, \quad \mathbf{w} = \begin{bmatrix} 1 \\ -2 \\ 3 \end{bmatrix},$$

the volume of the parallelepiped formed by these three vectors is

$$\left| \begin{bmatrix} 1 \\ -2 \\ 3 \end{bmatrix} \cdot \left(\begin{bmatrix} 1 \\ 3 \\ -2 \end{bmatrix} \times \begin{bmatrix} 0 \\ 4 \\ -1 \end{bmatrix} \right) \right| = \left| \begin{bmatrix} 1 \\ -2 \\ 3 \end{bmatrix} \cdot \begin{bmatrix} 5 \\ 1 \\ 4 \end{bmatrix} \right| = 15.$$

Exercises

1. Let $\mathbf{u} = \begin{bmatrix} 4 \\ 5 \\ 0 \end{bmatrix}$, $\mathbf{v} = \begin{bmatrix} -1 \\ -4 \\ 2 \end{bmatrix}$, and $\mathbf{w} = \begin{bmatrix} 0 \\ 3 \\ -7 \end{bmatrix}$. Compute if possible.

   (a) $\mathbf{u} \times \mathbf{v}$

   (b) $(\mathbf{w} \times \mathbf{u}) \cdot (\mathbf{w} - \mathbf{u})$

   (c) $(\mathbf{u} \times \mathbf{v}) \cdot \mathbf{w}$

   (d) $(\mathbf{v} \times \mathbf{w}) - (\mathbf{w} \times \mathbf{v})$

   (e) $(\mathbf{v} \cdot \mathbf{w}) \times \mathbf{w}$

   (f) $[(\mathbf{u} \cdot \mathbf{v})\mathbf{w}] \times [(\mathbf{u} \cdot \mathbf{w})\mathbf{v}]$

2. Find a unit vector orthogonal to $\begin{bmatrix} 7 \\ 5 \\ 3 \end{bmatrix}$ and $\begin{bmatrix} 2 \\ 4 \\ 6 \end{bmatrix}$.

3. Find the plane in $\mathbb{R}^3$ containing $\begin{bmatrix} 1 \\ 3 \\ 0 \end{bmatrix}$, $\begin{bmatrix} -2 \\ -2 \\ 4 \end{bmatrix}$, and $\begin{bmatrix} 2 \\ 0 \\ -3 \end{bmatrix}$.

4. Find the equation of the line that contains $\begin{bmatrix} 4 \\ 5 \\ 0 \end{bmatrix}$ and is orthogonal to the plane

   that is parallel to $\begin{bmatrix} -1 \\ -4 \\ 2 \end{bmatrix}$ and $\begin{bmatrix} 0 \\ 3 \\ -7 \end{bmatrix}$ and contains the zero vector.

5. Find the intersection of the planes $x + 5y - 2z = 3$ and $4x - 2y + 3z = 1$.

6. Let $\mathbf{u} = \begin{bmatrix} 7 \\ 2 \\ 1 \end{bmatrix}$, $\mathbf{v} = \begin{bmatrix} -6 \\ 4 \\ 2 \end{bmatrix}$, and $\mathbf{w} = \begin{bmatrix} 1 \\ 5 \\ 3 \end{bmatrix}$. Verify the cross product properties of
   Theorem 2.3.4 by computing and comparing results when appropriate.

   (a) $\mathbf{u} \times \mathbf{v}$ and $-(\mathbf{v} \times \mathbf{u})$ (d) $r(\mathbf{u} \times \mathbf{v})$, $r\mathbf{u} \times \mathbf{v}$, and $\mathbf{u} \times r\mathbf{v}$

   (b) $\mathbf{u} \times (\mathbf{v} + \mathbf{w})$ and $\mathbf{u} \times \mathbf{v} + \mathbf{u} \times \mathbf{w}$ (e) $\mathbf{u} \times \mathbf{0}$ and $\mathbf{0} \times \mathbf{u}$

   (c) $(\mathbf{u} + \mathbf{v}) \times \mathbf{w}$ and $\mathbf{u} \times \mathbf{w} + \mathbf{v} \times \mathbf{w}$ (f) $\mathbf{u} \times \mathbf{u} = \mathbf{0}$

7. Verify Lagrange's Identity (Theorem 2.3.5) using $\mathbf{u} = \begin{bmatrix} 3 \\ -2 \\ 1 \end{bmatrix}$ and $\mathbf{v} = \begin{bmatrix} -1 \\ 2 \\ -3 \end{bmatrix}$.

8. Let $\mathbf{u}$ and $\mathbf{v}$ be unit vectors parallel to the same plane.

   (a) What is the maximum of $\|\mathbf{u} \times \mathbf{v}\|$?

   (b) If $\|\mathbf{u} \times \mathbf{v}\| = 0$, what is true about $\mathbf{u}$ and $\mathbf{v}$?

9. Finish the proof of Theorem 2.3.2 by solving (2.40) for $a = 0$ and $ae - bd = 0$.

10. Prove Theorem 2.3.4 using Definition 2.3.1.

11. Let $\mathbf{u}, \mathbf{v}, \mathbf{w} \in \mathbb{R}^3$. Prove each **vector triple product**.

    (a) $\mathbf{u} \times (\mathbf{v} \times \mathbf{w}) = (\mathbf{u} \cdot \mathbf{w})\mathbf{v} - (\mathbf{u} \cdot \mathbf{v})\mathbf{w}$

    (b) $(\mathbf{u} \times \mathbf{v}) \times \mathbf{w} = (\mathbf{u} \cdot \mathbf{w})\mathbf{v} - (\mathbf{v} \cdot \mathbf{w})\mathbf{u}$

12. Let $\mathbf{u}, \mathbf{v}, \mathbf{w}, \mathbf{x} \in \mathbb{R}^3$. Prove that $(\mathbf{u} \times \mathbf{v}) \cdot (\mathbf{w} \times \mathbf{x}) = (\mathbf{u} \cdot \mathbf{w})(\mathbf{v} \cdot \mathbf{x}) - (\mathbf{v} \cdot \mathbf{w})(\mathbf{u} \cdot \mathbf{x})$.

13. Let $\mathbf{u}, \mathbf{v}, \mathbf{w} \in \mathbb{R}^3$ and $r \in \mathbb{R}$. Prove.

    (a) $(\mathbf{u} + r\mathbf{v}) \times \mathbf{v} = \mathbf{u} \times \mathbf{v}$

    (b) $\mathbf{u} \cdot (\mathbf{v} \times \mathbf{w}) = -(\mathbf{u} \times \mathbf{w}) \cdot \mathbf{v}$

14. Let $\mathbf{u} = \begin{bmatrix} 6 \\ 3 \\ 0 \end{bmatrix}$, $\mathbf{v} = \begin{bmatrix} 1 \\ 1 \\ 1 \end{bmatrix}$, and $\mathbf{w} = \begin{bmatrix} -3 \\ -2 \\ 1 \end{bmatrix}$. Compute.

    (a) The area of the parallelogram described by $\mathbf{u}$ and $\mathbf{v}$

    (b) The area of the triangle described by $\mathbf{u}$ and $\mathbf{w}$

    (c) The volume of the parallelepiped described by $\mathbf{u}$, $\mathbf{v}$, and $\mathbf{w}$

15. Use the triple scalar product to show that $\begin{bmatrix} 1 \\ 3 \\ 0 \end{bmatrix}$, $\begin{bmatrix} -2 \\ -2 \\ 4 \end{bmatrix}$, and $\begin{bmatrix} 0 \\ 4 \\ 4 \end{bmatrix}$ are in the same
    plane. Explain why the method works.

16. Prove: For all $\mathbf{u}, \mathbf{v} \in \mathbb{R}^3$ it follows that $\|\mathbf{u} + \mathbf{v}\| = \|\mathbf{u}\| + \|\mathbf{v}\|$ if and only if $\mathbf{u} = r\mathbf{v}$ for some $r \in \mathbb{R}$.

17. Using the right hand in its natural position, the **right-hand rule** states that if $\mathbf{u}$ is represented by the index finger pointing straight and $\mathbf{v}$ is represented by the middle finger pointing to the left of the index finger, $\mathbf{u} \times \mathbf{v}$ is represented by the thumb.

    (a) Use the right-hand rule to draw a figure representing $\mathbf{u} \times \mathbf{v}$ and $\mathbf{v} \times \mathbf{u}$.

    (b) Explain why the right-hand rule works.

CHAPTER 3

Transformations and Matrices

3.1 Linear Transformations

Consider the function $f : \mathbb{R} \times \mathbb{R} \to \mathbb{R} \times \mathbb{R}$ defined by $f(x, y) = (x + 3, y - 2)$. It can be regarded as the motion in the Cartesian plane that moves points three units to the right and two units down. It moves segments and polygons in such a way that they neither change size nor rotate during the move. This function can also be interpreted as transforming segments and polygons in the plane into other segments and polygons in the plane that are identical to the originals except for location. For example, if S represents the segment joining $(5, 7)$ and $(-1, 2)$, then $f[S]$ is the segment joining $(8, 5)$ and $(2, 0)$. Functions such as these are sometimes called **transformations**.

Transformations are good to use when the elements of the space need to be moved, but if the function also needs to take into consideration any algebra of the space, not just any transformation will do. For example, define $F : \mathbb{R}^2 \to \mathbb{R}^2$ by

$$F \begin{bmatrix} x \\ y \end{bmatrix} = \begin{bmatrix} x + 3 \\ y - 2 \end{bmatrix}, \tag{3.1}$$

Linear Algebra, First Edition. Michael L. O'Leary.
© 2021 John Wiley & Sons, Inc. Published 2021 by John Wiley & Sons, Inc.
Companion Website: www.wiley.com/go/o'leary/linearalgebra

99

where the parentheses are excluded for readability. Then, $F \begin{bmatrix} 0 \\ 0 \end{bmatrix} = \begin{bmatrix} 3 \\ -2 \end{bmatrix}$, so the image of the zero vector is not the zero vector, so although $\begin{bmatrix} 0 \\ 0 \end{bmatrix}$ is the additive identity in the domain of F, its image is not the additive identity in the range of F. Similarly, the sum of vectors in the domain is not the sum of the images of the vectors because, for instance, although

$$\begin{bmatrix} 3 \\ 6 \end{bmatrix} + \begin{bmatrix} 0 \\ -2 \end{bmatrix} = \begin{bmatrix} 3 \\ 4 \end{bmatrix},$$

it is nonetheless the case that

$$F \begin{bmatrix} 3 \\ 6 \end{bmatrix} + F \begin{bmatrix} 0 \\ -2 \end{bmatrix} = \begin{bmatrix} 6 \\ 4 \end{bmatrix} + \begin{bmatrix} 3 \\ -4 \end{bmatrix} \neq \begin{bmatrix} 6 \\ 2 \end{bmatrix} = F \begin{bmatrix} 3 \\ 4 \end{bmatrix}.$$

There are similar problems with scalar multiplication.

In order to transform a space in such a manner that operations on the vectors in the range give answers that align with the answers given by operations on the vectors in the domain, the function needs to satisfy the following definition, which is at the heart of linear algebra.

■ Definition 3.1.1

A function $T : \mathbb{R}^n \to \mathbb{R}^m$ is called a **linear transformation** when for all $\mathbf{u}, \mathbf{v} \in \mathbb{R}^n$ and $r \in \mathbb{R}$,

$$T(\mathbf{u} + \mathbf{v}) = T\mathbf{u} + T\mathbf{v} \tag{3.2}$$

and

$$T(r\mathbf{u}) = r(T\mathbf{u}). \tag{3.3}$$

(3.2) means that T is an **additive function**, and (3.3) means that T is a **homogeneous function**.

It is worth examining Definition 3.1.1 in detail. The two properties of a linear transformation copy the actions of the operations on the domain over to the range. In particular, because of the additive condition (3.2), the sum of vectors in the domain is mapped to the sum of their images in the range. The linear transformation is said to **preserve** vector addition (Figure 3.1). Also, because of the homogeneous condition (3.3), the result of multiplying a vector in the domain by a scalar is mapped to the product of the scalar and the image of the vector. The linear transformation is said to **preserve** scalar multiplication (Figure 3.2). Stated another way, a linear transformation pairs vectors in such a way that the operations on the range are copies of the operations on the domain.

It is important to realize that there are possibly two vector additions and two scalar multiplications because it is possible that $n \neq m$ in Definition 3.1.1. Specifically, the addition and scalar multiplication on the left-hand side of the equations

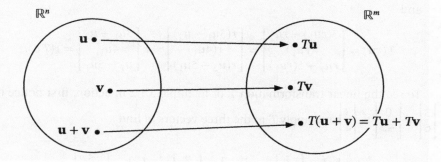

Figure 3.1   Linear transformations preserve addition.

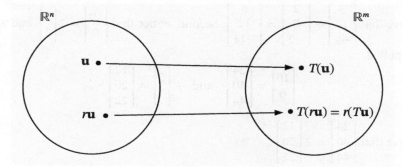

Figure 3.2   Linear transformations preserve scalar multiplication.

are the operations for $\mathbb{R}^n$, and the operations on the right-hand side are for $\mathbb{R}^m$. Despite this, the same notation is used on both sides to keep the notation from becoming cumbersome. The context makes it clear which operations are being used.

■ Example 3.1.2

Define $T : \mathbb{R}^2 \to \mathbb{R}^3$ by

$$T\begin{bmatrix} x \\ y \end{bmatrix} = \begin{bmatrix} 3x - y \\ 4x \\ y - 5x \end{bmatrix}. \tag{3.4}$$

To prove that T is a linear transformation, take $\mathbf{u}, \mathbf{v} \in \mathbb{R}^2$ and $r \in \mathbb{R}$ and write $\mathbf{u} = \begin{bmatrix} u_1 \\ u_2 \end{bmatrix}$ and $\mathbf{v} = \begin{bmatrix} v_1 \\ v_2 \end{bmatrix}$. Then,

$$T(\mathbf{u} + \mathbf{v}) = \begin{bmatrix} 3(u_1 + v_1) - (u_2 + v_2) \\ 4(u_1 + v_1) \\ u_2 + v_2 - 5(u_1 + v_1) \end{bmatrix} = \begin{bmatrix} 3u_1 - u_2 \\ 4u_1 \\ u_2 - 5u_1 \end{bmatrix} + \begin{bmatrix} 3v_1 - v_2 \\ 4v_1 \\ v_2 - 5v_2 \end{bmatrix} = T\mathbf{u} + T\mathbf{v},$$

and

$$T(r\mathbf{u}) = \begin{bmatrix} 3(ru_1) - ru_2 \\ 4(ru_1) \\ ru_2 - 5(ru_1) \end{bmatrix} = \begin{bmatrix} r(3u_1 - u_2) \\ r(4u_1) \\ r(u_2 - 5u_1) \end{bmatrix} = r \begin{bmatrix} 3u_1 - u_2 \\ 4u_1 \\ u_2 - 5u_1 \end{bmatrix} = r(T\mathbf{u}).$$

To see the linear transformation T of Example 3.1.2 in action, first notice that $\begin{bmatrix} 3 \\ 6 \end{bmatrix} + \begin{bmatrix} 0 \\ -2 \end{bmatrix} = \begin{bmatrix} 3 \\ 4 \end{bmatrix}$ and apply T to the three vectors to find

$$T \begin{bmatrix} 3 \\ 6 \end{bmatrix} = \begin{bmatrix} 3 \\ 12 \\ -9 \end{bmatrix}, \quad T \begin{bmatrix} 0 \\ -2 \end{bmatrix} = \begin{bmatrix} 2 \\ 0 \\ -2 \end{bmatrix}, \quad T \begin{bmatrix} 3 \\ 4 \end{bmatrix} = \begin{bmatrix} 5 \\ 12 \\ -11 \end{bmatrix}.$$

Observe that $\begin{bmatrix} 3 \\ 12 \\ -9 \end{bmatrix} + \begin{bmatrix} 2 \\ 0 \\ -2 \end{bmatrix} = \begin{bmatrix} 5 \\ 12 \\ -11 \end{bmatrix}$. Second, notice that $\begin{bmatrix} 10 \\ 6 \end{bmatrix} = 2 \begin{bmatrix} 5 \\ 3 \end{bmatrix}$, and when T is applied,

$$T \begin{bmatrix} 10 \\ 6 \end{bmatrix} = \begin{bmatrix} 24 \\ 40 \\ 44 \end{bmatrix} \quad \text{and} \quad T \begin{bmatrix} 5 \\ 3 \end{bmatrix} = \begin{bmatrix} 12 \\ 20 \\ 22 \end{bmatrix}.$$

Observe that $\begin{bmatrix} 24 \\ 40 \\ 44 \end{bmatrix} = 2 \begin{bmatrix} 12 \\ 20 \\ 22 \end{bmatrix}$.

■ **Example 3.1.3**

Let $0_{nm} : \mathbb{R}^n \to \mathbb{R}^m$ be defined for all $\mathbf{x} \in \mathbb{R}^n$ by $0_{nm}\mathbf{x} = \mathbf{0}_m$. This function is called a **zero transformation** (compare Example 1.4.40). To see that 0_{nm} is a linear transformation, let $\mathbf{u}, \mathbf{v} \in \mathbb{R}^n$ and $r \in \mathbb{R}$ and evaluate

$$0_{nm}(\mathbf{u} + \mathbf{v}) = \mathbf{0}_m = \mathbf{0}_m + \mathbf{0}_m = 0_{nm}\mathbf{u} + 0_{nm}\mathbf{v}$$

and

$$0_{nm}(r\mathbf{u}) = \mathbf{0}_m = r\mathbf{0}_m = r(0_{nm}\mathbf{u}).$$

When the domain and codomain are clear from context, the zero transformation can be written as 0.

■ **Example 3.1.4**

Similar to Example 1.4.13, define $T : \mathbb{R}^2 \to \mathbb{R}^3$ by $T \begin{bmatrix} x \\ y \end{bmatrix} = \begin{bmatrix} x \\ y \\ 0 \end{bmatrix}$. To prove that T is a linear transformation, let $\mathbf{u}, \mathbf{v} \in \mathbb{R}^2$ and $r \in \mathbb{R}$. Write $\mathbf{u} = \begin{bmatrix} u_1 \\ u_2 \end{bmatrix}$

and $\mathbf{v} = \begin{bmatrix} v_1 \\ v_2 \end{bmatrix}$. Then, $T(\mathbf{u} + \mathbf{v}) = T\mathbf{u} + T\mathbf{v}$ because

$$T\begin{bmatrix} u_1 + v_1 \\ u_2 + v_2 \end{bmatrix} = \begin{bmatrix} u_1 + v_1 \\ u_2 + v_2 \\ 0 \end{bmatrix} = \begin{bmatrix} u_1 \\ u_2 \\ 0 \end{bmatrix} + \begin{bmatrix} v_1 \\ v_2 \\ 0 \end{bmatrix} = T\begin{bmatrix} u_1 \\ u_2 \end{bmatrix} + T\begin{bmatrix} v_1 \\ v_2 \end{bmatrix},$$

and $T(r\mathbf{u}) = r(T\mathbf{u})$ because

$$T\begin{bmatrix} ru_1 \\ ru_2 \end{bmatrix} = \begin{bmatrix} ru_1 \\ ru_2 \\ 0 \end{bmatrix} = r\begin{bmatrix} u_1 \\ u_2 \\ 0 \end{bmatrix} = rT\begin{bmatrix} u_1 \\ u_2 \end{bmatrix}.$$

■ **Example 3.1.5**

Let $\mathbf{n} \in \mathbb{R}^n$. Define $T : \mathbb{R}^n \to \mathbb{R}^n$ by $T\mathbf{x} = \text{proj}_\mathbf{n}\mathbf{x}$. To see that T is a linear transformation, let $\mathbf{u}, \mathbf{v} \in \mathbb{R}^n$ and $r \in \mathbb{R}$. By Theorem 2.2.6(a) and (b),

$$\begin{aligned}
T(\mathbf{u} + \mathbf{v}) &= \text{proj}_\mathbf{n}(\mathbf{u} + \mathbf{v}) \\
&= \frac{(\mathbf{u} + \mathbf{v}) \cdot \mathbf{n}}{\|\mathbf{n}\|^2}\mathbf{n} \\
&= \frac{\mathbf{u} \cdot \mathbf{n}}{\|\mathbf{n}\|^2}\mathbf{n} + \frac{\mathbf{v} \cdot \mathbf{n}}{\|\mathbf{n}\|^2}\mathbf{n} \\
&= \text{proj}_\mathbf{n}\mathbf{u} + \text{proj}_\mathbf{n}\mathbf{v} = T\mathbf{u} + T\mathbf{v},
\end{aligned}$$

and by Theorem 2.2.6(c),

$$T(r\mathbf{u}) = \text{proj}_\mathbf{n}r\mathbf{u} = \frac{r\mathbf{u} \cdot \mathbf{n}}{\|\mathbf{n}\|^2}\mathbf{n} = \frac{r(\mathbf{u} \cdot \mathbf{n})}{\|\mathbf{n}\|^2}\mathbf{n} = r\text{proj}_\mathbf{n}\mathbf{u} = r(T\mathbf{u}).$$

Properties

The function defined by (3.1) does not satisfy either condition of Definition 3.1.1 as has been noted. There are other properties that every linear transformation possesses that when missing indicate that a function is not a linear transformation.

■ **Theorem 3.1.6**

Let $T : \mathbb{R}^n \to \mathbb{R}^m$ be a linear transformation.

(a) $T\mathbf{0}_n = \mathbf{0}_m$.

(b) $T(-\mathbf{u}) = -(T\mathbf{u})$ for all $\mathbf{u} \in \mathbb{R}^n$.

(c) $T(\mathbf{u} - \mathbf{v}) = T\mathbf{u} - T\mathbf{v}$ for all $\mathbf{u}, \mathbf{v} \in \mathbb{R}^n$.

Proof

This proof relies on Theorem 2.1.7. (a) is proved by

$$T\mathbf{0}_n = T(0 \cdot \mathbf{0}_n) = 0(T\mathbf{0}_n) = \mathbf{0}_m.$$

Now, let $\mathbf{u}, \mathbf{v} \in \mathbb{R}^n$. (b) is proved by noting that

$$T(-\mathbf{u}) = T(-1\mathbf{u}) = -1(T\mathbf{u}) = -(T\mathbf{u}),$$

and then (c) follows using (b) in

$$T(\mathbf{u} - \mathbf{v}) = T(\mathbf{u} + [-\mathbf{v}]) = T\mathbf{u} + T(-\mathbf{v}) = T\mathbf{u} + [-(T\mathbf{v})] = T\mathbf{u} - T\mathbf{v}. \blacksquare$$

Using the linear transformation T given by (3.4), $T\begin{bmatrix} x \\ y \end{bmatrix} = \begin{bmatrix} 3x - y \\ 4x \\ y - 5x \end{bmatrix}$, parts (a) and (b) of Theorem 3.1.6 are illustrated by

$$T\begin{bmatrix} 0 \\ 0 \end{bmatrix} = \begin{bmatrix} 3(0) - 0 \\ 4(0) \\ 0 - 5(0) \end{bmatrix} = \begin{bmatrix} 0 \\ 0 \\ 0 \end{bmatrix}$$

and

$$T\begin{bmatrix} -1 \\ -2 \end{bmatrix} = \begin{bmatrix} 3(-1) - (-2) \\ 4(-1) \\ -2 - 5(-1) \end{bmatrix} = -1\begin{bmatrix} 3(1) - (2) \\ 4(1) \\ 2 - 5(1) \end{bmatrix} = -1\left(T\begin{bmatrix} 1 \\ 2 \end{bmatrix}\right) = -\left(T\begin{bmatrix} 1 \\ 2 \end{bmatrix}\right).$$

However, the function F of (3.1) is not a linear transformation for various reasons, including the facts that $F\begin{bmatrix} 0 \\ 0 \end{bmatrix} = \begin{bmatrix} 3 \\ -2 \end{bmatrix}$ and $F\begin{bmatrix} -1 \\ -2 \end{bmatrix} = \begin{bmatrix} 2 \\ -4 \end{bmatrix} \neq \begin{bmatrix} -4 \\ 0 \end{bmatrix} = -1\left(F\begin{bmatrix} 1 \\ 2 \end{bmatrix}\right)$.

Let $L = S\left(\begin{bmatrix} 1 \\ 4 \end{bmatrix}\right)$. That is, L is the line containing $\begin{bmatrix} 0 \\ 0 \end{bmatrix}$ with direction vector $\begin{bmatrix} 1 \\ 4 \end{bmatrix}$. Continuing to let T be defined by (3.4),

$$T[L] = \left\{ T\begin{bmatrix} r \\ 4r \end{bmatrix} : r \in \mathbb{R} \right\} = \left\{ \begin{bmatrix} 3r - 4r \\ 4r \\ 4r - 5r \end{bmatrix} : r \in \mathbb{R} \right\} = S\left(\begin{bmatrix} -1 \\ 4 \\ -1 \end{bmatrix}\right).$$

Thus, $T[L]$ is the line through $\begin{bmatrix} 0 \\ 0 \\ 0 \end{bmatrix}$ with direction vector $T\begin{bmatrix} 1 \\ 4 \end{bmatrix} = \begin{bmatrix} -1 \\ 4 \\ -1 \end{bmatrix}$. This suggests the following result.

■ **Theorem 3.1.7**

If $T : \mathbb{R}^n \to \mathbb{R}^m$ is a linear transformation and $\mathbf{u}_1, \mathbf{u}_2, \dots, \mathbf{u}_k \in \mathbb{R}^n$,

$$T[S(\mathbf{u}_1, \mathbf{u}_2, \dots, \mathbf{u}_k)] = S(T\mathbf{u}_1, T\mathbf{u}_2, \dots, T\mathbf{u}_k).$$

Proof

Let $\mathbf{v} \in T[S(\mathbf{u}_1, \mathbf{u}_2, \ldots, \mathbf{u}_k)]$. This means there exist $r_1, r_2, \ldots, r_k \in \mathbb{R}$ so that

$$\mathbf{v} = T(r_1\mathbf{u}_1 + r_2\mathbf{u}_2 + \cdots + r_k\mathbf{u}_k).$$

Hence, $\mathbf{v} \in S(T\mathbf{u}_1, T\mathbf{u}_2, \ldots, T\mathbf{u}_k)$ because T is a linear transformation and

$$T(r_1\mathbf{u}_1 + r_2\mathbf{u}_2 + \cdots + r_k\mathbf{u}_k) = T(r_1\mathbf{u}_1) + T(r_2\mathbf{u}_2) + \cdots + T(r_k\mathbf{u}_k)$$
$$= r_1(T\mathbf{u}_1) + r_2(T\mathbf{u}_2) + \cdots + r_k(T\mathbf{u}_k).$$

The converse is proved by reversing the steps. ■

The theorem states that linear transformations map cosets containing **0** to cosets containing **0**. In general, such functions map cosets to cosets. The proof is left to Exercise 23.

■ **Theorem 3.1.8**

If $T : \mathbb{R}^n \to \mathbb{R}^m$ is a linear transformation and $\mathbf{u}, \mathbf{u}_1, \mathbf{u}_2, \ldots, \mathbf{u}_k \in \mathbb{R}^n$,

$$T[\mathbf{u} + S(\mathbf{u}_1, \mathbf{u}_2, \ldots, \mathbf{u}_k)] = T\mathbf{u} + S(T\mathbf{u}_1, T\mathbf{u}_2, \ldots, T\mathbf{u}_k).$$

■ **Example 3.1.9**

Let $T : \mathbb{R}^3 \to \mathbb{R}^3$ be the linear transformation defined by

$$T\begin{bmatrix} x \\ y \\ z \end{bmatrix} = \begin{bmatrix} 2x + z \\ 2z + y \\ 2y + x \end{bmatrix}.$$

Let P be the plane

$$\begin{bmatrix} 1 \\ 2 \\ 3 \end{bmatrix} + S\left(\begin{bmatrix} 1 \\ 0 \\ -5 \end{bmatrix}, \begin{bmatrix} 2 \\ -1 \\ -4 \end{bmatrix} \right).$$

Illustrate Theorem 3.1.8 by applying T to P to find

$$T[P] = \left\{ T\left(\begin{bmatrix} 1 \\ 2 \\ 3 \end{bmatrix} + r\begin{bmatrix} 1 \\ 0 \\ -5 \end{bmatrix} + s\begin{bmatrix} 2 \\ -1 \\ -4 \end{bmatrix} \right) : r, s \in \mathbb{R} \right\}$$

$$= \left\{ T\begin{bmatrix} 1 \\ 2 \\ 3 \end{bmatrix} + rT\begin{bmatrix} 1 \\ 0 \\ -5 \end{bmatrix} + sT\begin{bmatrix} 2 \\ -1 \\ -4 \end{bmatrix} : r, s \in \mathbb{R} \right\}$$

$$= T\begin{bmatrix} 1 \\ 2 \\ 3 \end{bmatrix} + S\left(\begin{bmatrix} -3 \\ -10 \\ 1 \end{bmatrix}, \begin{bmatrix} 0 \\ -9 \\ 0 \end{bmatrix} \right)$$

$$= \begin{bmatrix} 5 \\ 8 \\ 5 \end{bmatrix} + S\left(\begin{bmatrix} -3 \\ -10 \\ 1 \end{bmatrix}, \begin{bmatrix} 0 \\ -9 \\ 0 \end{bmatrix} \right).$$

Matrices

The subject **linear algebra** can be considered the study of linear transformations. To begin such a study, it is beneficial to represent linear transformations in a simple way so that they can be easily examined. Operations can then be defined on the representations, creating a system of algebra that mimics the algebra of functions defined in Section 1.4. To see how this is done, again consider the linear transformation $T : \mathbb{R}^2 \to \mathbb{R}^3$ defined by (3.4),

$$T \begin{bmatrix} x \\ y \end{bmatrix} = \begin{bmatrix} 3x - y \\ 4x \\ y - 5x \end{bmatrix}.$$

Write $T \begin{bmatrix} x \\ y \end{bmatrix}$ as a linear combination of vectors in $\mathbb{R}^3$ (Definition 2.1.18). In other words, for every $x, y \in \mathbb{R}$, write

$$T \begin{bmatrix} x \\ y \end{bmatrix} = \begin{bmatrix} 3x - y \\ 4x \\ y - 5x \end{bmatrix} = x \begin{bmatrix} 3 \\ 4 \\ -5 \end{bmatrix} + y \begin{bmatrix} -1 \\ 0 \\ 1 \end{bmatrix}. \tag{3.5}$$

What is important in the linear combination of (3.5) are the coordinates of the two vectors and the scalars that that multiply each of the vectors. To carefully study any linear transformation, a notation is needed to emphasize these two points, so represent $T \begin{bmatrix} x \\ y \end{bmatrix}$ by the notation

$$\begin{bmatrix} 3 & -1 \\ 4 & 0 \\ -5 & 1 \end{bmatrix} \begin{bmatrix} x \\ y \end{bmatrix}. \tag{3.6}$$

The array of real numbers on the left represents the set of the two vectors, and the vector on the right contains the arbitrary scalars. Now that this has been done for this example, it is time to generalize this for any linear transformation. The first step is to formalize the concept of an array of numbers.

■ **Definition 3.1.10**

Let $m, n \in \mathbb{Z}^+$ and $a_{ij} \in \mathbb{R}$ with $i = 1, 2, \ldots, m$, and $j = 1, 2, \ldots, n$. The array,

$$\begin{bmatrix} a_{11} & a_{12} & \cdots & a_{1n} \\ a_{21} & a_{22} & \cdots & a_{2n} \\ \vdots & \vdots & & \vdots \\ a_{m1} & a_{m2} & \cdots & a_{mn} \end{bmatrix},$$

is called a **matrix** and can be represented concisely as $[a_{ij}]$. Its **size** is $m \times n$ because the matrix has m rows and n columns. If the matrix is $[x_{ij}]$ with each

x_{ij} being a variable, the matrix is called a **variable matrix**. If the size of the matrix is 1×1, it is customary to remove the brackets from $[a]$ and simply write a. This suggests that the brackets in the matrix notation simply serve as grouping symbols for the array.

Definition 3.1.10 implies that the expression given in (3.6) has two matrices. They are the 3×2 matrix, $\begin{bmatrix} 3 & -1 \\ 4 & 0 \\ -5 & 1 \end{bmatrix}$, and the 2×1 variable matrix, $\begin{bmatrix} x \\ y \end{bmatrix}$. Because an $m \times 1$ matrix is a vector, it makes sense to use (3.6) to write

$$\begin{bmatrix} 3x - y \\ 4x \\ y - 5x \end{bmatrix} = \begin{bmatrix} 3 & -1 \\ 4 & 0 \\ -5 & 1 \end{bmatrix} \begin{bmatrix} x \\ y \end{bmatrix}, \tag{3.7}$$

which is simply an equation between two vectors from $\mathbb{R}^3$. The obvious generalization is made to define the equality of two matrices.

■ **Definition 3.1.11**

If A represents the matrix $[a_{ij}]$ and B represents $[b_{ij}]$, where both are $m \times n$ matrices, then A is **equal** to B (written as $A = B$) means that $a_{ij} = b_{ij}$ for all $i = 1, 2, \dots m$ and $j = 1, 2, \dots, n$. If A is not equal to B or A and B are of different sizes, write $A \neq B$.

■ **Example 3.1.12**

The $m \times n$ matrix $[a_{ij}]$ such that $a_{ij} = 0$ for all $i = 1, 2, \dots, m$ and $j = 1, 2, \dots, n$ is called the **zero matrix**. It is usually denoted by $0_{m,n}$, yet 0 or [0] is acceptable when its size is clear from context. To illustrate,

$$0_{3,4} = \begin{bmatrix} 0 & 0 & 0 & 0 \\ 0 & 0 & 0 & 0 \\ 0 & 0 & 0 & 0 \end{bmatrix}.$$

Since the columns of the 3×2 matrix of (3.6) come from the vectors in the linear combination of (3.5), it is important to be able to identify the columns of a matrix. For this reason the notation of the next definition is introduced. There are also times when the rows of a matrix need to be identified, so rows are included in the definition as well.

■ **Definition 3.1.13**

Let $A = [a_{ij}]$ be an $m \times n$ matrix. A **column** of A is an $m \times 1$ matrix

$$\mathbf{c}_j = \begin{bmatrix} a_{1j} \\ a_{2j} \\ \vdots \\ a_{mj} \end{bmatrix}$$

for some $j = 1, 2, \ldots n$, and a **row** of A is a $1 \times n$ matrix

$$\mathbf{r}_i = \begin{bmatrix} a_{i1} & a_{i2} & \cdots & a_{in} \end{bmatrix}$$

for some $i = 1, 2, \ldots m$. Observe that an $m \times 1$ column is an element of $\mathbb{R}^m$. Using these vectors, A can be represented by identifying its rows, as in

$$A = \begin{bmatrix} \mathbf{r}_1 \\ \mathbf{r}_2 \\ \vdots \\ \mathbf{r}_m \end{bmatrix},$$

or by identifying its columns, as in

$$A = \begin{bmatrix} \mathbf{c}_1 & \mathbf{c}_2 & \cdots & \mathbf{c}_n \end{bmatrix}.$$

■ **Example 3.1.14**

Consider the matrix $A = \begin{bmatrix} 1 & 2 & 3 & 4 \\ 5 & 6 & 7 & 8 \end{bmatrix}$. This matrix is a 2×4 matrix with rows,

$$\mathbf{r}_1 = \begin{bmatrix} 1 & 2 & 3 & 4 \end{bmatrix} \text{ and } \mathbf{r}_2 = \begin{bmatrix} 5 & 6 & 7 & 8 \end{bmatrix},$$

and columns,

$$\mathbf{c}_1 = \begin{bmatrix} 1 \\ 5 \end{bmatrix}, \quad \mathbf{c}_2 = \begin{bmatrix} 2 \\ 6 \end{bmatrix}, \quad \mathbf{c}_3 = \begin{bmatrix} 3 \\ 7 \end{bmatrix}, \quad \mathbf{c}_4 = \begin{bmatrix} 4 \\ 8 \end{bmatrix}.$$

This means that

$$A = \begin{bmatrix} \mathbf{r}_1 \\ \mathbf{r}_2 \end{bmatrix} = \begin{bmatrix} \mathbf{c}_1 & \mathbf{c}_2 & \mathbf{c}_3 & \mathbf{c}_4 \end{bmatrix}.$$

Returning to the computations that motivated Definition 3.1.10, from (3.5) and (3.7) it follows that

$$\begin{bmatrix} 3 & -1 \\ 4 & 0 \\ -5 & 1 \end{bmatrix} \begin{bmatrix} x \\ y \end{bmatrix} = x \begin{bmatrix} 3 \\ 4 \\ -5 \end{bmatrix} + y \begin{bmatrix} -1 \\ 0 \\ 1 \end{bmatrix}.$$

Since the term on the left side of this equation resembles multiplication, call it **matrix multiplication** and define it as follows.

■ **Definition 3.1.15 [Matrix Multiplication – Preliminary Form]**

Let A be an $m \times n$ matrix with columns $\mathbf{c}_1, \mathbf{c}_2, \ldots, \mathbf{c}_n$. Write $\mathbf{u} \in \mathbb{R}^n$ as

$$\mathbf{u} = \begin{bmatrix} u_1 \\ u_2 \\ \vdots \\ u_n \end{bmatrix}.$$

Then, $A\mathbf{u} = u_1\mathbf{c}_1 + u_2\mathbf{c}_2 + \cdots + u_n\mathbf{c}_n$.

Notice that the multiplication of Definition 3.1.15 requires that the number of columns in A equals the number of coordinates of $\mathbf{u}$. For example,

$$\begin{bmatrix} 1 & 2 & 3 & 4 \\ 5 & 6 & 7 & 8 \end{bmatrix} \begin{bmatrix} 9 \\ 1 \\ 2 \\ 3 \end{bmatrix} = 9\begin{bmatrix} 1 \\ 5 \end{bmatrix} + 1\begin{bmatrix} 2 \\ 6 \end{bmatrix} + 2\begin{bmatrix} 3 \\ 7 \end{bmatrix} + 3\begin{bmatrix} 4 \\ 8 \end{bmatrix} = \begin{bmatrix} 29 \\ 89 \end{bmatrix}, \tag{3.8}$$

but $\begin{bmatrix} 1 & 2 & 3 & 4 \\ 5 & 6 & 7 & 8 \end{bmatrix} \begin{bmatrix} 9 \\ 1 \\ 2 \end{bmatrix}$ is undefined.

Although $A\mathbf{u}$ is defined using the columns of A, it can also be computed using the rows of A, but a definition is needed first.

■ **Definition 3.1.16**

Let $A = [a_{ij}]$ be an $m \times n$ matrix. The **transpose** of A is the $n \times m$ matrix $A^T = [a_{ji}]$ such that for all $i = 1, 2, \ldots, m$ and $j = 1, 2, \ldots, n$, entry j of column i of A^T is a_{ij}.

Intuitively, the columns of A^T are the rows of A, and the rows of A^T are the columns of A. Therefore, it is straightforward to find a transpose, as with

$$\begin{bmatrix} 1 & 2 & 3 & 4 \\ 5 & 6 & 7 & 8 \end{bmatrix}^T = \begin{bmatrix} 1 & 5 \\ 2 & 6 \\ 3 & 7 \\ 4 & 8 \end{bmatrix},$$

and some basic properties of the transpose are found in Exercise 19. For now all that is needed is to work with columns and rows. Definition 3.1.16 simply states that the transpose of a column is the row consisting of the entries of the column and the transpose of a row is the column consisting of the entries of the row. For example,

$$\begin{bmatrix} 2 \\ 4 \\ 8 \end{bmatrix}^T = \begin{bmatrix} 2 & 4 & 8 \end{bmatrix} \text{ and } \begin{bmatrix} 1 & 3 & 5 & 7 \end{bmatrix}^T = \begin{bmatrix} 1 \\ 3 \\ 5 \\ 7 \end{bmatrix}.$$

The transpose is required for the next result because the dot product of two vectors (Definition 2.2.1) is only defined for vectors that are of the same dimension.

■ **Theorem 3.1.17**

Let A be an $m \times n$ matrix and $\mathbf{u} \in \mathbb{R}^n$. If $\mathbf{r}_1, \mathbf{r}_2, \ldots, \mathbf{r}_m$ are the rows of A, then

$$A\mathbf{u} = \begin{bmatrix} \mathbf{r}_1^T \cdot \mathbf{u} \\ \mathbf{r}_2^T \cdot \mathbf{u} \\ \vdots \\ \mathbf{r}_m^T \cdot \mathbf{u} \end{bmatrix}.$$

Proof

Write $A = [a_{ij}]$ with rows $\mathbf{r}_1, \mathbf{r}_2, \ldots, \mathbf{r}_m$ and let $u_1, u_2, \ldots, u_n$ be the coordinates

of $\mathbf{u}$. Then, $\mathbf{r}_i^T = \begin{bmatrix} a_{i1} \\ a_{i2} \\ \vdots \\ a_{in} \end{bmatrix}$ for $i = 1, 2, \ldots, m$, which implies that

$$\mathbf{r}_i^T \cdot \mathbf{u} = \begin{bmatrix} a_{i1} \\ a_{i2} \\ \vdots \\ a_{i,n} \end{bmatrix} \cdot \begin{bmatrix} u_1 \\ u_2 \\ \vdots \\ u_n \end{bmatrix} = a_{i1}u_1 + a_{i2}u_2 + \cdots + a_{in}u_n.$$

Now, compute using Definition 3.1.15,

$$A \begin{bmatrix} u_1 \\ u_2 \\ \vdots \\ u_n \end{bmatrix} = u_1 \begin{bmatrix} a_{11} \\ a_{21} \\ \vdots \\ a_{m1} \end{bmatrix} + u_2 \begin{bmatrix} a_{12} \\ a_{22} \\ \vdots \\ a_{m2} \end{bmatrix} + \cdots + u_n \begin{bmatrix} a_{1n} \\ a_{2n} \\ \vdots \\ a_{mn} \end{bmatrix}$$

$$= \begin{bmatrix} a_{11}u_1 + a_{12}u_2 + \cdots + a_{1n}u_n \\ a_{21}u_1 + a_{22}u_2 + \cdots + a_{2n}u_n \\ \vdots \\ a_{m1}u_1 + a_{m2}u_2 + \cdots + a_{mn}u_n \end{bmatrix} = \begin{bmatrix} \mathbf{r}_1^T \cdot \mathbf{u} \\ \mathbf{r}_2^T \cdot \mathbf{u} \\ \vdots \\ \mathbf{r}_m^T \cdot \mathbf{u} \end{bmatrix}. \blacksquare$$

For instance, recomputing the matrix multiplication of (3.8),

$$\begin{bmatrix} 1 & 2 & 3 & 4 \\ 5 & 6 & 7 & 8 \end{bmatrix} \begin{bmatrix} 9 \\ 1 \\ 2 \\ 3 \end{bmatrix} = \begin{bmatrix} 1(9) + 2(1) + 3(2) + 4(3) \\ 5(9) + 6(1) + 7(2) + 8(3) \end{bmatrix} = \begin{bmatrix} 29 \\ 89 \end{bmatrix}.$$

Theorem 3.1.17 can be used to show that the preliminary form of matrix multiplication is a linear transformation.

■ **Theorem 3.1.18**

Let A be an $m \times n$ matrix, $\mathbf{u}, \mathbf{v} \in \mathbb{R}^n$, and $s \in \mathbb{R}$.

(a) $A(\mathbf{u} + \mathbf{v}) = A\mathbf{u} + A\mathbf{v}$.

(b) $A(s\mathbf{u}) = s(A\mathbf{u})$.

Proof

Let $\mathbf{r}_i \in \mathbb{R}^n$ ($i = 1, 2, \ldots, m$) represent the rows of A. By Theorem 2.2.6(b), $A(\mathbf{u} + \mathbf{v}) = A\mathbf{u} + A\mathbf{v}$ because

$$\begin{bmatrix} \mathbf{r}_1^T \bullet (\mathbf{u}+\mathbf{v}) \\ \mathbf{r}_2^T \bullet (\mathbf{u}+\mathbf{v}) \\ \vdots \\ \mathbf{r}_m^T \bullet (\mathbf{u}+\mathbf{v}) \end{bmatrix} = \begin{bmatrix} \mathbf{r}_1^T \bullet \mathbf{u}+\mathbf{r}_1^T \bullet \mathbf{v} \\ \mathbf{r}_2^T \bullet \mathbf{u}+\mathbf{r}_2^T \bullet \mathbf{v} \\ \vdots \\ \mathbf{r}_m^T \bullet \mathbf{u}+\mathbf{r}_m^T \bullet \mathbf{v} \end{bmatrix} = \begin{bmatrix} \mathbf{r}_1^T \bullet \mathbf{u} \\ \mathbf{r}_2^T \bullet \mathbf{u} \\ \vdots \\ \mathbf{r}_m^T \bullet \mathbf{u} \end{bmatrix} + \begin{bmatrix} \mathbf{r}_1^T \bullet \mathbf{v} \\ \mathbf{r}_2^T \bullet \mathbf{v} \\ \vdots \\ \mathbf{r}_m^T \bullet \mathbf{v} \end{bmatrix},$$

and Theorem 2.2.6(c) implies that $A(s\mathbf{u}) = s(A\mathbf{u})$ because

$$\begin{bmatrix} \mathbf{r}_1^T \bullet s\mathbf{u} \\ \mathbf{r}_2^T \bullet s\mathbf{u} \\ \vdots \\ \mathbf{r}_m^T \bullet s\mathbf{u} \end{bmatrix} = \begin{bmatrix} s(\mathbf{r}_1^T \bullet \mathbf{u}) \\ s(\mathbf{r}_2^T \bullet \mathbf{u}) \\ \vdots \\ s(\mathbf{r}_m^T \bullet \mathbf{u}) \end{bmatrix} = s \begin{bmatrix} \mathbf{r}_1^T \bullet \mathbf{u} \\ \mathbf{r}_2^T \bullet \mathbf{u} \\ \vdots \\ \mathbf{r}_m^T \bullet \mathbf{u} \end{bmatrix}. \blacksquare$$

The purpose of matrices and matrix multiplication is to represent linear transformations. Theorem 3.1.18 plays an important role in finalizing this.

■ **Theorem 3.1.19**

Let T be a function $\mathbb{R}^n \to \mathbb{R}^m$. The following are equivalent.

  (a) T is a linear transformation.

  (b) There exists an $m \times n$ matrix A such that $T\mathbf{x} = A\mathbf{x}$ for all $\mathbf{x} \in \mathbb{R}^n$.

Proof

For the linear transformation $T : \mathbb{R}^n \to \mathbb{R}^m$ define $A = \begin{bmatrix} T\mathbf{e}_1 & T\mathbf{e}_2 & \cdots & T\mathbf{e}_n \end{bmatrix}$. Let $\mathbf{u} \in \mathbb{R}^n$ and write

$$\mathbf{u} = \begin{bmatrix} u_1 \\ u_2 \\ \vdots \\ u_n \end{bmatrix}.$$

Because T is a linear transformation and by Definition 3.1.15,

$$T\mathbf{u} = T(u_1\mathbf{e}_1 + u_2\mathbf{e}_2 + \cdots + u_n\mathbf{e}_n)$$
$$= u_1(T\mathbf{e}_1) + u_2(T\mathbf{e}_2) + \cdots + u_n(T\mathbf{e}_n) = A\mathbf{u}.$$

The converse is Theorem 3.1.18. ■

The proof of Theorem 3.1.19 shows how to find a matrix that describes a linear transformation.

■ **Definition 3.1.20**

Let $T : \mathbb{R}^n \to \mathbb{R}^m$ be a linear transformation and suppose that $\mathbf{e}_1, \mathbf{e}_2, \dots, \mathbf{e}_n$ are the standard vectors of $\mathbb{R}^n$. The **standard matrix** of T is

$$[T] = \begin{bmatrix} T\mathbf{e}_1 & T\mathbf{e}_2 & \cdots & T\mathbf{e}_n \end{bmatrix},$$

■ **Example 3.1.21**

As in Example 3.1.9, let $T : \mathbb{R}^3 \to \mathbb{R}^3$ be the linear transformation defined by

$$T \begin{bmatrix} x \\ y \\ z \end{bmatrix} = \begin{bmatrix} 2x + z \\ 2z + y \\ 2y + x \end{bmatrix}.$$

Because

$$T \begin{bmatrix} 1 \\ 0 \\ 0 \end{bmatrix} = \begin{bmatrix} 2 \\ 0 \\ 1 \end{bmatrix}, \quad T \begin{bmatrix} 0 \\ 1 \\ 0 \end{bmatrix} = \begin{bmatrix} 0 \\ 1 \\ 2 \end{bmatrix}, \quad T \begin{bmatrix} 0 \\ 0 \\ 1 \end{bmatrix} = \begin{bmatrix} 1 \\ 2 \\ 0 \end{bmatrix},$$

the standard matrix for T is

$$[T] = \begin{bmatrix} 2 & 0 & 1 \\ 0 & 1 & 2 \\ 1 & 2 & 0 \end{bmatrix}.$$

The linear transformation T applied to any vector $\mathbf{u}$ has the same result as $[T]$ times $\mathbf{u}$, such as when

$$T \begin{bmatrix} 1 \\ 2 \\ 3 \end{bmatrix} = \begin{bmatrix} 5 \\ 8 \\ 5 \end{bmatrix} \quad \text{and} \quad \begin{bmatrix} 2 & 0 & 1 \\ 0 & 1 & 2 \\ 1 & 2 & 0 \end{bmatrix} \begin{bmatrix} 1 \\ 2 \\ 3 \end{bmatrix} = \begin{bmatrix} 5 \\ 8 \\ 5 \end{bmatrix}.$$

The construction of the standard matrix suggests that $[T]$ is the only matrix that behaves as T. This is proved next.

■ **Theorem 3.1.22**

Let $T : \mathbb{R}^n \to \mathbb{R}^m$ be a linear transformation. If A is an $m \times n$ matrix such that $A\mathbf{u} = T\mathbf{u}$ for all $\mathbf{u} \in \mathbb{R}^n$, then $A = [T]$.

Proof

Suppose that A is an $m \times n$ matrix such that $A\mathbf{u} = T\mathbf{u} = [T]\mathbf{u}$ for every $\mathbf{u}$ in $\mathbb{R}^n$. Let $\mathbf{c}_1, \mathbf{c}_2, \dots, \mathbf{c}_n$ be the columns of A. Thus, for all real numbers $u_1, u_2, \dots, u_n$,

$$u_1\mathbf{c}_1 + u_2\mathbf{c}_2 + \cdots + u_n\mathbf{c}_n = u_1(T\mathbf{e}_1) + u_2(T\mathbf{e}_2) + \cdots + u_n(T\mathbf{e}_n).$$

Then, choosing any $i = 1, 2, \dots, n$ and letting $u_i = 1$ but $u_j = 0$ for $j \neq i$, it follows that $\mathbf{c}_i = T\mathbf{e}_i$. Hence, $A = [T]$. ■

Exercises

1. Prove that the given functions are linear transformations.

   (a) $T_1 : \mathbb{R} \to \mathbb{R}$, $T_1 x = 4x$

   (b) $T_2 : \mathbb{R}^3 \to \mathbb{R}^2$, $T_2 \begin{bmatrix} x \\ y \\ z \end{bmatrix} = \begin{bmatrix} 2x - 4y \\ 6z + 8x - y \end{bmatrix}$

   (c) $T_3 : \mathbb{R}^4 \to \mathbb{R}^3$, $T_3 \begin{bmatrix} x \\ y \\ z \\ w \end{bmatrix} = \begin{bmatrix} 0 \\ x + y + z + w \\ 0 \end{bmatrix}$

   (d) $T_4 : \mathbb{R}^4 \to \mathbb{R}^4$, $T_4 \begin{bmatrix} x \\ y \\ z \\ w \end{bmatrix} = \begin{bmatrix} x \\ 0 \\ 0 \\ 0 \end{bmatrix}$

   (e) $T_5 : \mathbb{R}^3 \to \mathbb{R}^2$, $T_5 \begin{bmatrix} x \\ y \\ z \end{bmatrix} = \begin{bmatrix} x \\ y \end{bmatrix}$

2. Find the standard matrix for each linear transformation in Exercise 1.

3. Prove that the given functions are not linear transformations.

   (a) $T : \mathbb{R} \to \mathbb{R}$, $Tx = 5x + 3$

   (b) $T : \mathbb{R}^3 \to \mathbb{R}$, $T \begin{bmatrix} x \\ y \\ z \end{bmatrix} = \sqrt{x + y + z}$

   (c) $T : \mathbb{R}^3 \to \mathbb{R}^3$, $T \begin{bmatrix} x \\ y \\ z \end{bmatrix} = \begin{bmatrix} xy \\ 0 \\ x + y + z \end{bmatrix}$

   (d) $T : \mathbb{R}^2 \to \mathbb{R}^5$, $T \begin{bmatrix} x \\ y \end{bmatrix} = \begin{bmatrix} 2x - 4z \\ x \\ y \\ x + 1 \\ y - 1 \end{bmatrix}$

4. Define $T : \mathbb{R}^2 \to \mathbb{R}^3$ by $T \begin{bmatrix} x \\ y \end{bmatrix} = \begin{bmatrix} 2x + 7y \\ x + y \\ 3x - y \end{bmatrix}$. Evaluate.

   (a) $T \begin{bmatrix} 4 \\ 6 \end{bmatrix}$ (b) $T \begin{bmatrix} 8 \\ 12 \end{bmatrix}$ (c) $T \begin{bmatrix} 0 \\ 0 \end{bmatrix}$ (d) $T \begin{bmatrix} 1 \\ 0 \end{bmatrix}$

5. Is the linear transformation T of Exercise 4 one-to-one or onto? Explain.

6. Define the linear transformation $T : \mathbb{R}^4 \to \mathbb{R}^2$ so that $[T] = \begin{bmatrix} 1 & 0 & 2 & -1 \\ 0 & 3 & 0 & 5 \end{bmatrix}$.
    Evaluate.

    (a) $T \begin{bmatrix} 0 \\ 1 \\ 0 \\ 0 \end{bmatrix}$
    (b) $T \begin{bmatrix} 0 \\ 0 \\ 0 \\ 1 \end{bmatrix}$
    (c) $T \begin{bmatrix} 1 \\ 2 \\ 3 \\ 4 \end{bmatrix}$
    (d) $T \begin{bmatrix} 0 \\ -4 \\ 3 \\ 8 \end{bmatrix}$

7. Is the linear transformation T of Exercise 6 one-to-one or onto? Explain.

8. Using the linear transformation T from Exercise 6, write $T \begin{bmatrix} 1 \\ 2 \\ 3 \\ 4 \end{bmatrix}$ and $T \begin{bmatrix} 0 \\ -4 \\ 3 \\ 8 \end{bmatrix}$ as
    linear combinations of the columns of $[T]$.

9. Let $T : \mathbb{R}^2 \to \mathbb{R}^2$ be a linear transformation so that T maps $\begin{bmatrix} 1 \\ 3 \end{bmatrix}$ to $\begin{bmatrix} 3 \\ -2 \end{bmatrix}$ and $\begin{bmatrix} 0 \\ 1 \end{bmatrix}$
    to $\begin{bmatrix} 2 \\ 7 \end{bmatrix}$. Evaluate.

    (a) $T \begin{bmatrix} 2 \\ 6 \end{bmatrix}$
    (b) $T \begin{bmatrix} 1 \\ 4 \end{bmatrix}$
    (c) $T \begin{bmatrix} 2 \\ 0 \end{bmatrix}$
    (d) $T \begin{bmatrix} -3 \\ -7 \end{bmatrix}$

10. Let T be the linear transformation such that $[T] = \begin{bmatrix} 1 & 0 & -2 & 3 \\ 0 & 4 & 1 & -1 \end{bmatrix}$.

    (a) Find the domain of T.
    (b) Find the codomain of T.
    (c) Find the range of T.
    (d) Find the images of the standard vectors of the domain of T.

11. Repeat Exercise 10 where T is the linear transformation with standard matrix
    $\begin{bmatrix} 1 & 0 & -2 & 3 \\ 0 & 4 & 1 & -1 \end{bmatrix}^T$.

12. Let $T : \mathbb{R}^2 \to \mathbb{R}^2$ be defined as $T \begin{bmatrix} x \\ y \end{bmatrix} = \begin{bmatrix} 7x + 3y \\ 2x - 5y \end{bmatrix}$. Let $\mathbf{u} = \begin{bmatrix} 7 \\ 2 \end{bmatrix}$ and $\mathbf{v} = \begin{bmatrix} -6 \\ 4 \end{bmatrix}$.
    Verify the properties of Theorem 3.1.6 by computing and comparing results
    when appropriate.

    (a) $T\mathbf{0}$ (c) $T(\mathbf{u} - \mathbf{v})$ and $T\mathbf{u} - T\mathbf{v}$
    (b) $T(-\mathbf{u})$ and $-(T\mathbf{u})$

13. Define $T : \mathbb{R}^2 \to \mathbb{R}^2$ by $T \begin{bmatrix} x \\ y \end{bmatrix} = \begin{bmatrix} 2x \\ x - 4y \end{bmatrix}$. Let $L = S\left(\begin{bmatrix} -1 \\ 2 \end{bmatrix}\right)$.

    (a) Show that T is a linear transformation.
    (b) Find $T[L]$ and explain why it is a line that does not contain the zero vector.

(c) Graph L and $T[L]$ on the same pair of axes.

(d) Find $T\left[\begin{bmatrix} 3 \\ 3 \end{bmatrix} + L\right]$ and explain why it is a line not containing $\mathbf{0}$.

(e) Graph $\begin{bmatrix} 3 \\ 3 \end{bmatrix} + L$ and $T\left[\begin{bmatrix} 3 \\ 3 \end{bmatrix} + L\right]$ on the same pair of axes.

14. Let $T : \mathbb{R}^4 \to \mathbb{R}^3$ be defined by

$$T\begin{bmatrix} x \\ y \\ z \\ w \end{bmatrix} = \begin{bmatrix} 0 \\ x + y + z + w \\ 0 \end{bmatrix}.$$

It is known that this is a linear transformation [Exercise 1(c)]. Define the plane

$$P = \begin{bmatrix} 1 \\ 0 \\ 3 \\ -1 \end{bmatrix} + S\left(\begin{bmatrix} 1 \\ 4 \\ 2 \\ 1 \end{bmatrix}, \begin{bmatrix} -1 \\ 0 \\ 5 \\ 2 \end{bmatrix}\right).$$

(a) Find $T[P]$.

(b) Is $T[P]$ a plane? Explain.

15. Let T be a linear transformation such that $[T] = \begin{bmatrix} 1 & -2 & 3 \\ 0 & 4 & 1 \\ -3 & 0 & 2 \end{bmatrix}$. Find the images of the xy-plane, xz-plane, and yz-plane under T. See Example 2.1.26.

16. Find a linear transformation $T : \mathbb{R}^2 \to \mathbb{R}^2$ such that the image of the x-axis is the y-axis and the image of the y-axis is the x-axis.

17. Find a linear transformation $T : \mathbb{R}^3 \to \mathbb{R}^3$ such that the image of the xy-plane is the xz-plane, the image of the xz-plane is the xy-plane, and the image of the yz-plane is the yz-plane.

18. Find a linear transformation $T : \mathbb{R}^3 \to \mathbb{R}^3$ such that the image of the xy-plane is the yz-plane, and the image of the xz-plane is the y-axis. Once T is found, determine the image of the yz-plane under T.

19. Let A and B be $m \times n$ matrices and $r \in \mathbb{R}$. Prove.

(a) $(A^T)^T = A$ (b) $(A+B)^T = A^T + B^T$ (c) $(rA)^T = rA^T$

20. Define $T : \mathbb{R}^n \to \mathbb{R}^m$ by $T\mathbf{u} = A\mathbf{u} + \mathbf{b}$ for some $m \times n$ matrix A and $\mathbf{b} \in \mathbb{R}^m$. Prove that T is a linear transformation if and only if $\mathbf{b} \neq \mathbf{0}$.

21. Prove that $T : \mathbb{R}^n \to \mathbb{R}^m$ is a linear transformation if and only if $T(r\mathbf{u} + s\mathbf{v}) = rT\mathbf{u} + sT\mathbf{v}$ for all $\mathbf{u}, \mathbf{v} \in \mathbb{R}^n$ and $r, s \in \mathbb{R}$.

22. Let $T : \mathbb{R}^n \to \mathbb{R}^m$ be a linear transformation. Prove that $T\mathbf{u} = T\mathbf{v}$ if and only if $T(\mathbf{u} - \mathbf{v}) = \mathbf{0}$ for all $\mathbf{u}, \mathbf{v} \in \mathbb{R}^n$.

23. Prove Theorem 3.1.8.

24. Prove by mathematical induction: If $T : \mathbb{R}^n \to \mathbb{R}^m$ is a linear transformation, then for all $\mathbf{u}_1, \mathbf{u}_2, \dots, \mathbf{u}_k \in \mathbb{R}^n$ and $r_1, r_2, \dots, r_k \in \mathbb{R}$,

$$T(r_1\mathbf{u}_1 + r_2\mathbf{u}_2 + \cdots + r_k\mathbf{u}_k) = r_1 T(\mathbf{u}_1) + r_2 T(\mathbf{u}_2) + \cdots + r_k T(\mathbf{u}_k).$$

25. Prove the converse of Exercise 24.

26. Let $\{\mathbf{u}_1, \mathbf{u}_2, \dots, \mathbf{u}_k\}$ be a spanning set for $\mathbb{R}^n$. Suppose that $T : \mathbb{R}^n \to \mathbb{R}^m$ is a linear transformation such that $T\mathbf{u}_i = \mathbf{0}$ for all $i = 1, 2, \dots, k$. Prove that T is a zero transformation. See Example 3.1.3.

3.2 Matrix Algebra

The initial definition of matrix multiplication (Definition 3.1.15) was motivated by an attempt to describe a linear transformation. This suggests that if an entire algebraic system is to be defined on matrices like that defined on the real numbers, it should flow from the study of linear transformations. Indeed, it will copy the algebra of functions.

Addition, Subtraction, and Scalar Multiplication

Define the linear transformations $f, g : \mathbb{R} \to \mathbb{R}$ by $f(x) = 4x$ and $g(x) = 7x$. The product, $(fg)(x) = 28x^2$, is not a linear transformation because, for example,

$$(fg)(1 + 3) = 448 \neq 280 = (fg)(1) + (fg)(3).$$

Also, the quotient f/g is not a linear transformation because $(f/g)(0)$ is undefined. However, $f + g$, $f - g$, and rf are all linear transformations and motive the next definition. Compare with Examples 1.4.37 and 1.4.43.

■ **Definition 3.2.1**

Let $T_1, T_2 : \mathbb{R}^n \to \mathbb{R}^m$ be linear transformations and r be a real number. Define $T_1 + T_2$, $T_1 - T_2$, and rT_1 as functions $\mathbb{R}^n \to \mathbb{R}^m$ such that for all $\mathbf{u} \in \mathbb{R}^n$,

(a) $(T_1 + T_2)\mathbf{u} = T_1\mathbf{u} + T_2\mathbf{u}$ (**addition**),

(b) $(T_1 - T_2)\mathbf{u} = T_1\mathbf{u} - T_2\mathbf{u}$ (**subtraction**),

(c) $rT_1\mathbf{u} = r(T_1\mathbf{u})$ (**scalar multiplication**).

■ **Example 3.2.2**

Define $T_1, T_2 : \mathbb{R}^2 \to \mathbb{R}^3$ by

$$T_1\begin{bmatrix} x \\ y \end{bmatrix} = \begin{bmatrix} 3x - y \\ 4x \\ -5x + y \end{bmatrix} \quad \text{and} \quad T_2\begin{bmatrix} x \\ y \end{bmatrix} = \begin{bmatrix} 3y \\ x + y \\ 3x + 6y \end{bmatrix}. \tag{3.9}$$

Definition 3.2.1 yields

$$(T_1 + T_2)\begin{bmatrix} x \\ y \end{bmatrix} = \begin{bmatrix} 3x + 2y \\ 5x + y \\ -2x + 7y \end{bmatrix}, \tag{3.10}$$

$$(T_1 - T_2)\begin{bmatrix} x \\ y \end{bmatrix} = \begin{bmatrix} 3x - 4y \\ 3x - y \\ -8x - 5y \end{bmatrix}, \tag{3.11}$$

and

$$7T_1\begin{bmatrix} x \\ y \end{bmatrix} = \begin{bmatrix} 21x - 7y \\ 28x \\ -35x + 7y \end{bmatrix}. \tag{3.12}$$

These operations on functions are familiar. They also yield linear transformations under the right conditions.

■ Theorem 3.2.3

If $T_1, T_2 : \mathbb{R}^n \to \mathbb{R}^m$ are linear transformations, $T_1 + T_2$, $T_1 - T_2$, and rT_1 for all $r \in \mathbb{R}$ are linear transformations.

Proof

Let $\mathbf{u}, \mathbf{v} \in \mathbb{R}^n$ and $r \in \mathbb{R}$. Then, $T_1 + T_2$ is a linear transformation because

$$
\begin{aligned}
(T_1 + T_2)(\mathbf{u} + \mathbf{v}) &= T_1(\mathbf{u} + \mathbf{v}) + T_2(\mathbf{u} + \mathbf{v}) \\
&= T_1\mathbf{u} + T_1\mathbf{v} + (T_2\mathbf{u} + T_2\mathbf{v}) \\
&= T_1\mathbf{u} + T_2\mathbf{u} + (T_1\mathbf{v} + T_2\mathbf{v}) = (T_1 + T_2)\mathbf{u} + (T_1 + T_2)\mathbf{v}
\end{aligned}
$$

and

$$
\begin{aligned}
(T_1 + T_2)(r\mathbf{u}) &= T_1(r\mathbf{u}) + T_2(r\mathbf{u}) \\
&= r(T_1\mathbf{u}) + r(T_2\mathbf{u}) \\
&= r(T_1\mathbf{u} + T_2\mathbf{u}) = r[(T_1 + T_2)\mathbf{u}].
\end{aligned}
$$

That $T_1 - T_2$ and rT_1 are linear transformations is left to Exercise 7. ■

■ Example 3.2.4

Let $T : \mathbb{R}^n \to \mathbb{R}^m$ be a linear transformation. Theorem 3.2.3 implies that $(-1)T$ is a linear transformation. Furthermore, by Theorem 2.1.7(a)

$$T\mathbf{u} + (-1)(T\mathbf{u}) = (1 - 1)(T\mathbf{u}) = \mathbf{0},$$

which implies that the additive inverse of T is $(-1)T$, in symbols $-T = (-1)T$, and $-T$ is a linear transformation.

The standard matrices for the linear transformations found in (3.9)–(3.12) are

$$[T_1] = \begin{bmatrix} 3 & -1 \\ 4 & 0 \\ -5 & 1 \end{bmatrix}, \ [T_2] = \begin{bmatrix} 0 & 3 \\ 1 & 1 \\ 3 & 6 \end{bmatrix},$$

$$[T_1 + T_2] = \begin{bmatrix} 3 & 2 \\ 5 & 1 \\ -2 & 7 \end{bmatrix}, \ [T_1 - T_2] = \begin{bmatrix} 3 & -4 \\ 3 & -1 \\ -8 & -5 \end{bmatrix}, \ [7T_1] = \begin{bmatrix} 21 & -7 \\ 28 & 0 \\ -35 & 7 \end{bmatrix}.$$

For example, because of the evaluation of $T_1 + T_2$ in (3.10),

$$(T_1 + T_2)\begin{bmatrix} 1 \\ 0 \end{bmatrix} = \begin{bmatrix} 3 \\ 5 \\ -2 \end{bmatrix} \ \text{and} \ (T_1 + T_2)\begin{bmatrix} 0 \\ 1 \end{bmatrix} = \begin{bmatrix} 2 \\ 1 \\ 7 \end{bmatrix},$$

so

$$[T_1 + T_2] = \begin{bmatrix} 3 & 2 \\ 5 & 1 \\ -2 & 7 \end{bmatrix}.$$

Hence, in this case the entries of $[T_1 + T_2]$ are the sums of the corresponding entries of $[T_1]$ and $[T_2]$. This suggests that matrix algebra should begin with the next definition.

■ **Definition 3.2.5**

Let $A = [a_{ij}]$ and $B = [b_{ij}]$ be $m \times n$ matrices. Let $r \in \mathbb{R}$.

(a) $A + B = [a_{ij} + b_{ij}]$.
(b) $A - B = [a_{ij} - b_{ij}]$.
(c) $rA = [ra_{ij}]$.

For example,

$$\begin{bmatrix} 3 & -1 \\ 4 & 0 \\ -5 & 1 \end{bmatrix} + \begin{bmatrix} 0 & 3 \\ 1 & 1 \\ 3 & 6 \end{bmatrix} = \begin{bmatrix} 3 & 2 \\ 5 & 1 \\ -2 & 7 \end{bmatrix}, \ \begin{bmatrix} 3 & -1 \\ 4 & 0 \\ -5 & 1 \end{bmatrix} - \begin{bmatrix} 0 & 3 \\ 1 & 1 \\ 3 & 6 \end{bmatrix} = \begin{bmatrix} 3 & -4 \\ 3 & -1 \\ -8 & -5 \end{bmatrix}$$

and

$$7\begin{bmatrix} 3 & -1 \\ 4 & 0 \\ -5 & 1 \end{bmatrix} = \begin{bmatrix} 21 & -7 \\ 28 & 0 \\ -35 & 7 \end{bmatrix}.$$

These computations suggest a close relationship between the operations on linear transformations and the corresponding operations on matrices. This is detailed in the next theorem.

■ **Theorem 3.2.6**

Let $T_1, T_2 : \mathbb{R}^n \to \mathbb{R}^m$ be linear transformations and let r be a real number.

(a) $[T_1 + T_2] = [T_1] + [T_2]$.

(b) $[T_1 - T_2] = [T_1] - [T_2]$.

(c) $[-T_1] = (-1)[T_1]$.

(d) $[rT_1] = r[T_1]$.

(e) $[0_{nm}] = 0_{m,n}$.

Proof

Let $\mathbf{u} \in \mathbb{R}^n$. To prove (a), use Theorem 3.1.18(a) to show that

$$(T_1 + T_2)(\mathbf{u}) = T_1\mathbf{u} + T_2\mathbf{u} = [T_1]\mathbf{u} + [T_2]\mathbf{u} = ([T_1] + [T_2])\mathbf{u},$$

so by the uniqueness of the standard matrix (Theorem 3.1.22),

$$[T_1 + T_2] = [T_1] + [T_2].$$

The proof of the remaining parts is left to Exercise 8. ■

Properties

Now that certain basic operations have been defined on both linear transformations and matrices, it is important to identify some properties that these operations satisfy. To begin, two definitions are required.

■ **Definition 3.2.7**

For all positive integers m and n,

$$L(\mathbb{R}^n, \mathbb{R}^m) = \{T : T \text{ is a linear transformation } \mathbb{R}^n \to \mathbb{R}^m\}.$$

For example, if $T : \mathbb{R}^5 \to \mathbb{R}^2$ is a linear transformation, $T \in L(\mathbb{R}^5, \mathbb{R}^2)$. Also, Theorem 3.2.3 implies that $L(\mathbb{R}^n, \mathbb{R}^m)$ is closed (Definitions 1.4.35 and 1.4.41) under the addition, subtraction, and scalar multiplication of Definition 3.2.1.

■ **Definition 3.2.8**

For all positive integers m and n,

$$\mathbb{R}^{m \times n} = \{A : A \text{ is an } m \times n \text{ matrix with real entries}\}.$$

For example,

$$\begin{bmatrix} 2 & 4 & 6 \\ 8 & 0 & 2 \end{bmatrix} \in \mathbb{R}^{2 \times 3}.$$

Furthermore, $\mathbb{R}^{m \times n}$ is seen to be closed under matrix addition, subtraction, and scalar multiplication by Definition 3.2.5.

The algebras that are being defined on $L(\mathbb{R}^n, \mathbb{R}^m)$ and $\mathbb{R}^{m \times n}$ resemble the algebra defined on $\mathbb{R}^n$. The main reason for this is that the linear transformation operations and the matrix operations have the same fundamental properties as the operations on $\mathbb{R}^n$, which are enumerated in Theorem 2.1.5.

■ **Theorem 3.2.9**

Let $R, S, T \in L(\mathbb{R}^n, \mathbb{R}^m)$. Let $r, s \in \mathbb{R}$.

(a) $S + T = T + S$ (**Commutative Law**).

(b) $R + S + T = R + (S + T)$ (**Associative Law**).

(c) $T + 0_{nm} = 0_{nm} + T = T$ (**Additive Identity Law**).

(d) $T + (-1)T = (-1)T + T = 0_{nm}$ (**Additive Inverse Law**).

(e) $rsT = r(sT)$ (**Associative Law**).

(f) $r(S + T) = rS + rT$ (**Distributive Law**).

(g) $(r + s)T = rT + sT$ (**Distributive Law**).

(h) $1T = T$ (**Multiplicative Identity**).

Proof

Let $\mathbf{u} \in \mathbb{R}^n$. Theorem 2.1.5(a) implies

$$(S + T)\mathbf{u} = S\mathbf{u} + T\mathbf{u} = T\mathbf{u} + S\mathbf{u} = (T + S)\mathbf{u}.$$

Thus, it follows that $S + T = T + S$, proving (a). Also, by Theorem 2.1.5(e), $(rsT)\mathbf{u} = r(sT)\mathbf{u}$, so $rsT = r(sT)$, which proves (e). The remaining parts can be proved using Theorem 2.1.5 as noted in Exercise 9. ■

As noted in Theorem 3.2.9, the additive identity of $L(\mathbb{R}^n, \mathbb{R}^m)$ is the zero function, and if $T \in L(\mathbb{R}^n, \mathbb{R}^m)$, the additive inverse of T is $(-1)T$. The additive identity of $\mathbb{R}^{m \times n}$ is $0_{m,n}$. To see this, let $A = [a_{ij}]$ be an $m \times n$ matrix. Then,

$$A + 0_{m,n} = [a_{ij}] + [0] = [a_{ij} + 0] = [a_{ij}] = A$$

and

$$0_{m,n} + A = [0] + [a_{ij}] = [0 + a_{ij}] = [a_{ij}] = A.$$

Also, the additive inverse of A is $(-1)A$ because

$$A + (-1)A = [a_{ij}] + [-a_{ij}] = [a_{ij} + (-a_{ij})] = [0]$$

and

$$(-1)A + A = [-a_{ij}] + [a_{ij}] = [-a_{ij} + a_{ij}] = [0],$$

so write $-A = (-1)A$. These and other properties of addition and scalar multiplication with matrices are listed in the next theorem, the results of which parallel Theorem 3.2.9.

■ **Theorem 3.2.10**

Let $A, B, C \in \mathbb{R}^{m \times n}$. Let $r, s \in \mathbb{R}$.

(a) $A + B = B + A$ (**Commutative Law**).

(b) $A + B + C = A + (B + C)$ (**Associative Law**).

(c) $A + 0_{mn} = 0_{mn} + A = A$ (**Additive Identity Law**).

(d) $A + (-1)A = (-1)A + A = 0_{mn}$ (**Additive Inverse Law**).

(e) $rsA = r(sA)$ (**Associative Law**).

(f) $r(A + B) = rA + rB$ (**Distributive Law**).

(g) $(r + s)A = rA + sA$ (**Distributive Law**).

(h) $1A = A$ (**Multiplicative Identity**).

Proof

Write $A = [a_{ij}]$, $B = [b_{ij}]$, and $C = [c_{ij}]$. Then, to prove (b),

$$
\begin{aligned}
A + B + C &= [a_{ij} + b_{ij}] + [c_{ij}] \\
&= [a_{ij} + b_{ij} + c_{ij}] \\
&= [a_{ij} + (b_{ij} + c_{ij})] \\
&= [a_{ij}] + [b_{ij} + c_{ij}] = A + (B + C).
\end{aligned}
$$

Also, to prove (f),

$$
\begin{aligned}
r(A + B) &= r[a_{ij} + b_{ij}] \\
&= [r(a_{ij} + b_{ij})] \\
&= [ra_{ij} + rb_{ij}] \\
&= [ra_{ij}] + [rb_{ij}] \\
&= r[a_{ij}] + r[b_{ij}] = rA + rB.
\end{aligned}
$$

The remaining proofs are left to Exercise 15. ∎

▪ Example 3.2.11

Let $A = \begin{bmatrix} 1 & 2 & 3 & 4 & 5 \\ 6 & 7 & 8 & 9 & 0 \\ 9 & 8 & 7 & 6 & 5 \\ 4 & 3 & 2 & 1 & 0 \end{bmatrix}$ and $B = \begin{bmatrix} 2 & 1 & 0 & 0 & 0 \\ 3 & 4 & 0 & 0 & 0 \\ 5 & 6 & 7 & 8 & 9 \\ 0 & 1 & 2 & 3 & 4 \end{bmatrix}$. Although the size of

these matrices is not large, it is sometimes helpful to write a larger matrix as a **block matrix**, sometimes called a **partitioned matrix**. For this example, define the **submatrices**

$$
A_{11} = \begin{bmatrix} 1 & 2 \\ 6 & 7 \end{bmatrix}, \quad A_{12} = \begin{bmatrix} 3 & 4 & 5 \\ 8 & 9 & 0 \end{bmatrix}, \quad A_{21} = \begin{bmatrix} 9 & 8 \\ 4 & 3 \end{bmatrix}, \quad A_{22} = \begin{bmatrix} 7 & 6 & 5 \\ 2 & 1 & 0 \end{bmatrix}
$$

and

$$
B_{11} = \begin{bmatrix} 2 & 1 \\ 3 & 4 \end{bmatrix}, \quad B_{12} = \begin{bmatrix} 0 & 0 & 0 \\ 0 & 0 & 0 \end{bmatrix}, \quad B_{21} = \begin{bmatrix} 5 & 6 \\ 0 & 1 \end{bmatrix}, \quad A_{22} = \begin{bmatrix} 7 & 8 & 9 \\ 2 & 3 & 4 \end{bmatrix}.
$$

Notice that B_{12} is a zero matrix. With these **blocks**, write

$$A = \begin{bmatrix} A_{11} & A_{12} \\ A_{21} & A_{22} \end{bmatrix} \text{ and } B = \begin{bmatrix} B_{11} & B_{12} \\ B_{21} & B_{22} \end{bmatrix} = \begin{bmatrix} B_{11} & 0 \\ B_{21} & B_{22} \end{bmatrix},$$

where 0 represents the appropriately sized zero matrix, which in this case is 2×3. Because relative blocks are the same size, matrix addition and subtraction can be done on blocks,

$$A + B = \begin{bmatrix} A_{11} + B_{11} & A_{12} + B_{12} \\ A_{21} + B_{21} & A_{22} + B_{22} \end{bmatrix}$$

and

$$A - B = \begin{bmatrix} A_{11} - B_{11} & A_{12} - B_{12} \\ A_{21} - B_{21} & A_{22} - B_{22} \end{bmatrix},$$

and using the entries of this example,

$$A + B = \left[\begin{array}{cc|ccc} 3 & 3 & 3 & 4 & 5 \\ 9 & 11 & 8 & 9 & 0 \\ \hline 14 & 14 & 14 & 14 & 14 \\ 4 & 4 & 4 & 4 & 4 \end{array} \right],$$

where the lines are included simply to highlight the blocks. Scalar multiplication works exactly as expected, namely for any $r \in \mathbb{R}$,

$$rA = \begin{bmatrix} rA_{11} & rA_{12} \\ rA_{21} & rA_{22} \end{bmatrix},$$

and using the entries of this example,

$$3A = \left[\begin{array}{cc|ccc} 3 & 6 & 9 & 12 & 15 \\ 18 & 21 & 24 & 27 & 0 \\ \hline 27 & 24 & 21 & 18 & 15 \\ 12 & 9 & 6 & 3 & 0 \end{array} \right].$$

Block matrices can have any number of blocks of any size. The **block arithmetic** can be done with block matrices provided the sizes of the relative blocks coincide.

Multiplication

As already noted, the product of linear transformations might not be a linear transformation. However, the composition of linear transformations will be a linear transformation.

■ Theorem 3.2.12

If $T_1 : \mathbb{R}^m \to \mathbb{R}^s$ and $T_2 : \mathbb{R}^n \to \mathbb{R}^m$ are linear transformations, the composition $T_1 \circ T_2 : \mathbb{R}^n \to \mathbb{R}^s$ is a linear transformation.

Proof

Let $\mathbf{u}, \mathbf{v} \in \mathbb{R}^n$ and $r \in \mathbb{R}$. Then,

$$\begin{aligned}(T_1 \circ T_2)(\mathbf{u} + \mathbf{v}) &= T_1(T_2[\mathbf{u} + \mathbf{v}]) \\ &= T_1(T_2\mathbf{u} + T_2\mathbf{v}) \\ &= T_1(T_2\mathbf{u}) + T_1(T_2\mathbf{v}) = (T_1 \circ T_2)\mathbf{u} + (T_1 \circ T_2)\mathbf{v}\end{aligned}$$

and

$$(T_1 \circ T_2)(r\mathbf{u}) = T_1(T_2[r\mathbf{u}]) = T_1(r[T_2\mathbf{u}]) = r(T_1[T_2\mathbf{u}]) = r(T_1 \circ T_2)\mathbf{u}. \blacksquare$$

The standard matrix for the sum, difference, and scalar multiple of linear transformations are the sum, difference, and scalar multiple of the standard matrices of the given linear transformations, but there is currently no good candidate for the standard matrix of the composition of linear transformations. To help find it, consider the functions $T_1 : \mathbb{R}^4 \to \mathbb{R}^3$ and $T_2 : \mathbb{R}^2 \to \mathbb{R}^4$ defined by

$$T_1 \begin{bmatrix} x \\ y \\ z \\ w \end{bmatrix} = \begin{bmatrix} x + y + z + w \\ x - 2y + 3z - 4w \\ 4x + 3w \end{bmatrix} \quad \text{and} \quad T_2 \begin{bmatrix} x \\ y \end{bmatrix} = \begin{bmatrix} 2x + y \\ 0 \\ -x + y \\ 3x - 2y \end{bmatrix}. \tag{3.13}$$

As noted in Exercise 6, the functions are linear transformations and their standard matrices are

$$[T_1] = \begin{bmatrix} 1 & 1 & 1 & 1 \\ 1 & -2 & 3 & -4 \\ 4 & 0 & 0 & 3 \end{bmatrix} \quad \text{and} \quad [T_2] = \begin{bmatrix} 2 & 1 \\ 0 & 0 \\ -1 & 1 \\ 3 & -2 \end{bmatrix}.$$

Using these matrices to evaluate $T_1 \circ T_2$ will help determine the standard matrix for the composition. Specifically, using Definition 3.1.15 and Theorem 3.1.18,

$$(T_1 \circ T_2) \begin{bmatrix} x \\ y \end{bmatrix} = \begin{bmatrix} 1 & 1 & 1 & 1 \\ 1 & -2 & 3 & -4 \\ 4 & 0 & 0 & 3 \end{bmatrix} \left(\begin{bmatrix} 2 & 1 \\ 0 & 0 \\ -1 & 1 \\ 3 & -2 \end{bmatrix} \begin{bmatrix} x \\ y \end{bmatrix} \right)$$

$$= \begin{bmatrix} 1 & 1 & 1 & 1 \\ 1 & -2 & 3 & -4 \\ 4 & 0 & 0 & 3 \end{bmatrix} \left(x \begin{bmatrix} 2 \\ 0 \\ -1 \\ 3 \end{bmatrix} + y \begin{bmatrix} 1 \\ 0 \\ 1 \\ -2 \end{bmatrix} \right)$$

Therefore,

$$(T_1 \circ T_2)\begin{bmatrix} x \\ y \end{bmatrix} = x \begin{bmatrix} 1 & 1 & 1 & 1 \\ 1 & -2 & 3 & -4 \\ 4 & 0 & 0 & 3 \end{bmatrix}\begin{bmatrix} 2 \\ 0 \\ -1 \\ 3 \end{bmatrix} + y \begin{bmatrix} 1 & 1 & 1 & 1 \\ 1 & -2 & 3 & -4 \\ 4 & 0 & 0 & 3 \end{bmatrix}\begin{bmatrix} 1 \\ 0 \\ 1 \\ -2 \end{bmatrix}.$$

Because

$$\begin{bmatrix} 1 & 1 & 1 & 1 \\ 1 & -2 & 3 & -4 \\ 4 & 0 & 0 & 3 \end{bmatrix}\begin{bmatrix} 2 \\ 0 \\ -1 \\ 3 \end{bmatrix} = 2\begin{bmatrix} 1 \\ 1 \\ 4 \end{bmatrix} + 0\begin{bmatrix} 1 \\ -2 \\ 0 \end{bmatrix} - 1\begin{bmatrix} 1 \\ 3 \\ 0 \end{bmatrix} + 3\begin{bmatrix} 1 \\ -4 \\ 3 \end{bmatrix}$$

and

$$\begin{bmatrix} 1 & 1 & 1 & 1 \\ 1 & -2 & 3 & -4 \\ 4 & 0 & 0 & 3 \end{bmatrix}\begin{bmatrix} 1 \\ 0 \\ 1 \\ -2 \end{bmatrix} = 1\begin{bmatrix} 1 \\ 1 \\ 4 \end{bmatrix} + 0\begin{bmatrix} 1 \\ -2 \\ 0 \end{bmatrix} + 1\begin{bmatrix} 1 \\ 3 \\ 0 \end{bmatrix} - 2\begin{bmatrix} 1 \\ -4 \\ 3 \end{bmatrix},$$

it follows that $(T_1 \circ T_2)\begin{bmatrix} x \\ y \end{bmatrix}$ equals

$$\begin{bmatrix} 2(1) + 0(1) - 1(1) + 3(1) & 1(1) + 0(1) + 1(1) - 2(1) \\ 2(1) + 0(-2) - 1(3) + 3(-4) & 1(1) + 0(-2) + 1(3) - 2(-4) \\ 2(4) + 0(0) - 1(0) + 3(3) & 1(4) + 0(0) + 1(0) - 2(3) \end{bmatrix}\begin{bmatrix} x \\ y \end{bmatrix}. \qquad (3.14)$$

Observe that the 3×2 matrix of (3.14) is the 3×2 matrix,

$$\begin{bmatrix} \begin{bmatrix} 1 & 1 & 1 & 1 \end{bmatrix}^T \cdot \begin{bmatrix} 2 \\ 0 \\ -1 \\ 3 \end{bmatrix} & \begin{bmatrix} 1 & 1 & 1 & 1 \end{bmatrix}^T \cdot \begin{bmatrix} 1 \\ 0 \\ 1 \\ -2 \end{bmatrix} \\ \begin{bmatrix} 1 & -2 & 3 & -4 \end{bmatrix}^T \cdot \begin{bmatrix} 2 \\ 0 \\ -1 \\ 3 \end{bmatrix} & \begin{bmatrix} 1 & -2 & 3 & -4 \end{bmatrix}^T \cdot \begin{bmatrix} 1 \\ 0 \\ 1 \\ -2 \end{bmatrix} \\ \begin{bmatrix} 4 & 0 & 0 & 3 \end{bmatrix}^T \cdot \begin{bmatrix} 2 \\ 0 \\ -1 \\ 3 \end{bmatrix} & \begin{bmatrix} 4 & 0 & 0 & 3 \end{bmatrix}^T \cdot \begin{bmatrix} 1 \\ 0 \\ 1 \\ -2 \end{bmatrix} \end{bmatrix}. \qquad (3.15)$$

The rows that appear in the entries of (3.15) are the rows of $[T_1]$ and the columns that appear in the entries of (3.15) are the columns of $[T_2]$. This work reveals what can be considered a generalization of Definition 3.1.15 and Theorem 3.1.17.

■ **Definition 3.2.13 [Matrix Multiplication]**

Let A be an $s \times m$ matrix and B be an $m \times n$ matrix. Let $\mathbf{r}_1, \mathbf{r}_2, \dots, \mathbf{r}_s$ be the rows of A and $\mathbf{c}_1, \mathbf{c}_2, \dots, \mathbf{c}_n$ be the columns of B. The product AB is the $s \times n$ matrix defined by

$$AB = \begin{bmatrix} \mathbf{r}_1^T \bullet \mathbf{c}_1 & \mathbf{r}_1^T \bullet \mathbf{c}_2 & \cdots & \mathbf{r}_1^T \bullet \mathbf{c}_n \\ \mathbf{r}_2^T \bullet \mathbf{c}_1 & \mathbf{r}_2^T \bullet \mathbf{c}_2 & \cdots & \mathbf{r}_2^T \bullet \mathbf{c}_n \\ \vdots & \vdots & & \vdots \\ \mathbf{r}_s^T \bullet \mathbf{c}_1 & \mathbf{r}_s^T \bullet \mathbf{c}_2 & \cdots & \mathbf{r}_s^T \bullet \mathbf{c}_n \end{bmatrix}.$$

■ **Theorem 3.2.14**

Let A be an $s \times m$ matrix and B be an $m \times n$ matrix.

(a) $AB = \begin{bmatrix} A\mathbf{c}_1 & A\mathbf{c}_2 & \dots & A\mathbf{c}_n \end{bmatrix}$ if $\mathbf{c}_1, \mathbf{c}_2, \dots, \mathbf{c}_n$ are the columns of B.

(b) $AB = \begin{bmatrix} \mathbf{r}_1 B \\ \mathbf{r}_2 B \\ \vdots \\ \mathbf{r}_s B \end{bmatrix}$ if $\mathbf{r}_1, \mathbf{r}_2, \dots, \mathbf{r}_s$ are the rows of A.

Proof

(a) follows by Theorem 3.1.17. To prove (b), first note that for $i = 1, 2, \dots, s$,

$$\mathbf{r}_i B = \begin{bmatrix} \mathbf{r}_i^T \bullet \mathbf{c}_1 & \mathbf{r}_i^T \bullet \mathbf{c}_2 & \cdots & \mathbf{r}_i^T \bullet \mathbf{c}_n \end{bmatrix},$$

from which $AB = \begin{bmatrix} \mathbf{r}_1 B \\ \mathbf{r}_2 B \\ \vdots \\ \mathbf{r}_m B \end{bmatrix}$ follows by Definition 3.2.13. ■

■ **Example 3.2.15**

Continue the study of block arithmetic started in Example 3.2.11. Multiplying block matrices follows the same pattern as Definition 3.2.13, but for this to work, the corresponding blocks must be of the right size. Using the notation of Example 3.2.11,

$$AB = \begin{bmatrix} A_{11} & A_{12} \\ A_{21} & A_{22} \end{bmatrix} \begin{bmatrix} B_{11} & B_{12} \\ B_{21} & B_{22} \end{bmatrix} = \begin{bmatrix} A_{11}B_{11} + A_{12}B_{21} & A_{11}B_{12} + A_{12}B_{22} \\ A_{21}B_{11} + A_{22}B_{21} & A_{21}B_{12} + A_{22}B_{22} \end{bmatrix},$$

provided that all of the block multiplications are legal. For instance, the multiplication,

$$\begin{bmatrix} 3 & 1 & 0 & 2 \\ 1 & 3 & 5 & 2 \\ \hline 2 & 4 & 1 & 3 \\ 0 & 3 & 0 & 5 \end{bmatrix} \begin{bmatrix} 1 & 2 & 4 & 2 \\ 0 & 6 & 8 & 1 \\ \hline 3 & 2 & 4 & 3 \\ 0 & 6 & 8 & 4 \end{bmatrix}, \tag{3.16}$$

can be done using blocks by computing

$$\begin{bmatrix} 3 & 1 \\ 1 & 3 \end{bmatrix}\begin{bmatrix} 1 & 2 \\ 0 & 6 \end{bmatrix} + \begin{bmatrix} 0 & 2 \\ 5 & 2 \end{bmatrix}\begin{bmatrix} 3 & 2 \\ 0 & 6 \end{bmatrix} = \begin{bmatrix} 3 & 12 \\ 1 & 20 \end{bmatrix} + \begin{bmatrix} 0 & 12 \\ 15 & 22 \end{bmatrix} = \begin{bmatrix} 3 & 24 \\ 16 & 42 \end{bmatrix},$$

$$\begin{bmatrix} 3 & 1 \\ 1 & 3 \end{bmatrix}\begin{bmatrix} 4 & 2 \\ 8 & 1 \end{bmatrix} + \begin{bmatrix} 0 & 2 \\ 5 & 2 \end{bmatrix}\begin{bmatrix} 4 & 3 \\ 8 & 4 \end{bmatrix} = \begin{bmatrix} 20 & 7 \\ 28 & 5 \end{bmatrix} + \begin{bmatrix} 16 & 8 \\ 36 & 23 \end{bmatrix} = \begin{bmatrix} 36 & 15 \\ 64 & 28 \end{bmatrix},$$

$$\begin{bmatrix} 2 & 4 \\ 0 & 3 \end{bmatrix}\begin{bmatrix} 1 & 2 \\ 0 & 6 \end{bmatrix} + \begin{bmatrix} 1 & 3 \\ 0 & 5 \end{bmatrix}\begin{bmatrix} 3 & 2 \\ 0 & 6 \end{bmatrix} = \begin{bmatrix} 2 & 28 \\ 0 & 18 \end{bmatrix} + \begin{bmatrix} 3 & 20 \\ 0 & 30 \end{bmatrix} = \begin{bmatrix} 5 & 48 \\ 0 & 48 \end{bmatrix},$$

$$\begin{bmatrix} 2 & 4 \\ 0 & 3 \end{bmatrix}\begin{bmatrix} 4 & 2 \\ 8 & 1 \end{bmatrix} + \begin{bmatrix} 1 & 3 \\ 0 & 5 \end{bmatrix}\begin{bmatrix} 4 & 3 \\ 8 & 4 \end{bmatrix} = \begin{bmatrix} 40 & 8 \\ 24 & 3 \end{bmatrix} + \begin{bmatrix} 28 & 15 \\ 40 & 20 \end{bmatrix} = \begin{bmatrix} 68 & 23 \\ 64 & 23 \end{bmatrix},$$

so the product (3.16) equals

$$\left[\begin{array}{cc|cc} 3 & 24 & 36 & 15 \\ 16 & 42 & 64 & 28 \\ \hline 5 & 48 & 68 & 23 \\ 0 & 48 & 64 & 23 \end{array}\right].$$

Another example is

$$\left[\begin{array}{cc|cc} 1 & 3 & 5 & 2 \\ \hline 2 & 4 & 1 & 3 \\ 0 & 3 & 0 & 5 \end{array}\right]\left[\begin{array}{c|c} 2 & 1 \\ 6 & 4 \\ \hline 3 & 4 \\ 5 & 8 \end{array}\right] = \left[\begin{array}{c|c} 46 & 49 \\ \hline 45 & 46 \\ 43 & 52 \end{array}\right].$$

■ Example 3.2.16

The ability to multiply two matrices is dependent on the sizes of the two matrices. In particular, if A is a 7×3 matrix and B is a 3×5 matrix, AB is a 7×5 matrix, but BA is not defined. This means that matrix multiplication is not commutative in general. Moreover,

$$\begin{bmatrix} 1 & 2 \\ 3 & 4 \end{bmatrix}\begin{bmatrix} 5 & 6 \\ 7 & 8 \end{bmatrix} = \begin{bmatrix} 19 & 22 \\ 43 & 50 \end{bmatrix},$$

but

$$\begin{bmatrix} 5 & 6 \\ 7 & 8 \end{bmatrix}\begin{bmatrix} 1 & 2 \\ 3 & 4 \end{bmatrix} = \begin{bmatrix} 23 & 34 \\ 31 & 46 \end{bmatrix},$$

so even for $n \times n$ matrices there is no guarantee that AB will equal BA. However, there are times when matrix multiplication does commute. One such example is when multiplying a matrix by itself. In this case, it is common to use exponents to represent the multiplication. The notation follows the same conventions as exponents on real numbers. For instance,

$$\begin{bmatrix} 1 & 3 \\ 0 & -2 \end{bmatrix}^3 = \begin{bmatrix} 1 & 3 \\ 0 & -2 \end{bmatrix}\begin{bmatrix} 1 & 3 \\ 0 & -2 \end{bmatrix}\begin{bmatrix} 1 & 3 \\ 0 & -2 \end{bmatrix} = \begin{bmatrix} 1 & -3 \\ 0 & 4 \end{bmatrix}\begin{bmatrix} 1 & 3 \\ 0 & -2 \end{bmatrix} = \begin{bmatrix} 1 & 9 \\ 0 & -8 \end{bmatrix}.$$

In general, letting k and l represent positive integers,

$$A^k = \underbrace{A \cdot A \cdots A}_{k \text{ times}}, \tag{3.17}$$

from which $A^k A^l = A^{k+l} = A^{l+k} = A^l A^k$ follows because

$$A^k A^l = \underbrace{A \cdot A \cdots A}_{k \text{ times}} \underbrace{A \cdot A \cdots A}_{l \text{ times}} = A^{k+l}.$$

Although matrix multiplication does not commute, it does have a very fundamental property that, if missing, would make matrix arithmetic very difficult.

■ **Theorem 3.2.17**

Matrix multiplication is associative. That is, $ABC = A(BC)$ for all $m \times n$ matrices A, all $n \times r$ matrices B, and all $r \times s$ matrices C.

Proof

Let $A = [a_{ij}]$ be an $m \times n$ matrix, $B = [b_{ij}]$ be an $n \times r$ matrix, and $C = [c_{ij}]$ be an $r \times s$ matrix. Compute ABC step-by-step. First, multiply A and B to obtain an $m \times r$ matrix,

$$\begin{bmatrix} a_{11} & a_{12} & \cdots & a_{1n} \\ a_{21} & a_{22} & \cdots & a_{2n} \\ \vdots & \vdots & & \vdots \\ a_{m1} & a_{m2} & \cdots & a_{mn} \end{bmatrix} \begin{bmatrix} b_{11} & b_{12} & \cdots & b_{1r} \\ b_{21} & b_{22} & \cdots & b_{2r} \\ \vdots & \vdots & & \vdots \\ b_{n1} & b_{n2} & \cdots & b_{nr} \end{bmatrix} = \begin{bmatrix} d_{11} & d_{12} & \cdots & d_{1r} \\ d_{21} & d_{22} & \cdots & d_{2r} \\ \vdots & \vdots & & \vdots \\ d_{m1} & d_{m2} & \cdots & d_{mr} \end{bmatrix},$$

where for $i = 1, 2, \ldots, m$ and $j = 1, 2, \ldots, r$,

$$d_{ij} = a_{i1}b_{1j} + a_{i2}b_{2j} + \cdots + a_{in}b_{nj}.$$

Second, multiply AB by C to obtain an $m \times s$ matrix,

$$\begin{bmatrix} d_{11} & d_{12} & \cdots & d_{1r} \\ d_{21} & d_{22} & \cdots & d_{2r} \\ \vdots & \vdots & & \vdots \\ d_{m1} & d_{m2} & \cdots & d_{mr} \end{bmatrix} \begin{bmatrix} c_{11} & c_{12} & \cdots & c_{1s} \\ c_{21} & c_{22} & \cdots & c_{2s} \\ \vdots & \vdots & & \vdots \\ c_{r1} & c_{r2} & \cdots & c_{rs} \end{bmatrix} = \begin{bmatrix} p_{11} & p_{12} & \cdots & p_{1s} \\ p_{21} & p_{22} & \cdots & p_{2s} \\ \vdots & \vdots & & \vdots \\ p_{m1} & p_{m2} & \cdots & p_{ms} \end{bmatrix},$$

where for $i = 1, 2, \ldots, m$ and $j = 1, 2, \ldots, s$,

$$\begin{aligned} p_{ij} &= d_{i1}c_{1j} + d_{i2}c_{2j} + \cdots + d_{ir}c_{rj} \\ &= (a_{i1}b_{11} + a_{i2}b_{21} + \cdots + a_{in}b_{n1})c_{1j} \\ &\quad + (a_{i1}b_{12} + a_{i2}b_{22} + \cdots + a_{in}b_{n2})c_{2j} \\ &\quad \vdots \\ &\quad + (a_{i1}b_{1r} + a_{i2}b_{2r} + \cdots + a_{in}b_{nr})c_{rj}. \end{aligned}$$

Lastly, gather all terms with a common a_{ij} factor and write

$$\begin{aligned}
p_{ij} = \; & a_{i1}(b_{11}c_{1j} + b_{12}c_{2j} + \cdots + b_{1r}c_{rj}) \\
& + a_{i2}(b_{21}c_{1j} + b_{22}c_{2j} + \cdots + b_{2r}c_{rj}) \\
& \;\; \vdots \\
& + a_{in}(b_{n1}c_{1r} + b_{n2}c_{2j} + \cdots + b_{nr}c_{rj}).
\end{aligned}$$

Next, compute $A(BC)$ step-by-step. First, multiply B and C to obtain an $n \times s$ matrix,

$$
\begin{bmatrix}
b_{11} & b_{12} & \cdots & b_{1r} \\
b_{21} & b_{22} & \cdots & b_{2r} \\
\vdots & \vdots & & \vdots \\
b_{n1} & b_{n2} & \cdots & b_{nr}
\end{bmatrix}
\begin{bmatrix}
c_{11} & c_{12} & \cdots & c_{1s} \\
c_{21} & c_{22} & \cdots & c_{2s} \\
\vdots & \vdots & & \vdots \\
c_{r1} & c_{r2} & \cdots & c_{rs}
\end{bmatrix}
=
\begin{bmatrix}
e_{11} & e_{12} & \cdots & e_{1r} \\
e_{21} & e_{22} & \cdots & e_{2r} \\
\vdots & \vdots & & \vdots \\
e_{n1} & e_{n2} & \cdots & e_{ns}
\end{bmatrix},
$$

where for $i = 1, 2, \ldots, n$ and $j = 1, 2, \ldots, s$,

$$e_{ij} = b_{i1}c_{1j} + b_{i2}c_{2j} + \cdots + b_{ir}c_{rj}.$$

Second, multiply A by BC to obtain an $m \times s$ matrix,

$$
\begin{bmatrix}
a_{11} & a_{12} & \cdots & a_{1n} \\
a_{21} & a_{22} & \cdots & a_{2n} \\
\vdots & \vdots & & \vdots \\
a_{m1} & a_{m2} & \cdots & a_{mn}
\end{bmatrix}
\begin{bmatrix}
e_{11} & e_{12} & \cdots & e_{1r} \\
e_{21} & e_{22} & \cdots & e_{2r} \\
\vdots & \vdots & & \vdots \\
e_{n1} & e_{n2} & \cdots & e_{ns}
\end{bmatrix}
=
\begin{bmatrix}
q_{11} & q_{12} & \cdots & q_{1s} \\
q_{21} & q_{22} & \cdots & q_{2s} \\
\vdots & \vdots & & \vdots \\
q_{m1} & q_{m2} & \cdots & q_{ms}
\end{bmatrix},
$$

where for $i = 1, 2, \ldots, m$ and $j = 1, 2, \ldots, s$,

$$\begin{aligned}
q_{ij} = \; & a_{i1}e_{1j} + a_{i2}e_{2j} + \cdots + a_{in}e_{nj} \\
= \; & a_{i1}(b_{11}c_{1j} + b_{12}c_{2j} + \cdots + b_{1r}c_{rj}) \\
& + a_{i2}(b_{21}c_{1j} + b_{22}c_{2j} + \cdots + b_{2r}c_{rj}) \\
& \;\; \vdots \\
& + a_{in}(b_{n1}c_{1r} + b_{n2}c_{2j} + \cdots + b_{nr}c_{rj}),
\end{aligned}$$

which implies that $p_{ij} = q_{ij}$. Therefore, $ABC = A(BC)$. $\blacksquare$

Using Definition 3.2.13 with T_1 and T_2 defined as in (3.13), the matrix product $[T_1][T_2]$ is

$$
\begin{bmatrix}
1 & 1 & 1 & 1 \\
1 & -2 & 3 & -4 \\
4 & 0 & 0 & 3
\end{bmatrix}
\begin{bmatrix}
2 & 1 \\
0 & 0 \\
-1 & 1 \\
3 & -2
\end{bmatrix}
=
\begin{bmatrix}
4 & 0 \\
-13 & 12 \\
17 & -2
\end{bmatrix},
$$

and this is the standard vector for $T_1 \circ T_2$.

■ **Theorem 3.2.18**

If $T_1 : \mathbb{R}^m \to \mathbb{R}^s$ and $T_2 : \mathbb{R}^n \to \mathbb{R}^m$ are linear transformations,

$$[T_1 \circ T_2] = [T_1][T_2].$$

Proof

First note that $[T_1]$ is an $s \times m$ matrix and $[T_2]$ is an $m \times n$ matrix. This implies that $[T_1][T_2]$ is an $s \times n$ matrix, which is the size of $[T_1 \circ T_2]$. Let $\mathbf{u} \in \mathbb{R}^m$. Because matrix multiplication is associative (Theorem 3.2.17),

$$[T_1][T_2]\mathbf{u} = [T_1]([T_2]\mathbf{u}) = [T_1](T_2\mathbf{u}) = T_1(T_2\mathbf{u}) = (T_1 \circ T_2)\mathbf{u}.$$

It follows that $[T_1 \circ T_2] = [T_1][T_2]$ because the standard matrix is unique by Theorem 3.1.22. ■

Identity Matrix

An $n \times n$ matrix is known as a **square matrix**. If $A = [a_{ij}]$ is a square matrix, the entries a_{ii} of A form the **main diagonal** of A. For example, the main diagonal of the 3×3 matrix,

$$\begin{bmatrix} 1 & 2 & 3 \\ 4 & 5 & 6 \\ 7 & 8 & 9 \end{bmatrix},$$

has entries 1, 5, and 9. The square matrices serve as the standard matrices of linear transformations of the form $\mathbb{R}^n \to \mathbb{R}^n$. These matrices play an important role in linear algebra.

The additive identity of the $n \times n$ matrices is the $n \times n$ zero matrix. There is also a multiplicative identity. In the 3×3 case,

$$\begin{bmatrix} 1 & 0 & 0 \\ 0 & 1 & 0 \\ 0 & 0 & 1 \end{bmatrix} \begin{bmatrix} 1 & 2 & 3 \\ 4 & 5 & 6 \\ 7 & 8 & 9 \end{bmatrix} = \begin{bmatrix} 1 & 2 & 3 \\ 4 & 5 & 6 \\ 7 & 8 & 9 \end{bmatrix},$$

and

$$\begin{bmatrix} 1 & 2 & 3 \\ 4 & 5 & 6 \\ 7 & 8 & 9 \end{bmatrix} \begin{bmatrix} 1 & 0 & 0 \\ 0 & 1 & 0 \\ 0 & 0 & 1 \end{bmatrix} = \begin{bmatrix} 1 & 2 & 3 \\ 4 & 5 & 6 \\ 7 & 8 & 9 \end{bmatrix}.$$

The matrix,

$$\begin{bmatrix} 1 & 0 & 0 \\ 0 & 1 & 0 \\ 0 & 0 & 1 \end{bmatrix},$$

appears to behave as the multiplicative identity, but this will need to be checked.

■ **Definition 3.2.19**

The $n \times n$ matrix with each entry of the main diagonal equal to 1 and every other entry equal to 0,

$$I_n = \begin{bmatrix} 1 & 0 & 0 & \cdots & 0 \\ 0 & 1 & 0 & \cdots & 0 \\ 0 & 0 & 1 & \cdots & 0 \\ \vdots & \vdots & \vdots & & \vdots \\ 0 & 0 & 0 & \cdots & 1 \end{bmatrix},$$

is called an **identity matrix**. That is, $I_n = [a_{ij}]$, where for all $i, j = 1, 2, \dots, n$, $a_{ii} = 1$ but $a_{ij} = 0$ if $i \neq j$. Moreover, letting $\mathbf{e}_1, \mathbf{e}_2, \dots, \mathbf{e}_n$ be the standard vectors of $\mathbb{R}^n$, the identity matrix can be written as

$$I_n = \begin{bmatrix} \mathbf{e}_1^T \\ \mathbf{e}_2^T \\ \vdots \\ \mathbf{e}_n^T \end{bmatrix} = \begin{bmatrix} \mathbf{e}_1 & \mathbf{e}_2 & \cdots & \mathbf{e}_n \end{bmatrix}.$$

To confirm that I_n is the multiplicative identity of $\mathbb{R}^{n \times n}$ for all positive integers n, start by multiplying the identity matrix by columns and rows.

■ **Lemma 3.2.20**

Let $\mathbf{u} \in \mathbb{R}^n$.

    (a) $I_n \mathbf{u} = \mathbf{u}$.

    (b) $\mathbf{u}^T I_n = \mathbf{u}^T$.

Proof

Let $\mathbf{u}$ have coordinates $u_1, u_2, \dots, u_n$. Observe that by the usual result illustrated in (2.28),

$$I_n \begin{bmatrix} u_1 \\ u_2 \\ \vdots \\ u_n \end{bmatrix} = \begin{bmatrix} \mathbf{e}_1 \cdot \mathbf{u} \\ \mathbf{e}_2 \cdot \mathbf{u} \\ \vdots \\ \mathbf{e}_n \cdot \mathbf{u} \end{bmatrix} = \begin{bmatrix} u_1 \\ u_2 \\ \vdots \\ u_n \end{bmatrix},$$

and

$$\begin{bmatrix} u_1 & u_2 & \cdots & u_n \end{bmatrix} I_n = \begin{bmatrix} \mathbf{u} \cdot \mathbf{e}_1 & \mathbf{u} \cdot \mathbf{e}_2 & \cdots & \mathbf{u} \cdot \mathbf{e}_n \end{bmatrix} = \begin{bmatrix} u_1 & u_2 & \cdots & u_n \end{bmatrix}. \blacksquare$$

Lemma 3.2.20 is used to show that I_n is a **left multiplicative identity** for all $n \times m$ matrices and a **right multiplicative identity** for all $m \times n$ matrices. For example,

$$\begin{bmatrix} 1 & 0 \\ 0 & 1 \end{bmatrix} \begin{bmatrix} 1 & 2 & 3 & 4 \\ 5 & 6 & 6 & 8 \end{bmatrix} = \begin{bmatrix} 1 & 2 & 3 & 4 \\ 5 & 6 & 6 & 8 \end{bmatrix} \quad \text{and} \quad \begin{bmatrix} 1 & 2 \\ 3 & 4 \\ 5 & 6 \\ 7 & 8 \end{bmatrix} \begin{bmatrix} 1 & 0 \\ 0 & 1 \end{bmatrix} = \begin{bmatrix} 1 & 2 \\ 3 & 4 \\ 5 & 6 \\ 7 & 8 \end{bmatrix}.$$

■ **Theorem 3.2.21**

$I_n A = A$ and $BI_n = B$ for all $n \times m$ matrices A and $m \times n$ matrices B.

Proof

Let $\mathbf{c}_i$ be the columns of A and $\mathbf{r}_i$ be the rows of B, where $i = 1, 2, \ldots, m$. Then, by Lemma 3.2.20(a),

$$I_n A = \begin{bmatrix} I_n \mathbf{c}_1 & I_n \mathbf{c}_2 & \cdots & I_n \mathbf{c}_n \end{bmatrix} = \begin{bmatrix} \mathbf{c}_1 & \mathbf{c}_2 & \cdots & \mathbf{c}_n \end{bmatrix} = A,$$

and by Lemma 3.2.20(b),

$$BI_n = \begin{bmatrix} \mathbf{r}_1 I_n \\ \mathbf{r}_2 I_n \\ \vdots \\ \mathbf{r}_n I_n \end{bmatrix} = \begin{bmatrix} \mathbf{r}_1 \\ \mathbf{r}_2 \\ \vdots \\ \mathbf{r}_n \end{bmatrix} = B. \ ∎$$

Of course, a matrix that is both a left multiplicative identity and a right multiplicative identity is simply a multiplicative identity.

■ **Corollary 3.2.22**

I_n is the multiplicative identity of $\mathbb{R}^{n \times n}$.

■ **Example 3.2.23**

Among real numbers, 0 and 1 are the only numbers that are **idempotents**, elements a such that $a^2 = a$. Notice that the square zero matrices and identity matrices are idempotents. For instance,

$$\begin{bmatrix} 0 & 0 \\ 0 & 0 \end{bmatrix}^2 = \begin{bmatrix} 0 & 0 \\ 0 & 0 \end{bmatrix} \quad \text{and} \quad \begin{bmatrix} 1 & 0 \\ 0 & 1 \end{bmatrix}^2 = \begin{bmatrix} 1 & 0 \\ 0 & 1 \end{bmatrix}.$$

However, unlike $\mathbb{R}$, there are other matrices in $\mathbb{R}^{n \times n}$ besides the additive and multiplicative identities that are idempotents. For example,

$$\begin{bmatrix} 1 & 0 \\ 0 & 0 \end{bmatrix}^2 = \begin{bmatrix} 1 & 0 \\ 0 & 0 \end{bmatrix} \quad \text{and} \quad \begin{bmatrix} 1/2 & 1/2 \\ 1/2 & 1/2 \end{bmatrix}^2 = \begin{bmatrix} 1/2 & 1/2 \\ 1/2 & 1/2 \end{bmatrix}.$$

As noted in Example 1.4.4, $i_{\mathbb{R}^n}$ serves as the identity with respect to composition on $\mathbb{R}^n \mathbb{R}^n$. Now restrict composition to linear transformations. Since $L(\mathbb{R}^n, \mathbb{R}^n)$ is closed under composition (Theorem 3.2.12) and the product of linear functions is not guaranteed to be a linear function, consider composition to be the multiplication of $L(\mathbb{R}^n, \mathbb{R}^n)$. With this understanding, $i_{\mathbb{R}^n}$ is the multiplicative identity of $L(\mathbb{R}^n, \mathbb{R}^n)$. Furthermore, because for all $\mathbf{x} \in \mathbb{R}^n$, $i_{\mathbb{R}^n}\mathbf{x} = \mathbf{x} = I_n\mathbf{x}$, conclude that $[i_{\mathbb{R}^n}] = I_n$. That is, that standard matrix of the multiplicative identity of $L(\mathbb{R}^n, \mathbb{R}^n)$ is the multiplicative identity of $\mathbb{R}^{n \times n}$.

Distributive Law

Now that there is a matrix addition and a matrix multiplication, it is important that they relate to each other. This is the purpose of the distributive law. Since matrix multiplication is not commutative, two versions are needed.

■ **Theorem 3.2.24 [Distributive Law]**

(a) $A(B + C) = AB + AC$ for all $m \times n$ matrices A and $n \times s$ matrices B, C.

(b) $(A + B)C = AC + BC$ for all $m \times n$ matrices A, B and $n \times s$ matrices C.

Proof

The proof of (b) is left to Exercise 30. To prove (a), let A be an $m \times n$ matrix. Let $\mathbf{b}_i, \mathbf{c}_i \in \mathbb{R}^n$ for $i = 1, 2, \ldots, s$ be the columns of the $n \times s$ matrices B and C, respectively. This implies that the columns of $B + C$ are $\mathbf{b}_i + \mathbf{c}_i$. Therefore, Theorem 3.1.18(a) implies

$$
\begin{aligned}
A(B + C) &= \begin{bmatrix} A(\mathbf{b}_1 + \mathbf{c}_1) & A(\mathbf{b}_2 + \mathbf{c}_2) & \cdots & A(\mathbf{b}_s + \mathbf{c}_s) \end{bmatrix} \\
&= \begin{bmatrix} A\mathbf{b}_1 + A\mathbf{c}_1 & A\mathbf{b}_2 + A\mathbf{c}_2 & \cdots & A\mathbf{b}_s + A\mathbf{c}_s \end{bmatrix} \\
&= \begin{bmatrix} A\mathbf{b}_1 & A\mathbf{b}_2 & \cdots & A\mathbf{b}_s \end{bmatrix} + \begin{bmatrix} A\mathbf{c}_1 & A\mathbf{c}_2 & \cdots & A\mathbf{c}_s \end{bmatrix} = AB + AC. \ \blacksquare
\end{aligned}
$$

For example,

$$
\begin{bmatrix} 1 & 2 \\ 3 & 4 \end{bmatrix} \left(\begin{bmatrix} 5 & 6 & 7 \\ 8 & 9 & 0 \end{bmatrix} + \begin{bmatrix} 1 & 2 & 3 \\ 4 & 5 & 6 \end{bmatrix} \right) = \begin{bmatrix} 1 & 2 \\ 3 & 4 \end{bmatrix} \begin{bmatrix} 6 & 8 & 10 \\ 12 & 14 & 6 \end{bmatrix} = \begin{bmatrix} 30 & 36 & 22 \\ 66 & 80 & 54 \end{bmatrix}.
$$

Using the Distributive Law [Theorem 3.2.24(a)] with the same matrices yields the same result,

$$
\begin{aligned}
\begin{bmatrix} 1 & 2 \\ 3 & 4 \end{bmatrix} \left(\begin{bmatrix} 5 & 6 & 7 \\ 8 & 9 & 0 \end{bmatrix} + \begin{bmatrix} 1 & 2 & 3 \\ 4 & 5 & 6 \end{bmatrix} \right) &= \begin{bmatrix} 1 & 2 \\ 3 & 4 \end{bmatrix} \begin{bmatrix} 5 & 6 & 7 \\ 8 & 9 & 0 \end{bmatrix} + \begin{bmatrix} 1 & 2 \\ 3 & 4 \end{bmatrix} \begin{bmatrix} 1 & 2 & 3 \\ 4 & 5 & 6 \end{bmatrix} \\
&= \begin{bmatrix} 21 & 24 & 7 \\ 47 & 54 & 21 \end{bmatrix} + \begin{bmatrix} 9 & 12 & 15 \\ 19 & 26 & 33 \end{bmatrix} \\
&= \begin{bmatrix} 30 & 36 & 22 \\ 66 & 80 & 54 \end{bmatrix}.
\end{aligned}
$$

Matrices and Polynomials

Because matrices can be added and subtracted, multiplied by a real number, and raised to powers when they are square, it is natural to define what it means to substitute a square matrix into a polynomial. Recall that $p(x)$ is a **polynomial with real coefficients** means that there exists $a_i \in \mathbb{R}$ for $i = 0, 1, \ldots, k$ such that

$$
p(x) = a_0 + a_1 x + \cdots + a_n x^k.
$$

The substitution is then defined as expected.

■ **Definition 3.2.25**

Let A be an $n \times n$ matrix and $p(x) = a_0 + a_1 x + \cdots + a_k x^k$. Define

$$p(A) = a_0 I_n + a_1 A + \cdots + a_k A^k.$$

To illustrate, let $A = \begin{bmatrix} 1 & -2 \\ 0 & 3 \end{bmatrix}$ and $p(x) = 7 - 6x + 3x^2$. Then,

$$p(A) = \begin{bmatrix} 7 & 0 \\ 0 & 7 \end{bmatrix} - 6 \begin{bmatrix} 1 & -2 \\ 0 & 3 \end{bmatrix} + 3 \begin{bmatrix} 1 & -2 \\ 0 & 3 \end{bmatrix}^2 = \begin{bmatrix} 4 & -12 \\ 0 & -16 \end{bmatrix}.$$

It is advantageous to use polynomials with matrices because it is a means to simplify linear combinations of powers of matrices. Although it might not seem to be the case at first, substitution of matrices into polynomials behaves similarly to substitution of real numbers as the next theorem shows.

■ **Theorem 3.2.26**

Let A be an $n \times n$ matrix. Let $f(x)$ and $g(x)$ be polynomials with real coefficients and $r \in \mathbb{R}$.

(a) $f(A)A = Af(A)$.

(b) $f(rA) = rf(A)$.

(c) $(f \pm g)(A) = f(A) \pm g(A)$.

(d) $(fg)(A) = f(A)g(A)$.

Proof

Let $f(x) = a_0 + a_1 x + \cdots + a_i x^k$ and $g(x) = b_0 + b_1 x + \cdots + b_i x^k$ be polynomials with coefficients from $\mathbb{R}$. A generalization of Theorem 3.1.18(b) proves (a) because

$$\begin{aligned} f(A)A &= (a_0 I + a_1 A + \cdots + a_k A^k)A \\ &= a_0 A + a_1 A^2 + \cdots + a_k A^{k+1} \\ &= A(a_0 I) + A(a_1 A) + \cdots + A(a_k A^k) \\ &= A(a_0 I + a_1 A + \cdots + a_k A^k) = Af(A). \end{aligned}$$

For (c), because scalar multiplication on matrices distributes and matrix addition is both commutative and associative,

$$\begin{aligned} (f \pm g)(A) &= (a_0 \pm b_0)I + (a_1 \pm b_1)A + \cdots + (a_k \pm b_k)A^k \\ &= (a_0 I \pm b_0 I) + (a_1 A \pm b_1 A) + \cdots + (a_k A^k \pm b_k A^k) \\ &= (a_0 I + a_1 A + \cdots + a_k A^k) \pm (b_0 I + b_1 A + \cdots + b_k A^k) \\ &= f(A) \pm g(A). \end{aligned}$$

Parts (b) and (d) are left to Exercise 36. ■

Exercises

1. Let $A = \begin{bmatrix} 1 & 2 & 3 \\ 4 & 5 & 6 \\ 7 & 8 & 9 \end{bmatrix}$, $B = \begin{bmatrix} 2 & -4 & 6 \\ -5 & -3 & 1 \\ 1 & 0 & 2 \end{bmatrix}$, and $C = \begin{bmatrix} 1 & 4 \\ -2 & 0 \\ 5 & 1 \\ 0 & 3 \end{bmatrix}$. Compute if possible.

    (a) $A + B$
    (b) $B - A$
    (c) $B + C$
    (d) $7C$
    (e) $A(7 + B)$
    (f) $C(3A + B)$
    (g) $3A + 5B - 7A$
    (h) $A3 + 5B$
    (i) $-2B - 3A$

2. Let A be an 8×3 matrix and B be a 3×5 matrix. If defined, what is the size of AB and what is the size of BA?

3. If AB is a 7×4 matrix, how many columns does B have?

4. Let $A = \begin{bmatrix} 1 & 2 & 0 & -1 \\ -2 & 3 & 1 & 2 \\ 4 & 2 & 0 & 2 \\ -2 & 4 & 0 & 1 \end{bmatrix}$, $B = \begin{bmatrix} 2 & 1 \\ 1 & -3 \\ 1 & -2 \\ 4 & 5 \end{bmatrix}$, and $C = \begin{bmatrix} 0 & -1 & 0 & 3 \\ 1 & 1 & -2 & 4 \end{bmatrix}$. Compute if possible.

    (a) AB
    (b) AC
    (c) BC
    (d) ABC
    (e) BAC
    (f) CAB

5. Find $x, y \in \mathbb{R}$ so that $AB - BA = B^2$, where

    $$A = \begin{bmatrix} 0 & 0 & 0 \\ 0 & 0 & 1 \\ 0 & 0 & 0 \end{bmatrix} \quad \text{and} \quad B = \begin{bmatrix} x-1 & 0 & y+1 \\ 2 & 0 & 0 \\ 2 & 0 & 3 \end{bmatrix}.$$

6. Define $T_1 : \mathbb{R}^3 \to \mathbb{R}^3$ and $T_2 : \mathbb{R}^3 \to \mathbb{R}^3$ by

    $$T_1 \begin{bmatrix} x \\ y \\ z \end{bmatrix} = \begin{bmatrix} x+y+z \\ x-2y+3z \\ 4x+3y \end{bmatrix} \quad \text{and} \quad T_2 \begin{bmatrix} x \\ y \\ z \end{bmatrix} = \begin{bmatrix} 2x+y \\ z+y-x \\ 3x-2y-z \end{bmatrix}.$$

    (a) Prove that T_1 and T_2 are linear transformations.
    (b) Find the standard matrices for $T_1 + T_2$, $T_1 - T_2$, $T_1 \circ T_2$, $T_2 \circ T_1$, $4T_1$, $3T_1 + 6T_2$, $(5T_1 - T_2) \circ (2T_1 + 7T_2)$, and $(T_1 \circ 3T_2) + (T_2 \circ 4T_2)$

7. Finish the proof of Theorem 3.2.3.

8. Prove the remaining parts of Theorem 3.2.6.

9. Finish the proof of Theorem 3.2.9.

10. Suppose that T_1, T_2, T_3, and T_4 are linear transformations $\mathbb{R}^n \to \mathbb{R}^m$. Let $r, s \in \mathbb{R}$. Prove or show false.

    (a) $T_1 + (T_2 + T_3) + T_4 = (T_3 + T_2) + (T_4 + T_1)$
    (b) $T_1 + 0_{nm} + T_2 = T_2 + T_1$

(c) $rs(T_1 + T_2) = rT_1 + sT_2$

(d) $r(T_1 + T_2 + T_3) = rT_1 + rT_2 + rT_3$

(e) $(r + s)(T_1 - T_2) = rT_1 + sT_1 - rT_2 - sT_2$

11. Suppose that A, B, C, and D are $m \times n$ matrices. Let $r, s \in \mathbb{R}$. Use Definition 3.2.5 to prove or show false.

   (a) $A + (B + C) + D = (C + B) + (D + A)$

   (b) $A + 0_{mn} + B = B + A$

   (c) $rs(A + B) = rA + sB$

   (d) $r(A + B + C) = rA + rB + rC$

   (e) $(r + s)(A - B) = rA + sA - rB - sB$

12. Explain how to answer Exercise 10 using the standard matrix of a linear transformation.

13. Let $f : \mathbb{R} \to \mathbb{R}$ be defined by $f(x) = 5x + b$. Prove that $f \in L(\mathbb{R}, \mathbb{R})$ if and only if $b = 0$.

14. Define $\mathbb{C}^{m \times n}$ to be the set of $n \times m$ matrices with complex entries. Prove.

   (a) $\mathbb{R}^{m \times n} \subset \mathbb{C}^{m \times n}$.

   (b) Theorem 3.2.10 holds true if $\mathbb{R}^{m \times n}$ is replaced with $\mathbb{C}^{m \times n}$.

15. Complete the proof of Theorem 3.2.10.

16. Let k be a positive integer and $A_1, A_2, \dots, A_k$ be $n \times n$ matrices. Use mathematical induction to prove that $A_1 A_2 \cdots A_k = A_1(A_2 \cdots A_k)$.

17. Let $T : \mathbb{R}^2 \to \mathbb{R}^2$ be the linear transformation defined by $T \begin{bmatrix} x \\ y \end{bmatrix} = \begin{bmatrix} T_1(x) \\ T_2(y) \end{bmatrix}$, where $T_1, T_2 : \mathbb{R} \to \mathbb{R}$ are linear transformations. Find the standard matrix of T.

18. Let $T : \mathbb{R}^3 \to \mathbb{R}^3$ be the linear transformation with $Te_1 = e_2$, $Te_2 = e_3$, and $Te_3 = \mathbf{0}$. Show that $T \neq 0$, $T \circ T \neq 0$, but $T \circ T \circ T = 0$.

19. Find nonzero $n \times n$ matrices A, B, and C such that $AC = BC$ but $A \neq B$.

20. Find nonzero $n \times n$ matrices A and B such that $AB = 0_{nn}$.

21. Prove or show false: There exists a 2×2 matrix $A \neq I_2$ such that $AB = B$ for all 2×2 matrices B.

22. Let A and B be $n \times n$ matrices. Let $r \in \mathbb{R}$. Prove $A(rB) = rAB$.

23. Suppose that $A = \begin{bmatrix} 1 & 3 \\ 0 & -2 \end{bmatrix}$ and $AB = \begin{bmatrix} 7 & 4 \\ -2 & -3 \end{bmatrix}$. Find B.

24. Let $A = \begin{bmatrix} a & 1 \\ a^2 & a \end{bmatrix}$ with $a \in \mathbb{R}$. Find A^n and prove the formula.

25. Let $B = \begin{bmatrix} b^2 & b \\ b & 1 \end{bmatrix}$ with $b \in \mathbb{R}$. Find B^n and prove the formula.

26. Let $A = \begin{bmatrix} 1 & 3 \\ a & b \end{bmatrix}$. If possible, find $a, b \in \mathbb{R}$ so that A satisfies the given equations.

    (a) $A^2 = 0$ (b) $A^2 = -I_2$

27. Find an example of a 3×3 idempotent matrix.

28. If possible, find a 2×2 idempotent matrix with exactly one nonzero entry.

29. Let A and B be $n \times n$ idempotent matrices. Find r when $A = rB$.

30. Prove Theorem 3.2.24(b).

31. Let $A = \begin{bmatrix} A_{11} & A_{12} \\ A_{21} & A_{22} \\ A_{31} & A_{32} \end{bmatrix}$ and $B = \begin{bmatrix} B_{11} & B_{12} & B_{13} \\ B_{21} & B_{22} & B_{32} \end{bmatrix}$. Suppose each A_{ij} is a 3×5 matrix for $i = 1, 2$ and $j = 1, 2, 3$. If AB is a 9×24 matrix, what is the size of each B_{ij} with $i = 1, 2, 3$ and $j = 1, 2$?

32. Letting 0 represent a zero matrix and I represent an identity matrix, perform the indicated block arithmetic calculations. Assume that all blocks are the correct size.

    (a) $\begin{bmatrix} A_{11} & A_{12} \\ A_{21} & A_{22} \end{bmatrix} + \begin{bmatrix} B_{11} & B_{12} \\ B_{21} & B_{22} \end{bmatrix}$ (d) $\begin{bmatrix} A_{11} & A_{12} & A_{13} \\ A_{21} & A_{22} & A_{23} \end{bmatrix} \begin{bmatrix} B_{11} & B_{12} \\ B_{21} & B_{22} \\ B_{31} & B_{32} \end{bmatrix}$

    (b) $\begin{bmatrix} A_{11} & A_{12} \\ A_{21} & A_{22} \end{bmatrix} - \begin{bmatrix} 0 & A_{12} \\ A_{21} & 0 \end{bmatrix}$ (e) $\begin{bmatrix} A_{11} & 0 \\ 0 & A_{22} \end{bmatrix} \begin{bmatrix} B_{11} & 0 \\ 0 & B_{22} \end{bmatrix}$

    (c) $\begin{bmatrix} A_{11} & A_{12} \\ A_{21} & A_{22} \end{bmatrix} \begin{bmatrix} I & 0 \\ 0 & I \end{bmatrix}$ (f) $\begin{bmatrix} A_{11} & A_{12} \\ 0 & A_{22} \end{bmatrix} \begin{bmatrix} B_{11} & B_{12} \\ 0 & B_{22} \end{bmatrix}$

33. Let A be an $m \times n$ matrix and $B_1, B_2, \dots, B_k$ be $n \times s$ matrices. Prove that

$$A(B_1 + B_2 + \cdots + B_k) = AB_1 + AB_2 + \cdots + AB_k.$$

34. For $n \times n$ matrices A and B, when can $A^2 + 2AB + B^2$ be factored?

35. Let A be a 2×2 matrix and $f(x) = -5 - 4x + x^2$.

    (a) If $A = \begin{bmatrix} 3 & 4 \\ 1 & -1 \end{bmatrix}$, what is $f(A)$?

    (b) Can $-5I_2 - 4A + A^2$ be factored to find solutions to this equation? Do the factors commute?

    (c) Show that $A = \begin{bmatrix} 2 & 3a \\ 3/a & 2 \end{bmatrix}$ is a solution to $f(x) = 0$ if $a \neq 0$.

    (d) Find two solutions to $f(A) = 0$ not of the form from Exercise 35(c).

36. Complete the proofs of (b) and (d) from Theorem 3.2.26.

3.3 Linear Operators

When looking for good examples of linear transformations, restricting them to functions of the form $\mathbb{R}^2 \to \mathbb{R}^2$ or $\mathbb{R}^3 \to \mathbb{R}^3$ make the linear transformations easy to visualize. This can aid in finding their standard matrices. A linear transformation $\mathbb{R}^n \to \mathbb{R}^n$ is called a **linear operator**.

Reflections

Take a line ℓ in a plane. Let P be a point on one side of ℓ as in Figure 3.3. The point P' is the **reflection** of P through ℓ if the line segment PP' is perpendicular to ℓ and the midpoint M of PP' lies on ℓ. Using the tools of basic geometry, given a line ℓ and a point P, it is easy to construct the reflection P' of P through ℓ.

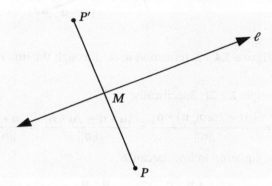

Figure 3.3   A reflection through the line ℓ.

The problem of finding a reflection can be translated to $\mathbb{R}^2$. Let $\mathbf{u} \in \mathbb{R}^2$. Interpret this vector as an arrow originating at the origin. Let L be a line containing $\mathbf{0}$ with normal vector $\mathbf{n}$. Project $\mathbf{u}$ onto $\mathbf{n}$. This is illustrated in Figure 3.4. Let

$$\mathbf{u}' = \mathbf{u} - 2\text{proj}_{\mathbf{n}}\mathbf{u} = (\mathbf{u} - \text{proj}_{\mathbf{n}}\mathbf{u}) - \text{proj}_{\mathbf{n}}\mathbf{u}. \qquad (3.18)$$

In other words, $\mathbf{u}'$ is the sum of the vector component of $\mathbf{u}$ orthogonal to $\mathbf{n}$ (Definition 2.2.18) and the vector that is the opposite of $\text{proj}_{\mathbf{n}}\mathbf{u}$. For $\mathbf{u}'$ to be the reflection of $\mathbf{u}$ through L, two statements must be true:

- The line containing $\mathbf{u}$ and $\mathbf{u}'$ is orthogonal to L.

- $\mathbf{u}$ and $\mathbf{u}'$ are equidistant to L.

The first is true because

$$\mathbf{u} - \mathbf{u}' = \mathbf{u} - (\mathbf{u} - 2\text{proj}_{\mathbf{n}}\mathbf{u}) = 2\text{proj}_{\mathbf{n}}\mathbf{u}, \qquad (3.19)$$

which is a multiple of $\mathbf{n}$. The second is seen to be true by showing that the orthogonal distance from $\mathbf{u}$ to L equals the orthogonal distance from $\mathbf{u}'$ to L using the

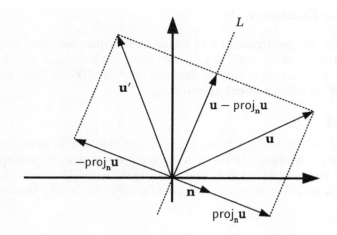

Figure 3.4 A reflection in $\mathbb{R}^2$ through the line L.

technique of Example 2.2.21. Specifically,

$$\frac{|\mathbf{u}' \cdot \mathbf{n}|}{\|\mathbf{n}\|} = \frac{|(\mathbf{u} - 2\text{proj}_\mathbf{n}\mathbf{u}) \cdot \mathbf{n}|}{\|\mathbf{n}\|} = \frac{|\mathbf{u} \cdot \mathbf{n} - 2\mathbf{u} \cdot \mathbf{n}|}{\|\mathbf{n}\|} = \frac{|\mathbf{u} \cdot \mathbf{n}|}{\|\mathbf{n}\|}, \tag{3.20}$$

where the second equation follows because

$$\text{proj}_\mathbf{n}\mathbf{u} \cdot \mathbf{n} = \left(\frac{\mathbf{u} \cdot \mathbf{n}}{\|\mathbf{n}\|^2}\mathbf{n}\right) \cdot \mathbf{n} = \frac{\mathbf{u} \cdot \mathbf{n}}{\|\mathbf{n}\|^2}(\mathbf{n} \cdot \mathbf{n}) = \mathbf{u} \cdot \mathbf{n}.$$

Both (3.19) and (3.20) imply that (3.18) can be used to find the reflection of a vector through a line. Start with the 2-dimensional case.

■ **Definition 3.3.1**

The linear operator $T : \mathbb{R}^2 \to \mathbb{R}^2$ is a **reflection** through the line containing $\mathbf{0}$ with normal $\begin{bmatrix} a \\ b \end{bmatrix}$ if for all $x, y \in \mathbb{R}$,

$$T\begin{bmatrix} x \\ y \end{bmatrix} = \begin{bmatrix} x \\ y \end{bmatrix} - \frac{ax + by}{a^2 + b^2}\begin{bmatrix} 2a \\ 2b \end{bmatrix}.$$

■ **Example 3.3.2**

Let L be the line in $\mathbb{R}^2$ given by the equation $y = x$. This means that $\mathbf{n} = \begin{bmatrix} 1 \\ -1 \end{bmatrix}$ is a normal vector to L. By Definition 3.3.1, the reflection T_1 through L is

$$T_1\begin{bmatrix} x \\ y \end{bmatrix} = \begin{bmatrix} x \\ y \end{bmatrix} - \begin{bmatrix} x - y \\ y - x \end{bmatrix} = \begin{bmatrix} y \\ x \end{bmatrix}. \tag{3.21}$$

In particular, $T_1 \begin{bmatrix} 4 \\ 1 \end{bmatrix} = \begin{bmatrix} 1 \\ 4 \end{bmatrix}$, which is the known reflection of $\begin{bmatrix} 4 \\ 1 \end{bmatrix}$ through $y = x$.

■ **Example 3.3.3**

The vector $\mathbf{n} = \begin{bmatrix} 0 \\ 1 \end{bmatrix}$ is a normal to the x-axis. Definition 3.3.1 implies that the function that describes reflections through the x-axis is

$$T_2 \begin{bmatrix} x \\ y \end{bmatrix} = \begin{bmatrix} x \\ y \end{bmatrix} - \begin{bmatrix} 0 \\ 2y \end{bmatrix} = \begin{bmatrix} x \\ -y \end{bmatrix}. \tag{3.22}$$

Similarly,

$$T_3 \begin{bmatrix} x \\ y \end{bmatrix} = \begin{bmatrix} -x \\ y \end{bmatrix} \tag{3.23}$$

is the function that describes reflections through the y-axis.

It is now required to determine if a reflection is always a linear operator. If it is, it has a standard matrix.

■ **Theorem 3.3.4**

A reflection through a line containing the zero vector is a linear operator.

Proof

Take $\mathbf{n} \in \mathbb{R}^n$. Let $T : \mathbb{R}^n \to \mathbb{R}^n$ be defined by $T\mathbf{x} = \mathbf{x} - 2\text{proj}_\mathbf{n}\mathbf{x}$. It is known that the functions $\mathbf{y} = \mathbf{x}$ and $\mathbf{y} = \text{proj}_\mathbf{n}\mathbf{x}$ are a linear operators (Example 3.1.5). Therefore, T is a linear operator by Theorem 3.2.3. ■

■ **Example 3.3.5**

Because of Theorem 3.3.4, standard matrices describe the linear operators of Examples 3.3.2 and 3.3.3. Using the operator T_1 of (3.21),

$$T_1 \begin{bmatrix} 1 \\ 0 \end{bmatrix} = \begin{bmatrix} 0 \\ 1 \end{bmatrix} \quad \text{and} \quad T_1 \begin{bmatrix} 0 \\ 1 \end{bmatrix} = \begin{bmatrix} 1 \\ 0 \end{bmatrix},$$

so the standard matrix of the reflection through $y = x$ is

$$[T_1] = \begin{bmatrix} 0 & 1 \\ 1 & 0 \end{bmatrix}. \tag{3.24}$$

Using the operator T_2 of (3.22),

$$T_2 \begin{bmatrix} 1 \\ 0 \end{bmatrix} = \begin{bmatrix} 1 \\ 0 \end{bmatrix} \quad \text{and} \quad T_2 \begin{bmatrix} 0 \\ 1 \end{bmatrix} = \begin{bmatrix} 0 \\ -1 \end{bmatrix},$$

so the standard matrix of the reflection through the x-axis is

$$[T_2] = \begin{bmatrix} 1 & 0 \\ 0 & -1 \end{bmatrix}.$$

Lastly, using the operator T_3 of (3.23),

$$T_3 \begin{bmatrix} 1 \\ 0 \end{bmatrix} = \begin{bmatrix} -1 \\ 0 \end{bmatrix} \quad \text{and} \quad T_3 \begin{bmatrix} 0 \\ 1 \end{bmatrix} = \begin{bmatrix} 0 \\ 1 \end{bmatrix},$$

so the standard matrix of the reflection through the y-axis is

$$[T_3] = \begin{bmatrix} -1 & 0 \\ 0 & 1 \end{bmatrix}.$$

Reflections can also be visualized in $\mathbb{R}^3$. Let $\mathbf{u} \in \mathbb{R}^3$ and consider the reflection of $\mathbf{u}$ through the xy-plane. Since the reflection of a vector already in the xy-plane is simply the vector itself, suppose that $\mathbf{u}$ is not in the xy-plane. Describing this reflection requires finding $\mathbf{u}'$ such that $\mathbf{u}$ and $\mathbf{u}'$ are equidistant from the xy-plane but on opposite sides as illustrated in Figure 3.5. Since $\mathbf{u}$ and $\mathbf{u}'$ lie in the same plane, this problem is equivalent to the two-dimensional case that has already been examined of reflecting through a line. This illustrates that Definition 3.3.1 can be copied for $\mathbb{R}^3$, provided certain modifications are made.

■ **Definition 3.3.6**

The linear operator $T : \mathbb{R}^3 \to \mathbb{R}^3$ is a **reflection** through the plane containing $\mathbf{0}$ with normal $\begin{bmatrix} a \\ b \\ c \end{bmatrix}$ if for all $x, y, z \in \mathbb{R}$,

$$T \begin{bmatrix} x \\ y \\ z \end{bmatrix} = \begin{bmatrix} x \\ y \\ z \end{bmatrix} - \frac{ax + by + cz}{a^2 + b^2 + c^2} \begin{bmatrix} 2a \\ 2b \\ 2c \end{bmatrix}.$$

■ **Example 3.3.7**

The vector $\mathbf{n} = \begin{bmatrix} 0 \\ 0 \\ 1 \end{bmatrix}$ is normal to the xy-plane. Let $T_4 : \mathbb{R}^3 \to \mathbb{R}^3$ be the reflection through the xy-plane. It follows from Definition 3.3.6 that

$$T_4 \begin{bmatrix} 1 \\ 2 \\ 3 \end{bmatrix} = \begin{bmatrix} 1 \\ 2 \\ 3 \end{bmatrix} - 2 \begin{bmatrix} 0 \\ 0 \\ 3 \end{bmatrix} = \begin{bmatrix} 1 \\ 2 \\ -3 \end{bmatrix}.$$

Moreover, because

$$T_4 \begin{bmatrix} 1 \\ 0 \\ 0 \end{bmatrix} = \begin{bmatrix} 1 \\ 0 \\ 0 \end{bmatrix}, \quad T_4 \begin{bmatrix} 0 \\ 1 \\ 0 \end{bmatrix} = \begin{bmatrix} 0 \\ 1 \\ 0 \end{bmatrix}, \quad T_4 \begin{bmatrix} 0 \\ 0 \\ 1 \end{bmatrix} = \begin{bmatrix} 0 \\ 0 \\ -1 \end{bmatrix},$$

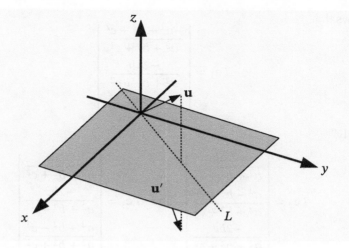

Figure 3.5 A reflection in $\mathbb{R}^3$ through the xy-plane.

the standard matrix for the reflection through the xy-plane is

$$[T_4] = \begin{bmatrix} 1 & 0 & 0 \\ 0 & 1 & 0 \\ 0 & 0 & -1 \end{bmatrix}.$$

Also, if the linear operator $T_5 : \mathbb{R}^3 \to \mathbb{R}^3$ is the reflection through the xz-plane and $T_6 : \mathbb{R}^3 \to \mathbb{R}^3$ is the reflection through the yz-plane,

$$[T_5] = \begin{bmatrix} 1 & 0 & 0 \\ 0 & -1 & 0 \\ 0 & 0 & 1 \end{bmatrix} \quad \text{and} \quad [T_6] = \begin{bmatrix} -1 & 0 & 0 \\ 0 & 1 & 0 \\ 0 & 0 & 1 \end{bmatrix}.$$

The standard matrix for the reflection in $\mathbb{R}^2$ is found in Exercise 1, and the standard matrix T that is a reflection through a plane P in $\mathbb{R}^3$ is derived as follows. Let $\mathbf{n} = \begin{bmatrix} a \\ b \\ c \end{bmatrix}$ be a normal vector of P. Because

$$T\mathbf{e}_1 = \begin{bmatrix} 1 \\ 0 \\ 0 \end{bmatrix} - \frac{2}{a^2 + b^2 + c^2} \begin{bmatrix} a^2 \\ ab \\ ac \end{bmatrix},$$

it follows that

$$T\mathbf{e}_1 = \begin{bmatrix} \dfrac{-a^2 + b^2 + c^2}{a^2 + b^2 + c^2} \\[2mm] \dfrac{-2ab}{a^2 + b^2 + c^2} \\[2mm] \dfrac{-2ac}{a^2 + b^2 + c^2} \end{bmatrix}.$$

Similarly,

$$T\mathbf{e}_2 = \begin{bmatrix} \dfrac{-2ab}{a^2 + b^2 + c^2} \\[2mm] \dfrac{a^2 - b^2 + c^2}{a^2 + b^2 + c^2} \\[2mm] \dfrac{-2bc}{a^2 + b^2 + c^2} \end{bmatrix} \quad \text{and} \quad T\mathbf{e}_3 = \begin{bmatrix} \dfrac{-2ac}{a^2 + b^2 + c^2} \\[2mm] \dfrac{-2bc}{a^2 + b^2 + c^2} \\[2mm] \dfrac{a^2 + b^2 - c^2}{a^2 + b^2 + c^2} \end{bmatrix}.$$

This implies that the standard matrix for a reflection through a plane with normal vector $\begin{bmatrix} a \\ b \\ c \end{bmatrix}$ is

$$[T] = \frac{1}{a^2 + b^2 + c^2} \begin{bmatrix} -a^2 + b^2 + c^2 & -2ab & -2ac \\ -2ab & a^2 - b^2 + c^2 & -2bc \\ -2ac & -2bc & a^2 + b^2 - c^2 \end{bmatrix}.$$

Rotations

Choose distinct points O and P in a plane. Join O and P with a segment. Choose a magnitude θ. Using O as the center of a circle, rotate the radius OP counterclockwise, turning P to P' so that $\angle POP'$ has measure θ and $OP = OP'$ (Figure 3.6). The function just described that maps P to P' is called a **rotation**.

As with reflections, this problem can be translated to $\mathbb{R}^2$. Choose $r, \theta \in \mathbb{R}$. Let $\mathbf{u}$ be the angle $\mathbb{R}^2$ such that $\|\mathbf{u}\| = r$ and the angle between $\mathbf{u}$ and $\mathbf{e}_1$ has measure α. Find the vector $\mathbf{u}'$ with the properties $\|\mathbf{u}'\| = r$ and the angle between $\mathbf{u}$ and $\mathbf{u}'$ has measure θ (Figure 3.7). This means that $\mathbf{u}'$ can be viewed as the image of $\mathbf{u}$ rotated through an angle of measure θ, which is counterclockwise if $\theta > 0$. Since the vector is viewed as an arrow originating at the origin, it is not crucial to note that the rotation is centered at the zero vector. Let $R : \mathbb{R}^2 \to \mathbb{R}^2$ be this rotation.

To find the formula for $R\mathbf{u}$, first write the polar forms for $\mathbf{u}$ and $\mathbf{u}'$,

$$\mathbf{u} = \begin{bmatrix} r\cos\alpha \\ r\sin\alpha \end{bmatrix} \quad \text{and} \quad \mathbf{u}' = \begin{bmatrix} r\cos(\theta + \alpha) \\ r\sin(\theta + \alpha) \end{bmatrix}.$$

By the addition formulas from trigonometry,

$$r\cos(\theta + \alpha) = r\cos\theta\cos\alpha - r\sin\theta\sin\alpha,$$

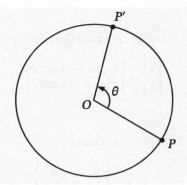

Figure 3.6 A rotation of θ centered at O.

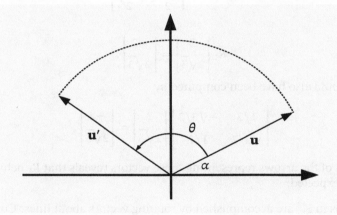

Figure 3.7 A rotation in $\mathbb{R}^2$ through θ.

and

$$r\sin(\theta + \alpha) = r\sin\theta\cos\alpha + r\cos\theta\sin\alpha.$$

Therefore,

$$\mathbf{u}' = R\mathbf{u} = \begin{bmatrix} \cos\theta & -\sin\theta \\ \sin\theta & \cos\theta \end{bmatrix} \begin{bmatrix} r\cos\alpha \\ r\sin\alpha \end{bmatrix}.$$

This shows that R is a linear operator (Theorem 3.1.19), and its standard matrix is

$$[R] = \begin{bmatrix} \cos\theta & -\sin\theta \\ \sin\theta & \cos\theta \end{bmatrix}. \tag{3.25}$$

The typical polar conversion equations of $x = r\cos\theta$ and $y = r\sin\theta$ give the definition of the rotation.

■ **Definition 3.3.8**

The linear operator $R : \mathbb{R}^2 \to \mathbb{R}^2$ is a **rotation** through θ if for all $x, y \in \mathbb{R}$,

$$R \begin{bmatrix} x \\ y \end{bmatrix} = \begin{bmatrix} x \cos \theta - y \sin \theta \\ x \sin \theta + y \cos \theta \end{bmatrix}.$$

■ **Example 3.3.9**

Let R_1 be the rotation through $\pi/3$. Then,

$$R_1 \begin{bmatrix} x \\ y \end{bmatrix} = \begin{bmatrix} \dfrac{1}{2}x - \dfrac{\sqrt{3}}{2}y \\ \dfrac{\sqrt{3}}{2}x + \dfrac{1}{2}y \end{bmatrix}.$$

Also,

$$R_1 \begin{bmatrix} 2 \\ 2\sqrt{3} \end{bmatrix} = \begin{bmatrix} -2 \\ 2\sqrt{3} \end{bmatrix},$$

which could also have been computed by

$$\begin{bmatrix} 1/2 & -\sqrt{3}/2 \\ \sqrt{3}/2 & 1/2 \end{bmatrix} \begin{bmatrix} 2 \\ 2\sqrt{3} \end{bmatrix} = \begin{bmatrix} -2 \\ 2\sqrt{3} \end{bmatrix}.$$

A sketch of the arrows representing these vectors reveals that R_1 behaves exactly as expected.

Rotations in $\mathbb{R}^3$ are accomplished by rotating vectors about lines. Consider the rotation R_2 about the z-axis through an angle of measure θ (Figure 3.8). Let $\mathbf{u} \in \mathbb{R}^3$. Observe that when the arrow representing $\mathbf{u}$ is rotated counterclockwise around the z-axis stopping at the arrow representing $\mathbf{u}'$, the terminating points of both arrows lie in a plane parallel to the xy-plane. This means that

$$\mathbf{u} = \begin{bmatrix} x \\ y \\ z \end{bmatrix} \quad \text{and} \quad \mathbf{u}' = \begin{bmatrix} x' \\ y' \\ z \end{bmatrix},$$

for some $x, x', y, y', z \in \mathbb{R}$. For this reason it is best to view this rotation from above the xy-plane (Figure 3.9), that is, from some point with a positive z-coordinate. When this is done, the problem is transformed to the two-dimensional case, leading to the equation,

$$R_2 \begin{bmatrix} x \\ y \\ z \end{bmatrix} = \begin{bmatrix} x \cos \theta - y \sin \theta \\ x \sin \theta + y \cos \theta \\ z \end{bmatrix} = \begin{bmatrix} \cos \theta & -\sin \theta & 0 \\ \sin \theta & \cos \theta & 0 \\ 0 & 0 & 1 \end{bmatrix} \begin{bmatrix} x \\ y \\ z \end{bmatrix}.$$

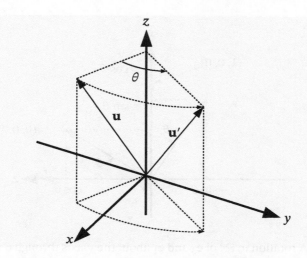

Figure 3.8 A rotation in $\mathbb{R}^3$ about the z-axis through θ.

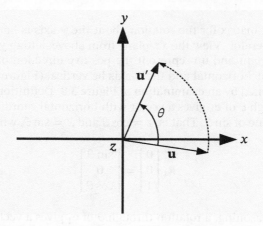

Figure 3.9 A rotation in $\mathbb{R}^3$ about the z-axis through θ viewed from above the xy-plane.

Thus, R_2 is a linear operator (Theorem 3.1.19), and its standard matrix is

$$[R_2] = \begin{bmatrix} \cos\theta & -\sin\theta & 0 \\ \sin\theta & \cos\theta & 0 \\ 0 & 0 & 1 \end{bmatrix}. \tag{3.26}$$

Using similar arguments, letting R_3 be the rotation in $\mathbb{R}^3$ about the x-axis, its standard matrix is

$$[R_3] = \begin{bmatrix} 1 & 0 & 0 \\ 0 & \cos\theta & -\sin\theta \\ 0 & \sin\theta & \cos\theta \end{bmatrix}. \tag{3.27}$$

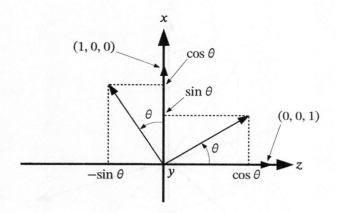

Figure 3.10 A rotation in $\mathbb{R}^3$ of $\mathbf{e}_3$ and $\mathbf{e}_1$ about the y-axis through θ viewed from above the xz-plane.

The standard matrix for the rotation about the y-axis is harder to see. Let R_4 be this linear operator. View the xz-plane from above, letting $y > 0$. Drawing the picture so that right and up represent the positive direction of the axes requires that the z-axis be horizontal and the x-axis be vertical (Figure 3.10). That this is correct is confirmed by an examination of Figure 3.8. Definition 3.3.8 implies that a rotation through θ of $\mathbf{e}_3$ gives a vector with horizontal coordinate of $\cos\theta$ and a vertical coordinate of $\sin\theta$. That is, $z = \cos\theta$ and $x = \sin\theta$, which implies that

$$R_4 \begin{bmatrix} 0 \\ 0 \\ 1 \end{bmatrix} = \begin{bmatrix} \sin\theta \\ 0 \\ \cos\theta \end{bmatrix}.$$

Using similar reasoning, a rotation through θ of $\mathbf{e}_1$ gives a vector with horizontal coordinate $-\sin\theta$ and a vertical coordinate of $\cos\theta$, so

$$R_4 \begin{bmatrix} 1 \\ 0 \\ 0 \end{bmatrix} = \begin{bmatrix} \cos\theta \\ 0 \\ -\sin\theta \end{bmatrix}.$$

Therefore, since $R_4 \begin{bmatrix} 0 \\ 1 \\ 0 \end{bmatrix} = \begin{bmatrix} 0 \\ 1 \\ 0 \end{bmatrix}$,

$$[R_4] = \begin{bmatrix} \cos\theta & 0 & \sin\theta \\ 0 & 1 & 0 \\ -\sin\theta & 0 & \cos\theta \end{bmatrix}. \tag{3.28}$$

Isometries

Both reflections and rotations have the property that they preserve distance. What this means is that the distance between the images of two vectors equals the distance between the vectors themselves. Writing this with the distance function, if T preserves distance, then for all $\mathbf{u}$ and $\mathbf{v}$ in the domain of T,

$$d(T\mathbf{u}, T\mathbf{v}) = d(\mathbf{u}, \mathbf{v}).$$

Since distance in $\mathbb{R}^n$ can be defined using the norm (Theorem 2.1.14), the next definition names such functions and the theorem that follows essentially states that an isometry preserves the geometry of $\mathbb{R}^n$.

■ **Definition 3.3.10**

A function $f : \mathbb{R}^n \to \mathbb{R}^n$ is an **isometry** when for all $\mathbf{u}, \mathbf{v} \in \mathbb{R}^n$,

$$\|f(\mathbf{u}) - f(\mathbf{v})\| = \|\mathbf{u} - \mathbf{v}\|.$$

■ **Theorem 3.3.11**

Isometries $\mathbb{R}^n \to \mathbb{R}^n$ preserve length, dot product, and angle measure.

Proof

Let $f : \mathbb{R}^n \to \mathbb{R}^n$ be an isometry. Because the length of a vector is nothing more than its distance to $\mathbf{0}$, it follows that f preserves length. This with Exercise 19 of Section 2.2 shows that isometries preserve the dot product. Lastly, to see that f preserves angle measure, let $\mathbf{u}$ and $\mathbf{v}$ be nonzero vectors in $\mathbb{R}^n$. Let θ be the measure of the angle between $\mathbf{u}$ and $\mathbf{v}$, and also let θ' be the measure of the angle between $f(\mathbf{u})$ and $f(\mathbf{v})$. Using Definition 2.2.2 and the previous two results,

$$\theta = \cos^{-1} \frac{\mathbf{u} \cdot \mathbf{v}}{\|\mathbf{u}\| \, \|\mathbf{v}\|} = \cos^{-1} \frac{f(\mathbf{u}) \cdot f(\mathbf{v})}{\|f(\mathbf{u})\| \, \|f(\mathbf{v})\|} = \theta'. \ ■$$

The two types of linear operators already considered are examples of isometries. First, reflect the vectors $\mathbf{u}, \mathbf{v} \in \mathbb{R}^2$ through a given line L that contains the zero vector. Suppose $\mathbf{n}$ is a normal vector to L. Let the reflections of $\mathbf{u}$ and $\mathbf{v}$ be represented by $\mathbf{u}'$ and $\mathbf{v}'$. It appears in Figure 3.11 that the distance between $\mathbf{u}$ and $\mathbf{v}$ equals the distance between $\mathbf{u}'$ and $\mathbf{v}'$. The next theorem proves this. The key to understanding its proof is to realize that $\|\text{proj}_{\mathbf{n}}\mathbf{u}\|$ is the distance of $\mathbf{u}$ to L (Figure 3.4) so that a line segment drawn from the terminal point of $\mathbf{u}$ that is parallel to $\mathbf{n}$ will have $\mathbf{u}'$ at its other endpoint if the segment intersects L and has length of $2\|\text{proj}_{\mathbf{n}}\mathbf{u}\|$.

■ **Theorem 3.3.12**

A reflection in $\mathbb{R}^2$ through a line containing the zero vector is an isometry.

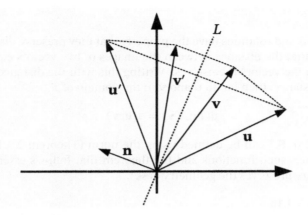

Figure 3.11   A reflection is an isometry.

Proof

Let L be a line containing $\mathbf{0}$ in $\mathbb{R}^2$ with normal vector $\mathbf{n}$. Let T be the reflection through L. Take $\mathbf{u}, \mathbf{v} \in \mathbb{R}^2$. Observe that

$$[(\mathbf{u} - \mathbf{v}) - \text{proj}_\mathbf{n}(\mathbf{u} - \mathbf{v})] \cdot \text{proj}_\mathbf{n}(\mathbf{u} - \mathbf{v}) = 0$$

because $(\mathbf{u} - \mathbf{v}) - \text{proj}_\mathbf{n}(\mathbf{u} - \mathbf{v})$ is the vector component of $\mathbf{u} - \mathbf{v}$ orthogonal to $\mathbf{n}$ (Definition 2.2.18). As noted in Example 3.1.5, the orthogonal projection onto $\mathbf{n}$ is a linear operator, so by Theorem 3.1.6(c),

$$\text{proj}_\mathbf{n}(\mathbf{u} - \mathbf{v}) = \text{proj}_\mathbf{n}\mathbf{u} - \text{proj}_\mathbf{n}\mathbf{v}.$$

Therefore, Theorem 2.2.6(c) implies that

$$(\mathbf{u} - \mathbf{v} - \text{proj}_\mathbf{n}\mathbf{u} + \text{proj}_\mathbf{n}\mathbf{v}) \cdot (\text{proj}_\mathbf{n}\mathbf{v} - \text{proj}_\mathbf{n}\mathbf{u}) = 0,$$

so by Exercise 18 in Section 2.2,

$$\|T\mathbf{u} - T\mathbf{v}\| = \|\mathbf{u} - \mathbf{v} - \text{proj}_\mathbf{n}\mathbf{u} + \text{proj}_\mathbf{n}\mathbf{v} + (\text{proj}_\mathbf{n}\mathbf{v} - \text{proj}_\mathbf{n}\mathbf{u})\|$$
$$= \|\mathbf{u} - \mathbf{v} - \text{proj}_\mathbf{n}\mathbf{u} + \text{proj}_\mathbf{n}\mathbf{v} - (\text{proj}_\mathbf{n}\mathbf{v} - \text{proj}_\mathbf{n}\mathbf{u})\| = \|\mathbf{u} - \mathbf{v}\|. \blacksquare$$

Definitions 3.3.1 and 3.3.6 state that the equations for a reflection in $\mathbb{R}^2$ through a line and a reflection in $\mathbb{R}^3$ through a plane are the same, except for the dimension, so the proof of Theorem 3.3.12 is essentially the proof of the next result.

■ Theorem 3.3.13

A reflection in $\mathbb{R}^3$ through a plane containing the zero vector is an isometry.

Second, rotate the vectors $\mathbf{u}, \mathbf{v} \in \mathbb{R}^2$ to the vectors $\mathbf{u}'$ and $\mathbf{v}'$, and then compare the distance between $\mathbf{u}$ and $\mathbf{v}$ to the distance between $\mathbf{u}'$ and $\mathbf{v}'$. Figure 3.12 suggests that the distance are equal.

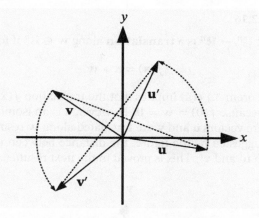

Figure 3.12 A rotation is an isometry.

■ Theorem 3.3.14

A rotation in $\mathbb{R}^2$ is an isometry.

Proof

Fix an angle measure θ. Let $\begin{bmatrix} x \\ y \end{bmatrix}, \begin{bmatrix} a \\ b \end{bmatrix} \in \mathbb{R}^2$. The rotation through θ is an isometry because

$$\left\| \begin{bmatrix} x - a \\ y - b \end{bmatrix} \right\| = \left\| \begin{bmatrix} (x - a)\cos\theta + (b - y)\sin\theta \\ (x - a)\sin\theta + (y - b)\cos\theta \end{bmatrix} \right\|.$$

This follows because

$$(x - a)^2 \cos^2\theta + (x - a)^2 \sin^2\theta = (x - a)^2,$$
$$(b - y)^2 \sin^2\theta + (y - b)^2 \cos^2\theta = (y - b)^2,$$

and

$$2(x - a)(b - y)\cos\theta\sin\theta + 2(x - a)(y - b)\sin\theta\cos\theta = 0. \ ■$$

Theorem 3.3.14 and the standard matrices for rotations in $\mathbb{R}^3$ given by (3.26), (3.27), and (3.28) can be used to prove that rotations about one of the axes in three dimensions are also isometries (Exercise 16).

■ Theorem 3.3.15

A rotation about an axis in $\mathbb{R}^3$ is an isometry.

Here is an example of an isometry that is not a linear operator. Take a random point (x, y) in $\mathbb{R} \times \mathbb{R}$. To shift this point five units to the right and three units down, use the function $f(x, y) = (x + 5, x - 3)$. For example, the result of shifting $(3, 7)$ like this is $f(3, 7) = (8, 4)$. This can be generalized to vectors of any dimension.

■ Definition 3.3.16

A function $f : \mathbb{R}^n \to \mathbb{R}^n$ is a **translation** along $\mathbf{w} \in \mathbb{R}^n$ if for all $\mathbf{x} \in \mathbb{R}^n$,

$$f(\mathbf{x}) = \mathbf{x} + \mathbf{w}.$$

If $\mathbf{w} \neq \mathbf{0}$, Theorem 3.1.6(a) implies that the translation $f(\mathbf{x}) = \mathbf{x} + \mathbf{w}$ is not a linear operator because $f(\mathbf{0}) = \mathbf{w} \neq \mathbf{0}$. However, it is an isometry. This is seen in Figure 3.13, where vectors $\mathbf{u}$ and $\mathbf{v}$ are translated along $\mathbf{w}$ resulting in the vectors $\mathbf{u}'$ and $\mathbf{v}'$. As suggested by the figure, the distance between $\mathbf{u}$ and $\mathbf{v}$ equals the distance between $\mathbf{u}'$ and $\mathbf{v}'$. This is proved in the next result.

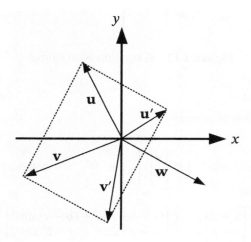

Figure 3.13   A translation is an isometry.

■ Theorem 3.3.17

A translation is an isometry.

Proof

Take $\mathbf{w} \in \mathbb{R}^n$ and let $f : \mathbb{R}^n \to \mathbb{R}^n$ be the translation along $\mathbf{w}$. Let the vectors $\mathbf{u}, \mathbf{v}$ be elements of $\mathbb{R}^n$. Then,

$$\|f(\mathbf{u}) - f(\mathbf{v})\| = \|(\mathbf{u} + \mathbf{w}) - (\mathbf{v} + \mathbf{w})\| = \|\mathbf{u} - \mathbf{v}\|. \quad \blacksquare$$

Contractions, Dilations, and Shears

A function that takes line segments and expands them by a fixed factor is called a **dilation,** while a function that takes lines segments and shrinks them by a fixed factor is called a **contraction**. Such functions are easily generalized to $\mathbb{R}^n$ where segments are replaced by vectors.

■ **Definition 3.3.18**

Let $k \in \mathbb{R}$ such that $k > 0$ and $k \neq 1$. The function $T : \mathbb{R}^n \to \mathbb{R}^n$ such that $T\mathbf{u} = k\mathbf{u}$ for all $\mathbf{u} \in \mathbb{R}^n$ is a **dilation** if $k > 1$ and is a **contraction** if $k < 1$.

It should be noted that dilations and contractions are not isometries. For example, consider the dilation $T : \mathbb{R}^2 \to \mathbb{R}^2$ defined by $T\begin{bmatrix} x \\ y \end{bmatrix} = \begin{bmatrix} 4x \\ 4y \end{bmatrix}$. Although the distance between the vectors $\begin{bmatrix} 1 \\ 0 \end{bmatrix}$ and $\begin{bmatrix} 0 \\ 1 \end{bmatrix}$ is $\sqrt{2}$, the distance between $\begin{bmatrix} 4 \\ 0 \end{bmatrix}$ and $\begin{bmatrix} 0 \\ 4 \end{bmatrix}$ is $4\sqrt{2}$, so a dilation (and also a contraction) is not an isometry. However, both of these types of functions are linear operators.

■ **Theorem 3.3.19**

Dilations and contractions are linear operators.

Proof

Fix $k \in \mathbb{R}$ and define $T : \mathbb{R}^n \to \mathbb{R}^n$ by $T\mathbf{x} = k\mathbf{x}$ for all $\mathbf{x} \in \mathbb{R}^n$. Then, letting $\mathbf{u}, \mathbf{v} \in \mathbb{R}^n$ and $r \in \mathbb{R}$, $T(\mathbf{u} + \mathbf{v}) = k(\mathbf{u} + \mathbf{v}) = k\mathbf{u} + k\mathbf{v} = T\mathbf{u} + T\mathbf{v}$, and $T(r\mathbf{u}) = k(r\mathbf{u}) = r(k\mathbf{u}) = r(T\mathbf{u})$. ■

Because dilations and contractions are linear operators, they have standard matrices, which is, if $T : \mathbb{R}^2 \to \mathbb{R}^2$ is the dilation or contraction with scalar k,

$$[T] = \begin{bmatrix} k & 0 \\ 0 & k \end{bmatrix}$$

because

$$T\begin{bmatrix} 1 \\ 0 \end{bmatrix} = \begin{bmatrix} k \\ 0 \end{bmatrix} \quad \text{and} \quad T\begin{bmatrix} 0 \\ 1 \end{bmatrix} = \begin{bmatrix} 0 \\ k \end{bmatrix}.$$

Another linear operator that is not an isometry is a **shear**. Take a line ℓ in a plane. Let P and Q be points on one side of ℓ. The shear maps P to P' and Q to Q' so that the segments joining both P and P' and both Q and Q' are parallel to ℓ but the lengths PP' and QQ' are proportional to the distance of the points to ℓ. For example, in Figure 3.14, the distance that P is from ℓ is $k > 0$ times the distance that Q is from ℓ. Since the figure represents a shear, $PP' = kQQ'$.

In $\mathbb{R}^2$, it is best to first examine a **horizontal shear**, one in which the line is the x-axis. Let k be a positive real number called the **shear factor**. Define $T : \mathbb{R}^2 \to \mathbb{R}^2$ by

$$T\begin{bmatrix} x \\ y \end{bmatrix} = \begin{bmatrix} x \\ y \end{bmatrix} + k\begin{bmatrix} y \\ 0 \end{bmatrix}.$$

Observe that because T is the sum of two linear operators, T is a linear operator (Theorem 3.2.3). This means that since

$$T\begin{bmatrix} 1 \\ 0 \end{bmatrix} = \begin{bmatrix} 1 \\ 0 \end{bmatrix} \quad \text{and} \quad T\begin{bmatrix} 0 \\ 1 \end{bmatrix} = \begin{bmatrix} k \\ 1 \end{bmatrix},$$

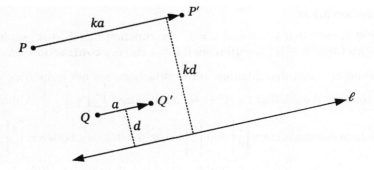

Figure 3.14   A shear along L.

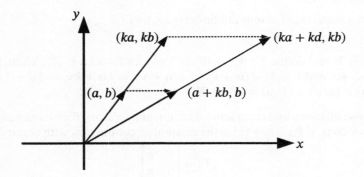

Figure 3.15   A horizontal shear with shear factor k.

it follows that

$$[T] = \begin{bmatrix} 1 & k \\ 0 & 1 \end{bmatrix}. \tag{3.29}$$

To see that this is a shear, let $\begin{bmatrix} a \\ b \end{bmatrix}, \begin{bmatrix} ka \\ kb \end{bmatrix} \in \mathbb{R}^2$. The arrows that represent these vectors and their terminal points are in Figure 3.15. Compute

$$\left\| T\begin{bmatrix} a \\ b \end{bmatrix} - \begin{bmatrix} a \\ b \end{bmatrix} \right\| = \left\| \begin{bmatrix} a + kb \\ b \end{bmatrix} - \begin{bmatrix} a \\ b \end{bmatrix} \right\| = \left\| \begin{bmatrix} kb \\ 0 \end{bmatrix} \right\| = k|b|$$

and

$$\left\| T\begin{bmatrix} ka \\ kb \end{bmatrix} - \begin{bmatrix} ka \\ kb \end{bmatrix} \right\| = \left\| \begin{bmatrix} ka + k^2b \\ kb \end{bmatrix} - \begin{bmatrix} ka \\ kb \end{bmatrix} \right\| = \left\| \begin{bmatrix} k^2b \\ 0 \end{bmatrix} \right\| = k^2|b|.$$

Therefore,

$$\frac{\left\| T\begin{bmatrix} ka \\ kb \end{bmatrix} - \begin{bmatrix} ka \\ kb \end{bmatrix} \right\|}{\left\| T\begin{bmatrix} a \\ b \end{bmatrix} - \begin{bmatrix} a \\ b \end{bmatrix} \right\|} = k.$$

That is, the distance of a point to its image under a horizontal shear is proportional its distance to the x-axis. The equation for a **vertical shear** is similar.

■ Definition 3.3.20

Let $T : \mathbb{R}^2 \to \mathbb{R}^2$ be a linear operator. Let $k \in \mathbb{R}$.

(a) If $T\begin{bmatrix} x \\ y \end{bmatrix} = \begin{bmatrix} x + ky \\ y \end{bmatrix}$, then T is a horizontal shear.

(b) If $T\begin{bmatrix} x \\ y \end{bmatrix} = \begin{bmatrix} x \\ y + kx \end{bmatrix}$, then T is a vertical shear.

Exercises

1. Find the standard matrix for a reflection in $\mathbb{R}^2$ through the line with normal vector $\begin{bmatrix} a \\ b \end{bmatrix}$.

2. Find the standard matrix for each reflection.

   (a) In $\mathbb{R}^2$ through $y = 3x$.

   (b) Maps $\begin{bmatrix} -2 \\ 4 \end{bmatrix}$ to $\begin{bmatrix} -4 \\ -2 \end{bmatrix}$.

   (c) Maps $5\mathbf{e}_3$ to $-5\mathbf{e}_3$.

   (d) In $\mathbb{R}^3$ through $x + y + z = 0$.

3. Find the standard matrix for each rotation.

   (a) In $\mathbb{R}^2$ through $\theta = 3\pi/4$.

   (b) A counterclockwise quarter turn in $\mathbb{R}^2$.

   (c) In $\mathbb{R}^3$ about the y-axis through $\theta = \pi/6$.

   (d) Maps $\begin{bmatrix} 4 \\ 3 \\ 1 \end{bmatrix}$ to $\begin{bmatrix} 4 \\ 0 \\ \sqrt{10} \end{bmatrix}$.

4. Find the standard matrix for each contraction.

   (a) In $\mathbb{R}^2$ with $k = 1/3$.

   (b) Maps $\begin{bmatrix} 3 \\ 8 \end{bmatrix}$ to $\begin{bmatrix} 12 \\ 32 \end{bmatrix}$.

5. Find the standard matrix for each dilation.

    (a) In $\mathbb{R}^3$ with $k = 4$.

    (b) Maps $\begin{bmatrix} 2 \\ 4 \\ 10 \end{bmatrix}$ to $\begin{bmatrix} 1 \\ 2 \\ 5 \end{bmatrix}$.

6. Find the standard matrix for each shear.

    (a) Horizontal with shear factor 4.

    (b) Vertical with shear factor 2.

    (c) Along the line $y = x$ with shear factor 3.

7. Find the standard matrix for the given linear operators $\mathbb{R}^2 \rightarrow \mathbb{R}^2$.

    (a) The $\pi/3$ rotation followed by the reflection through the y-axis.

    (b) The reflection through the x-axis followed by the vertical shear with shear factor $k = 4$.

    (c) The reflection through the y-axis followed by a π rotation followed by the reflection through the x-axis.

    (d) Reflection through the line containing $\mathbf{0}$ that is orthogonal to $4x - 2y = 7$.

8. Let S be the unit square in $\mathbb{R}^2$, which is the parallelogram described by $\mathbf{e}_1$ and $\mathbf{e}_2$. Find the image of S using the linear operators of Exercise 7.

9. Apply each of the linear operators in Exercise 7 to the vector $\begin{bmatrix} 3 \\ 5 \end{bmatrix}$ and give a geometric explanation for why each result makes sense.

10. Find the standard matrix for the given linear operators $\mathbb{R}^3 \rightarrow \mathbb{R}^3$.

    (a) The reflection through the xy-plane followed by the reflection through the yz-plane followed by the reflection through the xz-plane.

    (b) The $\pi/3$ rotation about the x-axis followed by a $\pi/6$ rotation about the y-axis followed by a reflection through the xz-plane.

    (c) Rotation in $\mathbb{R}^3$ about the line $S\left(\begin{bmatrix} 1 \\ 2 \\ 3 \end{bmatrix} \right)$.

11. Let C be the unit cube in $\mathbb{R}^3$, which is the parallelepiped described by $\mathbf{e}_1, \mathbf{e}_2$, and $\mathbf{e}_3$. Find the image of C using the linear operators of Exercise 10.

12. Apply each linear operator in Exercise 10 to the vector $\begin{bmatrix} 1 \\ 2 \\ 5 \end{bmatrix}$ and give a geometric explaining for why each result makes sense.

13. Let $T_1, T_2 : \mathbb{R}^2 \rightarrow \mathbb{R}^2$ be linear operators. Prove.

    (a) If T_1 and T_2 are rotations, $T_1 \circ T_2$ is a rotation.

    (b) If T_1 and T_2 are reflections, $T_1 \circ T_2$ is a rotation.

    (c) If T_1 is a reflection and T_2 is a rotation, $T_1 \circ T_2$ and $T_1 \circ T_2$ are reflections.

14. Assume that $T : \mathbb{R}^3 \to \mathbb{R}^3$ is a translation that maps $\begin{bmatrix} 4 \\ 0 \\ 1 \end{bmatrix}$ to $\begin{bmatrix} -7 \\ 3 \\ -2 \end{bmatrix}$. Find T.

15. Let $T : \mathbb{R}^2 \to \mathbb{R}^2$ be the translation along $\begin{bmatrix} 7 \\ -2 \end{bmatrix}$.

    (a) Verify that T is an isometry using the vectors $\begin{bmatrix} 8 \\ -7 \end{bmatrix}$ and $\begin{bmatrix} -3 \\ 2 \end{bmatrix}$.

    (b) Although T does not have a standard matrix because it is not a linear transformation, T can be defined using a matrix. Find this definition.

16. Prove Theorem 3.3.15.

17. Let $f, g : \mathbb{R}^n \to \mathbb{R}^n$ be isometries. Prove that $f \circ g$ is an isometry. Also, prove that f^{-1} is an isometry if f is invertible.

18. Let $T : \mathbb{R}^n \to \mathbb{R}^n$ be an isometry that is onto and such that $T\mathbf{0} = \mathbf{0}$. Prove that T is a linear operator.

19. Let $T : \mathbb{R}^2 \to \mathbb{R}^2$ be an isometry that is also a linear operator. Prove that T is a rotation or a reflection through a line containing the zero vector.

20. Definition 3.3.20 defines only two types of shears, and they are both in $\mathbb{R}^2$.

    (a) Find the standard matrix of the shear in $\mathbb{R}^2$ along the line $y = mx + b$.

    (b) Define a shear in $\mathbb{R}^3$ and find its standard matrix.

3.4 Injections and Surjections

Linear transformations can be one-to-one and onto like other functions, but they have an advantage in that there are means by which a linear transformation can be shown to be one-to-one or onto besides simply relying on the definition.

Kernel

Consider $T : \mathbb{R}^3 \to \mathbb{R}^2$ defined by

$$T \begin{bmatrix} x \\ y \\ z \end{bmatrix} = \begin{bmatrix} 2x + 4y + 3z \\ 5x + 8y + 8z \end{bmatrix}. \tag{3.30}$$

This function is not one-to-one because, for instance,

$$T \begin{bmatrix} 0 \\ 0 \\ 0 \end{bmatrix} = T \begin{bmatrix} -8 \\ 1 \\ 4 \end{bmatrix} = \begin{bmatrix} 0 \\ 0 \end{bmatrix}. \tag{3.31}$$

Looking at pre-images of the zero vector seems like a good place to check when trying to determine whether a linear transformation is one-to-one. It is already known that $T\mathbf{0} = \mathbf{0}$, so if there is another vector in the domain that has the zero vector as its image, T is not one-to-one. However, if T was defined differently so that the only vector that maps to the zero vector in the range is the zero vector of the domain, does it follow that T is one-to-one? To investigate this, a definition is needed.

■ **Definition 3.4.1**

The **kernel** of the linear transformation $T : \mathbb{R}^n \to \mathbb{R}^m$ is the set

$$\ker T = \{\mathbf{x} \in \mathbb{R}^n : T\mathbf{x} = \mathbf{0}\}.$$

Therefore, continuing the definition of T given in (3.30), the equations of (3.31) show that $\begin{bmatrix} 0 \\ 0 \\ 0 \end{bmatrix}, \begin{bmatrix} -8 \\ 1 \\ 4 \end{bmatrix} \in \ker T$. To find the entire kernel of T, it is required to find all vectors of $\mathbb{R}^3$ such that

$$T \begin{bmatrix} x \\ y \\ z \end{bmatrix} = \begin{bmatrix} 0 \\ 0 \end{bmatrix}.$$

To accomplish this, it suffices to solve the system of equations,

$$\begin{array}{rrrrrrl} 2x & + & 4y & + & 3z & = & 0, \\ 5x & + & 8y & + & 8z & = & 0. \end{array} \tag{3.32}$$

By basic algebra, the solution set of (3.32) in parametric form is the line,

$$x = -8t,$$
$$y = t,$$
$$z = 4t,$$

so $\ker T = \mathsf{S} \left(\begin{bmatrix} -8 \\ 1 \\ 4 \end{bmatrix} \right)$.

The kernel can be used to determine whether a linear transformation is one-to-one. Some fundamental results about the kernel are needed before the main theorem.

■ **Theorem 3.4.2**

Let $T : \mathbb{R}^n \to \mathbb{R}^m$ be a linear transformation.

(a) $\mathbf{0} \in \ker T$.

(b) $\mathbf{u} + \mathbf{v} \in \ker T$ for all $\mathbf{u}, \mathbf{v} \in \ker T$.

(c) $r\mathbf{u} \in \ker T$ for all $\mathbf{u} \in \ker T$ and $r \in \mathbb{R}$.

Proof

Because $T\mathbf{0} = \mathbf{0}$ by Theorem 3.1.6(a), the zero vector of $\mathbb{R}^n$ is an element of the kernel of T, proving part (a). To prove (b), take $\mathbf{u}, \mathbf{v} \in \ker T$. This implies that $T\mathbf{u} = \mathbf{0}$ and $T\mathbf{v} = \mathbf{0}$. Since $T(\mathbf{u} + \mathbf{v}) = T(\mathbf{u}) + T(\mathbf{v}) = \mathbf{0}$, it is the case that $\mathbf{u} + \mathbf{v} \in \ker T$. Part (c) is proved similarly (Exercise 4). ∎

■ **Theorem 3.4.3**

The linear transformation $T : \mathbb{R}^n \to \mathbb{R}^m$ is one-to-one if and only if the kernel of T equals $\{\mathbf{0}\}$.

Proof

Suppose that T is one-to-one. By Theorem 3.4.2(a), $\mathbf{0} \in \ker T$. Assume that $\mathbf{u} \in \ker T$. Therefore, $T\mathbf{0} = T\mathbf{u} = \mathbf{0}$. Hence, $\mathbf{u} = \mathbf{0}$ by supposition, which implies that $\ker T = \{\mathbf{0}\}$. To prove the converse, suppose that the only vector in $\ker T$ is the zero vector. Let $\mathbf{u}_1, \mathbf{u}_2 \in \mathbb{R}^n$ and assume that $T\mathbf{u}_1 = T\mathbf{u}_2$. Then, by Theorem 3.1.6(c), $T(\mathbf{u}_1 - \mathbf{u}_2) = T\mathbf{u}_1 - T\mathbf{u}_2 = \mathbf{0}$, so $\mathbf{u}_1 - \mathbf{v}_2 \in \ker T$. Hence, $\mathbf{u}_1 = \mathbf{u}_2$. ∎

■ **Example 3.4.4**

Demonstrate that $T : \mathbb{R}^2 \to \mathbb{R}^4$ is one-to-one, where

$$[T] = \begin{bmatrix} 2 & -4 \\ 6 & -1 \\ -5 & 3 \\ 1 & 1 \end{bmatrix}.$$

Since $\{\mathbf{0}\} \subseteq \ker T$ by Theorem 3.4.2(a), take $\begin{bmatrix} x \\ y \end{bmatrix} \in \ker(T)$. This means that

$$\begin{bmatrix} 2 & -4 \\ 6 & -1 \\ -5 & 3 \\ 1 & 1 \end{bmatrix} \begin{bmatrix} x \\ y \end{bmatrix} = \begin{bmatrix} 0 \\ 0 \\ 0 \\ 0 \end{bmatrix}.$$

Solving this equation is equivalent to solving the system of linear equations,

$$\begin{array}{rcrcl} 2x & - & 4y & = & 0, \\ 6x & - & y & = & 0, \\ -5x & + & 3y & = & 0, \\ x & + & y & = & 0. \end{array}$$

Adding the second equation to the last equation, replacing the last equation, gives

$$\begin{array}{rcrcl} 2x & - & 4y & = & 0, \\ 6x & - & y & = & 0, \\ -5x & + & 3y & = & 0, \\ 7x & & & = & 0. \end{array}$$

Hence, $x = 0$, and then by substituting $x = 0$ into any of the other equations yields $y = 0$. This implies that $\ker T = \{\mathbf{0}\}$, so T is one-to-one by Theorem 3.4.3.

Range

When checking whether a linear transformation is a surjection, it is helpful to know exactly the vectors in the range of the function. The next results gives that information, and follows directly from Definition 3.1.15 (Exercise 5).

■ Theorem 3.4.5

If $T : \mathbb{R}^n \rightarrow \mathbb{R}^m$ is a linear transformation and $\mathbf{c}_1, \mathbf{c}_2, \ldots, \mathbf{c}_n$ are the columns of $[T]$, then $\operatorname{ran} T = \mathsf{S}(\mathbf{c}_1, \mathbf{c}_2, \ldots, \mathbf{c}_n)$.

■ Example 3.4.6

Notice that Theorem 3.4.5 implies that the range of the linear transformation T from Example 3.4.4 is

$$
\operatorname{ran} T = \mathsf{S}\left(\begin{bmatrix} 2 \\ 6 \\ -5 \\ 1 \end{bmatrix}, \begin{bmatrix} -4 \\ -1 \\ 3 \\ 1 \end{bmatrix} \right),
$$

and this means that

$$
\operatorname{ran} T = \left\{ \begin{bmatrix} 2x - 4y \\ 6x - y \\ -5x + 3y \\ x + y \end{bmatrix} : x, y \in \mathbb{R} \right\}.
$$

Since the codomain of T is $\mathbb{R}^4$, it appears that since $\operatorname{ran} T$ can be described with only two parameters, T is not onto. To confirm this, let $\mathbf{u} \in \mathbb{R}^4$ with coordinates u_1, u_2, u_3, u_4. To determine if $\mathbf{u}$ is in the range, try to find real numbers x and y such that

$$
\begin{array}{rcrcl}
2x & - & 4y & = & u_1, \\
6x & - & y & = & u_2, \\
-5x & + & 3y & = & u_3, \\
x & + & y & = & u_4.
\end{array}
$$

There are many ways to solve this system. One way is to start by moving the last equation to the beginning of the system.

$$
\begin{array}{rcrcl}
x & + & y & = & u_4, \\
2x & - & 4y & = & u_1, \\
6x & - & y & = & u_2, \\
-5x & + & 3y & = & u_3.
\end{array}
$$

This makes the next step easier to see, which is to multiply the new first equation by -2, -6, and then 5 and add the results to the second, third, and fourth equations, respectively, to obtain

$$
\begin{aligned}
x \ + \ \ y &= u_4, \\
-6y &= u_1 - 2u_4, \\
-7y &= u_2 - 6u_4, \\
8y &= u_3 + 5u_4.
\end{aligned}
$$

Next, divide the second equation by -6 to find

$$
\begin{aligned}
x \ + \ \ y &= u_4, \\
y &= \frac{-1}{6}u_1 + \frac{1}{3}u_4, \\
-7y &= u_2 - 6u_4, \\
8y &= u_3 + 5u_4.
\end{aligned}
$$

Lastly, multiply the second equation by -1, 7, and -8 and add the results to the first, third, and fourth equations, respectively, to obtain

$$
\begin{aligned}
x &= \frac{1}{6}u_1 + \frac{2}{3}u_4, \\
y &= \frac{-1}{6}u_1 + \frac{1}{3}u_4, \\
0 &= -\frac{7}{6}u_1 + u_2 - \frac{11}{3}u_4, \\
0 &= \frac{4}{3}u_1 + u_3 + \frac{7}{3}u_4.
\end{aligned}
$$

Therefore, any **u** with coordinates such that the right-hand side of the third or fourth equation is not zero will not be an element of the range. For example, $\mathbf{e}_4 \notin \operatorname{ran} T$. This means that T is not onto.

■ **Example 3.4.7**

Show that the linear transformation $T : \mathbb{R}^3 \to \mathbb{R}^2$ is onto, where

$$
[T] = \begin{bmatrix} 3 & -1 & 5 \\ 1 & 4 & 4 \end{bmatrix}.
$$

To see this, take $\begin{bmatrix} u_1 \\ u_2 \end{bmatrix} \in \mathbb{R}^2$ and solve the system of linear equations

$$
\begin{aligned}
3x \ - \ \ y \ + \ 5z &= u_1, \\
x \ + \ 4y \ + \ 4z &= u_2.
\end{aligned}
$$

Using the method of Example 3.4.6, first exchange the equations to obtain the system,

$$
\begin{aligned}
x \ + \ 4y \ + \ 4z &= 0, \\
3x \ - \ \ y \ + \ 5z &= 0.
\end{aligned}
$$

Multiply the first equation by -3 and add to the second equation, giving

$$
\begin{aligned}
x \;+\; 4y \;+\; 4z &= u_2, \\
-13y \;-\; 7z &= u_1 - 3u_2.
\end{aligned}
$$

Now, multiply the second equation by $-1/13$. This gives

$$
\begin{aligned}
x \;+\; 4y \;+\; 4z &= u_2, \\
y \;+\; \frac{7}{13}z &= -\frac{1}{13}u_1 + \frac{3}{13}u_2.
\end{aligned}
$$

Finally, multiply the second equation by -4 and add to the first to find

$$
\begin{aligned}
x \;+\; \frac{24}{13}z &= \frac{4}{13}u_1 + \frac{1}{13}u_2, \\
y \;+\; \frac{7}{13}z &= -\frac{1}{13}u_1 + \frac{3}{13}u_2.
\end{aligned}
$$

Therefore,

$$
\begin{aligned}
x &= -\frac{24}{13}r + \frac{4u_1 + u_2}{13}, \\
y &= -\frac{7}{13}r + \frac{-u_1 + 3u_2}{13}, \\
z &= r,
\end{aligned}
\tag{3.33}
$$

which has a solution for all $u_1, u_2 \in \mathbb{R}$.

■ **Example 3.4.8**

Continuing Example 3.4.7, show that the linear transformation T is not one-to-one. To find the kernel of T, let $u_1 = u_2 = 0$ in (3.33). This implies that

$$
\ker T = S\left(\begin{bmatrix} -24/13 \\ -7/13 \\ 1 \end{bmatrix} \right) = S\left(\begin{bmatrix} 24 \\ 7 \\ -13 \end{bmatrix} \right).
$$

For example, $\begin{bmatrix} 24 \\ 7 \\ -13 \end{bmatrix} \in \ker T$, so T is not one-to-one.

Exercises

1. Find the kernel of each linear transformation T and determine whether it is one-to-one.

    (a) $T : \mathbb{R}^2 \to \mathbb{R}^2$, $T\begin{bmatrix} x \\ y \end{bmatrix} = \begin{bmatrix} 4x + 2y \\ x - 5y \end{bmatrix}$

    (b) $T : \mathbb{R}^3 \to \mathbb{R}^4$, $T\begin{bmatrix} x \\ y \\ z \end{bmatrix} = \begin{bmatrix} 2x - 5y + z \\ x - 2y + z \\ 4x - y \\ 4x - 3y + 2z \end{bmatrix}$

(c) $T : \mathbb{R}^3 \to \mathbb{R}^3$, $T \begin{bmatrix} x \\ y \\ z \end{bmatrix} = \begin{bmatrix} 7x + 3y + 4z \\ 3x + 4y - z \\ 6x + 2y - 4z \end{bmatrix}$

(d) $T : \mathbb{R}^4 \to \mathbb{R}^2$, $T \begin{bmatrix} x_1 \\ x_2 \\ x_3 \\ x_4 \end{bmatrix} = \begin{bmatrix} 2x_1 - 4x_2 \\ 5x_3 - 2x_4 \end{bmatrix}$

2. Find the kernel and range of the identity transformation $\mathbb{R}^n \to \mathbb{R}^n$.

3. Find the kernel and range of the zero transformation $\mathbb{R}^n \to \mathbb{R}^m$.

4. Let $T : \mathbb{R}^n \to \mathbb{R}^m$ be a linear transformation. Prove that $r\mathbf{u} \in \ker(T)$ for all $\mathbf{u} \in \ker(T)$ and $r \in \mathbb{R}$.

5. Prove Theorem 3.4.5.

6. Find the range of each linear transformation T in Exercise 1 and determine whether it is onto.

7. Can a linear transformation $\mathbb{R}^n \to \mathbb{R}^m$ be one-to-one if $n > m$? Explain.

8. Can a linear transformation $\mathbb{R}^n \to \mathbb{R}^m$ onto if $n < m$? Explain.

9. Prove or show false: $T_1 + T_2$ is one-to-one if and only if the linear transformations $T_1 : \mathbb{R}^n \to \mathbb{R}^m$ and $T_2 : \mathbb{R}^n \to \mathbb{R}^m$ are one-to-one.

10. Prove or show false: $T_1 + T_2$ is onto if and only if the linear transformations $T_1 : \mathbb{R}^n \to \mathbb{R}^m$ and $T_2 : \mathbb{R}^n \to \mathbb{R}^m$ are onto.

11. Let $T : \mathbb{R}^2 \to \mathbb{R}^2$ be defined by $T \begin{bmatrix} x \\ y \end{bmatrix} = \begin{bmatrix} T_1(x) \\ T_2(y) \end{bmatrix}$ where $T_1, T_2 : \mathbb{R} \to \mathbb{R}$ are linear operators.

    (a) Find the kernel of T. When is T one-to-one?
    (b) Find the range of T. When is T onto?

12. Let $T_1, T_2 : \mathbb{R}^3 \to \mathbb{R}^3$ be linear operators.

    (a) Find the kernel of $T_1 \circ T_2$. When is this function one-to-one?
    (b) Fine the range of $T_1 \circ T_2$. When is this function onto?

13. Let $T : \mathbb{R}^n \to \mathbb{R}^m$ be a linear transformation. Given the standard matrix for T, determine whether T is one-to-one or onto by finding its kernel and range.

    (a) $[T] = \begin{bmatrix} 4 & 8 \\ -3 & -6 \end{bmatrix}$

    (b) $[T] = \begin{bmatrix} 2 & -7 & -1 \\ 5 & 0 & -6 \\ -2 & 9 & -1 \end{bmatrix}$

    (c) $[T] = \begin{bmatrix} 1 & 8 \\ 0 & 5 \end{bmatrix}$

    (d) $[T] = \begin{bmatrix} 2 & 3 & 7 \\ 3 & -5 & 1 \\ 0 & 1 & 1 \end{bmatrix}$

    (e) $[T] = \begin{bmatrix} 7 & 3 \\ 3 & 4 \\ 6 & 2 \\ 1 & 0 \end{bmatrix}$

    (f) $[T] = \begin{bmatrix} -1 & 2 & 0 & -1 \\ 0 & 1 & -1 & 2 \\ 3 & 0 & 2 & 0 \end{bmatrix}$

14. Redo Exercise 13 using the definition of one-to-one (Definition 1.4.11) and the definition of onto (Definition 1.4.15).

3.5 Gauss–Jordan Elimination

As seen in Section 3.4, determining whether a linear transformation is one-to-one or onto involves solving a system of linear equations. What was also evident in that same section was that the number of equations and the number of variables in the system are dependent on the dimensions of the domain and codomain of the transformation. This means that the system can become unwieldy. What would be helpful is an algorithm that would aid in solving such systems. Looking at the work in Examples 3.4.4, 3.4.6, 3.4.7, and 3.4.8, it appears that the systems could have been solved by simply doing work on the coefficients with the variables essentially serving only as placeholders. With this in mind, make the next definition, which will serve as the starting point of the algorithm.

■ **Definition 3.5.1**

Let $a_{ij} \in \mathbb{R}$ for $i = 1, 2, \ldots, m$ and $j = 1, 2, \ldots, n$. The **coefficient matrix** of the system of linear equations

$$
\begin{array}{ccccccccc}
a_{11}x_1 & + & a_{12}x_2 & + & \cdots & + & a_{1n}x_n & = & b_1 \\
a_{21}x_1 & + & a_{22}x_2 & + & \cdots & + & a_{2n}x_n & = & b_2 \\
\vdots & & \vdots & & & & \vdots & & \vdots \\
a_{m1}x_1 & + & a_{m2}x_2 & + & \cdots & + & a_{mn}x_n & = & b_m
\end{array}
$$

is

$$
\begin{bmatrix}
a_{11} & a_{12} & \cdots & a_{1n} \\
a_{21} & a_{22} & \cdots & a_{2n} \\
\vdots & \vdots & & \vdots \\
a_{m1} & a_{m2} & \cdots & a_{mn}
\end{bmatrix},
$$

and its **augmented coefficient matrix** is

$$
\left[
\begin{array}{cccc|c}
a_{11} & a_{12} & \cdots & a_{1n} & b_1 \\
a_{21} & a_{22} & \cdots & a_{2n} & b_2 \\
\vdots & \vdots & & \vdots & \vdots \\
a_{m1} & a_{m2} & \cdots & a_{mn} & b_m
\end{array}
\right].
$$

■ **Example 3.5.2**

The coefficient matrix for the system of linear equations,

$$
\begin{array}{ccccccc}
2x_1 & + & 5x_2 & + & x_3 & = & 1, \\
x_1 & - & x_2 & + & 6x_3 & = & 4, \\
-2x_1 & + & 2x_2 & + & 4x_3 & = & 2,
\end{array}
\tag{3.34}
$$

is the 3×3 matrix,

$$A = \begin{bmatrix} 2 & 5 & 1 \\ 1 & -1 & 6 \\ -2 & 2 & 4 \end{bmatrix},$$

and its augmented coefficient matrix is the 3×4 matrix,

$$\begin{bmatrix} 2 & 5 & 1 & 1 \\ 1 & -1 & 6 & 4 \\ -2 & 2 & 4 & 2 \end{bmatrix},$$

which can be written using $\mathbf{b} = \begin{bmatrix} 1 \\ 4 \\ 2 \end{bmatrix}$ as $[A \mid \mathbf{b}]$. The purpose of the augmentation is to identify which entries represent the coefficients on the variables and which entries represent the constants. Otherwise, the matrix is treated as any other matrix as if the line was not present.

The system of equations (3.34) can be interpreted as three dot product equations, namely

$$\begin{bmatrix} 2 \\ 5 \\ 1 \end{bmatrix} \cdot \begin{bmatrix} x_1 \\ x_2 \\ x_3 \end{bmatrix} = 1, \quad \begin{bmatrix} 1 \\ -1 \\ 6 \end{bmatrix} \cdot \begin{bmatrix} x_1 \\ x_2 \\ x_3 \end{bmatrix} = 4, \quad \begin{bmatrix} -2 \\ 2 \\ 4 \end{bmatrix} \cdot \begin{bmatrix} x_1 \\ x_2 \\ x_3 \end{bmatrix} = 2.$$

This can be said about every system of linear equations. Because the dot product equations involve the columns of the coefficient matrix, every linear system of equations can be written as a matrix equation of the form

$$A\mathbf{x} = \mathbf{b}, \tag{3.35}$$

where A is the $m \times n$ coefficient matrix of the system, $\mathbf{x}$ is a variable vector from $\mathbb{R}^n$, and $\mathbf{b} \in \mathbb{R}^m$. This means that familiar terminology regarding systems of equations can be restated using (3.35).

■ Definition 3.5.3

Let $A\mathbf{x} = \mathbf{b}$ be a given system of linear equations with A being an $m \times n$ matrix, $\mathbf{x}$ being an $n \times 1$ variable matrix, and $\mathbf{b}$ being an $m \times 1$ matrix.

(a) A vector $\mathbf{u} \in \mathbb{R}^n$ such that $A\mathbf{u} = \mathbf{b}$ is called a **solution** of the system.

(b) The **solution set** of the system is $\{\mathbf{x} : A\mathbf{x} = \mathbf{b}\}$.

(c) Systems of linear equations with equal solution sets are **equivalent**.

(d) If the system has a solution, it is **consistent**. Otherwise, the system is **inconsistent**.

(e) The system is **homogeneous** if $\mathbf{b} = \mathbf{0}$.

(f) The vector $\mathbf{0}$ is called the **trivial solution** of $A\mathbf{x} = \mathbf{0}$ because $A\mathbf{0} = \mathbf{0}$.

Elementary Row Operations

In many of the examples of Section 3.3, sequences of systems of equations were found that transformed the given system to a system in which the solution was easily obtained. Each system in the sequence was equivalent to the previous system. This was guaranteed by the algebra performed at each step. A close inspection reveals that there were only three types of moves performed to transform one system into another. They were as follows:

- Equations were reordered.

- An equation was multiplied by a nonzero constant on both sides of the equation.

- An equation was multiplied by a constant on both sides of the equation and the resulting equation was added to another equation of the system.

To prove that these three algebraic steps result in a system of linear equations that is equivalent to the given system, take them one at a time. Let $\mathbf{u} \in \mathbb{R}^n$ with coordinates $u_1, u_2, \ldots, u_n$.

- Certainly, the order in which the equations are written has no affect on the solution set.

- Take the equation from a given system of linear equations,

$$a_1 x_1 + a_2 x_2 + \cdots + a_n x_n = b. \tag{3.36}$$

Let $r \in \mathbb{R}$ but $r \neq 0$ and define a new system with the same equations as the given system except that (3.36) has been replaced with

$$r a_1 x_1 + r a_2 x_2 + \cdots + r a_n x_n = rb. \tag{3.37}$$

To show that the systems are equivalent, it suffices to prove that (3.36) is equivalent to (3.37). Since it is clear that every solution of (3.36) is a solution of (3.37), assume that $\mathbf{u}$ is a solution to (3.37). Hence,

$$r a_1 u_1 + r a_2 u_2 + \cdots + r a_n u_n = rb. \tag{3.38}$$

Since $r \neq 0$, divide both sides of (3.38) to obtain

$$a_1 u_1 + a_2 u_2 + \cdots + a_n u_n = b.$$

- Suppose (3.36) is in a given system of linear equations and let

$$c_1 x_1 + c_2 x_2 + \cdots + c_n x_n = d \tag{3.39}$$

be an equation in the system different from (3.36). Write another system that has the same equations as the given system except that (3.39) is replaced with

$$(r a_1 + c_1) x_1 + (r a_2 + c_2) x_2 + \cdots + (r a_n + c_n) x_n = rb + d. \tag{3.40}$$

Suppose that **u** is a solution to the given system. This implies that

$$a_1 u_1 + a_2 u_2 + \cdots + a_n u_n = b \tag{3.41}$$

and

$$c_1 u_1 + c_2 u_2 + \cdots + c_n u_n = d. \tag{3.42}$$

Then,

$$(ra_1 + c_1)u_1 + (ra_2 + c_2)u_2 + \cdots + (ra_n + c_n)u_n = rb + d, \tag{3.43}$$

which implies that **u** is a solution to the second system. Conversely, if both (3.41) and (3.43) hold,

$$(ra_1 + c_1)u_1 - ra_1 u_1 + \cdots + (ra_n + c_n)u_n - ra_n u_n = rb + d - rb$$

also holds, which means that (3.42) follows.

Notice that the three steps that can be performed on a system of linear equations again focuses on the coefficients of the equations. For example, starting with the coefficient matrix, the work of Example 3.4.7 with $u_1 = u_2 = 0$ could be written as

$$\begin{bmatrix} 3 & -1 & 5 \\ 1 & 4 & 4 \end{bmatrix} \longrightarrow \begin{bmatrix} 1 & 4 & 4 \\ 3 & -1 & 5 \end{bmatrix} \longrightarrow \begin{bmatrix} 1 & 4 & 4 \\ 0 & -13 & -7 \end{bmatrix}$$

$$\longrightarrow \begin{bmatrix} 1 & 4 & 4 \\ 0 & 1 & 7/13 \end{bmatrix} \longrightarrow \begin{bmatrix} 1 & 0 & 24/13 \\ 0 & 1 & 7/13 \end{bmatrix}.$$

This observation motivates the next definition. It simply copies the three algebraic steps done to a system of linear equations.

■ **Definition 3.5.4**

Let $A = [A_{ij}]$ be an $m \times n$ matrix and take $s \in \mathbb{R}$. The **elementary row operations** of a A are certain modifications of the rows of A that are used to define a new matrix A'. The following give notation for each of the operations along with a description of the operation.

(a) $R_i \leftrightarrow R_j$
Rows i and j of A are exchanged, resulting in A'.

(b) $sR_i \ (s \neq 0)$
The new matrix A' is defined so that it is equal to the A except that entry k of row i is sa_{ik} for all $k = 1, 2, \ldots, n$.

(c) $sR_i + R_j$
The matrix A' is A except that entry k of row j is $sa_{ik} + a_{jk}$ for all $k = 1, 2, \ldots, n$.

If the nonzero condition on s is removed from (b), these operations are simply called **row operations**.

■ **Example 3.5.5**

Let A be the matrix $\begin{bmatrix} 1 & 2 & 3 & 4 \\ 5 & 6 & 7 & 8 \end{bmatrix}$. Let A' be the matrix obtained from A using the given elementary row operations.

- $R_1 \leftrightarrow R_2$: $A' = \begin{bmatrix} 5 & 6 & 7 & 8 \\ 1 & 2 & 3 & 4 \end{bmatrix}$.

- $5R_2$: $A' = \begin{bmatrix} 1 & 2 & 3 & 4 \\ 25 & 30 & 35 & 40 \end{bmatrix}$.

- $-3R_2 + R_1$: $A' = \begin{bmatrix} -14 & -16 & -18 & -20 \\ 5 & 6 & 7 & 8 \end{bmatrix}$.

To develop a technique that will simplify the study of systems of linear equations, begin by rewriting Definition 3.5.3(c) in terms of matrices.

■ **Definition 3.5.6**

The $m \times n$ matrices A and B are **row equivalent** if A can be transformed into B using the elementary row operations.

Suppose that $[A \mid \mathbf{c}]$ is row equivalent to $[B \mid \mathbf{d}]$. Because each elementary row operation used to transform $[A \mid \mathbf{c}]$ into $[B \mid \mathbf{d}]$ corresponds to an algebraic step, the solution set of the system of linear equations described by $[A \mid \mathbf{c}]$ equals the solution set of the system of linear equations described by $[B \mid \mathbf{d}]$. For this reason, it is said that the elementary row operations **preserve** the solution set. Conversely, if $A\mathbf{x} = \mathbf{c}$ is equivalent to $B\mathbf{x} = \mathbf{d}$, the one system can be algebraically transformed into the other, meaning that $[A \mid \mathbf{c}]$ is row equivalent to $[B \mid \mathbf{d}]$ because each algebraic step corresponds to an elementary row operation. Hence, the next theorem can be stated.

■ **Theorem 3.5.7**

Let A and B be $m \times n$ matrices and $\mathbf{c}, \mathbf{d} \in \mathbb{R}^n$. Then, $[A \mid \mathbf{c}]$ is row equivalent to $[B \mid \mathbf{d}]$ if and only if $A\mathbf{x} = \mathbf{c}$ is equivalent to $B\mathbf{x} = \mathbf{d}$.

The strategy that will be developed here to solve the system of linear equations, $A\mathbf{x} = \mathbf{c}$, starts by writing the system's augmented coefficient matrix, $[A \mid \mathbf{c}]$. Then, the strategy transforms the matrix using the elementary row operations into a matrix $[B \mid \mathbf{d}]$, which will be row equivalent to $[A \mid \mathbf{c}]$. Because of Theorem 3.5.7, the solution to $B\mathbf{x} = \mathbf{d}$ will also be a solution to $A\mathbf{x} = \mathbf{c}$. The goal will be to make sure that the solution to the system $B\mathbf{x} = \mathbf{d}$ is easily read from $[B \mid \mathbf{d}]$. What it means to be easily read is defined next.

■ **Definition 3.5.8**

A matrix is in **reduced row echelon form** when all of the following hold.

(a) The first nonzero entry of a row from the left is 1 (called a **leading 1**).

(b) Every leading 1 is to the right of the leading one in any row above it.

(c) All other entries in a column with a leading 1 are 0.

(d) There is no row consisting of all zero entries (called a **zero row**) above a row with a nonzero entry.

■ **Example 3.5.9**

The matrices,

$$\begin{bmatrix} 1 & 0 & 0 \\ 0 & 1 & 0 \\ 0 & 0 & 1 \end{bmatrix}, \quad \begin{bmatrix} 1 & 5 & 0 & 3 \\ 0 & 0 & 1 & 9 \end{bmatrix}, \quad \left[\begin{array}{cc|c} 1 & 0 & 2 \\ 0 & 1 & 3 \\ 0 & 0 & 0 \\ 0 & 0 & 0 \end{array}\right],$$

are in reduced row echelon form, but these matrices,

$$\begin{bmatrix} 1 & 0 & 0 \\ 0 & 5 & 0 \\ 0 & 0 & 1 \end{bmatrix}, \quad \begin{bmatrix} 0 & 1 & 0 \\ 1 & 0 & 0 \\ 0 & 0 & 1 \end{bmatrix},$$

$$\left[\begin{array}{cc|c} 1 & 0 & 0 \\ 2 & 1 & 0 \\ 3 & 0 & 1 \end{array}\right], \quad \left[\begin{array}{cc|c} 0 & 0 & 0 \\ 0 & 1 & 0 \\ 0 & 0 & 1 \end{array}\right],$$

are not in reduced row echelon form.

Square Matrices

Every matrix A is row equivalent to a matrix B in reduced row echelon form, and the method used to find B involves the elementary row operations. The technique is credited to Johann Carl Friedrich Gauss and Wilhelm Jordan.

■ **Definition 3.5.10**

The process of transforming a matrix to reduced row echelon form using only the elementary row operations is called **Gauss–Jordan elimination**.

■ **Example 3.5.11**

Solve the homogeneous system of linear equations,

$$\begin{aligned} 2x_1 + \quad\quad\; + \; x_3 &= 0, \\ x_1 - \; x_2 + 6x_3 &= 0, \\ 2x_2 + 4x_3 &= 0. \end{aligned} \tag{3.44}$$

The augmented coefficient matrix for this system of linear equations is

$$\left[\begin{array}{ccc|c} 2 & 0 & 1 & 0 \\ 1 & -1 & 6 & 0 \\ 0 & 2 & 4 & 0 \end{array}\right].$$

Since the elementary row operations will not change the last column because all of the entries are 0, the coefficient matrix,

$$\begin{bmatrix} 2 & 0 & 1 \\ 1 & -1 & 6 \\ 0 & 2 & 4 \end{bmatrix},$$

can be used instead. Following the technique of Example 3.4.6, Gauss-Jordan elimination will lead to the reduced row echelon form of the coefficient matrix. Do this by working column by column. Start by obtaining a 1 in the first entry of the first column. This could be done by multiplying the first row by 1/2. However, in this case a simple exchange of rows gives the desired outcome. The remaining entries of the first column become 0 when the appropriate scalar multiple of the first row is added to the other two rows. In notation, what happens is this:

$$\begin{bmatrix} 2 & 0 & 1 \\ 1 & -1 & 6 \\ 0 & 2 & 4 \end{bmatrix} \xrightarrow{R_1 \leftrightarrow R_2} \begin{bmatrix} 1 & -1 & 6 \\ 2 & 0 & 1 \\ 0 & 2 & 4 \end{bmatrix} \xrightarrow{-2R_1+R_2} \begin{bmatrix} 1 & -1 & 6 \\ 0 & 2 & -11 \\ 0 & 2 & 4 \end{bmatrix}$$

The first column is done. Next, it is desired to have a 1 in the second position of the second column. Although, this can be done by multiplying row 2 by 1/2, to avoid a fraction at this point, switch rows 2 and 3, and then multiply row 2 by 1/2. The -1 in the first row is then eliminated by a simple addition, and the 2 in the third row is eliminated as usual:

$$\xrightarrow{R_2 \leftrightarrow R_3} \begin{bmatrix} 1 & -1 & 6 \\ 0 & 2 & 4 \\ 0 & 2 & -11 \end{bmatrix} \xrightarrow{\frac{1}{2}R_2} \begin{bmatrix} 1 & -1 & 6 \\ 0 & 1 & 2 \\ 0 & 2 & -11 \end{bmatrix}$$

$$\xrightarrow{R_2+R_1} \begin{bmatrix} 1 & 0 & 8 \\ 0 & 1 & 2 \\ 0 & 2 & -11 \end{bmatrix} \xrightarrow{-2R_2+R_3} \begin{bmatrix} 1 & 0 & 8 \\ 0 & 1 & 2 \\ 0 & 0 & -15 \end{bmatrix}.$$

Lastly, it is easy to transform the third column to reach the reduced row echelon form:

$$\xrightarrow{-\frac{1}{15}R_3} \begin{bmatrix} 1 & 0 & 8 \\ 0 & 1 & 2 \\ 0 & 0 & 1 \end{bmatrix} \xrightarrow{-8R_3+R_1} \begin{bmatrix} 1 & 0 & 0 \\ 0 & 1 & 2 \\ 0 & 0 & 1 \end{bmatrix} \xrightarrow{-2R_3+R_2} \begin{bmatrix} 1 & 0 & 0 \\ 0 & 1 & 0 \\ 0 & 0 & 1 \end{bmatrix}.$$

Therefore, the **elementary row operation sequence**,

$$R_1 \leftrightarrow R_2, \quad -2R_1 + R_2, \quad R_2 \leftrightarrow R_3, \quad \frac{1}{2}R_2, \quad R_2 + R_1,$$

$$-2R_2 + R_3, \quad -\frac{1}{15}R_3, \quad -8R_3 + R_1, \quad -2R_3 + R_2,$$

shows that $\begin{bmatrix} 2 & 0 & 1 \\ 1 & -1 & 6 \\ 0 & 2 & 4 \end{bmatrix}$ is row equivalent to $\begin{bmatrix} 1 & 0 & 0 \\ 0 & 1 & 0 \\ 0 & 0 & 1 \end{bmatrix}$. Because the 3×3

identity matrix corresponds to the system of linear equations,

$$x_1 = 0,$$
$$x_2 = 0,$$
$$x_3 = 0,$$

the solution set to (3.44) is $\left\{ \begin{bmatrix} 0 \\ 0 \\ 0 \end{bmatrix} \right\}$.

■ **Example 3.5.12**

Change the system of linear equations found in Example 3.5.11 to the non-homogeneous system

$$\begin{array}{rcrcrcl} 2x_1 & + & & + & x_3 & = & 1, \\ x_1 & - & x_2 & + & 6x_3 & = & 4, \\ & & 2x_2 & + & 4x_3 & = & 2. \end{array} \qquad (3.45)$$

The reduced row echelon form of its augmented coefficient matrix is

$$\begin{bmatrix} 1 & 0 & 0 & | & 1/5 \\ 0 & 1 & 0 & | & -1/5 \\ 0 & 0 & 1 & | & 3/5 \end{bmatrix}$$

and is found using the same elementary row operations as in Example 3.5.11.

Thus, the solution set to (3.45) is $\left\{ \begin{bmatrix} 1/5 \\ -1/5 \\ 3/5 \end{bmatrix} \right\}$, which in parametric form is

$$x_1 = 1/5,$$
$$x_2 = -1/5,$$
$$x_3 = 3/5.$$

■ **Example 3.5.13**

The system of linear equations,

$$\begin{array}{rcrcr} 4x & - & 8y & = & 12, \\ -12x & + & 24y & = & -36, \end{array} \qquad (3.46)$$

has infinitely many solutions because the two equations represent the same line. A matrix can be used to find this line. The augmented coefficient matrix of (3.46) is

$$\begin{bmatrix} 4 & -8 & | & 12 \\ -12 & 24 & | & -36 \end{bmatrix},$$

which has reduced row echelon form,

$$\left[\begin{array}{cc|c} 1 & -2 & 3 \\ 0 & 0 & 0 \end{array}\right].$$

This corresponds to the system of linear equations,

$$\begin{array}{rcl} x - 2y &=& 3, \\ 0x + 0y &=& 0. \end{array}$$

Since all vectors of $\mathbb{R}^2$ satisfy $0x + 0y = 0$, the system can be rewritten as

$$x - 2y = 3.$$

The parametric equations for this line are

$$x = 2r + 3,$$
$$y = r,$$

which means that the solution set to (3.46) is $\begin{bmatrix} 3 \\ 0 \end{bmatrix} + S\left(\begin{bmatrix} 2 \\ 1 \end{bmatrix}\right)$.

■ **Example 3.5.14**

The next system of linear equations is a slight modification of (3.46):

$$\begin{array}{rcl} 4x - 8y &=& 12, \\ -12x + 24y &=& 36. \end{array} \tag{3.47}$$

The reduced row echelon form of its augmented coefficient matrix is

$$\left[\begin{array}{cc|c} 1 & -2 & 0 \\ 0 & 0 & 1 \end{array}\right].$$

Because the last row of this matrix corresponds to the equation $0x + 0y = 1$, the solution set to (3.47) is $\emptyset$.

■ **Example 3.5.15**

As in Example 3.5.13, the following system of linear equations has infinitely many solutions:

$$\begin{array}{rcl} 2x_1 + 6x_2 - 2x_3 &=& 0, \\ x_1 + 5x_2 + x_3 &=& 0, \\ x_1 + 9x_2 + 5x_3 &=& 0. \end{array} \tag{3.48}$$

The reduced row echelon form of the coefficient matrix of (3.48) is

$$\begin{bmatrix} 1 & 0 & -4 \\ 0 & 1 & 1 \\ 0 & 0 & 0 \end{bmatrix}.$$

This implies that the solution set of (3.48) is, $S\left(\begin{bmatrix} 4 \\ -1 \\ 1 \end{bmatrix}\right)$, which is the line

containing **0** with parametric equations,

$$x = 4r,$$
$$y = -r,$$
$$z = r.$$

The solutions sets of the systems in Examples 3.5.13 and 3.5.15 are lines. This need not always be the case for systems that have infinitely many solutions. For example, if the reduced row echelon form of a coefficient matrix is

$$\begin{bmatrix} 1 & -5 & 9 \\ 0 & 0 & 0 \\ 0 & 0 & 0 \end{bmatrix},$$

the solution set of the corresponding homogeneous system of linear equations is the plane $x - 5y + 9z = 0$. Nonetheless, this solution set contains a line, which will always be the case when there are infinitely many solutions. This fact is crucial to the proof of a famous result.

■ **Theorem 3.5.16**

A system of linear equations has either no solutions, exactly one solution, or infinitely many solutions.

Proof

Let $A\mathbf{x} = \mathbf{b}$ represent a system of linear of equations with distinct solutions $\mathbf{u}_1$ and $\mathbf{u}_2$. The vectors in the line containing $\mathbf{u}_1$ and $\mathbf{u}_2$,

$$\mathbf{u}_1 + S(\mathbf{u}_2 - \mathbf{u}_1),$$

are also solutions to this system. To confirm this, take $r \in \mathbb{R}$ and evaluate,

$$A[\mathbf{u}_1 + r(\mathbf{u}_2 - \mathbf{u}_1)] = A\mathbf{u}_1 + Ar\mathbf{u}_2 - Ar\mathbf{u}_1$$
$$= A\mathbf{u}_1 + r(A\mathbf{u}_2 - A\mathbf{u}_1)$$
$$= \mathbf{b} + r(\mathbf{b} - \mathbf{b}) = \mathbf{b}. \blacksquare$$

Nonsquare Matrices

It is important to emphasize that Theorem 3.5.16 is not restricted to systems of linear equations with square coefficient matrices. It applies to all systems, including those that have more variables than equations (**underdetermined**) and those that have more equations than variables (**overdetermined**).

The coefficient matrix of an underdetermined system of linear equations will have more columns than rows. The solution set for such a system is guaranteed to have infinitely many solutions if it has any. For example, if the coefficient matrix is 3×4 and the system is homogeneous, the reduced row echelon form of such a matrix takes one of the following eight forms, where $*$ represents an arbitrary real number. Each of the associated systems of linear equations will have infinitely many solutions with the number of parameters being indicated for each matrix.

$$\begin{bmatrix} 1 & 0 & 0 & * \\ 0 & 1 & 0 & * \\ 0 & 0 & 1 & * \end{bmatrix} \text{ 1 parameter} \qquad \begin{bmatrix} 1 & 0 & * & 0 \\ 0 & 1 & * & 0 \\ 0 & 0 & 0 & 1 \end{bmatrix} \text{ 1 parameter}$$

$$\begin{bmatrix} 1 & 0 & * & * \\ 0 & 1 & * & * \\ 0 & 0 & 0 & 0 \end{bmatrix} \text{ 2 parameters} \qquad \begin{bmatrix} 1 & * & 0 & * \\ 0 & 0 & 1 & * \\ 0 & 0 & 0 & 0 \end{bmatrix} \text{ 2 parameters}$$

$$\begin{bmatrix} 1 & * & * & 0 \\ 0 & 0 & 0 & 1 \\ 0 & 0 & 0 & 0 \end{bmatrix} \text{ 2 parameters} \qquad \begin{bmatrix} 1 & * & * & * \\ 0 & 0 & 0 & 0 \\ 0 & 0 & 0 & 0 \end{bmatrix} \text{ 3 parameters}$$

$$\begin{bmatrix} 0 & 1 & * & * \\ 0 & 0 & 0 & 0 \\ 0 & 0 & 0 & 0 \end{bmatrix} \text{ 3 parameters} \qquad \begin{bmatrix} 0 & 0 & 1 & * \\ 0 & 0 & 0 & 0 \\ 0 & 0 & 0 & 0 \end{bmatrix} \text{ 3 parameters}$$

$$\begin{bmatrix} 0 & 0 & 0 & 1 \\ 0 & 0 & 0 & 0 \\ 0 & 0 & 0 & 0 \end{bmatrix} \text{ 3 parameters} \qquad \begin{bmatrix} 0 & 0 & 0 & 0 \\ 0 & 0 & 0 & 0 \\ 0 & 0 & 0 & 0 \end{bmatrix} \text{ 4 parameters}$$

■ Example 3.5.17

Consider the underdetermined system of linear equations,

$$\begin{array}{rcrcrcrcl} x_1 & + & 2x_2 & + & 3x_3 & + & 4x_4 & = & 0, \\ 5x_1 & + & 6x_2 & + & 7x_3 & + & 8x_3 & = & 0, \\ 9x_1 & & & + & x_3 & + & 2x_4 & = & 0. \end{array} \qquad (3.49)$$

Its coefficient matrix is $\begin{bmatrix} 1 & 2 & 3 & 4 \\ 5 & 6 & 7 & 8 \\ 9 & 0 & 1 & 2 \end{bmatrix}$. Using Gauss–Jordan elimination:

$$\begin{bmatrix} 1 & 2 & 3 & 4 \\ 5 & 6 & 7 & 8 \\ 9 & 0 & 1 & 2 \end{bmatrix} \xrightarrow{-5R_1+R_2} \begin{bmatrix} 1 & 2 & 3 & 4 \\ 0 & -4 & -8 & -12 \\ 9 & 0 & 1 & 2 \end{bmatrix}$$

$$\xrightarrow{-9R_1+R_3} \begin{bmatrix} 1 & 2 & 3 & 4 \\ 0 & -4 & -8 & -12 \\ 0 & -18 & -26 & -34 \end{bmatrix} \xrightarrow{-\frac{1}{4}R_2} \begin{bmatrix} 1 & 2 & 3 & 4 \\ 0 & 1 & 2 & 3 \\ 0 & -18 & -26 & -34 \end{bmatrix}$$

$$\xrightarrow{-2R_2+R_1} \begin{bmatrix} 1 & 0 & -1 & -2 \\ 0 & 1 & 2 & 3 \\ 0 & -18 & -26 & -34 \end{bmatrix} \xrightarrow{18R_2+R_3} \begin{bmatrix} 1 & 0 & -1 & -2 \\ 0 & 1 & 2 & 3 \\ 0 & 0 & 10 & 20 \end{bmatrix}$$

$$\xrightarrow{\frac{1}{10}R_3} \begin{bmatrix} 1 & 0 & -1 & -2 \\ 0 & 1 & 2 & 3 \\ 0 & 0 & 1 & 2 \end{bmatrix} \xrightarrow{R_3+R_1} \begin{bmatrix} 1 & 0 & 0 & 0 \\ 0 & 1 & 2 & 3 \\ 0 & 0 & 1 & 2 \end{bmatrix}$$

$$\xrightarrow{-2R_3+R_2} \begin{bmatrix} 1 & 0 & 0 & 0 \\ 0 & 1 & 0 & -1 \\ 0 & 0 & 1 & 2 \end{bmatrix}.$$

Therefore, the elementary row operation sequence,

$$-5R_1 + R_2, \quad -9R_1 + R_3, \quad -\frac{1}{4}R_2, \quad -2R_2 + R_1,$$

$$18R_2 + R_3, \quad \frac{1}{10}R_3, \quad R_3 + R_1, \quad -2R_3 + R_2,$$

shows that $\begin{bmatrix} 1 & 2 & 3 & 4 \\ 5 & 6 & 7 & 8 \\ 9 & 0 & 1 & 2 \end{bmatrix}$ is row equivalent to $\begin{bmatrix} 1 & 0 & 0 & 0 \\ 0 & 1 & 0 & -1 \\ 0 & 0 & 1 & 2 \end{bmatrix}$. The solution to (3.49) in parametric form is

$$x_1 = 0,$$
$$x_2 = r,$$
$$x_3 = -2r,$$
$$x_4 = r.$$

This means that the solution set to (3.49) is $S\left(\begin{bmatrix} 0 \\ 1 \\ -2 \\ 1 \end{bmatrix}\right)$.

■ **Example 3.5.18**

Removing the last equation from (3.49) gives the system,

$$\begin{array}{ccccccccc} x_1 & + & 2x_2 & + & 3x_3 & + & 4x_4 & = & 0, \\ 5x_1 & + & 6x_2 & + & 7x_3 & + & 8x_3 & = & 0. \end{array} \qquad (3.50)$$

This has

$$\begin{bmatrix} 1 & 2 & 3 & 4 \\ 5 & 6 & 7 & 8 \end{bmatrix}$$

as its coefficient matrix, which in reduced row echelon form is

$$\begin{bmatrix} 1 & 0 & -1 & -2 \\ 0 & 1 & 2 & 3 \end{bmatrix}.$$

Therefore, the parametric equations for the solution of (3.50) are

$$x_1 = r + 2s,$$
$$x_2 = -2r - 3s,$$
$$x_3 = r,$$
$$x_4 = s,$$

and its solution set is the plane containing the zero vector,

$$S\left(\begin{bmatrix} 1 \\ -2 \\ 1 \\ 0 \end{bmatrix}, \begin{bmatrix} 2 \\ -3 \\ 0 \\ 1 \end{bmatrix}\right).$$

If the constants of (3.50) are changed creating, say, the system,

$$\begin{array}{ccccccccc} x_1 & + & 2x_2 & + & 3x_3 & + & 4x_4 & = & 1, \\ 5x_1 & + & 6x_2 & + & 7x_3 & + & 8x_3 & = & 9, \end{array} \tag{3.51}$$

with

$$\left[\begin{array}{cccc|c} 1 & 2 & 3 & 4 & 1 \\ 5 & 6 & 7 & 8 & 9 \end{array}\right]$$

as its augmented coefficient matrix, its reduced row echelon form is

$$\left[\begin{array}{cccc|c} 1 & 0 & -1 & -2 & 3 \\ 0 & 1 & 2 & 3 & -1 \end{array}\right].$$

Hence, the solution set of (3.51) is the coset,

$$\begin{bmatrix} 3 \\ -1 \\ 0 \\ 0 \end{bmatrix} + S\left(\begin{bmatrix} 1 \\ -2 \\ 1 \\ 0 \end{bmatrix}, \begin{bmatrix} 2 \\ -3 \\ 0 \\ 1 \end{bmatrix}\right). \tag{3.52}$$

Every nonempty solution set of a system of linear equations has this form.

■ **Theorem 3.5.19**

For all $m \times n$ matrices A and $\mathbf{b} \in \mathbb{R}^m$, if there exists $\mathbf{u} \in \mathbb{R}^n$ such that $A\mathbf{u} = \mathbf{b}$,

$$\{\mathbf{x} : A\mathbf{x} = \mathbf{b}\} = \{\mathbf{u} + \mathbf{x} : A\mathbf{x} = \mathbf{0}\}.$$

Proof

Let $M = \{\mathbf{x} : A\mathbf{x} = \mathbf{b}\}$ and $N = \{\mathbf{u} + \mathbf{x} : A\mathbf{x} = \mathbf{0}\}$. Suppose $\mathbf{v} \in M$. This means that $A\mathbf{v} = \mathbf{b}$. Because $\mathbf{v} = \mathbf{u} + (\mathbf{v} - \mathbf{u})$ and $A(\mathbf{v} - \mathbf{u}) = \mathbf{b} - \mathbf{b} = \mathbf{0}$, it follows that $\mathbf{v} \in N$. Conversely, let $\mathbf{v} \in N$, so $\mathbf{v} = \mathbf{u} + \mathbf{w}$, where $A\mathbf{w} = \mathbf{0}$. Hence, $\mathbf{v} \in M$ since $A\mathbf{v} = A(\mathbf{u} + \mathbf{w}) = A\mathbf{u} + A\mathbf{w} = \mathbf{b} + \mathbf{0} = \mathbf{b}$. ■

The coefficient matrix of an overdetermined system of linear equations will have more rows than columns. It is possible for such a system to have exactly one solution in addition to the other two possibilities of Theorem 3.5.16. To illustrate the possibilities of an overdetermined system, consider a homogeneous system of linear equations with a 4×2 coefficient matrix. The reduced row echelon form of such a matrix takes one of the following seven forms.

$$\begin{bmatrix} 1 & 0 & 0 \\ 0 & 1 & 0 \\ 0 & 0 & 1 \\ 0 & 0 & 0 \end{bmatrix} \text{ 0 parameters} \qquad \begin{bmatrix} 1 & 0 & * \\ 0 & 1 & * \\ 0 & 0 & 0 \\ 0 & 0 & 0 \end{bmatrix} \text{ 1 parameter}$$

$$\begin{bmatrix} 1 & * & 0 \\ 0 & 0 & 1 \\ 0 & 0 & 0 \\ 0 & 0 & 0 \end{bmatrix} \text{ 1 parameter} \qquad \begin{bmatrix} 1 & * & * \\ 0 & 0 & 0 \\ 0 & 0 & 0 \\ 0 & 0 & 0 \end{bmatrix} \text{ 2 parameters}$$

$$\begin{bmatrix} 0 & 1 & * \\ 0 & 0 & 0 \\ 0 & 0 & 0 \\ 0 & 0 & 0 \end{bmatrix} \text{ 2 parameters} \qquad \begin{bmatrix} 0 & 0 & 1 \\ 0 & 0 & 0 \\ 0 & 0 & 0 \\ 0 & 0 & 0 \end{bmatrix} \text{ 2 parameters}$$

$$\begin{bmatrix} 0 & 0 & 0 \\ 0 & 0 & 0 \\ 0 & 0 & 0 \\ 0 & 0 & 0 \end{bmatrix} \text{ 3 parameters}$$

■ **Example 3.5.20**

The reduced row echelon form of the augmented coefficient matrix of the overdetermined system of linear equations,

$$\begin{aligned} x - 3y &= 1, \\ x + 2y &= 2, \\ 4x + y &= 1, \\ -x - y &= 4, \end{aligned} \tag{3.53}$$

is

$$\begin{bmatrix} 1 & 0 & 0 \\ 0 & 1 & 0 \\ 0 & 0 & 1 \\ 0 & 0 & 0 \end{bmatrix}.$$

Therefore, (3.53) has no solution. However, the reduced row echelon form of the coefficient matrix for the homogeneous system

$$\begin{aligned}
x - 3y &= 0, \\
x + 2y &= 0, \\
4x + y &= 0, \\
-x - y &= 0,
\end{aligned} \tag{3.54}$$

is $\begin{bmatrix} 1 & 0 \\ 0 & 1 \\ 0 & 0 \\ 0 & 0 \end{bmatrix}$, which has $\left\{ \begin{bmatrix} 0 \\ 0 \end{bmatrix} \right\}$ as its solution set.

The examples starting with Example 3.5.11 and ending with Example 3.5.20 appear to be working under the assumption that the reduced row echelon matrix found is unique to the initial matrix, that it is a type of characteristic of the matrix. This is indeed the case.

■ Theorem 3.5.21

Every matrix is row equivalent to exactly one matrix in reduced row echelon form.

Proof

Let A be an $m \times n$ matrix. Proceed by induction on n. The basis case of $n = 1$ is left to Exercise 12, so let $n > 1$. Suppose that A is row equivalent to B and C that are in reduced row echelon form. Let A', B', and C' be the matrices that result from deleting column n from A, B, and C, respectively. Also, let $\mathbf{a}$, $\mathbf{b}$, and $\mathbf{c}$ be column n for A, B, and C, respectively. This means that A can be viewed as an augmented matrix representing the system $A'\mathbf{x} = \mathbf{a}$. Likewise, B represents $B'\mathbf{x} = \mathbf{b}$, and C represents $C'\mathbf{x} = \mathbf{c}$. Since B' and C' are the reduced row echelon forms of A', it follows that $B' = C'$ by induction. There are two cases to consider.

- The last column of B has a leading 1. This implies that the other entries in column n of B are 0 and B is of the form

$$B = \begin{bmatrix}
\cdots & 1 & \cdots & 0 & \cdots & 0 \\
\cdots & 0 & \cdots & 1 & \cdots & 0 \\
& \vdots & & \vdots & & \vdots \\
\cdots & 0 & \cdots & 0 & \cdots & 1 \\
\cdots & 0 & \cdots & 0 & \cdots & 0 \\
& \vdots & & \vdots & & \vdots
\end{bmatrix}.$$

Then, $B'\mathbf{x} = \mathbf{b}$ has no solution, so $A'\mathbf{x} = \mathbf{a}$ has no solution. Therefore, $C'\mathbf{x} = \mathbf{c}$ has no solution because C is row equivalent to A. Hence, column n of C must have a leading 1. Since $B' = C'$, the leading 1 in the last columns of B and C must be in the same row because all zero rows are at the bottom of each matrix. Therefore, $B = C$.

- The last column of B does not have a leading 1. This implies that the system $B'\mathbf{x} = \mathbf{b}$ has a solution $\mathbf{u}$. Hence, $C'\mathbf{u} = \mathbf{c}$ by Theorem 3.5.7. Because $B' = C'$ it must be the case that $\mathbf{b} = \mathbf{c}$, which implies that $B = C$. ∎

Gaussian Elimination

Although Guass-Jordan elimination results in a unique reduced row echelon form matrix, it does have the disadvantage of having many steps that are sometimes inefficient. If speed is an issue or the existence of a solution is all that is needed, the following technique can be used.

■ **Definition 3.5.22**

A matrix A is in **row echelon form** when both of the following hold.

(a) The first nonzero entry of a row (called a **pivot**) is to the right of the pivot in any row above it.

(b) There is no zero row above a row with a nonzero entry.

The process of transforming a matrix into row echelon form using only the elementary row operations is called **Gaussian elimination**.

■ **Example 3.5.23**

Consider the system of linear equations,

$$
\begin{array}{rcrcrcr}
2x & - & y & - & 2z & = & 1, \\
x & + & y & + & 3z & = & 3, \\
-x & + & 5y & + & 2z & = & -2.
\end{array}
\qquad (3.55)
$$

The augmented coefficient matrix of (3.55) is

$$
\left[\begin{array}{ccc|c}
2 & -1 & -2 & 1 \\
1 & 1 & 3 & 3 \\
-1 & 5 & 2 & -2
\end{array}\right].
$$

Use Gaussian elimination to put this matrix into row echelon form, taking two extra steps to rationalize the entries:

$$
\left[\begin{array}{ccc|c}
2 & -1 & -2 & 1 \\
1 & 1 & 3 & 3 \\
-1 & 5 & 2 & -2
\end{array}\right]
\xrightarrow{R_3+R_2}
\left[\begin{array}{ccc|c}
2 & -1 & -2 & 1 \\
0 & 6 & 5 & 1 \\
-1 & 5 & 2 & -2
\end{array}\right]
$$

$$
\xrightarrow{\frac{1}{2}R_1+R_3}
\left[\begin{array}{ccc|c}
2 & -1 & -2 & 1 \\
0 & 6 & 5 & 1 \\
0 & 9/2 & 1 & -3/2
\end{array}\right]
\xrightarrow{2R_3}
\left[\begin{array}{ccc|c}
2 & -1 & -2 & 1 \\
0 & 6 & 5 & 1 \\
0 & 9 & 2 & -3
\end{array}\right]
$$

$$
\xrightarrow{-\frac{3}{2}R_2+R_3}
\left[\begin{array}{ccc|c}
2 & -1 & -2 & 1 \\
0 & 6 & 5 & 1 \\
0 & 0 & -11/2 & -9/2
\end{array}\right]
\xrightarrow{2R_3}
\left[\begin{array}{ccc|c}
2 & -1 & -2 & 1 \\
0 & 6 & 5 & 1 \\
0 & 0 & -11 & -9
\end{array}\right]
$$

At this point since the coefficient matrix is square, it is apparent that the system has a unique solution. So, if this knowledge is all that is required, the work can stop, but if the solution is required, use a method called **back substitution**. Write the system given by the row echelon matrix,

$$
\begin{array}{rcrcrcl}
2x & - & y & - & 2z & = & 1, \\
 & & 6y & + & 5z & = & 1, \\
 & & & - & 11z & = & -9.
\end{array}
\tag{3.56}
$$

Then,

$$
z = \frac{9}{11},
$$

and

$$
y = -\frac{17}{33} \quad \text{because} \quad 6y + 5\left(\frac{9}{11}\right) = 1,
$$

and

$$
x = \frac{35}{33} \quad \text{because} \quad 2x + \frac{17}{33} - 2\left(\frac{9}{11}\right) = 1.
$$

In Example 3.5.23, both

$$
\left[\begin{array}{ccc|c}
2 & -1 & -2 & 1 \\
0 & 6 & 5 & 1 \\
0 & 0 & -11/2 & -9/2
\end{array}\right]
\quad \text{and} \quad
\left[\begin{array}{ccc|c}
2 & -1 & -2 & 1 \\
0 & 6 & 5 & 1 \\
0 & 0 & -11 & -9
\end{array}\right]
$$

are row equivalent matrices in row echelon form, proving that the echelon form of a matrix is not unique.

■ **Example 3.5.24**

Gaussian elimination with back substitution can also be used on more complicated systems. To solve the overdetermined system,

$$
\begin{array}{rcrcrcl}
\dfrac{\sqrt{6}}{6}x & + & \dfrac{\sqrt{3}}{6}y & - & \dfrac{\sqrt{3}}{6}z & = & 0, \\[2mm]
 & & -\dfrac{\sqrt{3}}{2}y & + & \dfrac{\sqrt{3}}{6}z & = & 1, \\[2mm]
\dfrac{\sqrt{6}}{6}x & + & \dfrac{\sqrt{3}}{6}y & + & \dfrac{\sqrt{3}}{2}z & = & 2, \\[2mm]
\dfrac{\sqrt{6}}{3}x & - & \dfrac{\sqrt{3}}{6}y & - & \dfrac{\sqrt{3}}{6}z & = & 1,
\end{array}
$$

first clear the denominators to find the system's augmented coefficient matrix,

$$
A = \left[\begin{array}{ccc|c}
\sqrt{6} & \sqrt{3} & -\sqrt{3} & 0 \\
0 & -3\sqrt{3} & \sqrt{3} & 6 \\
\sqrt{6} & \sqrt{3} & 3\sqrt{3} & 12 \\
2\sqrt{6} & -\sqrt{3} & -\sqrt{3} & 6
\end{array}\right]
$$

A row echelon form of A is

$$\left[\begin{array}{ccc|c} \sqrt{6} & \sqrt{3} & -\sqrt{3} & 0 \\ 0 & -3\sqrt{3} & \sqrt{3} & 6 \\ 0 & 0 & 4\sqrt{3} & 12 \\ 0 & 0 & 0 & 0 \end{array}\right],$$

and back substitution yields the solution set, $\left\{\begin{bmatrix} 4/\sqrt{6} \\ -1/\sqrt{3} \\ 3/\sqrt{3} \end{bmatrix}\right\}$.

Exercises

1. List all of the reduced row echelon form possibilities for matrices of the given sizes.

    (a) 3×3 (b) 2×5 (c) 1×6 (d) 4×2

2. Solve the given systems of linear equations using Gauss–Jordan elimination.

    (a) $\begin{aligned} 4x + y &= 5 \\ 3x + 7y &= 0 \end{aligned}$

    (b) $\begin{aligned} 4x + y &= 5 \\ 3x + 7y &= 0 \end{aligned}$

    (c) $\begin{aligned} 4x + y &= 5 \\ 3x + 7y &= 0 \end{aligned}$

    (d) $\begin{aligned} x + 4y + 3z &= 0 \\ x + 6y + 7z &= 0 \\ -x - 2y + z &= 0 \end{aligned}$

    (e) $\begin{aligned} 2x - 3y + 4z &= -5 \\ -6x + 7y - 8z &= 9 \\ 2x - 6y + 8z &= 4 \end{aligned}$

    (f) $\begin{aligned} 5x + 2y - 3z &= 7 \\ x + 10y - 5z &= 8 \\ 2x - 4y + z &= 1 \end{aligned}$

    (g) $\begin{aligned} x_1 + 2x_2 + 4x_3 + 8x_4 &= 0 \\ x_1 - 4x_2 + 16x_3 - 64x_4 &= 0 \\ x_1 - 3x_2 + 9x_3 - 27x_4 &= 0 \\ x_1 + 5x_2 + 25x_3 + 125x_4 &= 0 \end{aligned}$

    (h) $\begin{aligned} x - 2y + 3z &= -4 \\ -5x + 6y - 7z &= 8 \end{aligned}$

    (i) $\begin{aligned} -x_1 + 2x_2 + x_3 - 2x_4 &= -2 \\ -x_1 + x_2 - 3x_3 + 3x_4 &= 3 \\ x_1 + x_2 - 3x_3 + x_4 &= 1 \end{aligned}$

$$
\begin{array}{rrrrrr}
x_1 & + & 4x_2 & - & 3x_3 & + & 2x_4 & + & 2x_5 & = & -1 \\
3x_1 & - & 2x_2 & + & 2x_3 & + & x_4 & + & x_5 & = & 2 \\
2x_1 & + & x_2 & - & 4x_3 & + & 2x_4 & + & 2x_5 & = & 5
\end{array}
$$

(j)

$$
\begin{array}{rrrrr}
2x_1 & - & 4x_2 & + & 2x_3 & & & + & 4x_5 & = & 0 \\
& & 3x_2 & + & 3x_3 & + & x_4 & + & 2x_5 & = & 0 \\
& & & & 4x_3 & + & 3x_4 & - & 2x_5 & = & 0 \\
& & & & & & 2x_4 & + & 6x_5 & = & 0
\end{array}
$$

(k)

$$
\begin{array}{rrr}
9x & + & 8y & = & 7 \\
4x & + & 5y & = & 6 \\
3x & + & 2y & = & 1
\end{array}
$$

(l)

$$
\begin{array}{rrrrr}
-2x_1 & + & 2x_2 & + & x_3 & + & 5x_4 & = & 0 \\
6x_1 & + & 4x_2 & - & x_3 & + & x_4 & = & 0 \\
3x_1 & + & 2x_2 & - & 4x_3 & - & 2x_4 & = & 0 \\
-2x_1 & + & x_2 & - & x_3 & + & 3x_4 & = & 0 \\
x_1 & - & 2x_2 & + & 5x_3 & + & x_4 & = & 0
\end{array}
$$

(m)

3. Use Gauss–Jordan elimination to find the kernel of the given linear operators.

   (a) $T\begin{bmatrix} x \\ y \end{bmatrix} = \begin{bmatrix} 4x + 2y \\ x - 5y \end{bmatrix}$

   (b) $T\begin{bmatrix} x \\ y \\ z \end{bmatrix} = \begin{bmatrix} x + 4y + 3z \\ x + 6y + 7z \\ -x - 2y + z \end{bmatrix}$

   (c) $T\begin{bmatrix} x \\ y \\ z \end{bmatrix} = \begin{bmatrix} x + y - z \\ 3x + 6y + 9z \\ 4x + 7y + 8z \end{bmatrix}$

4. Use Gauss–Jordan elimination to determine whether the given vector is in the range of the given linear operator.

   (a) $\begin{bmatrix} 5 \\ 0 \end{bmatrix}$, $T\begin{bmatrix} x \\ y \end{bmatrix} = \begin{bmatrix} 4x + y \\ 3x + 7y \end{bmatrix}$

   (b) $\begin{bmatrix} 5 \\ 9 \\ 4 \end{bmatrix}$, $T\begin{bmatrix} x \\ y \\ z \end{bmatrix} = \begin{bmatrix} 2x - 3y + 4z \\ 6x + 7y - 8z \\ 2x - 6y + 8z \end{bmatrix}$

   (c) $\begin{bmatrix} 7 \\ 8 \\ 1 \end{bmatrix}$, $T\begin{bmatrix} x \\ y \\ z \end{bmatrix} = \begin{bmatrix} 5x + 2y - 3z \\ x + 10y - 5z \\ 2x - 4y + z \end{bmatrix}$

5. Use Gauss–Jordan elimination to find the range of the given linear operators.

   (a) $T\begin{bmatrix} x \\ y \end{bmatrix} = \begin{bmatrix} 4x + 2y \\ x - 5y \end{bmatrix}$

   (b) $T\begin{bmatrix} x \\ y \\ z \end{bmatrix} = \begin{bmatrix} x + 4y + 3z \\ x + 6y + 7z \\ -x - 2y + z \end{bmatrix}$

   (c) $T\begin{bmatrix} x \\ y \\ z \end{bmatrix} = \begin{bmatrix} x + y - z \\ 3x + 6y + 9z \\ 4x + 7y + 8z \end{bmatrix}$

6. Let $\mathbf{r}_1, \mathbf{r}_2, \dots, \mathbf{r}_m \in \mathbb{R}^n$. Prove the following.

   (a) If s is a nonzero scalar, $S(\mathbf{r}_1, \mathbf{r}_2, \dots, \mathbf{r}_m) = S(s\mathbf{r}_1, \mathbf{r}_2, \dots, \mathbf{r}_m)$.

   (b) If $s \in \mathbb{R}$, then $S(\mathbf{r}_1, \mathbf{r}_2, \dots, \mathbf{r}_m) = S(s\mathbf{r}_i + \mathbf{r}_1, \mathbf{r}_2, \dots, \mathbf{r}_m)$, for all $i = 1, 2, \dots m$.

7. Find the coefficients in the equation of the parabola $y = ax^2 + bx + c$ that passes through the points $(1, -1)$, $(2, 3)$, and $(3, 13)$.

8. Find the coefficients in the equation of the cubic $y = ax^3 + bx^2 + cx + d$ that passes through the points $(-2, -35)$, $(-1, -7)$, $(1, 5)$, and $(2, 11)$.

9. Find angles θ_1, θ_2, and θ_3 such that $0 \le \theta_1, \theta_2, \theta_3 \le 2\pi$, and

$$\begin{array}{rcrcrcr}
\cos\theta_1 & + & 2\sin\theta_2 & + & \tan\theta_3 & = & 1, \\
\cos\theta_1 & - & 4\sin\theta_2 & - & 3\tan\theta_3 & = & -6, \\
3\cos\theta_1 & + & 2\sin\theta_2 & + & 2\tan\theta_3 & = & 0.
\end{array}$$

10. Find all values of a and b so that the given system of equations has infinitely many solutions, no solutions, and a unique solution.

$$\begin{array}{rcrcrcr}
ax & + & by & + & z & = & 9, \\
ax & + & (2b+1)y & + & 2z & = & 4, \\
ax & + & by & + & (b+1)z & = & 5b+1.
\end{array}$$

11. Find all values of a so that the given system of equations has infinitely many solutions, no solutions, and a unique solution.

$$\begin{array}{rcrcrcr}
x & + & 4y & - & 2z & = & -6, \\
2x & + & y & + & 3z & = & 9, \\
3x & + & y & + & (a^2-5)z & = & 4a+2.
\end{array}$$

12. Confirm the basis case in the proof of Theorem 3.5.21.

13. Suppose that A is an $m \times n$ matrix and $\mathbf{b} \in \mathbb{R}^m$. Prove or show false: if $\mathbf{u}$ and $\mathbf{v}$ are solutions to $A\mathbf{x} = \mathbf{b}$, then $r\mathbf{u} + s\mathbf{v}$ is a solution to $A\mathbf{x} = \mathbf{b}$ for all $r, s \in \mathbb{R}$.

14. Let A be a matrix. Prove that the solution set of $A\mathbf{x} = \mathbf{0}$ is the set of all vectors that are orthogonal to the columns of A.

15. Let A be an $m \times n$ matrix and $\mathbf{b} \in \mathbb{R}^m$. Prove that the following are equivalent.

- $A\mathbf{x} = \mathbf{b}$ is consistent.

- $\mathbf{b}$ is a linear combination of the columns of A.

- The reduced row echelon form of the augmented matrix $[A \mid \mathbf{b}]$ does not have a row of the form $[\,0 \ 0 \ \cdots \ 0 \mid a\,]$ with $a \ne 0$.

16. Use Gaussian elimination to solve the given systems of linear equations.

(a)
$$\begin{array}{rcrcr}
6x & - & 4y & = & 6 \\
2x & - & y & = & -8
\end{array}$$

(b)
$$\begin{array}{rcrcr}
5x & - & 3y & = & 2 \\
-10x & + & 6y & = & -4
\end{array}$$

(c)
$$\begin{array}{rcrcrcr}
-3x & + & 2y & - & 8z & = & 0 \\
-8x & + & & & 8z & = & 0 \\
6x & + & 2y & & & = & 0
\end{array}$$

(d)
$$\begin{array}{rcrcrcr}
5x & + & 4y & + & 7z & = & 0 \\
3x & + & 6y & + & 13z & = & -2 \\
-2x & + & 2y & + & 6z & = & -2
\end{array}$$

(e)
$$
\begin{aligned}
-2x_1 + x_2 - 7x_3 - 4x_4 &= 0 \\
3x_1 + 6x_3 + x_4 &= 0 \\
8x_1 + 5x_2 - 7x_3 - 10x_4 &= 0 \\
-9x_1 - 7x_2 + x_3 - 7x_4 &= 0
\end{aligned}
$$

(f)
$$
\begin{aligned}
6x_1 + 4x_2 + x_3 + 5x_4 &= 5 \\
3x_1 - 2x_2 - 7x_3 - 6x_4 &= 2 \\
12x_1 - 13x_3 - 7x_4 &= 9 \\
3x_1 + 6x_2 + 8x_3 + 11x_4 &= 3
\end{aligned}
$$

CHAPTER 4

Invertibility

4.1 Invertible Matrices

Consider the linear operator $T : \mathbb{R} \to \mathbb{R}$ defined by $Tx = 4x$. It is a bijection, so its inverse is a function $\mathbb{R} \to \mathbb{R}$ (Theorem 1.4.25), which is

$$T^{-1}x = \frac{1}{4}x.$$

To prove that T^{-1} is a linear operator, take $u_1, u_2, r \in \mathbb{R}$. Then,

$$T^{-1}(u_1 + u_2) = \frac{1}{4}(u_1 + u_2) = \frac{1}{4}u_1 + \frac{1}{4}u_2 = T^{-1}u_1 + T^{-1}u_2,$$

and

$$T^{-1}(ru_1) = \frac{1}{4}(ru_1) = r\left(\frac{1}{4}u_1\right) = r(T^{-1}u_1).$$

This simple example suggests a general result.

Linear Algebra, First Edition. Michael L. O'Leary.
© 2021 John Wiley & Sons, Inc. Published 2021 by John Wiley & Sons, Inc.
Companion Website: www.wiley.com/go/o'leary/linearalgebra

183

■ Theorem 4.1.1

If $T : \mathbb{R}^n \to \mathbb{R}^m$ is an invertible linear transformation, $T^{-1} : \mathbb{R}^m \to \mathbb{R}^n$ is an invertible linear transformation.

Proof

Assume that $T : \mathbb{R}^n \to \mathbb{R}^m$ is a linear transformation that is invertible. Recall that T^{-1} is invertible because T is invertible. To prove that T^{-1} is a linear transformation, let $\mathbf{v}_1, \mathbf{v}_2 \in \mathbb{R}^m$. Because T is onto by Theorem 1.4.25, there are $\mathbf{u}_1, \mathbf{u}_2 \in \mathbb{R}^n$ such that $T\mathbf{u}_1 = \mathbf{v}_1$ and $T\mathbf{u}_2 = \mathbf{v}_2$. Thus, $T^{-1}\mathbf{v}_1 = \mathbf{u}_1$ and $T^{-1}\mathbf{v}_2 = \mathbf{u}_2$. Relying on the fact that $T(\mathbf{u}_1 + \mathbf{u}_2) = T\mathbf{u}_1 + T\mathbf{u}_2 = \mathbf{v}_1 + \mathbf{v}_2$, it follows that

$$T^{-1}(\mathbf{v}_1 + \mathbf{v}_2) = \mathbf{u}_1 + \mathbf{u}_2 = T^{-1}\mathbf{v}_1 + T^{-1}\mathbf{v}_2.$$

In addition, take $r \in \mathbb{R}$. Because $T(r\mathbf{u}_1) = r(T\mathbf{u}_1) = r\mathbf{v}_1$,

$$T^{-1}(r\mathbf{v}_1) = r\mathbf{u}_1 = r(T^{-1}\mathbf{v}_1). \blacksquare$$

Since every linear transformation can be described by a matrix as noted in Theorem 3.1.19, the next goal should be to determine the standard matrix for the inverse of an invertible linear transformation. Since the inverse function is defined in terms of the given function, the sought-after standard matrix should also be found using the standard matrix of the given linear transformation.

Let $T : \mathbb{R}^n \to \mathbb{R}^m$ be an invertible linear transformation. Write $A = [T]$ and $B = [T^{-1}]$. Theorem 1.4.28 implies that for all $\mathbf{u} \in \mathbb{R}^n$,

$$\mathbf{u} = (T \circ T^{-1})\mathbf{u} = T(B\mathbf{u}) = A(B\mathbf{u}) = AB\mathbf{u}, \tag{4.1}$$

and

$$\mathbf{u} = (T^{-1} \circ T)\mathbf{u} = T^{-1}(A\mathbf{u}) = B(A\mathbf{u}) = BA\mathbf{u}. \tag{4.2}$$

Using the standard vectors $\mathbf{e}_1, \mathbf{e}_2, \dots, \mathbf{e}_n$ of $\mathbb{R}^n$, an appeal to (4.1) gives

$$(AB)\mathbf{e}_i = \mathbf{e}_i$$

for all $i = 1, 2, \dots, n$. The columns of I_n are $\mathbf{e}_1, \mathbf{e}_2, \dots, \mathbf{e}_n$, so Theorem 3.2.14(a) implies that

$$AB = (AB)I_n = I_n.$$

Similarly, (4.2) implies that $BA = I_n$. Therefore, B is the multiplicative inverse of A, so A must be a square matrix and $n = m$. All of this leads to the next definition.

■ Definition 4.1.2

The $n \times n$ matrix A is **invertible** (or **nonsingular**) if there exists an $n \times n$ matrix B such that

$$AB = BA = I_n.$$

The matrix B is the **(multiplicative) inverse** of A (written $B = A^{-1}$). If A is not invertible, it is called **noninvertible** (or **singular**).

Continuing the work before Definition 4.1.2, suppose that $[T]$ is invertible. Take $\mathbf{u}_1, \mathbf{u}_2 \in \mathbb{R}^n$ such that $T\mathbf{u}_1 = T\mathbf{u}_2$. This implies that

$$[T]\mathbf{u}_1 = [T]\mathbf{u}_2$$
$$[T]^{-1}[T]\mathbf{u}_1 = [T]^{-1}[T]\mathbf{u}_2$$
$$\mathbf{u}_1 = \mathbf{u}_2,$$

so T is one-to-one. Also, $T[T]^{-1}\mathbf{u} = [T][T]^{-1}\mathbf{u} = \mathbf{u}$, for any $\mathbf{u} \in \mathbb{R}^n$, so T is onto. Hence, combining the work with Theorem 1.4.25 yields the next important result connecting invertible linear operators and invertible matrices.

■ **Theorem 4.1.3**

Let $T : \mathbb{R}^n \to \mathbb{R}^n$ be a linear operator.

   (a) T is invertible if and only if $[T]$ is invertible.

   (b) $[T^{-1}] = [T]^{-1}$ if T is invertible.

■ **Example 4.1.4**

Not every nonzero linear operator $\mathbb{R}^n \to \mathbb{R}^n$ is invertible. For example, $T : \mathbb{R}^2 \to \mathbb{R}^2$ defined by

$$T\begin{bmatrix} x \\ y \end{bmatrix} = \begin{bmatrix} x \\ 0 \end{bmatrix}$$

is not invertible because it is not a bijection (Theorem 1.4.25). This suggests that its standard matrix,

$$[T] = \begin{bmatrix} 1 & 0 \\ 0 & 0 \end{bmatrix},$$

should not have a multiplicative inverse. This is confirmed by observing that the matrix equation, $\begin{bmatrix} 1 & 0 \\ 0 & 0 \end{bmatrix}\begin{bmatrix} a & b \\ c & d \end{bmatrix} = \begin{bmatrix} 1 & 0 \\ 0 & 1 \end{bmatrix}$, has no solution.

■ **Example 4.1.5**

Consider the 1×1 matrix $[a]$. If $a \neq 0$, then $[a]^{-1} = [1/a]$ because both $[a][1/a] = [1]$ and $[1/a][a] = [1]$. This is not surprising because $\mathbb{R}^{1\times 1}$ is essentially $\mathbb{R}$.

■ **Example 4.1.6**

Consider the 2×2 matrix $A = \begin{bmatrix} a & b \\ c & d \end{bmatrix}$. To find A^{-1}, it is required to demonstrate two matrix equations as in Example 4.1.5. First, solve

$$\begin{bmatrix} a & b \\ c & d \end{bmatrix}\begin{bmatrix} x & y \\ z & w \end{bmatrix} = \begin{bmatrix} 1 & 0 \\ 0 & 1 \end{bmatrix}.$$

This is done by solving the system,

$$
\begin{array}{rcl}
ax + bz &=& 1, \\
ay + bw &=& 0, \\
cx + dz &=& 0, \\
cy + dw &=& 1.
\end{array}
\tag{4.3}
$$

By a careful use of Gauss–Jordan elimination, the solution to this system of linear equations is seen to be

$$
x = \frac{d}{ad - bc}, \quad y = \frac{-b}{ad - bc}, \quad z = \frac{-c}{ad - bc}, \quad w = \frac{a}{ad - bc},
$$

provided that $ad - bc \neq 0$. Second, solve the matrix equation,

$$
\begin{bmatrix} x & y \\ z & w \end{bmatrix} \begin{bmatrix} a & b \\ c & d \end{bmatrix} = \begin{bmatrix} 1 & 0 \\ 0 & 1 \end{bmatrix}.
$$

This yields a system similar to (4.3) and has an identical solution. Therefore, if $ad - bc \neq 0$,

$$
\begin{bmatrix} a & b \\ c & d \end{bmatrix}^{-1} = \frac{1}{ad - bc} \begin{bmatrix} d & -b \\ -c & a \end{bmatrix}.
$$

Since the work in Example 4.1.6 is significantly more complicated than that of Example 4.1.5, it is probable that finding the inverse of an invertible 3×3 matrix by generalizing the strategy of Example 4.1.6 will be overly complicated. For this reason, it is advantageous to seek another method.

Elementary Matrices

Gauss–Jordan elimination uses the elementary row operations that are given in Definition 3.5.4 to find the reduced row echelon form of a matrix. Starting with a matrix, a sequence of matrices is found, each matrix being the result of an elementary row operation on the previous matrix, culminating in the reduced row echelon form of the original matrix. It would be preferred, however, if Gauss–Jordan elimination resulted in an equation instead of a sequence. For example, if A is a given $n \times n$ matrix and its reduced row echelon form is R, it would be advantageous if a matrix B could be found such that $BA = R$.

Take the matrix $A = \begin{bmatrix} 1 & 2 \\ 3 & 4 \end{bmatrix}$ and perform elementary row operations on it. For example, $R_1 \leftrightarrow R_2$ gives $A_1 = \begin{bmatrix} 3 & 4 \\ 1 & 2 \end{bmatrix}$, $5R_2$ gives $A_2 = \begin{bmatrix} 1 & 2 \\ 15 & 20 \end{bmatrix}$, and $-4R_2 + R_1$ gives $A_3 = \begin{bmatrix} -11 & -14 \\ 3 & 4 \end{bmatrix}$. It should be noted that each of A_1, A_2, and A_3 is the product of a 2×2 matrix with A. Specifically,

$$
A_1 = \begin{bmatrix} 0 & 1 \\ 1 & 0 \end{bmatrix} \begin{bmatrix} 1 & 2 \\ 3 & 4 \end{bmatrix}, \quad A_2 = \begin{bmatrix} 1 & 0 \\ 0 & 5 \end{bmatrix} \begin{bmatrix} 1 & 2 \\ 3 & 4 \end{bmatrix}, \quad A_3 = \begin{bmatrix} 1 & -4 \\ 0 & 1 \end{bmatrix} \begin{bmatrix} 1 & 2 \\ 3 & 4 \end{bmatrix}.
$$

Let

$$E_1 = \begin{bmatrix} 0 & 1 \\ 1 & 0 \end{bmatrix}, \quad E_2 = \begin{bmatrix} 1 & 0 \\ 0 & 5 \end{bmatrix}, \quad E_3 = \begin{bmatrix} 1 & -4 \\ 0 & 1 \end{bmatrix}. \tag{4.4}$$

Notice that E_1 is obtained from I_2 using $R_1 \leftrightarrow R_2$, E_2 is obtained from I_2 using $5R_2$, and E_3 is obtained from I_2 using $-4R_2 + R_1$. That is, the elementary row operations used on A to obtain A_1, A_2, and A_3 are the same operations used on I_n to obtain E_1, E_2, and E_3. Furthermore, multiplying A on the left by E_1, E_2, and E_3 had the same result as performing the corresponding elementary row operation on A. All of this is now generalized.

■ Definition 4.1.7

The $n \times n$ matrix E is an **elementary matrix** if E is obtained from I_n using exactly one row operation, which is denoted by op E.

Using the notation of Definition 3.5.4, this means that E_1, E_2, and E_3 from (4.4) are elementary matrices, and

$$\text{op}\, E_1 = R_1 \leftrightarrow R_2, \quad \text{op}\, E_2 = 5R_2, \quad \text{op}\, E_3 = -4R_2 + R_1.$$

■ Example 4.1.8

Let $s \in \mathbb{R}$. The elementary 3×3 matrices are

$$E_1 = \begin{bmatrix} 0 & 1 & 0 \\ 1 & 0 & 0 \\ 0 & 0 & 1 \end{bmatrix}, \quad E_2 = \begin{bmatrix} 0 & 0 & 1 \\ 0 & 1 & 0 \\ 1 & 0 & 0 \end{bmatrix}, \quad E_3 = \begin{bmatrix} 1 & 0 & 0 \\ 0 & 0 & 1 \\ 0 & 1 & 0 \end{bmatrix},$$

which are obtained using $R_i \leftrightarrow R_j$,

$$E_4 = \begin{bmatrix} s & 0 & 0 \\ 0 & 1 & 0 \\ 0 & 0 & 1 \end{bmatrix}, \quad E_5 = \begin{bmatrix} 1 & 0 & 0 \\ 0 & s & 0 \\ 0 & 0 & 1 \end{bmatrix}, \quad E_6 = \begin{bmatrix} 1 & 0 & 0 \\ 0 & 1 & 0 \\ 0 & 0 & s \end{bmatrix},$$

which are obtained using sR_i, and

$$E_7 = \begin{bmatrix} 1 & 0 & 0 \\ s & 1 & 0 \\ 0 & 0 & 1 \end{bmatrix}, \quad E_8 = \begin{bmatrix} 1 & 0 & 0 \\ 0 & 1 & 0 \\ s & 0 & 1 \end{bmatrix}, \quad E_9 = \begin{bmatrix} 1 & s & 0 \\ 0 & 1 & 0 \\ 0 & 0 & 1 \end{bmatrix},$$

$$E_{10} = \begin{bmatrix} 1 & 0 & 0 \\ 0 & 1 & 0 \\ 0 & s & 1 \end{bmatrix}, \quad E_{11} = \begin{bmatrix} 1 & 0 & s \\ 0 & 1 & 0 \\ 0 & 0 & 1 \end{bmatrix}, \quad E_{12} = \begin{bmatrix} 1 & 0 & 0 \\ 0 & 1 & s \\ 0 & 0 & 1 \end{bmatrix},$$

which are obtained using $sR_i + R_j$. Multiplication of any of these matrices on the left of any $3 \times n$ matrix is the same as performing the corresponding row operation. This is illustrated by

$$\begin{bmatrix} 0 & 1 & 0 \\ 1 & 0 & 0 \\ 0 & 0 & 1 \end{bmatrix} \begin{bmatrix} a_{11} & a_{12} & \cdots & a_{1n} \\ a_{21} & a_{22} & \cdots & a_{2n} \\ a_{31} & a_{32} & \cdots & a_{3n} \end{bmatrix} = \begin{bmatrix} a_{21} & a_{22} & \cdots & a_{2n} \\ a_{11} & a_{12} & \cdots & a_{1n} \\ a_{31} & a_{32} & \cdots & a_{3n} \end{bmatrix},$$

$$\begin{bmatrix} 1 & 0 & 0 \\ 0 & 1 & 0 \\ 0 & 0 & s \end{bmatrix}\begin{bmatrix} a_{11} & a_{12} & \cdots & a_{1n} \\ a_{21} & a_{22} & \cdots & a_{2n} \\ a_{31} & a_{32} & \cdots & a_{3n} \end{bmatrix} = \begin{bmatrix} a_{11} & a_{12} & \cdots & a_{1n} \\ a_{21} & a_{22} & \cdots & a_{2n} \\ sa_{31} & sa_{32} & \cdots & sa_{3n} \end{bmatrix},$$

and

$$\begin{bmatrix} 1 & 0 & 0 \\ 0 & 1 & 0 \\ s & 0 & 1 \end{bmatrix}\begin{bmatrix} a_{11} & a_{12} & \cdots & a_{1n} \\ a_{21} & a_{22} & \cdots & a_{2n} \\ a_{31} & a_{32} & \cdots & a_{3n} \end{bmatrix} = \begin{bmatrix} a_{11} & \cdots & a_{1n} \\ a_{21} & \cdots & a_{2n} \\ sa_{11} + a_{31} & \cdots & sa_{1n} + a_{3n} \end{bmatrix}.$$

■ Theorem 4.1.9

Let E be an $m \times m$ matrix and A be an $m \times n$ matrix. If E is an elementary matrix, then EA is obtained from A using op E.

Proof

Let $\mathbf{c}_1, \mathbf{c}_2, \ldots, \mathbf{c}_n \in \mathbb{R}^m$ be the columns of $A = [a_{ij}]$. Suppose that E is an elementary matrix. To begin, let op $E = R_i \leftrightarrow R_j$ with $i < j$. Letting $\mathbf{e}_i$ be a standard matrix of $\mathbb{R}^m$ for $i = 1, 2, \ldots, m$,

$$E = \begin{bmatrix} \mathbf{e}_1^T \\ \vdots \\ \mathbf{e}_j^T \\ \vdots \\ \mathbf{e}_i^T \\ \vdots \\ \mathbf{e}_m^T \end{bmatrix}.$$

Using Definition 3.2.13,

$$EA = \begin{bmatrix} \mathbf{e}_1^T \\ \vdots \\ \mathbf{e}_j^T \\ \vdots \\ \mathbf{e}_i^T \\ \vdots \\ \mathbf{e}_m^T \end{bmatrix}\begin{bmatrix} \mathbf{c}_1 & \mathbf{c}_2 & \cdots & \mathbf{c}_n \end{bmatrix} = \begin{bmatrix} \mathbf{e}_1 \bullet \mathbf{c}_1 & \mathbf{e}_1 \bullet \mathbf{c}_2 & \cdots & \mathbf{e}_1 \bullet \mathbf{c}_n \\ \vdots & \vdots & & \vdots \\ \mathbf{e}_j \bullet \mathbf{c}_1 & \mathbf{e}_j \bullet \mathbf{c}_2 & \cdots & \mathbf{e}_j \bullet \mathbf{c}_n \\ \vdots & \vdots & & \vdots \\ \mathbf{e}_i \bullet \mathbf{c}_1 & \mathbf{e}_i \bullet \mathbf{c}_2 & \cdots & \mathbf{e}_i \bullet \mathbf{c}_n \\ \vdots & \vdots & & \vdots \\ \mathbf{e}_m \bullet \mathbf{c}_1 & \mathbf{e}_m \bullet \mathbf{c}_2 & \cdots & \mathbf{e}_m \bullet \mathbf{c}_n \end{bmatrix}.$$

Using (2.28) from Example 2.2.5,

$$\mathbf{e}_k \bullet \mathbf{c}_l = \mathbf{e}_k \bullet \begin{bmatrix} a_{1l} \\ \vdots \\ a_{kl} \\ \vdots \\ a_{ml} \end{bmatrix} = a_{kl} \tag{4.5}$$

for all $k = 1, 2, \ldots, m$ and $l = 1, 2, \ldots, n$, so it follows that

$$
EA = \begin{bmatrix} a_{11} & a_{12} & \cdots & a_{in} \\ \vdots & \vdots & & \vdots \\ a_{j1} & a_{j2} & \cdots & a_{jn} \\ \vdots & \vdots & \cdots & \vdots \\ a_{i1} & a_{i2} & \cdots & a_{in} \\ \vdots & \vdots & & \vdots \\ a_{m1} & a_{m2} & \cdots & a_{mn} \end{bmatrix},
$$

which is A with rows i and j exchanged.

Next, suppose that op $E = r\mathrm{R}_i$ for some scalar r. This implies that

$$
E = \begin{bmatrix} \mathbf{e}_1^T \\ \vdots \\ r\mathbf{e}_i^T \\ \vdots \\ \mathbf{e}_m^T \end{bmatrix}.
$$

Thus, $(r\mathbf{e}_i^T)^T = r\mathbf{e}_i$ by Exercises 19(a) and 19(c) of Section 3.1, so

$$
EA = \begin{bmatrix} \mathbf{e}_1 \bullet \mathbf{c}_1 & \mathbf{e}_1 \bullet \mathbf{c}_2 & \cdots & \mathbf{e}_1 \bullet \mathbf{c}_n \\ \vdots & \vdots & & \vdots \\ r\mathbf{e}_i \bullet \mathbf{c}_1 & r\mathbf{e}_i \bullet \mathbf{c}_2 & \cdots & r\mathbf{e}_i \bullet \mathbf{c}_n \\ \vdots & \vdots & & \vdots \\ \mathbf{e}_m \bullet \mathbf{c}_1 & \mathbf{e}_m \bullet \mathbf{c}_2 & \cdots & \mathbf{e}_m \bullet \mathbf{c}_n \end{bmatrix}.
$$

By Theorem 2.2.6(c) and (4.5), $r\mathbf{e}_i \bullet \mathbf{c}_j = r(\mathbf{e}_i \bullet \mathbf{c}_j) = ra_{ij}$, so EA is A with its ith row multiplied by r.

Lastly, let op $E = \mathrm{R}_i + r\mathrm{R}_j$ with $r \in \mathbb{R}$, so

$$
E = \begin{bmatrix} \mathbf{e}_1^T \\ \vdots \\ \mathbf{e}_i^T + r\mathbf{e}_j^T \\ \vdots \\ \mathbf{e}_j^T \\ \vdots \\ \mathbf{e}_m^T \end{bmatrix}.
$$

Therefore,

$$
EA = \begin{bmatrix} \mathbf{e}_1 \bullet \mathbf{c}_1 & \mathbf{e}_1 \bullet \mathbf{c}_2 & \cdots & \mathbf{e}_1 \bullet \mathbf{c}_n \\ \vdots & \vdots & & \vdots \\ (\mathbf{e}_i + r\mathbf{e}_j) \bullet \mathbf{c}_1 & (\mathbf{e}_i + r\mathbf{e}_j) \bullet \mathbf{c}_2 & \cdots & (\mathbf{e}_i + r\mathbf{e}_j) \bullet \mathbf{c}_n \\ \vdots & \vdots & & \vdots \\ \mathbf{e}_j \bullet \mathbf{c}_1 & \mathbf{e}_j \bullet \mathbf{c}_2 & \cdots & \mathbf{e}_j \bullet \mathbf{c}_n \\ \vdots & \vdots & & \vdots \\ \mathbf{e}_m \bullet \mathbf{c}_1 & \mathbf{e}_m \bullet \mathbf{c}_2 & \cdots & \mathbf{e}_m \bullet \mathbf{c}_n \end{bmatrix},
$$

which follows because $(\mathbf{e}_i^T + r\mathbf{e}_j^T)^T = \mathbf{e}_i + r\mathbf{e}_j$ by Exercises 19(a) and 19(b) of Section 3.1. Theorem 2.2.6 and (4.5) imply that for $k = 1, 2, \dots, n$,

$$(\mathbf{e}_i + r\mathbf{e}_j) \bullet \mathbf{c}_k = \mathbf{e}_i \bullet \mathbf{c}_k + r(\mathbf{e}_j \bullet \mathbf{c}_k) = a_{ik} + ra_{jk},$$

and this implies that EA is obtained from A by multiplying row j of A by r and adding it to row i of A. ∎

■ **Corollary 4.1.10**

For all row equivalent $m \times n$ matrices A and B, there exist elementary $m \times m$ matrices $E_1, E_2, \dots, E_k$ such that $B = E_k \cdots E_2 E_1 A$.

■ **Example 4.1.11**

Example 3.5.11 showed that the reduced row echelon form of

$$A = \begin{bmatrix} 2 & 0 & 1 \\ 1 & -1 & 6 \\ 0 & 2 & 4 \end{bmatrix}$$

is the identity matrix, and this was obtained using the elementary row operation sequence,

$$R_1 \leftrightarrow R_2, \quad -2R_1 + R_2, \quad R_2 \leftrightarrow R_3, \quad \frac{1}{2}R_2, \quad R_2 + R_1,$$

$$-2R_2 + R_3, \quad -\frac{1}{15}R_3, \quad -8R_3 + R_1, \quad -2R_3 + R_2.$$

Each of these elementary row operations corresponds to one of the following elementary matrices, in order:

$$F_1 = \begin{bmatrix} 0 & 1 & 0 \\ 1 & 0 & 0 \\ 0 & 0 & 1 \end{bmatrix}, \qquad F_2 = \begin{bmatrix} 1 & 0 & 0 \\ -2 & 1 & 0 \\ 0 & 0 & 1 \end{bmatrix}, \quad F_3 = \begin{bmatrix} 1 & 0 & 0 \\ 0 & 0 & 1 \\ 0 & 1 & 0 \end{bmatrix},$$

$$F_4 = \begin{bmatrix} 1 & 0 & 0 \\ 0 & 1/2 & 0 \\ 0 & 0 & 1 \end{bmatrix}, \qquad F_5 = \begin{bmatrix} 1 & 1 & 0 \\ 0 & 1 & 0 \\ 0 & 0 & 1 \end{bmatrix}, \qquad F_6 = \begin{bmatrix} 1 & 0 & 0 \\ 0 & 1 & 0 \\ 0 & -2 & 1 \end{bmatrix},$$

$$F_7 = \begin{bmatrix} 1 & 0 & 0 \\ 0 & 1 & 0 \\ 0 & 0 & -1/15 \end{bmatrix}, \quad F_8 = \begin{bmatrix} 1 & 0 & -8 \\ 0 & 1 & 0 \\ 0 & 0 & 1 \end{bmatrix}, \quad F_9 = \begin{bmatrix} 1 & 0 & 0 \\ 0 & 1 & -2 \\ 0 & 0 & 1 \end{bmatrix}.$$

A check of the multiplication reveals that

$$F_9 F_8 F_7 F_6 F_5 F_4 F_3 F_2 F_1 A = I_3.$$

■ **Example 4.1.12**

The elementary matrices from Example 4.1.8 are invertible. To find the inverse of an elementary matrix E, what needs to be found is the elementary row operation that reverses op E. This will reveal the inverse of E. Consider the three types of elementary matrices. If op $E = R_i \leftrightarrow R_j$, then $E^{-1} = E$ because to return E to I_3, simply switch the rows again. Therefore,

$$E_1^{-1} = E_1, \quad E_2^{-1} = E_2, \quad E_3^{-1} = E_3.$$

If op $E = sR_i$ and $s \neq 0$, simply multiply the row by $1/s$ to return E to the identity. This means that

$$E_4^{-1} = \begin{bmatrix} 1/s & 0 & 0 \\ 0 & 1 & 0 \\ 0 & 0 & 1 \end{bmatrix}, \quad E_5^{-1} = \begin{bmatrix} 1 & 0 & 0 \\ 0 & 1/s & 0 \\ 0 & 0 & 1 \end{bmatrix}, \quad E_6^{-1} = \begin{bmatrix} 1 & 0 & 0 \\ 0 & 1 & 0 \\ 0 & 0 & 1/s \end{bmatrix}.$$

Lastly, if op $E = sR_i + R_j$, multiply row i by $-s$ and add to row j so that the (i, j) entry of the matrix is returned to 0. This means that

$$E_7^{-1} = \begin{bmatrix} 1 & 0 & 0 \\ -s & 1 & 0 \\ 0 & 0 & 1 \end{bmatrix}, \quad E_8^{-1} = \begin{bmatrix} 1 & 0 & 0 \\ 0 & 1 & 0 \\ -s & 0 & 1 \end{bmatrix}, \quad E_9^{-1} = \begin{bmatrix} 1 & -s & 0 \\ 0 & 1 & 0 \\ 0 & 0 & 1 \end{bmatrix},$$

$$E_{10}^{-1} = \begin{bmatrix} 1 & 0 & 0 \\ 0 & 1 & 0 \\ 0 & -s & 1 \end{bmatrix}, \quad E_{11}^{-1} = \begin{bmatrix} 1 & 0 & -s \\ 0 & 1 & 0 \\ 0 & 0 & 1 \end{bmatrix}, \quad E_{12}^{-1} = \begin{bmatrix} 1 & 0 & 0 \\ 0 & 1 & -s \\ 0 & 0 & 1 \end{bmatrix}.$$

As a confirmation,

$$\begin{bmatrix} 1 & s & 0 \\ 0 & 1 & 0 \\ 0 & 0 & 1 \end{bmatrix}^{-1} = \begin{bmatrix} 1 & -s & 0 \\ 0 & 1 & 0 \\ 0 & 0 & 1 \end{bmatrix}$$

because

$$\begin{bmatrix} 1 & -s & 0 \\ 0 & 1 & 0 \\ 0 & 0 & 1 \end{bmatrix}\begin{bmatrix} 1 & s & 0 \\ 0 & 1 & 0 \\ 0 & 0 & 1 \end{bmatrix} = \begin{bmatrix} 1 & 0 & 0 \\ 0 & 1 & 0 \\ 0 & 0 & 1 \end{bmatrix}$$

and

$$\begin{bmatrix} 1 & s & 0 \\ 0 & 1 & 0 \\ 0 & 0 & 1 \end{bmatrix}\begin{bmatrix} 1 & -s & 0 \\ 0 & 1 & 0 \\ 0 & 0 & 1 \end{bmatrix} = \begin{bmatrix} 1 & 0 & 0 \\ 0 & 1 & 0 \\ 0 & 0 & 1 \end{bmatrix}.$$

Notice that each of the inverses in Example 4.1.12 is an elementary matrix. This illustrates the next theorem. Its proof is in the spirit of the proof of Theorem 4.1.9 and is left to Exercise 21.

■ **Theorem 4.1.13**

For every elementary matrix E, if op E is an elementary row operation, E is invertible and its inverse is also an elementary matrix.

Finding the Inverse of a Matrix

The elementary matrices will play an important role in discovering a method to compute the inverse of an invertible matrix. To start, it will be required to find the inverse of the product of invertible matrices. When there are only two matrices, the inverse of the product of the invertible $n \times n$ matrices A and B is given by

$$(AB)^{-1} = B^{-1}A^{-1}.$$

This is because

$$AB(B^{-1}A^{-1}) = A[B(B^{-1}A^{-1})] = A(BB^{-1}A^{-1}) = A(I_nA^{-1}) = AA^{-1} = I_n,$$

and, by a similar argument,

$$B^{-1}A^{-1}(AB) = I_n.$$

This is generalized with the next theorem.

■ **Theorem 4.1.14**

If $A_1, A_2, \dots, A_k$ are invertible $n \times n$ matrices, $A_1 A_2 \cdots A_k$ is invertible and

$$(A_1 A_2 \cdots A_k)^{-1} = A_k^{-1} \cdots A_2^{-1} A_1^{-1}.$$

Proof

Proceed by induction on k. The case when $k = 1$ is trivial and $k = 2$ has already been done, so let $k > 2$. Then, by induction and Exercise 16 from Section 3.2,

$$\begin{aligned}
A_1 \cdots A_k A_{k+1}(A_{k+1}^{-1} A_k^{-1} \cdots A_1^{-1}) &= A_1 \cdots A_k[A_{k+1}(A_{k+1}^{-1} A_k^{-1} \cdots A_1^{-1})] \\
&= A_1 \cdots A_k[A_{k+1}(A_{k+1}^{-1}[A_k^{-1} \cdots A_1^{-1}])] \\
&= A_1 \cdots A_k[A_{k+1} A_{k+1}^{-1}(A_k^{-1} \cdots A_1^{-1})] \\
&= A_1 \cdots A_k(A_k^{-1} \cdots A_1^{-1}) \\
&= A_1 \cdots A_k(A_1 \cdots A_k)^{-1} = I_n,
\end{aligned}$$

and, similarly, $A_{k+1}^{-1} A_k^{-1} \cdots A_1^{-1}(A_1 \cdots A_k A_{k+1}) = A_{k+1}^{-1} A_{k+1} = I_n$, where the details are left to Exercise 22. ■

Theorem 4.1.14 along with the fact that most elementary matrices are invertible provide a method by which the inverse of an invertible matrix can be computed.

■ **Theorem 4.1.15**

If the reduced row echelon form of a given matrix is the identity matrix, the given matrix is invertible.

Proof

Let A be an $n \times n$ matrix such that its reduced row echelon form is I_n. By Corollary 4.1.10, there are elementary matrices $E_1, E_2, \dots, E_k$ such that

$$E_k \cdots E_2 E_1 A = I_n.$$

Elementary matrices defined by elementary row operations are invertible by Theorem 4.1.13. Hence, by multiplying on the left,

$$A = E_1^{-1} E_2^{-1} \cdots E_k^{-1},$$

and then by multiplying on the right,

$$A E_k \cdots E_2 E_1 = I_n$$

This demonstrates that A is invertible. ■

An examination of the proof of Theorem 4.1.15 justifies the corollary.

■ **Corollary 4.1.16**

If A is an $n \times n$ matrix so that there exists elementary matrices $E_1, E_2, \dots, E_k$ such that $E_k \cdots E_2 E_1 A = I_n$, then A is invertible and $A^{-1} = E_k \cdots E_2 E_1$.

■ **Example 4.1.17**

Find the inverse of $A = \begin{bmatrix} 1 & 0 & 2 \\ 2 & -1 & 4 \\ 0 & 2 & 4 \end{bmatrix}$ by finding the elementary matrices that

when multiplied on the left of A yield I_3. Augment A with I_3 on the right and put the matrix into reduced row echelon form. The work given here has only one elementary row operation per step so that the corresponding elementary matrices can be found:

$$\left[\begin{array}{ccc|ccc} 1 & 0 & 2 & 1 & 0 & 0 \\ 2 & -1 & 4 & 0 & 1 & 0 \\ 0 & 2 & 4 & 0 & 0 & 1 \end{array}\right] \xrightarrow{-2R_1 + R_2} \left[\begin{array}{ccc|ccc} 1 & 0 & 2 & 1 & 0 & 0 \\ 0 & -1 & 0 & -2 & 1 & 0 \\ 0 & 2 & 4 & 0 & 0 & 1 \end{array}\right]$$

$$\xrightarrow{-R_2} \left[\begin{array}{ccc|ccc} 1 & 0 & 2 & 1 & 0 & 0 \\ 0 & 1 & 0 & 2 & -1 & 0 \\ 0 & 2 & 4 & 0 & 0 & 1 \end{array}\right] \xrightarrow{-2R_2 + R_3} \left[\begin{array}{ccc|ccc} 1 & 0 & 2 & 1 & 0 & 0 \\ 0 & 1 & 0 & 2 & -1 & 0 \\ 0 & 0 & 4 & -4 & 2 & 1 \end{array}\right]$$

$$\xrightarrow{\frac{1}{4}R_3} \left[\begin{array}{ccc|ccc} 1 & 0 & 2 & 1 & 0 & 0 \\ 0 & 1 & 0 & 2 & -1 & 0 \\ 0 & 0 & 1 & -1 & 1/2 & 1/4 \end{array}\right]$$

$$\xrightarrow{-2R_3+R_1} \left[\begin{array}{ccc|ccc} 1 & 0 & 0 & 3 & -1 & -1/2 \\ 0 & 1 & 0 & 2 & -1 & 0 \\ 0 & 0 & 1 & -1 & 1/2 & 1/4 \end{array}\right]$$

The elementary row operation sequence is

$$-2R_1 + R_2, \quad -R_2, \quad -2R_2 + R_3, \quad \tfrac{1}{4}R_3, \quad -2R_3 + R_1,$$

which means that the elementary matrices used in the Gauss–Jordan elimination are

$$E_1 = \begin{bmatrix} 1 & 0 & 0 \\ -2 & 1 & 0 \\ 0 & 0 & 1 \end{bmatrix}, E_2 = \begin{bmatrix} 1 & 0 & 0 \\ 0 & -1 & 0 \\ 0 & 0 & 1 \end{bmatrix}, E_3 = \begin{bmatrix} 1 & 0 & 0 \\ 0 & 1 & 0 \\ 0 & -2 & 1 \end{bmatrix},$$

$$E_4 = \begin{bmatrix} 1 & 0 & 0 \\ 0 & 1 & 0 \\ 0 & 0 & 1/4 \end{bmatrix}, E_5 = \begin{bmatrix} 1 & 0 & -2 \\ 0 & 1 & 0 \\ 0 & 0 & 1 \end{bmatrix}.$$

Therefore,

$$E_5 E_4 E_3 E_2 E_1 A = I_3, \tag{4.6}$$

so by Corollary 4.1.16,

$$A^{-1} = E_5 E_4 E_3 E_2 E_1 = \begin{bmatrix} 3 & -1 & -1/2 \\ 2 & -1 & 0 \\ -1 & 1/2 & 1/4 \end{bmatrix},$$

which is the matrix augmented to the identity matrix in the last step of the Gauss–Jordan elimination.

Systems of Linear Equations

Recall the homogeneous system of linear equations from Example 3.5.11,

$$\begin{array}{rcrcrcl} 2x_1 & + & & + & x_3 & = & 0, \\ x_1 & - & x_2 & + & 6x_3 & = & 0, \\ & & 2x_2 & + & 4x_3 & = & 0. \end{array} \tag{4.7}$$

The given system can be written as the matrix equation,

$$\begin{bmatrix} 2 & 0 & 1 \\ 1 & -1 & 6 \\ 0 & 2 & 4 \end{bmatrix} \begin{bmatrix} x_1 \\ x_2 \\ x_3 \end{bmatrix} = \begin{bmatrix} 0 \\ 0 \\ 0 \end{bmatrix}. \tag{4.8}$$

Let $A = \begin{bmatrix} 2 & 0 & 1 \\ 1 & -1 & 6 \\ 0 & 2 & 4 \end{bmatrix}$, $\mathbf{x} = \begin{bmatrix} x_1 \\ x_2 \\ x_3 \end{bmatrix}$, and $\mathbf{0} = \begin{bmatrix} 0 \\ 0 \\ 0 \end{bmatrix}$ so that (4.8) becomes $A\mathbf{x} = \mathbf{0}$. By Example 4.1.17, the matrix A is invertible and shown to be

$$A^{-1} = \begin{bmatrix} 8/15 & -1/15 & -1/30 \\ 2/15 & -4/15 & 11/30 \\ -1/15 & 2/15 & 1/15 \end{bmatrix}.$$

Use A^{-1} to solve (4.8):

$$A^{-1}(Ax) = A^{-1}0,$$
$$A^{-1}Ax = 0,$$
$$I_n x = 0, \tag{4.9}$$
$$x = 0.$$

Therefore, the solution set to (4.7) is $\{0\}$.

Now, rewrite the system of linear equations so that it is no longer homogeneous. For example, write

$$
\begin{array}{rcrcrcl}
2x_1 & + & & + & x_3 & = & 15, \\
x_1 & - & x_2 & + & 6x_3 & = & 45, \\
& & 2x_2 & + & 4x_4 & = & 30.
\end{array}
\tag{4.10}
$$

Letting $\mathbf{b} = \begin{bmatrix} 15 \\ 45 \\ 30 \end{bmatrix}$, the linear system (4.10) can be represented as $Ax = \mathbf{b}$, and, again, use A^{-1} to solve (4.10):

$$A^{-1}(Ax) = A^{-1}\mathbf{b},$$
$$A^{-1}Ax = A^{-1}\mathbf{b},$$
$$I_n x = A^{-1}\mathbf{b}, \tag{4.11}$$
$$x = A^{-1}\mathbf{b}.$$

Thus, $\left\{ \begin{bmatrix} 4 \\ 1 \\ 7 \end{bmatrix} \right\}$ is the solution set for (4.10) because

$$
x = \begin{bmatrix} 8/15 & -1/15 & -1/30 \\ 2/15 & -4/15 & 11/30 \\ -1/15 & 2/15 & 1/15 \end{bmatrix} \begin{bmatrix} 15 \\ 45 \\ 30 \end{bmatrix} = \begin{bmatrix} 4 \\ 1 \\ 7 \end{bmatrix}.
$$

The work of (4.9) and (4.11) prove the next theorem.

■ **Theorem 4.1.18**

Let A be an invertible $n \times n$ matrix and $\mathbf{x} \in \mathbb{R}^n$ be a variable vector.

(a) $Ax = \mathbf{0}$ has only the trivial solution.

(b) For all $\mathbf{b} \in \mathbb{R}^n$, the solution set of $Ax = \mathbf{b}$ is $\{A^{-1}\mathbf{b}\}$.

■ **Example 4.1.19**

Define the linear operator $T : \mathbb{R}^3 \to \mathbb{R}^3$ by

$$
T\begin{bmatrix} x_1 \\ x_2 \\ x_3 \end{bmatrix} = \begin{bmatrix} 2x_1 + x_3 \\ x_1 - x_2 + 6x_3 \\ 2x_2 + 4x_4 \end{bmatrix}.
$$

Because $[T]$ is invertible, Theorem 4.1.18(a) implies that the system of equations (4.7) only has the trivial solution, so $\ker T = \{0\}$ and T is one-to-one. Also, because $[T]$ is invertible, T is onto since $[T]\mathbf{x} = \mathbf{b}$ has a solution for every $\mathbf{b} \in \mathbb{R}^3$ by Theorem 4.1.18(b). All of this is confirmed with the work solving the linear systems (4.7) and (4.10).

Exercises

1. Prove that each linear operator T is invertible without using a matrix.

   (a) $T : \mathbb{R} \to \mathbb{R}$, $Tx = 5x$

   (b) $T : \mathbb{R}^2 \to \mathbb{R}^2$, $T\begin{bmatrix} x \\ y \end{bmatrix} = \begin{bmatrix} 4x + 2y \\ x - 7y \end{bmatrix}$

   (c) $T : \mathbb{R}^3 \to \mathbb{R}^3$, $T\begin{bmatrix} x \\ y \\ z \end{bmatrix} = \begin{bmatrix} x + y + z \\ 2x + 5y \\ 4x \end{bmatrix}$

2. Find the inverse of each matrix if possible.

   (a) $[6]$

   (b) $\begin{bmatrix} 6 & 3 \\ 8 & 4 \end{bmatrix}$

   (c) $\begin{bmatrix} 4 & 1 \\ 3 & 2 \end{bmatrix}$

   (d) $\begin{bmatrix} -4 & 0 & -3 \\ 5 & 2 & 1 \\ 2 & 2 & 0 \end{bmatrix}$

   (e) $\begin{bmatrix} 3 & 0 & 1 \\ 5 & 2 & 1 \\ 7 & 4 & 1 \end{bmatrix}$

   (f) $\begin{bmatrix} 0 & 2 & -1 & 4 \\ 2 & 1 & -1 & -2 \\ 0 & 1 & 1 & 2 \\ 1 & 3 & 3 & 2 \end{bmatrix}$

3. Assume that the standard matrix of the linear operator $T : \mathbb{R} \to \mathbb{R}$ is the matrix $\begin{bmatrix} 1 & 2 & 3 \\ -1 & 0 & 1 \\ 0 & 1 & -2 \end{bmatrix}$. Prove that T is invertible.

4. Prove that $\begin{bmatrix} a & b \\ c & d \end{bmatrix}$ is not invertible if $ad - bc = 0$.

5. Find the inverses of the standard matrices for the given linear operators.

   (a) Reflection through the x-axis in $\mathbb{R}^2$.

   (b) Reflection through the yz-plane in $\mathbb{R}^3$.

   (c) Rotation in $\mathbb{R}^2$ through θ.

   (d) Rotation in $\mathbb{R}^3$ about the z-axis through θ.

   (e) Horizontal shear in $\mathbb{R}^2$ with $k = 5$.

   (f) The composition of the rotation through θ and the reflection through the y-axis in $\mathbb{R}^2$.

   (g) The composition of the rotation about the x-axis through θ, the rotation about the y-axis through θ, and the rotation about the z-axis through θ in $\mathbb{R}^3$.

6. Find the corresponding elementary 4×4 matrix for each row operation.

   (a) $R_2 \leftrightarrow R_3$ (c) $3R_3$ (e) $-1R_1 + R_2$

   (b) $7R_1 + R_4$ (d) $R_1 \leftrightarrow R_4$ (f) $-3R_4$

7. Find the inverse of each elementary matrix in Exercise 6.

8. Find the reduced row echelon form R of each matrix A and find a sequence of elementary matrices $E_1, E_2, \ldots, E_k$ such that $R = E_k \cdots E_2 E_1 A$.

   (a) $A = \begin{bmatrix} 3 & 7 \\ 7 & 3 \end{bmatrix}$

   (b) $A = \begin{bmatrix} -4 & 6 & 2 & 3 \\ -1 & 5 & 1 & 0 \\ -7 & 7 & 3 & 6 \\ -3 & 1 & 1 & 3 \end{bmatrix}$

   (c) $A = \begin{bmatrix} 6 & -1 & 7 \\ 2 & 1 & 4 \\ 2 & -3 & -1 \end{bmatrix}$

   (d) $A = \begin{bmatrix} 1 & 2 & 2 \\ 1 & 4 & -3 \\ -2 & 0 & 1 \end{bmatrix}$

9. Write each invertible matrix from Exercise 2 as the product of elementary matrices.

10. Let $A = BC$ where A, B, and C are $n \times n$ matrices and B is invertible. Prove that every sequence of elementary row operations that reduce B to I_n also reduces A to C.

11. Prove that the converse of Exercise 10 is false.

12. If possible, use the inverse of the coefficient matrix to find the solution to each system of linear equations. Whenever a coefficient matrix is not invertible, prove that it is not.

   (a) $\begin{aligned} -3x &+ 6y &= 13 \\ x &+ 5y &= 42 \end{aligned}$

   (b) $\begin{aligned} 5x &+ 2y &- 3z &= 1 \\ x &+ 10y &- 5z &= 2 \\ 2x &- 6y &+ z &= 3 \end{aligned}$

   (c) $\begin{aligned} 6x &- 5y &- 6z &= 3 \\ 5x &- 3y &- 7z &= 9 \\ x &- 2y &+ z &= 0 \end{aligned}$

   (d) $\begin{aligned} x_1 & & &+ 3x_3 &+ x_4 &= 5 \\ 2x_1 &+ x_2 &+ 2x_3 &+ 3x_4 &= -7 \\ x_1 &+ x_2 &- x_3 &- x_4 &= -3 \\ 2x_1 & & &+ 2x_3 &+ x_4 &= 13 \end{aligned}$

13. The method used in Exercise 12 enables systems of linear equations with the same coefficient matrix to be solved quickly when their constants are changed.
   Let $A = \begin{bmatrix} 5 & 2 & -3 \\ 1 & 10 & -5 \\ 2 & -6 & 1 \end{bmatrix}$. Solve $Ax = \mathbf{b}$ for each given $\mathbf{b} \in \mathbb{R}^3$.

(a) $\mathbf{b} = \begin{bmatrix} 0 \\ 0 \\ 0 \end{bmatrix}$ (b) $\mathbf{b} = \begin{bmatrix} 4 \\ -1 \\ 6 \end{bmatrix}$ (c) $\mathbf{b} = \begin{bmatrix} -6 \\ 2 \\ 0 \end{bmatrix}$

14. What can be said about the number of solutions to a system of linear equations when the coefficient matrix is not invertible? Give examples.

15. Prove that a linear operator $T : \mathbb{R}^n \to \mathbb{R}^n$ is a bijection if and only if $[T]$ is invertible.

16. Let A be an invertible matrix. Prove that A^{-1} is invertible and $(A^{-1})^{-1} = A$.

17. Let A, B, and P be $n \times n$ matrices with P being invertible. Solve for B given $A = P^{-1}BP$.

18. Find A given that $B^{-1} = \begin{bmatrix} 2 & 3 \\ 7 & -1 \end{bmatrix}$ and $AB = \begin{bmatrix} -3 & 6 \\ -2 & 4 \end{bmatrix}$.

19. Take $A = \begin{bmatrix} 2 & 4 & 3 \\ r & 1 & 5 \\ 2 & 5 & 6 \end{bmatrix}$. For what values of r is the matrix A noninvertible?

20. Let A be an $m \times m$ matrix. Let B and C be $m \times n$ matrices. Prove.

    (a) If A is invertible and $AB = AC$, then $B = C$.

    (b) If A is not invertible, it is possible for $AB = AC$ but $B \neq C$.

21. Prove Theorem 4.1.13.

22. Provide the finishing details to the proof of Theorem 4.1.14.

23. Generalize Example 4.1.19 by proving that T is one-to-one if and only if T is onto for all linear operators $T : \mathbb{R}^n \to \mathbb{R}^n$.

24. If A is an $n \times n$ matrix such that $I_n - A$ is invertible, prove that

$$A(I_n - A)^{-1} = (I_n - A)^{-1}A.$$

25. Let A be an invertible $n \times n$ matrix, show that $(I_n + A)^{-1} = I_n - \left(A^{-1} + I_n\right)^{-1}$.

4.2 Determinants

If A is a square matrix, the equation $A\mathbf{x} = \mathbf{b}$ will have a unique solution when A is invertible (Theorem 4.1.18). For this reason it is desirable to have a test on A that determines whether A is invertible or not. In Example 4.1.5, it was found that

$$[a] \text{ is invertible if and only if } a \neq 0, \tag{4.12}$$

and in Example 4.1.6, it was determined that

$$\begin{bmatrix} a & b \\ c & d \end{bmatrix} \text{ is invertible if and only if } ad - bc \neq 0. \tag{4.13}$$

Finding the test involves extending these two results to a function that when given a matrix it returns a nonzero real number if and only if the given matrix is invertible. The sought-after function will be defined recursively and requires some introductory terminology.

■ Definition 4.2.1

Let $A = [a_{ij}]$ be an $n \times n$ matrix. The **(k, l)-submatrix** of A is the matrix A_{kl} formed by removing row k and column l from A. In notation,

$$A_{kl} = [a_{ij} : i \neq k \text{ and } j \neq l].$$

For example, when

$$A = \begin{bmatrix} 1 & 2 & 3 & 4 \\ 5 & 6 & 7 & 8 \\ 9 & 0 & 1 & 2 \\ 3 & 4 & 5 & 6 \end{bmatrix}, \tag{4.14}$$

to find A_{23}, delete row 2 and column 3 from A. This results is the (2, 3)-submatrix,

$$A_{23} = \begin{bmatrix} 1 & 2 & 4 \\ 9 & 0 & 2 \\ 3 & 4 & 6 \end{bmatrix}.$$

These submatrices are used in the definition of the sought-after function.

■ Definition 4.2.2

The **determinant** is a function

$$\det : \bigcup_{i=1}^{\infty} \mathbb{R}^{i \times i} \to \mathbb{R}$$

such that for any $n \times n$ matrix $A = [a_{ij}]$, if $n = 1$, then $\det A = a_{11}$, and if $n > 1$,

$$\det A = \sum_{j=1}^{n} (-1)^{1+j} a_{1j} \det A_{1j}.$$

Alternately, write

$$\det A = \begin{vmatrix} a_{11} & a_{12} & \cdots & a_{1n} \\ a_{21} & a_{22} & \cdots & a_{2n} \\ \vdots & \vdots & & \vdots \\ a_{n1} & a_{n2} & \cdots & a_{nn} \end{vmatrix}.$$

Because $\det [a] = a$, (4.12) can be rewritten as

$$[a] \text{ is invertible if and only if } |a| \neq 0. \tag{4.15}$$

Also, if $A = \begin{bmatrix} a & b \\ c & d \end{bmatrix}$, then $A_{11} = [d]$ and $A_{12} = [c]$, so

$$\det A = ad - bc. \tag{4.16}$$

This implies that (4.13) can be rewritten as

$$\begin{bmatrix} a & b \\ c & d \end{bmatrix} \text{ is invertible if and only if } \begin{vmatrix} a & b \\ c & d \end{vmatrix} \neq 0. \tag{4.17}$$

■ **Example 4.2.3**

Let $T : \mathbb{R}^2 \to \mathbb{R}^2$ be a linear operator. If T is a rotation through θ, then by Definition 3.3.8,

$$\det [T] = \begin{vmatrix} \cos\theta & -\sin\theta \\ \sin\theta & \cos\theta \end{vmatrix} = \cos^2\theta + \sin^2\theta = 1.$$

Suppose, instead, that T is a reflection through a line L containing the zero vector. Let θ be the measure of the angle that L forms with $\mathbf{e}_1$. This implies that T is the composition of the rotation through θ, the reflection through the x-axis, and the inverse of the first rotation. Because cosine is an even function and sine is an odd function, Example 3.3.5 gives

$$\begin{aligned}
[T] &= \begin{bmatrix} \cos\theta & -\sin\theta \\ \sin\theta & \cos\theta \end{bmatrix} \begin{bmatrix} 1 & 0 \\ 0 & -1 \end{bmatrix} \begin{bmatrix} \cos\theta & \sin\theta \\ -\sin\theta & \cos\theta \end{bmatrix} \\
&= \begin{bmatrix} \cos\theta & -\sin\theta \\ \sin\theta & \cos\theta \end{bmatrix} \begin{bmatrix} \cos\theta & \sin\theta \\ \sin\theta & -\cos\theta \end{bmatrix} \\
&= \begin{bmatrix} \cos^2\theta - \sin^2\theta & 2\cos\theta\sin\theta \\ 2\cos\theta\sin\theta & \sin^2\theta - \cos^2\theta \end{bmatrix} \\
&= \begin{bmatrix} \cos 2\theta & \sin 2\theta \\ \sin 2\theta & -\cos 2\theta \end{bmatrix}.
\end{aligned}$$

Hence, $\det [T] = -\cos^2 2\theta - \sin^2 2\theta = -1$. Summarizing, the determinant of the standard matrix of a rotation in $\mathbb{R}^2$ equals 1, and the determinant of the standard matrix of a reflection in $\mathbb{R}^2$ through a line containing the zero vector is -1. The same result holds for rotations and reflections in $\mathbb{R}^3$ (Exercise 19).

■ **Example 4.2.4**

To find the determinant of the 3×3 matrix A given in (4.14), find the determinants of the submatrices,

$$\det A_{11} = 0, \quad \det A_{12} = -20, \quad \det A_{13} = -40, \quad \det A_{14} = -20.$$

In particular, $\det A_{12}$ is computed by finding the determinants of the appropriate submatrices of

$$B = A_{12} = \begin{bmatrix} 5 & 7 & 8 \\ 9 & 1 & 2 \\ 3 & 5 & 6 \end{bmatrix}.$$

They are

$$\det B_{11} = 1(6) - 2(5) = -4,$$
$$\det B_{12} = 9(6) - 2(3) = 48,$$
$$\det B_{13} = 9(5) - 1(3) = 42.$$

This is because the determinant of a 2×2 matrix uses (4.16). Then,

$$\det B = 5(-4) - 7(48) + 8(42) = -20.$$

Finally, finishing the computation,

$$\det A = 1(0) - 2(-20) + 3(-40) - 4(-20) = 0,$$

or using the alternate notation,

$$\begin{vmatrix} 1 & 2 & 3 & 4 \\ 5 & 6 & 7 & 8 \\ 9 & 0 & 1 & 2 \\ 3 & 4 & 5 & 6 \end{vmatrix} = 0. \tag{4.18}$$

■ **Example 4.2.5**

Find the determinants of

$$A = \begin{bmatrix} 4 & 0 & 0 \\ 2 & 5 & 0 \\ 1 & 2 & -3 \end{bmatrix}, \quad B = \begin{bmatrix} 7 & 1 & 3 \\ 0 & -1 & 8 \\ 0 & 0 & 2 \end{bmatrix}, \quad C = \begin{bmatrix} 5 & 0 & 0 \\ 0 & -2 & 0 \\ 0 & 0 & 6 \end{bmatrix}.$$

First,

$$\det A = 4 \begin{vmatrix} 5 & 0 \\ 2 & -3 \end{vmatrix} = 4(-15) = -60.$$

Observe that $-60 = (4)(5)(-3)$. Second,

$$\det B = 7 \begin{vmatrix} -1 & 8 \\ 0 & 2 \end{vmatrix} - \begin{vmatrix} 0 & 8 \\ 0 & 2 \end{vmatrix} + 3 \begin{vmatrix} 0 & -1 \\ 0 & 0 \end{vmatrix} = 7(-2) - 0 + 3(0) = -14.$$

Observe that $-14 = (7)(-1)(2)$. Third,

$$\det C = 5 \begin{vmatrix} -2 & 0 \\ 0 & 6 \end{vmatrix} = 5(-12) = -60,$$

and observe that $-60 = (5)(-2)(6)$. It appears that if the matrix is just right, its determinant is easy to find.

■ **Definition 4.2.6**

Let A be an $n \times n$ matrix.

(a) A is **lower triangular** if $a_{ij} = 0$ for all i and j such that $i < j$.

(b) A is **upper triangular** if $a_{ij} = 0$ for all i and j such that $i > j$.

(c) A is **diagonal** if $a_{ij} = 0$ for all i and j such that $i \neq j$.

A is **triangular** if A is lower or upper triangular, and note that every diagonal matrix is both lower and upper triangular.

The next theorem confirms what was illustrated in Example 4.2.5 about lower triangular matrices. The result concerning upper triangle matrices will wait.

■ **Theorem 4.2.7**

If A is an $n \times n$ lower triangular matrix with $a_{11}, a_{22}, \ldots, a_{nn}$ as the entries of its main diagonal, $\det A = a_{11} a_{22} \cdots a_{nn}$.

Proof

Let A be an $n \times n$ matrix. Prove this by induction on n. The basis case follows because $\det [a_{11}] = a_{11}$. For the induction step, write

$$
A = \begin{bmatrix}
a_{11} & 0 & 0 & \cdots & 0 \\
a_{21} & a_{22} & 0 & \cdots & 0 \\
a_{31} & a_{32} & a_{33} & \cdots & 0 \\
\vdots & \vdots & \vdots & & \vdots \\
a_{n+1,1} & a_{n+1,2} & a_{n+1,3} & \cdots & a_{n+1,n+1}
\end{bmatrix},
$$

so using the induction hypothesis for the second equality,

$$
\det A = a_{11} \begin{vmatrix}
a_{22} & 0 & \cdots & 0 \\
a_{32} & a_{33} & \cdots & 0 \\
\vdots & \vdots & & \vdots \\
a_{n+1,2} & a_{n+1,3} & \cdots & a_{n+1,n+1}
\end{vmatrix} = a_{11} a_{22} \cdots a_{n+1,n+1}. \blacksquare
$$

■ **Example 4.2.8**

It is a straightforward exercise to show that the set of diagonal matrices is closed under the operations of addition, multiplication, and scalar multiplication [Exercise 17(a)]. The same can be said for the set of lower triangular matrices and the set of upper triangular matrices (Exercise 18). Similarly, following Example 3.2.15, define the block matrices,

$$
A = \begin{bmatrix} A_{11} & 0 \\ 0 & A_{22} \end{bmatrix} \quad \text{and} \quad B = \begin{bmatrix} B_{11} & 0 \\ 0 & B_{22} \end{bmatrix}.
$$

Each is an example of a **block diagonal matrix**. With the continued assumption that the blocks are of the right size,

$$
AB = \begin{bmatrix} A_{11} B_{11} & 0 \\ 0 & A_{22} B_{22} \end{bmatrix}.
$$

Also, the identity matrix can be viewed as a block diagonal matrix, such as

$$I_7 = \begin{bmatrix} I_3 & 0 \\ 0 & I_4 \end{bmatrix} = \begin{bmatrix} I_2 & 0 & 0 \\ 0 & I_3 & 0 \\ 0 & 0 & I_2 \end{bmatrix}.$$

Lastly, if A_{11} is an invertible 3×3 matrix and A_{22} is an invertible 6×6 matrix,

$$A^{-1} = \begin{bmatrix} A_{11}^{-1} & 0 \\ 0 & A_{22}^{-1} \end{bmatrix}$$

because

$$\begin{bmatrix} A_{11} & 0 \\ 0 & A_{22} \end{bmatrix} \begin{bmatrix} A_{11}^{-1} & 0 \\ 0 & A_{22}^{-1} \end{bmatrix} = \begin{bmatrix} A_{11}A_{11}^{-1} & 0 \\ 0 & A_{22}A_{22}^{-1} \end{bmatrix} = \begin{bmatrix} I_3 & 0 \\ 0 & I_6 \end{bmatrix} = I_9$$

and

$$\begin{bmatrix} A_{11}^{-1} & 0 \\ 0 & A_{22}^{-1} \end{bmatrix} \begin{bmatrix} A_{11} & 0 \\ 0 & A_{22} \end{bmatrix} = \begin{bmatrix} A_{11}^{-1}A_{11} & 0 \\ 0 & A_{22}^{-1}A_{22} \end{bmatrix} = \begin{bmatrix} I_3 & 0 \\ 0 & I_6 \end{bmatrix} = I_9.$$

There are other matrices that have easily computed determinants. Consider $A = \begin{bmatrix} 5 & 3 & 1 \\ 0 & 0 & 0 \\ 2 & 9 & 4 \end{bmatrix}$. Then, $\det A = 5 \begin{vmatrix} 0 & 0 \\ 9 & 4 \end{vmatrix} - 3 \begin{vmatrix} 0 & 0 \\ 2 & 4 \end{vmatrix} + \begin{vmatrix} 0 & 0 \\ 2 & 9 \end{vmatrix} = 0$, which suggests the following.

■ Theorem 4.2.9

If the square matrix A has a row of zeros, $\det A = 0$.

Proof

If the first row of the $n \times n$ matrix A consists entirely of zeros, then Definition 4.2.2 quickly implies that $\det A = 0$. Suppose, instead, that the entries of row $i > 1$ are all 0 and proceed by induction on n. The basis case of $n = 1$ is trivial. When A is an $(n+1) \times (n+1)$ matrix, row $i-1$ of A_{1j} consists entirely of zeros for all $j = 1, 2, \ldots, n + 1$, so $\det A_{1j} = 0$ by induction. Therefore, $\det A = 0$. ■

Multiplying a Row by a Scalar

The determinant has a close relationship with the row operations given in Definition 3.5.4. Consider them each in turn, starting with the result of multiplying a row of a matrix by a scalar. Let $s \in \mathbb{R}$. For the 1×1 case, multiplying the only row of $A = [a]$ by s results in $A' = [sa]$, so

$$\det A' = s \det A. \tag{4.19}$$

To prove that (4.19) holds for all matrices under this row operation, let

$$A = \begin{bmatrix} a_{11} & a_{12} & \cdots & a_{1,n+1} \\ \vdots & \vdots & & \vdots \\ a_{i1} & a_{i2} & \cdots & a_{i,n+1} \\ \vdots & \vdots & & \vdots \\ a_{n+1,1} & a_{n+1,2} & \cdots & a_{n+1,n+1} \end{bmatrix}$$

and suppose that E is the $(n+1) \times (n+1)$ elementary matrix such that op $E = sR_i$. That is,

$$EA = \begin{bmatrix} a_{11} & a_{12} & \cdots & a_{1,n+1} \\ \vdots & \vdots & & \vdots \\ sa_{i1} & sa_{i2} & \cdots & sa_{i,n+1} \\ \vdots & \vdots & & \vdots \\ a_{n+1,1} & a_{n+1,2} & \cdots & a_{n+1,n+1} \end{bmatrix}.$$

If $i \neq 1$, by induction,

$$\det EA = \sum_{j=1}^{n+1} (-1)^{1+j} a_{1j} \begin{vmatrix} a_{21} & \cdots & a_{2,j-1} & a_{2,j+1} & \cdots & a_{2,n+1} \\ \vdots & & \vdots & \vdots & & \vdots \\ sa_{i1} & \cdots & sa_{i,j-1} & sa_{i,j+1} & \cdots & sa_{i,n+1} \\ \vdots & & \vdots & \vdots & & \vdots \\ a_{n+1,1} & \cdots & a_{n+1,j-1} & a_{n+1,j+1} & \cdots & a_{n+1,n+1} \end{vmatrix}$$

$$= \sum_{j=1}^{n+1} (-1)^{1+j} a_{1j} s \begin{vmatrix} a_{21} & \cdots & a_{2,j-1} & a_{2,j+1} & \cdots & a_{2,n+1} \\ \vdots & & \vdots & \vdots & & \vdots \\ a_{i1} & \cdots & a_{i,j-1} & a_{i,j+1} & \cdots & a_{i,n+1} \\ \vdots & & \vdots & \vdots & & \vdots \\ a_{n+1,1} & \cdots & a_{n+1,j-1} & a_{n+1,j+1} & \cdots & a_{n+1,n+1} \end{vmatrix}$$

$$= s \sum_{j=1}^{n+1} (-1)^{1+j} a_{1j} \begin{vmatrix} a_{21} & \cdots & a_{2,j-1} & a_{2,j+1} & \cdots & a_{2,n+1} \\ \vdots & & \vdots & \vdots & & \vdots \\ a_{i1} & \cdots & a_{i,j-1} & a_{i,j+1} & \cdots & a_{i,n+1} \\ \vdots & & \vdots & \vdots & & \vdots \\ a_{n+1,1} & \cdots & a_{n+1,j-1} & a_{n+1,j+1} & \cdots & a_{n+1,n+1} \end{vmatrix}$$

$$= s \det A.$$

However, if $n = 1$,

$$\det EA = \sum_{j=1}^{n+1} (-1)^{1+j} sa_{1j} \begin{vmatrix} a_{21} & \cdots & a_{2,j-1} & a_{2,j+1} & \cdots & a_{2,n+1} \\ a_{31} & \cdots & a_{3,j-1} & a_{3,j+1} & \cdots & a_{3,n+1} \\ \vdots & & \vdots & \vdots & & \vdots \\ a_{n+1,1} & \cdots & a_{n+1,j-1} & a_{n+1,j+1} & \cdots & a_{n+1,n} \end{vmatrix}$$

$$= s \det A.$$

Therefore,

$$\text{if op } E = sR_i, \text{ then } \det EA = s \det A. \tag{4.20}$$

■ **Example 4.2.10**

Let $A = \begin{bmatrix} 4 & 2 & -1 \\ -3 & 5 & 2 \\ 1 & 0 & -2 \end{bmatrix}$. Then,

$$\det A = 4\begin{vmatrix} 5 & 2 \\ 0 & -2 \end{vmatrix} - 2\begin{vmatrix} -3 & 2 \\ 1 & -2 \end{vmatrix} - \begin{vmatrix} -3 & 5 \\ 1 & 0 \end{vmatrix} = -43,$$

and if E is the 3×3 elementary matrix such that op $E = 6R_2$,

$$\det EA = \begin{vmatrix} 4 & 2 & -1 \\ -18 & 30 & 12 \\ 1 & 0 & -2 \end{vmatrix} = -258 = 6(-43).$$

Adding a Multiple of a Row to Another Row

To find how the next row operation affects the determinant, some background is needed. Usually, if A and B are square matrices of the same size, it is expected that $\det(A + B) \neq \det A + \det B$, but under the right circumstances the equation holds.

■ **Lemma 4.2.11**

For any integer i such that $1 \leq i \leq n$,

$$\begin{vmatrix} a_{11} & \cdots & a_{1n} \\ \vdots & & \vdots \\ a_{i1} + b_{i1} & \cdots & a_{in} + b_{in} \\ \vdots & & \vdots \\ a_{n,1} & \cdots & a_{n,n} \end{vmatrix} = \begin{vmatrix} a_{11} & \cdots & a_{1n} \\ \vdots & & \vdots \\ a_{i1} & \cdots & a_{in} \\ \vdots & & \vdots \\ a_{n1} & \cdots & a_{nn} \end{vmatrix} + \begin{vmatrix} a_{11} & \cdots & a_{1n} \\ \vdots & & \vdots \\ b_{i1} & \cdots & b_{in} \\ \vdots & & \vdots \\ a_{n1} & \cdots & a_{nn} \end{vmatrix}.$$

Proof

This is proved by induction on n. The basis case holds because

$$\left| a_{11} + b_{11} \right| = \left| a_{11} \right| + \left| b_{11} \right|.$$

The induction step has two cases. Write

$$C = \begin{bmatrix} a_{11} & \cdots & a_{1,n+1} \\ \vdots & & \vdots \\ a_{i1} + b_{i1} & \cdots & a_{i,n+1} + b_{i,n+1} \\ \vdots & & \vdots \\ a_{n+1,1} & \cdots & a_{n+1,n+1} \end{bmatrix},$$

$$A = \begin{bmatrix} a_{11} & \cdots & a_{1,n+1} \\ \vdots & & \vdots \\ a_{i1} & \cdots & a_{i,n+1} \\ \vdots & & \vdots \\ a_{n+1,1} & \cdots & a_{n+1,n+1} \end{bmatrix}, \quad \text{and} \quad B = \begin{bmatrix} a_{11} & \cdots & a_{1,n+1} \\ \vdots & & \vdots \\ b_{i1} & \cdots & b_{i,n+1} \\ \vdots & & \vdots \\ a_{n+1,1} & \cdots & a_{n+1,n+1} \end{bmatrix}.$$

If the addition occurs in the first row ($i = 1$), for all $j = 1, 2, \ldots, n + 1$,

$$A_{1j} = B_{1j} = C_{1j}.$$

Therefore,

$$\det C = \sum_{j=1}^{n+1}(-1)^{1+j}(a_{1j} + b_{1j})\det C_{1j}$$

$$= \sum_{j=1}^{n+1}(-1)^{1+j}a_{1j}\det A_{1j} + \sum_{j=1}^{n+1}(-1)^{1+j}b_{1j}\det B_{1j} = \det A + \det B.$$

However, if the addition occurs in row $i > 1$,

$$\det C = \sum_{j=1}^{n+1}(-1)^{1+j}a_{1j}\det C_{1j},$$

where $\det C_{1j}$ is

$$\begin{vmatrix} a_{21} & \cdots & a_{2,j-1} & a_{2,j+1} & \cdots & a_{2,n+1} \\ \vdots & & \vdots & \vdots & & \vdots \\ a_{i1} + b_{i1} & \cdots & a_{i,j-1} + b_{i,j-1} & a_{i,j+1} + b_{i,j+1} & \cdots & a_{i,n+1} + b_{i,n+1} \\ \vdots & & \vdots & \vdots & & \vdots \\ a_{n+1,1} & \cdots & a_{n+1,j-1} & a_{n+1,j+1} & \cdots & a_{n+1,n+1} + b_{n+1,n+1} \end{vmatrix},$$

by induction $\det C_{1j} = \det A_{1j} + \det B_{1j}$. It follows that

$$\det C = \sum_{j=1}^{n+1}(-1)^{1+j}a_{1j}(\det A_{1j} + \det B_{1j})$$

$$= \sum_{j=1}^{n+1}(-1)^{1+j}a_{1j}\det A_{1j} + \sum_{j=1}^{n+1}(-1)^{1+j}a_{1j}\det B_{1j} = \det A + \det B. \blacksquare$$

Since Lemma 4.2.11 allows the determinant of matrix C to be split into the sum of the determinants of A and B based on the simple fact that a row of C is the sum of the corresponding rows of A and B with all other corresponding entries being equal, the determinant of a matrix can be decomposed into simpler matrices. For example,

$$\begin{vmatrix} 1 & 2 \\ 3 & 4 \end{vmatrix} = \begin{vmatrix} 1 & 0 \\ 3 & 4 \end{vmatrix} + \begin{vmatrix} 0 & 2 \\ 3 & 4 \end{vmatrix}$$

because $[1 \ 2] = [1 \ 0] + [0 \ 2]$, and then

$$\begin{vmatrix} 1 & 0 \\ 3 & 4 \end{vmatrix} = \begin{vmatrix} 1 & 0 \\ 3 & 0 \end{vmatrix} + \begin{vmatrix} 1 & 0 \\ 0 & 4 \end{vmatrix} \quad \text{and} \quad \begin{vmatrix} 0 & 2 \\ 3 & 4 \end{vmatrix} = \begin{vmatrix} 0 & 2 \\ 3 & 0 \end{vmatrix} + \begin{vmatrix} 0 & 2 \\ 0 & 4 \end{vmatrix}$$

because $[3 \ 4] = [3 \ 0] + [0 \ 4]$. Therefore,

$$\begin{vmatrix} 1 & 2 \\ 3 & 4 \end{vmatrix} = \begin{vmatrix} 1 & 0 \\ 3 & 0 \end{vmatrix} + \begin{vmatrix} 1 & 0 \\ 0 & 4 \end{vmatrix} + \begin{vmatrix} 0 & 2 \\ 3 & 0 \end{vmatrix} + \begin{vmatrix} 0 & 2 \\ 0 & 4 \end{vmatrix}.$$

However, the first and last determinants equal zero, so

$$\begin{vmatrix} 1 & 2 \\ 3 & 4 \end{vmatrix} = \begin{vmatrix} 1 & 0 \\ 0 & 4 \end{vmatrix} + \begin{vmatrix} 0 & 2 \\ 3 & 0 \end{vmatrix}. \tag{4.21}$$

Think of the two matrices of this sum (4.21) as being formed by first hiding all of the entries of the given matrix with a 0 and then revealing original entries in such a way that every row and every column has only one entry revealed. The two matrices of (4.21) represent all possibilities. That is, either positions $(1, 1)$ and $(2, 2)$ are revealed or positions $(1, 2)$ and $(2, 1)$ are revealed. This can be generalized to a matrix of any size, but it needs a function to identify all of the possible patterns.

■ Definition 4.2.12

Let $n \in \mathbb{Z}$. A **permutation** of $\{1, 2, \dots, n\}$ is a bijection

$$\sigma : \{1, 2, \dots, n\} \to \{1, 2, \dots, n\}.$$

Furthermore, $S_n = \{\sigma : \sigma \text{ is a permutation of } \{1, 2, \dots, n\}\}$.

There is a standard notation used to represent a permutation. Take S_2, for instance. If $\sigma \in S_2$, then either $\sigma(1) = 1$ and $\sigma(2) = 2$, or $\sigma(1) = 2$ and $\sigma(2) = 1$. Represent these two functions by

$$\begin{pmatrix} 1 & 2 \\ 1 & 2 \end{pmatrix} \text{ and } \begin{pmatrix} 1 & 2 \\ 2 & 1 \end{pmatrix},$$

where the first row contains the numbers in the domain of σ and the second row are the numbers in its range. Tying this back to (4.21), consider the first row as indicating the row of the matrix and the second row as indicating the entry that will be revealed in that row. For example,

$$\sigma = \begin{pmatrix} 1 & 2 \\ 2 & 1 \end{pmatrix}$$

reveals the second entry of the first row because $\sigma(1) = 2$ and reveals the first entry of the second row because $\sigma(2) = 1$.

Now, do this for a 3×3 matrix. Let

$$A = \begin{bmatrix} 4 & 2 & -1 \\ -3 & 5 & 2 \\ 1 & 7 & -2 \end{bmatrix}. \tag{4.22}$$

The elements of S_3 are

$$\begin{pmatrix} 1 & 2 & 3 \\ 1 & 2 & 3 \end{pmatrix}, \begin{pmatrix} 1 & 2 & 3 \\ 1 & 3 & 2 \end{pmatrix}, \begin{pmatrix} 1 & 2 & 3 \\ 2 & 1 & 3 \end{pmatrix},$$

$$\begin{pmatrix} 1 & 2 & 3 \\ 2 & 3 & 1 \end{pmatrix}, \begin{pmatrix} 1 & 2 & 3 \\ 3 & 1 & 2 \end{pmatrix}, \begin{pmatrix} 1 & 2 & 3 \\ 3 & 2 & 1 \end{pmatrix}.$$

(4.23)

Notice that there are $6 = 3!$ elements of S_3. This is because S_n has $n!$ permutations. Each of these permutations define a matrix from A by revealing certain entries. Following the order of (4.23), the matrices are

$$\begin{bmatrix} 4 & 0 & 0 \\ 0 & 5 & 0 \\ 0 & 0 & -2 \end{bmatrix}, \begin{bmatrix} 4 & 0 & 0 \\ 0 & 0 & 2 \\ 0 & 7 & 0 \end{bmatrix}, \begin{bmatrix} 0 & 2 & 0 \\ -3 & 0 & 0 \\ 0 & 0 & -2 \end{bmatrix}$$

$$\begin{bmatrix} 0 & 2 & 0 \\ 0 & 0 & 2 \\ 1 & 0 & 0 \end{bmatrix}, \begin{bmatrix} 0 & 0 & -1 \\ -3 & 0 & 0 \\ 0 & 7 & 0 \end{bmatrix}, \begin{bmatrix} 0 & 0 & -1 \\ 0 & 5 & 0 \\ 1 & 0 & 0 \end{bmatrix}.$$

(4.24)

This is generalized using the next definition, which in turn generalizes (4.21) using the subsequent theorem.

■ Definition 4.2.13

Let $A = [a_{ij}]$ be an $n \times n$ matrix and let σ be a permutation of $\{1, 2, \dots, n\}$. The **refinement** of A using σ is the matrix A^σ such that every entry of A^σ is 0 except that entry $\sigma(i)$ of row i is $a_{i\sigma(i)}$.

For example, let $\sigma = \begin{pmatrix} 1 & 2 & 3 \\ 3 & 1 & 2 \end{pmatrix}$. Continuing to use matrix A of (4.22), since $\sigma(1) = 3, \sigma(2) = 1$, and $\sigma(3) = 2$,

$$A^\sigma = \begin{bmatrix} 0 & 0 & -1 \\ -3 & 0 & 0 \\ 0 & 7 & 0 \end{bmatrix}.$$

■ Theorem 4.2.14

For every $n \times n$ matrix A, $\det A = \sum_{\sigma \in S_n} \det A^\sigma$.

Proof

Proceed by induction on n. Since the basis case is trivial, assume that the equation holds for all $n \times n$ matrices. Take $A = [a_{ij}]$ to be an $(n+1) \times (n+1)$ matrix. Multiple applications of Lemma 4.2.11 implies that $\det A$ equals

$$\begin{vmatrix} a_{11} & 0 & \dots & 0 \\ a_{21} & a_{22} & \dots & a_{2,n+1} \\ \vdots & \vdots & & \vdots \\ a_{n+1,1} & a_{n+1,2} & \dots & a_{n+1,n+1} \end{vmatrix} + \dots + \begin{vmatrix} 0 & 0 & \dots & a_{1,n+1} \\ a_{21} & a_{22} & \dots & a_{2,n+1} \\ \vdots & \vdots & & \vdots \\ a_{n+1,1} & a_{n+1,2} & \dots & a_{n+1,n+1} \end{vmatrix}.$$

Observe that by induction, for $j = 1, 2, \ldots, n + 1$,

$$\det A_{1j} = \sum_{\sigma \in S_n} \det A_{1j}^{\sigma}.$$

Therefore, $\det A = \sum_{j=1}^{n+1} \left(a_{1j} \sum_{\sigma \in S_n} \det A_{1j}^{\sigma} \right) = \sum_{\sigma \in S_{n+1}} \det A^{\sigma}.$ ∎

■ **Example 4.2.15**

Using the determinants of the matrices given in (4.24),

$$\begin{vmatrix} 4 & 2 & -1 \\ -3 & 5 & 2 \\ 1 & 7 & -2 \end{vmatrix} = -40 - 56 - 12 + 4 + 21 + 5 = -78.$$

Use Theorem 4.2.14 to show the second determinant equation involving a row operation. Note that if A is a 1×1 matrix, it is true that whenever a multiple of another row is added to the first row resulting in the matrix A', $\det A' = \det A$. This is because there is no other row. This proves the basis case, so for the induction step, let E be the $(n + 1) \times (n + 1)$ elementary matrix such that op $E = s R_i + R_j$. That is,

$$EA = \begin{bmatrix} a_{11} & a_{12} & \cdots & a_{1,n+1} \\ \vdots & \vdots & & \vdots \\ a_{i1} & a_{i2} & \cdots & a_{i,n+1} \\ \vdots & \vdots & & \vdots \\ sa_{i1} + a_{j1} & sa_{i2} + a_{j2} & \cdots & sa_{i,n+1} + a_{j,n+1} \\ \vdots & \vdots & & \vdots \\ a_{n+1,1} & a_{n+1,2} & \cdots & a_{n+1,n+1} \end{bmatrix}.$$

As with the demonstration of (4.20), it is a straight-forward exercise to use induction to show $\det EA = \det A$ when $i \neq 1$ and $j \neq 1$ (Exercise 12). There are two other cases.

For the first, assume that op $E = s R_i + R_1$. By Theorem 4.2.14,

$$\det A = \sum_{\sigma \in S_{n+1}} \det A^{\sigma}. \tag{4.25}$$

To help visualize what will happen, for all $\sigma \in S_n$, write

$$A^{\sigma} = \begin{bmatrix} \cdots & 0 & \cdots & 0 & \cdots & a_{1\sigma(1)} & \cdots \\ \cdots & 0 & \cdots & a_{2\sigma(2)} & \cdots & 0 & \cdots \\ & \vdots & & \vdots & & \vdots & \\ \cdots & a_{i\sigma(i)} & \cdots & 0 & \cdots & 0 & \cdots \\ & \vdots & & \vdots & & \vdots & \end{bmatrix}. \tag{4.26}$$

The notation of (4.26) is intended to represent a refinement of A. In such a matrix all of its entries are zero except possibly for a few entries that are located such that every row and every column have at most one nonzero entry. Then,

$$
\det EA = \sum_{\sigma \in S_{n+1}}
\begin{vmatrix}
\cdots & sa_{i\sigma(i)} & \cdots & 0 & \cdots & a_{1\sigma(1)} & \cdots \\
\cdots & 0 & \cdots & a_{2\sigma(2)} & \cdots & 0 & \cdots \\
& \vdots & & \vdots & & \vdots & \\
\cdots & a_{i\sigma(i)} & \cdots & 0 & \cdots & 0 & \cdots \\
& \vdots & & \vdots & & \vdots &
\end{vmatrix}.
$$

By Lemma 4.2.11, each term of this sum can be written as

$$
\begin{vmatrix}
\cdots & 0 & \cdots & 0 & \cdots & a_{1\sigma(1)} & \cdots \\
\cdots & 0 & \cdots & a_{2\sigma(2)} & \cdots & 0 & \cdots \\
& \vdots & & \vdots & & \vdots & \\
\cdots & a_{i\sigma(i)} & \cdots & 0 & \cdots & 0 & \cdots \\
& \vdots & & \vdots & & \vdots &
\end{vmatrix}
+
\begin{vmatrix}
\cdots & sa_{i\sigma(i)} & \cdots & 0 & \cdots & 0 & \cdots \\
\cdots & 0 & \cdots & a_{2\sigma(2)} & \cdots & 0 & \cdots \\
& \vdots & & \vdots & & \vdots & \\
\cdots & a_{i\sigma(i)} & \cdots & 0 & \cdots & 0 & \cdots \\
& \vdots & & \vdots & & \vdots &
\end{vmatrix}.
$$

However, the submatrix of

$$
\begin{bmatrix}
\cdots & sa_{i\sigma(i)} & \cdots & 0 & \cdots & 0 & \cdots \\
\cdots & 0 & \cdots & a_{2\sigma(2)} & \cdots & 0 & \cdots \\
& \vdots & & \vdots & & \vdots & \\
\cdots & a_{i\sigma(i)} & \cdots & 0 & \cdots & 0 & \cdots \\
& \vdots & & \vdots & & \vdots &
\end{bmatrix}
$$

formed by deleting the first row and column $\sigma(i)$ has a row of zeroes, which implies that the determinant of this matrix is zero by Theorem 4.2.9. From this conclude that $\det EA = \det A$.

For the second case, let op $E = sR_1 + R_i$. Again, write $\det A$ as in (4.25) and write A^σ as in (4.26). This implies that

$$
\det EA = \sum_{\sigma \in S_{n+1}}
\begin{vmatrix}
\cdots & 0 & \cdots & 0 & \cdots & a_{1\sigma(1)} & \cdots \\
\cdots & 0 & \cdots & a_{2\sigma(2)} & \cdots & 0 & \cdots \\
& \vdots & & \vdots & & \vdots & \\
\cdots & a_{i\sigma(i)} & \cdots & 0 & \cdots & sa_{1\sigma(1)} & \cdots \\
& \vdots & & \vdots & & \vdots &
\end{vmatrix}.
$$

$$
= \sum_{\sigma \in S_{n+1}} a_{1\sigma(1)}
\begin{vmatrix}
\cdots & 0 & \cdots & a_{2\sigma(2)} & \cdots \\
& \vdots & & \vdots & \\
\cdots & a_{i\sigma(i)} & \cdots & 0 & \cdots \\
& \vdots & & \vdots &
\end{vmatrix}
= \det A.
$$

All of this work shows that

$$
\text{if op } E = sR_i + R_j, \text{ then } \det EA = \det A. \tag{4.27}
$$

Switching Rows

For the final elementary row operation, take the $n \times n$ matrix A and exchange rows i and j. That is, consider EA, where E is an $n \times n$ elementary matrix such that op $E = R_i \leftrightarrow R_j$. This switch can be obtained using the other two elementary row operations. Let $E_1, E_2, E_3, E_4, E_5, E_6$ be $n \times n$ elementary matrices such that

$$\begin{aligned}
&\text{op } E_1 = -R_i, &&\text{op } E_2 = R_i + R_j, &&\text{op } E_3 = -R_j, \\
&\text{op } E_4 = R_j + R_i, &&\text{op } E_5 = -R_i, &&\text{op } E_6 = R_i + R_j.
\end{aligned} \tag{4.28}$$

Then, $EA = E_6 E_5 E_4 E_3 E_2 E_1 A$. For example, use these six row operations to switch the first and third rows,

$$
\begin{bmatrix} 1 & 2 & 3 \\ 4 & 5 & 6 \\ 7 & 8 & 9 \end{bmatrix}
\xrightarrow{-R_1}
\begin{bmatrix} -1 & -2 & -3 \\ 4 & 5 & 6 \\ 7 & 8 & 9 \end{bmatrix}
\xrightarrow{R_1+R_3}
\begin{bmatrix} -1 & -2 & -3 \\ 4 & 5 & 6 \\ 6 & 6 & 6 \end{bmatrix}
$$

$$
\xrightarrow{-R_3}
\begin{bmatrix} -1 & -2 & -3 \\ 4 & 5 & 6 \\ -6 & -6 & -6 \end{bmatrix}
\xrightarrow{R_3+R_1}
\begin{bmatrix} -7 & -8 & -9 \\ 4 & 5 & 6 \\ -6 & -6 & -6 \end{bmatrix}
$$

$$
\xrightarrow{-R_1}
\begin{bmatrix} 7 & 8 & 9 \\ 4 & 5 & 6 \\ -6 & -6 & -6 \end{bmatrix}
\xrightarrow{R_1+R_3}
\begin{bmatrix} 7 & 8 & 9 \\ 4 & 5 & 6 \\ 1 & 2 & 3 \end{bmatrix}.
$$

The proof that the sequence of elementary matrices given in (4.28) will always exchange two rows of a matrix is left to Exercise 13.

The elementary matrices of (4.28) can now be used to compute the determinant of A. Namely,

$$\begin{aligned}
\det A &= -\det E_1 A \\
&= -\det E_2 E_1 A \\
&= \det E_3 E_2 E_1 A \\
&= \det E_4 E_3 E_2 E_1 A \\
&= -\det E_5 E_4 E_3 E_2 E_1 A \\
&= -\det E_6 E_5 E_4 E_3 E_2 E_1 A = -\det EA.
\end{aligned}$$

Therefore,

$$\text{if op } E = R_i \leftrightarrow R_k, \text{ then } \det EA = -\det A. \tag{4.29}$$

Lastly, it should be noted that

$$\det I_n = 1. \tag{4.30}$$

This is proved by induction on n. First, $\det[1] = 1$, showing the basis case. Next,

$$\det I_{n+1} = 1\det I_n = 1,$$

where the first equality is Definition 4.2.2 and the second equality follows by the induction hypothesis.

The results found in (4.20), (4.27), (4.29), and (4.30) regarding the relationship between the row operations and the determinant are summarized in the next theorem. The two corollaries are its direct results.

■ **Theorem 4.2.16**

Let A be an $n \times n$ matrix and E be an $n \times n$ elementary matrix. Let $s \in \mathbb{R}$.

(a) $\det I_n = 1$.

(b) If $\mathrm{op}\, E = \mathrm{R}_i \leftrightarrow \mathrm{R}_k$, then $\det EA = -\det A$.

(c) If $\mathrm{op}\, E = s\mathrm{R}_i$, then $\det EA = s \det A$.

(d) If $\mathrm{op}\, E = s\mathrm{R}_i + \mathrm{R}_k$, then $\det EA = \det A$.

■ **Corollary 4.2.17**

Let E be an elementary matrix.

(a) If $\mathrm{op}\, E = \mathrm{R}_i \leftrightarrow \mathrm{R}_k$, then $\det E = -1$.

(b) If $\mathrm{op}\, E = s\mathrm{R}_i$, then $\det E = s$.

(c) If $\mathrm{op}\, E = s\mathrm{R}_i + \mathrm{R}_k$, then $\det E = 1$.

Proof

All four parts of Theorem 4.2.16 are required to prove this corollary. There are three cases to consider.

- If $\mathrm{op}\, E = \mathrm{R}_i \leftrightarrow \mathrm{R}_k$, then $\det E = -\det I_n = -1$.
- If $\mathrm{op}\, E = s\mathrm{R}_i$, then $\det E = s \det I_n = s$.
- If $\mathrm{op}\, E = s\mathrm{R}_i + \mathrm{R}_k$, then $\det E = \det I_n = 1$. ■

■ **Corollary 4.2.18**

Let A and E be $n \times n$ matrices. If E is an invertible elementary matrix,

$$\det EA = \det E \det A.$$

Proof

Let E be an elementary matrix. As in Corollary 4.2.17, there are three cases to consider.

- If $\mathrm{op}\, E = \mathrm{R}_i \leftrightarrow \mathrm{R}_k$, then $\det E = -1$, so

$$\det EA = -\det A = (-1)\det A = \det E \det A.$$

- If $\mathrm{op}\, E = s\mathrm{R}_i$, then $\det E = s$, so $\det EA = s \det A = \det E \det A$.
- If $\mathrm{op}\, E = s\mathrm{R}_i + \mathrm{R}_k$, then $\det E = 1$, so $\det EA = \det A = \det E \det A$. ■

■ **Corollary 4.2.19**

If a square matrix A has two identical rows, $\det A = 0$.

Proof

If a matrix A has two identical rows and B is obtained from A by switching those rows, $A = B$, resulting in $\det A = -\det A$ by Theorem 4.2.16(b). ■

■ **Example 4.2.20**

Use Theorem 4.2.16 and its corollary to find the determinant of the matrix

$$A = \begin{bmatrix} 2 & 5 & 1 \\ 0 & -2 & 6 \\ 1 & 0 & 4 \end{bmatrix}.$$

By Gauss–Jordan elimination:

$$\begin{bmatrix} 2 & 5 & 1 \\ 0 & -2 & 6 \\ 1 & 0 & 4 \end{bmatrix} \xrightarrow{R_1 \leftrightarrow R_3} \begin{bmatrix} 1 & 0 & 4 \\ 0 & -2 & 6 \\ 2 & 5 & 1 \end{bmatrix} \xrightarrow{-2R_1 + R_3} \begin{bmatrix} 1 & 0 & 4 \\ 0 & -2 & 6 \\ 0 & 5 & -7 \end{bmatrix}$$

$$\xrightarrow{-\frac{1}{2}R_2} \begin{bmatrix} 1 & 0 & 4 \\ 0 & 1 & -3 \\ 0 & 5 & -7 \end{bmatrix} \xrightarrow{-5R_2 + R_3} \begin{bmatrix} 1 & 0 & 4 \\ 0 & 1 & -3 \\ 0 & 0 & 8 \end{bmatrix}$$

$$\xrightarrow{\frac{1}{8}R_3} \begin{bmatrix} 1 & 0 & 4 \\ 0 & 1 & -3 \\ 0 & 0 & 1 \end{bmatrix} \xrightarrow{3R_3 + R_2} \begin{bmatrix} 1 & 0 & 4 \\ 0 & 1 & 0 \\ 0 & 0 & 1 \end{bmatrix}$$

$$\xrightarrow{-4R_3 + R_1} \begin{bmatrix} 1 & 0 & 0 \\ 0 & 1 & 0 \\ 0 & 0 & 1 \end{bmatrix}$$

Therefore, $\det I = -\left(-\frac{1}{2}\right)\left(\frac{1}{8}\right) \det A$, which implies that $\det A = 16$.

Exercises

1. Find the determinants of the given matrices using only Definition 4.2.2.

   (a) $\begin{bmatrix} 6 & -2 \\ 4 & 1 \end{bmatrix}$

   (c) $\begin{bmatrix} 1 & 2 & -3 \\ 0 & 0 & 0 \\ 5 & 3 & 1 \end{bmatrix}$

   (b) $\begin{bmatrix} 9 & 1 & -4 & -5 \\ 5 & -4 & 9 & -6 \\ -9 & 4 & -8 & 0 \\ -10 & -6 & 1 & 0 \end{bmatrix}$

   (d) $\begin{bmatrix} 3 & -3 & 2 \\ 1 & -1 & 2 \\ 5 & -5 & -1 \end{bmatrix}$

2. Find the determinants of the given matrices by finding the reduced row echelon form of the matrix.

   (a) $\begin{bmatrix} 0 & 3 & 2 \\ -3 & -1 & -5 \\ 0 & -4 & -6 \end{bmatrix}$

   (c) $\begin{bmatrix} 5 & 0 & 0 \\ 0 & 6 & 0 \\ 0 & 0 & -3 \end{bmatrix}$

   (b) $\begin{bmatrix} 3 & 6 & -9 \\ 0 & 0 & -2 \\ -2 & 1 & 5 \end{bmatrix}$

   (d) $\begin{bmatrix} -5 & 4 & 3 \\ 3 & -2 & -2 \\ -2 & 2 & 1 \end{bmatrix}$

3. Find the determinants of the given matrices using any method.

(a) $\begin{bmatrix} 3 & 2 & 1 \\ 0 & 2 & -5 \\ 3 & 2 & 1 \end{bmatrix}$ (c) $\begin{bmatrix} 4 & -1 & -2 \\ 0 & -5 & 1 \\ 0 & 0 & 2 \end{bmatrix}$ (e) $\begin{bmatrix} 7 & 0 & 0 \\ 0 & 0 & 4 \\ 0 & 2 & 0 \end{bmatrix}$

(b) $\begin{bmatrix} 1 & 1 & 1 \\ 2 & -3 & 4 \\ -5 & 6 & -7 \end{bmatrix}$ (d) $\begin{bmatrix} 3 & 6 & 9 \\ 8 & 4 & 2 \\ 0 & 0 & 0 \end{bmatrix}$ (f) $\begin{bmatrix} -1 & 0 & 0 \\ 0 & 2 & 0 \\ 0 & 0 & 4 \end{bmatrix}$

4. Find $\begin{vmatrix} 2 & -1 & 0 & 0 & 0 \\ -1 & 2 & -1 & 0 & 0 \\ 0 & -1 & 2 & -1 & 0 \\ 0 & 0 & -1 & 2 & -1 \\ 0 & 0 & 0 & -1 & 2 \end{vmatrix}$.

5. Take $a, b \in \mathbb{R}$ and let $A = \begin{bmatrix} a & b & b & \cdots & b \\ b & a & b & \cdots & b \\ b & b & a & \cdots & b \\ \vdots & \vdots & \vdots & & \vdots \\ b & b & b & \cdots & a \end{bmatrix}$ so that size of A is $n \times n$. Then,

$\det A = (a - b)^{n-1}[a + (n - 1)b]$. Prove this using the following techniques.

(a) Use Definition 4.2.2.

(b) Use the elementary row operations.

6. Find all permutations in S_4.

7. Compute the given determinants using Theorem 4.2.14.

(a) $\begin{vmatrix} 8 & -3 \\ 7 & 4 \end{vmatrix}$ (b) $\begin{vmatrix} -4 & 2 & 1 \\ 0 & 0 & 0 \\ 1 & -4 & 5 \end{vmatrix}$ (c) $\begin{vmatrix} 2 & 1 & 3 \\ 6 & 4 & 2 \\ 1 & 1 & 5 \end{vmatrix}$

8. Solve for x given $\begin{vmatrix} 1 & 0 & -2 \\ 3 & 2x & -6 \\ 1 & 2 & x - 4 \end{vmatrix} = \begin{vmatrix} x & -2 \\ 5 & x - 1 \end{vmatrix}$.

9. Use $\begin{vmatrix} 2 & 1 & 4 \\ 1 & 2 & 3 \\ 0 & 4 & 1 \end{vmatrix}$ and $\begin{vmatrix} 2 & 1 & 4 \\ 5 & -2 & 1 \\ 0 & 4 & 1 \end{vmatrix}$ to compute $\begin{vmatrix} 2 & 1 & 4 \\ 6 & 0 & 4 \\ 0 & 4 & 1 \end{vmatrix}$.

10. Prove Theorem 4.3.6(a).

11. Prove (4.31) from the proof of Theorem 4.3.6.

12. Prove that op $E = sR_i + R_k$ implies $\det EA = \det A$ when $i \neq 1$ and $k \neq 1$.

13. Let A be an $n \times n$ matrix and $E_1, E_2, E_3, E_4, E_5, E_6$ be the elementary $n \times n$ matrices defined using (4.28). Prove that $E_6 E_5 E_4 E_3 E_2 E_1 A$ equals the matrix obtained by exchanging rows $i \neq j$ in A.

14. Let A and B be $n \times n$ matrices such that $AB = 0_n$. Show that

$$\det (I_n - aA) \det (I_n - bB) = \det (I_n - aA - bB).$$

15. Let $a_1, a_2, \ldots, a_n \in \mathbb{R}$. Define

$$A = \begin{bmatrix} 1 & a_1 & a_1^2 & \cdots & a_1^{n-1} \\ 1 & a_2 & a_2^2 & \cdots & a_2^{n-1} \\ \vdots & \vdots & \vdots & & \vdots \\ 1 & a_n & a_n^2 & \cdots & a_n^{n-1} \end{bmatrix}.$$

This is a **Vandermonde matrix**. Find $\det A$ without using Theorem 4.3.6.

16. Let A and B be square matrices, not necessarily of the same size.

   (a) Prove that $\det \begin{bmatrix} A & 0 \\ 0 & B \end{bmatrix} = \det A \det B.$

   (b) Use Exercise 16(a) to evaluate $\begin{vmatrix} 5 & 2 & 0 & 0 & 0 \\ -1 & 3 & 0 & 0 & 0 \\ 0 & 0 & 4 & -2 & 5 \\ 0 & 0 & 1 & 2 & 3 \\ 0 & 0 & 3 & -1 & 1 \end{vmatrix}.$

   (c) Prove that $\begin{bmatrix} A & 0 \\ 0 & B \end{bmatrix}$ is invertible if and only if A and B are invertible.

17. Let A and B be $n \times n$ matrices.

   (a) Show that if A and B are diagonal matrices, $A + B$, AB, and rA for $r \in \mathbb{R}$ are diagonal matrices.

   (b) For all positive integers n, evaluate A^n when A is a diagonal matrix. Prove the result.

   (c) Find an example of a linear operator T such that $[T]$ is a diagonal matrix.

18. Prove that both the set of lower triangular matrices and the set of upper triangular matrices are closed under addition, multiplication, and scalar multiplication.

19. Prove that the determinant of the standard matrix of a rotation in $\mathbb{R}^3$ about an axis equals 1 and that the determinant of the standard matrix of a reflection in $\mathbb{R}^3$ about a plane is -1.

4.3   Inverses and Determinants

There is a close relationship between determinants and the inverses of square matrices. Specifically, it is indicated in (4.15) for the 1×1 case and in (4.17) for the 2×2 case that the determinant is nonzero for invertible matrices of these two sizes. That this can be generalized to every invertible square matrix, along with other connections, will be shown soon. First, however, it should be proved that the determinant has a very important property.

Uniqueness of the Determinant

Knowing that the determinant has the four properties of Theorem 4.2.16, it is natural to ask whether it is the only function $\bigcup_{i=1}^{\infty} \mathbb{R}^{i \times i} \to \mathbb{R}$ that satisfies these four properties. The answer to this question is yes, and a proof of this starts by naming such functions and ends by showing that any such function is actually the determinant of Definition 4.2.2.

■ Definition 4.3.1

A function

$$d : \bigcup_{i=1}^{\infty} \mathbb{R}^{i \times i} \to \mathbb{R}$$

is a **pseudo-determinant** if d satisfies the following properties for all $n \times n$ matrices A and elementary E:

(a) $d(I_n) = 1$.
(b) If op $E = s R_i$, then $d(EA) = s d(A)$.
(c) If op $E = s R_i + R_k$, then $d(EA) = d(A)$.

Definition 4.3.1 has only three properties because (b) and (c) can be used to exchange two rows of a matrix. For this reason, this property is listed in the next theorem along with the corresponding result for Theorem 4.2.9.

■ Theorem 4.3.2

Let d be a pseudo-determinant.

(a) Let E be an elementary matrix that is the same size as the matrix A. If op $E = R_i \leftrightarrow R_j$, then $d(EA) = -d(A)$.
(b) If A is a matrix with a row of zeros, $d(A) = 0$.
(c) If A is a matrix in which one row is a multiple of another row, then $d(A) = 0$.

Proof

Let E be an elementary matrix that has the same size as the $n \times n$ matrix A. Since the proof of Theorem 4.2.16(b) relies only on parts (b) and (c) of Definition 4.3.1, (a) is true. Next, suppose that all entries of row i in A are zero. Assume that op $E = 0 R_i$. Then, $A = EA$, so by Definition 4.3.1(b), $d(A) = d(EA) = 0$. This proves (b). Lastly, write $A = [a_{ij}]$ and suppose that for rows $i \neq j$, there exists $s \in \mathbb{R}$ such that $a_{ik} = s a_{jk}$ for all $k = 1, 2, \ldots, n$. Assume that op $E = -s R_j + R_i$. Then, EA has a row of zeros, so by (b), $d(EA) = 0$, and the theorem is proved by Definition 4.3.1(c). ■

The work leading to (4.20), (4.27), and (4.30) show that pseudo-determinants might be an appropriate set of functions that can be used to show the uniqueness of the determinant. Indeed, Theorem 4.2.16 implies the next result.

■ **Theorem 4.3.3**

The determinant is a pseudo-determinant.

Completing the demonstration of the uniqueness of the determinant requires the converse of Theorem 4.3.3, and this requires a lemma.

■ **Lemma 4.3.4**

Let $A, E_1, E_2, \ldots, E_k$ be $n \times n$ matrices such that $E_1, E_2, \ldots E_k$ are invertible and elementary. There exists a nonzero $r \in \mathbb{R}$ so that $d(E_k \cdots E_2 E_1 A) = rd(A)$ for all pseudo-determinants d.

Proof

First, define for all invertible elementary matrices E,

$$r_E = \begin{cases} -1 & \text{if op } E = \text{R}_i \leftrightarrow \text{R}_j, \\ s & \text{if op } E = s\text{R}_i \text{ with } s \neq 0, \\ 1 & \text{if op } E = s\text{R}_i + \text{R}_k. \end{cases}$$

Prove the result by induction on k. Definition 4.3.1, Theorem 4.3.2(a) and the definition of r_E imply that the basis case holds. Next, let $E_1, E_2, \ldots, E_{k+1}$ be $n \times n$ invertible elementary matrices. By induction, there is a nonzero $r_1 \in \mathbb{R}$ such that $f(E_k \cdots E_2 E_1 A) = r_1 f(A)$ for all pseudo-determinants f. Choose $r = r_{E_{k+1}} r_1 \neq 0$. Let d be an arbitrary pseudo-determinant and compute,

$$d(E_{k+1} E_k \cdots E_2 E_1 A) = r_{E_{k+1}} d(E_k \cdots E_2 E_1 A) = r_{E_{k+1}} r_1 d(A) = rd(A). \blacksquare$$

■ **Theorem 4.3.5**

A pseudo-determinant is the determinant.

Proof

Assume that d is a pseudo-determinant. To prove that $d = \det$, let A be an $n \times n$ matrix. Let R be the reduced row echelon form of A and suppose that $E_1, E_2, \ldots, E_k$ are invertible elementary matrices such that

$$R = E_k \cdots E_2 E_1 A.$$

This implies that $R = I_n$ or R has a row of zeros by Exercise 8. If $R = I_n$, by Definition 4.3.1(a) and Theorem 4.2.16(a),

$$d(R) = 1 = \det R.$$

Otherwise, because of Theorems 4.2.9 and 4.3.2(b),

$$d(R) = 0 = \det R.$$

In either case, appealing to Lemma 4.3.4, since the determinant is a pseudo-determinant (Theorem 4.3.3), there exists a nonzero real number r such that

$$rd(A) = d(E_k \cdots E_2 E_1 A) = d(R) = \det R = \det E_k \cdots E_2 E_1 A = r \det A.$$

Hence, $d(A) = \det A$. ■

Theorem 4.3.5 means that any function that takes a matrix as an input, returns a real number, and satisfies the three properties of Definition 4.3.1 is, in fact, the determinant. This has significant consequences.

■ **Theorem 4.3.6**

Let $A = [a_{ij}]$ be an $n \times n$ matrix with $n > 1$.

(a) The determinant can be calculated by expanding on any row. That is, for any $i = 1, 2, \ldots, n$,

$$\det A = \sum_{j=1}^{n} (-1)^{i+j} a_{ij} \det A_{ij}.$$

(b) The determinant can be calculated by expanding on any column. That is, for any $j = 1, 2, \ldots, n$,

$$\det A = \sum_{i=1}^{n} (-1)^{i+j} a_{ij} \det A_{ij}.$$

Proof

The proof of (a) is left to Exercise 10. To prove (b), it suffices to show that expanding on columns satisfies the three properties of pseudo-determinants given in Definition 4.3.1. That $d(I_n) = 1$ is proved by induction in essentially the same manner as (4.30), while the proof of

$$\text{if op } E = sR_i + R_k, \text{ then } d(EA) = d(A) \tag{4.31}$$

is left to Exercise 11. The proof of (b) is by induction on the size of A. That (b) holds for a 1×1 matrix is trivial, so let A be an $(n+1) \times (n+1)$ matrix. Chose an integer j such that $1 \leq j \leq n$. Define

$$d(A) = \sum_{i=1}^{n} (-1)^{i+j} a_{ij} \det A_{ij}.$$

Next, let E be an $(n+1) \times (n+1)$ elementary matrix with op $E = sR_i$. Thus,

$$EA = \begin{bmatrix} a_{11} & a_{12} & \cdots & a_{1,n+1} \\ \vdots & \vdots & & \vdots \\ sa_{i1} & sa_{i2} & \cdots & sa_{i,n+1} \\ \vdots & \vdots & & \vdots \\ a_{n+1,1} & a_{n+1,2} & \cdots & a_{n+1,n+1} \end{bmatrix},$$

which implies

$$d(EA) = (-1)^{i+j} sa_{ij} \det (EA)_{ij} + \sum_{k \neq i} (-1)^{k+j} a_{kj} \det (EA)_{kj}.$$

When $k \neq i$,

$$\det(EA)_{kj} = \begin{vmatrix} a_{11} & a_{12} & \cdots & a_{1,j-1} & a_{1,j+1} & \cdots & a_{1,n+1} \\ \vdots & \vdots & & \vdots & \vdots & & \vdots \\ sa_{i1} & sa_{i2} & \cdots & sa_{i,j-1} & sa_{i,j+1} & \cdots & sa_{i,n+1} \\ \vdots & \vdots & & \vdots & \vdots & & \vdots \\ a_{k-1,1} & a_{k-1,2} & \cdots & a_{k-1,j-1} & a_{k-1,j+1} & \cdots & a_{k-1,n+1} \\ a_{k+1,1} & a_{k+1,2} & \cdots & a_{k+1,j-1} & a_{k+1,j+1} & \cdots & a_{k+1,n+1} \\ \vdots & \vdots & & \vdots & \vdots & & \vdots \\ a_{n+1,1} & a_{n+1,2} & \cdots & a_{n+1,j-1} & a_{n+1,j+1} & \cdots & a_{n+1,n+1} \end{vmatrix},$$

so by induction, $\det(EA)_{kj} = s \det A_{kj}$. Because, $\det(EA)_{ij} = \det A_{ij}$, it follows that

$$d(EA) = (-1)^{i+j} sa_{ij} \det A_{ij} + \sum_{k \neq i} (-1)^{k+j} a_{kj} s \det A_{kj} = sd(A).$$

All of this means that d is a pseudo-determinant. Hence, $d = \det$ by Theorem 4.3.5, so determinants can be found using columns instead of rows. ∎

■ Corollary 4.3.7

Let $A = [a_{ij}]$ be an $n \times n$ matrix.

(a) $\det A = \det A^T$.

(b) If A is a triangular matrix, $\det(A) = a_{11}a_{22} \cdots a_{nn}$.

(c) If A is a matrix in which one row is a multiple of another row or one column is a multiple of another column, $\det A = 0$.

■ Example 4.3.8

Since the determinant can be calculated by expanding on any row or column by Theorem 4.3.6, expand on the first column to compute

$$\begin{vmatrix} 2 & 5 & 1 \\ 0 & -2 & 6 \\ 1 & 0 & 4 \end{vmatrix} = 2\begin{vmatrix} -2 & 6 \\ 0 & 4 \end{vmatrix} + \begin{vmatrix} 5 & 1 \\ -2 & 6 \end{vmatrix} = 2(-8) + (32) = 16,$$

expand on the second column to compute

$$\begin{vmatrix} 1 & 1 & 2 \\ 3 & 1 & 4 \\ 5 & 1 & 6 \end{vmatrix} = -\begin{vmatrix} 3 & 4 \\ 5 & 6 \end{vmatrix} + \begin{vmatrix} 1 & 2 \\ 5 & 6 \end{vmatrix} - \begin{vmatrix} 1 & 2 \\ 3 & 4 \end{vmatrix} = 2 - 4 + 2 = 0,$$

and expand on the second row to compute

$$\begin{vmatrix} 1 & 4 & -2 \\ 0 & 0 & 3 \\ -3 & 1 & 4 \end{vmatrix} = -3\begin{vmatrix} 1 & 4 \\ -3 & 1 \end{vmatrix} = -3(13) = -39.$$

Also, by Corollary 4.3.7,

$$\begin{vmatrix} 1 & 0 & -3 \\ 4 & 0 & 1 \\ -2 & 3 & 4 \end{vmatrix} = -39, \quad \begin{vmatrix} 1 & 3 & 5 \\ 0 & 2 & 1 \\ 0 & 0 & 6 \end{vmatrix} = 12, \quad \begin{vmatrix} 1 & 3 & 5 \\ 1 & 2 & 3 \\ 2 & 4 & 6 \end{vmatrix} = 0, \quad \begin{vmatrix} 1 & 3 & 6 \\ 1 & 1 & 2 \\ 1 & 2 & 4 \end{vmatrix} = 0.$$

Equivalents to Invertibility

The goal of the definition of the determinant is to generalize the invertibility tests of (4.15) and (4.17) to arbitrary square matrices. It is now the time to prove this and some other common equivalences of invertibility.

■ **Theorem 4.3.9**

The square matrix A is invertible if and only if the reduced row echelon form of A is the identity matrix.

Proof

Let A be an invertible $n \times n$ matrix. Suppose that R, the reduced row echelon form of A, is not the identity. This implies that R has a row of zeros (Exercise 8), so the solution set of $R\mathbf{x} = \mathbf{0}$ is described using a parameter. Therefore, $A\mathbf{x} = \mathbf{0}$ has infinitely many solutions by Theorem 3.5.7, but this contradicts Theorem 4.1.18. Thus, the reduced row echelon form of A must be I_n. The converse is Theorem 4.1.15. ■

■ **Example 4.3.10**

The matrix,

$$\begin{bmatrix} 2 & 5 & 1 \\ 0 & -2 & 6 \\ 1 & 0 & 4 \end{bmatrix},$$

is invertible because its reduced row echelon form is I_3, but

$$\begin{bmatrix} 1 & 2 & 3 & 4 \\ 5 & 6 & 7 & 8 \\ 9 & 0 & 1 & 2 \\ 3 & 4 & 5 & 6 \end{bmatrix}$$

is not invertible because its reduced row echelon form is

$$\begin{bmatrix} 1 & 0 & 0 & 0 \\ 0 & 1 & 0 & -1 \\ 0 & 0 & 1 & 2 \\ 0 & 0 & 0 & 0 \end{bmatrix}.$$

■ **Theorem 4.3.11**

The square matrix A is invertible if and only if $\det A \neq 0$.

Proof

Let R be the reduced row echelon form of A. Suppose that A is invertible. By Theorem 4.3.9, R is the identity matrix, so $\det R \neq 0$. By Lemma 4.3.4 and Theorem 4.3.5, $\det A \neq 0$. To prove the converse, suppose that A is not invertible. This implies that R is not the identity matrix, so R has a row of zeros (Exercise 8). Hence, $\det A = 0$ by Theorem 4.2.9. ∎

■ **Example 4.3.12**

The matrix,

$$A = \begin{bmatrix} 2 & 5 & 1 \\ 0 & -2 & 6 \\ 1 & 0 & 4 \end{bmatrix},$$

is invertible because $\det A = 16$ (Example 4.2.20), but

$$B = \begin{bmatrix} 1 & 2 & 3 & 4 \\ 5 & 6 & 7 & 8 \\ 9 & 0 & 1 & 2 \\ 3 & 4 & 5 & 6 \end{bmatrix}$$

is not invertible because $\det B = 0$ by (4.18).

■ **Theorem 4.3.13**

The square matrix A is invertible if and only if A is the product of elementary matrices.

Proof

Let A be an $n \times n$ matrix. Assume that A is invertible. By Theorem 4.3.9, there exists elementary matrices $E_1, E_2, \ldots, E_k$ such that $E_k \cdots E_2 E_1 A = I_n$. These elementary matrices are invertible, and their inverses are also elementary (Theorem 4.1.13). Hence, $A = E_1^{-1} E_2^{-1} \cdots E_k^{-1}$, showing that A is the product of elementary matrices. The converse follows from Corollary 4.1.16. ∎

■ **Example 4.3.14**

Continuing Example 4.1.17, where $A = \begin{bmatrix} 1 & 0 & 2 \\ 2 & -1 & 4 \\ 0 & 2 & 4 \end{bmatrix}$, it follows from (4.6) that

$$A = E_1^{-1} E_2^{-1} E_3^{-1} E_4^{-1} E_5^{-1}, \tag{4.32}$$

where, as in Example 4.1.12,

$$E_1^{-1} = \begin{bmatrix} 1 & 0 & 0 \\ -2 & 1 & 0 \\ 0 & 0 & 1 \end{bmatrix}^{-1} = \begin{bmatrix} 1 & 0 & 0 \\ 2 & 1 & 0 \\ 0 & 0 & 1 \end{bmatrix},$$

$$E_2^{-1} = \begin{bmatrix} 1 & 0 & 0 \\ 0 & -1 & 0 \\ 0 & 0 & 1 \end{bmatrix}^{-1} = \begin{bmatrix} 1 & 0 & 0 \\ 0 & -1 & 0 \\ 0 & 0 & 1 \end{bmatrix},$$

$$E_3^{-1} = \begin{bmatrix} 1 & 0 & 0 \\ 0 & 1 & 0 \\ 0 & -2 & 1 \end{bmatrix}^{-1} = \begin{bmatrix} 1 & 0 & 0 \\ 0 & 1 & 0 \\ 0 & 2 & 1 \end{bmatrix},$$

$$E_4^{-1} = \begin{bmatrix} 1 & 0 & 0 \\ 0 & 1 & 0 \\ 0 & 0 & 1/4 \end{bmatrix}^{-1} = \begin{bmatrix} 1 & 0 & 0 \\ 0 & 1 & 0 \\ 0 & 0 & 4 \end{bmatrix},$$

$$E_5^{-1} = \begin{bmatrix} 1 & 0 & -2 \\ 0 & 1 & 0 \\ 0 & 0 & 1 \end{bmatrix}^{-1} = \begin{bmatrix} 1 & 0 & 2 \\ 0 & 1 & 0 \\ 0 & 0 & 1 \end{bmatrix}.$$

It is left to Exercise 2 to confirm (4.32).

■ **Theorem 4.3.15**

Let A be an $n \times n$ matrix and $\mathbf{x} \in \mathbb{R}^n$ be a variable vector. The following are equivalent.

(a) A is invertible.

(b) $A\mathbf{x} = \mathbf{0}$ has only the trivial solution.

(c) The solution set of $A\mathbf{x} = \mathbf{b}$ has exactly one solution for all $\mathbf{b} \in \mathbb{R}^n$.

Proof

Because of Theorem 4.1.18, (a) implies both (b) and (c). Since (c) includes the case when $\mathbf{b} = \mathbf{0}$, (c) proves (b). It remains to prove that (b) implies (c) and that (c) implies (a).

First, assume that $A\mathbf{u}_1 = \mathbf{b}$ and $A\mathbf{u}_2 = \mathbf{b}$ for some $\mathbf{u}_1, \mathbf{u}_2, \mathbf{b} \in \mathbb{R}^n$ such that $\mathbf{u}_1 \neq \mathbf{u}_2$. Then, $A\mathbf{x} = \mathbf{0}$ has a nontrivial solution because

$$A(\mathbf{u}_1 - \mathbf{u}_2) = A\mathbf{u}_1 - A\mathbf{u}_2 = \mathbf{0}.$$

Second, suppose that A is not invertible. This implies that the reduced row echelon form of A has a row of zeros (Theorem 4.3.9), which implies that for any $\mathbf{b} \in \mathbb{R}^n$, the reduced row echelon form of $[A \mid \mathbf{b}]$ has a leading 1 in the last column (no solutions) or the solution set of $A\mathbf{x} = \mathbf{b}$ requires at least one parameter (infinitely many solutions). ■

Products

Given an operation $*$ on a set A, for $a \in A$ to have an inverse a', it must be the case that both $a * a'$ and $a' * a$ equal the identity of A with respect to $*$. This is Definition 1.4.38(d). If $*$ is commutative, only one of these computations need to be checked. However, many times in algebra the operation represented by $*$ will not be commutative, so both $a * a'$ and $a' * a$ will need to be confirmed to equal the identity. Because matrix multiplication is not commutative, both computations were checked in Examples 4.1.5 and 4.1.6. However, opposed to what is typical in noncommutative algebra, only one of the computations is actually needed for matrix multiplication. This nice result is the next theorem.

■ **Theorem 4.3.16**

For all $n \times n$ matrices A and B, if $AB = I_n$, then A and B are invertible and $A^{-1} = B$.

Proof

Let $AB = I_n$. Suppose that $B\mathbf{x} = \mathbf{0}$. Then,

$$A(B\mathbf{x}) = A\mathbf{0},$$
$$AB\mathbf{x} = \mathbf{0},$$
$$I_n\mathbf{x} = \mathbf{0},$$
$$\mathbf{x} = \mathbf{0},$$

so B is invertible by Theorem 4.3.15. Furthermore, $A^{-1} = B$ because

$$AB = I_n$$
$$ABB^{-1} = I_nB^{-1},$$
$$A(BB^{-1}) = B^{-1},$$
$$AI_n = B^{-1},$$
$$A = B^{-1}. ■$$

Because of Theorem 4.3.16,

$$\begin{bmatrix} 1 & 0 & 0 \\ 0 & 1 & 0 \\ 0 & -2 & 1 \end{bmatrix} \begin{bmatrix} 1 & 0 & 0 \\ 0 & 1 & 0 \\ 0 & 2 & 1 \end{bmatrix} = \begin{bmatrix} 1 & 0 & 0 \\ 0 & 1 & 0 \\ 0 & 0 & 1 \end{bmatrix} \tag{4.33}$$

is enough to prove that

$$\begin{bmatrix} 1 & 0 & 0 \\ 0 & 1 & 0 \\ 0 & -2 & 1 \end{bmatrix}^{-1} = \begin{bmatrix} 1 & 0 & 0 \\ 0 & 1 & 0 \\ 0 & 2 & 1 \end{bmatrix}.$$

It should also be noted that since the identity matrix is invertible, (4.33) is also sufficient to prove that both factors of (4.33) are invertible. This is the next result.

■ **Theorem 4.3.17**

For all $n \times n$ matrices A and B, if AB is invertible, then A and B are invertible.

Proof

Let A and B be square matrices of the same size such that AB is invertible. Suppose that $B\mathbf{x} = \mathbf{0}$. Then, $AB\mathbf{x} = \mathbf{0}$, so $\mathbf{x} = \mathbf{0}$ by Theorem 4.3.15. Hence, B is invertible. In addition, $A[B(AB)^{-1}] = AB(AB)^{-1} = I_n$, which implies that A is invertible by Theorem 4.3.16. ∎

Under the right conditions, the determinant has an additive property as illustrated in Lemma 4.2.11. In general, however, the determinant always is multiplicative. What this means is seen in the theorem after the required lemma.

■ **Lemma 4.3.18**

For all $n \times n$ matrices $E_1, E_2, \ldots, E_k, B$ such that the matrices $E_1, E_2, \ldots, E_k$ are elementary, $\det E_1 E_2 \cdots E_k B = \det E_1 \det E_2 \cdots \det E_k \det B$.

Proof

Proceed by induction on the number k of elementary matrices in the product. The basis case is Corollary 4.2.18. For the induction step, let $k > 1$. Again, by Corollary 4.2.18 and the induction hypothesis,

$$\det E_1 E_2 \cdots E_k E_{k+1} B = \det E_1(E_2 \cdots E_k E_{k+1} B)$$
$$= \det E_1 \det E_2 \cdots E_k E_{k+1} B$$
$$= \det E_1 \det E_2 \cdots \det E_k \det E_{k+1} \det B. \ ∎$$

■ **Theorem 4.3.19**

For all $n \times n$ matrices A and B, $\det AB = \det A \det B$.

Proof

There are two cases to consider. First, suppose that A is not invertible. This implies that $\det A = 0$ by Theorem 4.3.11. Also, AB cannot be invertible by Theorem 4.3.17, so, again, $\det AB = 0$. Therefore, $\det AB = \det A \det B$.

Second, assume that A is invertible. This means that there exists elementary matrices $E_1, E_2, \ldots, E_k$ such that $A = E_1, E_2, \ldots, E_k$ by Theorem 4.3.13. Appealing to Lemma 4.3.18,

$$\det AB = \det E_1 \cdots E_k B$$
$$= \det E_1 \det E_2 \cdots \det E_k \det B$$
$$= \det E_1 E_2 \cdots E_k \det B = \det A \det B. \ ∎$$

To illustrate Theorem 4.3.19,

$$\begin{bmatrix} 1 & 2 \\ 3 & 4 \end{bmatrix} \begin{bmatrix} 5 & 6 \\ 7 & 8 \end{bmatrix} = \begin{bmatrix} 19 & 22 \\ 43 & 50 \end{bmatrix},$$

and

$$\begin{vmatrix} 1 & 2 \\ 3 & 4 \end{vmatrix} = -2, \quad \begin{vmatrix} 5 & 6 \\ 7 & 8 \end{vmatrix} = -2, \quad \begin{vmatrix} 19 & 22 \\ 43 & 50 \end{vmatrix} = 4.$$

Another example is

$$\begin{vmatrix} 1 & 0 & 0 \\ 0 & 3 & 0 \\ 0 & 0 & 1 \end{vmatrix} = 3, \quad \begin{vmatrix} 1 & 0 & 0 \\ 0 & 1/3 & 0 \\ 0 & 0 & 1 \end{vmatrix} = \frac{1}{3}, \quad \begin{vmatrix} 1 & 0 & 0 \\ 0 & 1 & 0 \\ 0 & 0 & 1 \end{vmatrix} = 1,$$

which illustrates the next theorem.

■ **Theorem 4.3.20**

For all square matrices A, if A is invertible, $\det A^{-1} = \dfrac{1}{\det A}$.

Proof

Assume that A is an invertible $n \times n$ matrix, so $\det A \neq 0$ by Theorem 4.3.11. Then, Theorem 4.3.19 implies that

$$1 = \det I_n = \det AA^{-1} = \det A \det A^{-1},$$

and this gives $\det A^{-1} = \dfrac{1}{\det A}$. ■

Exercises

1. Find the determinants of the given matrices.

   (a) $\begin{bmatrix} 1 & 2 & 3 \\ 4 & 5 & 6 \\ 0 & 1 & 0 \end{bmatrix}$ (b) $\begin{bmatrix} 1 & 2 & 3 \\ 0 & 1 & 0 \\ 4 & 5 & 6 \end{bmatrix}$ (c) $\begin{bmatrix} 1 & 0 & 4 \\ 2 & 1 & 5 \\ 3 & 0 & 6 \end{bmatrix}$

2. Confirm (4.32) in Example 4.3.14.

3. Use $A = \begin{bmatrix} 2 & 0 & 4 \\ 1 & 2 & 0 \\ 1 & 2 & 1 \end{bmatrix}$ and $B = \begin{bmatrix} 1/2 & 2 & -2 \\ -1/4 & -1/2 & 1 \\ 0 & -1 & 1 \end{bmatrix}$ to verify Theorem 4.3.16.

4. Determine which matrices from Exercises 1 to 3 in Section 4.2 are invertible.

5. Take $A = \begin{bmatrix} 2 & 4 & 3 \\ r & 1 & 5 \\ 2 & 5 & 6 \end{bmatrix}$. For what value of r is the matrix A noninvertible?

6. What must be said about the entries of an upper (lower) triangle matrix to guarantee that it is invertible?

7. If A is an invertible $n \times n$ matrix and $\mathbf{u} \neq \mathbf{v}$ are vectors in $\mathbb{R}^n$, prove that $A\mathbf{u} \neq A\mathbf{v}$.

8. Prove that if A is a square matrix in reduced row echelon form, A is the identity matrix or A has a row of zeros.

9. Prove that the square matrix A is invertible if and only if A^T is invertible.

10. Let A be an $n \times n$ matrix. Prove that the following are equivalent.
    - The system $A\mathbf{x} = \mathbf{b}$ has exactly one solution for all $\mathbf{b} \in \mathbb{R}^n$.
    - For all positive integers k, the system of linear equations $A^k\mathbf{x} = \mathbf{b}$ has exactly one solution for all $\mathbf{b} \in \mathbb{R}^n$.
    - For every invertible $n \times n$ matrix B, the system $BA\mathbf{x} = \mathbf{b}$ has exactly one solution for all $\mathbf{b} \in \mathbb{R}^n$.

11. Confirm that $C = \begin{bmatrix} 2 & 2 & 9 \\ 4 & 1 & 4 \\ 1 & 6 & 7 \end{bmatrix}$ is invertible. Then, find 3×3 nonidentity matrices A and B such that $C = AB$ and confirm that A and B are invertible, thus verifying Theorem 4.3.17.

12. Use $A = \begin{bmatrix} -6 & -3 & 5 \\ -3 & 1 & -8 \\ 6 & 2 & -6 \end{bmatrix}$ and $B = \begin{bmatrix} -2 & 1 & 0 \\ -2 & 6 & -2 \\ 9 & 0 & 9 \end{bmatrix}$ to verify Theorem 4.3.19.

13. Check that $A = \begin{bmatrix} 0 & -2 & 1 \\ 1 & 0 & 3 \\ 2 & 1 & 0 \end{bmatrix}$ is invertible and use it to verify Theorem 4.3.20.

14. Let A and B be 3×3 matrices such that $\det A = 4$ and $\det B = 5$. Evaluate.
    (a) $\det A^{-1}$ (c) $\det 3A$ (e) $\det (BA)^{-1}$
    (b) $\det AB$ (d) $\det A^3 B^2$ (f) $\det (5B)^{-1}$

15. Let $A_1, A_2, \dots, A_k$ be $n \times n$ matrices. Use mathematical induction to prove that
$$\det A_1 A_2 \cdots A_k = \det A_1 \det A_2 \cdots \det A_k.$$

16. Assume that A and B are $n \times n$ matrices. If $\det A = 5$ and $\det B = 2$, what is the determinant of $A^{-1}B$? What would happen if $\det A = 0$?

17. Prove that $\det AB = \det BA$ for all $n \times n$ matrices A and B.

18. Show that for all $n \times n$ matrices A and P, if P is invertible, $\det P^{-1}AP = \det A$.

19. Let A be an $n \times n$ matrix such that $AA^T = rI_n$ for some $r > 0$. Find $\det A$.

20. Prove that if A is an invertible matrix, AA^T and $A^T A$ are also invertible.

21. Suppose for the square matrix A that there exists a positive integer n such that $\det A^n = 0$. Prove that A is not invertible.

22. Let A and B be $n \times n$ matrices. Prove that if $AB = I_n$, then $AB = BA$.

23. Let A, B, C be appropriately sized for the given block matrices. Prove.

    (a) $\det \begin{bmatrix} A & B \\ 0 & C \end{bmatrix} = \det A \det C.$

    (b) $\det \begin{bmatrix} A & 0 \\ B & C \end{bmatrix} = \det A \det C.$ Can this be proved using Exercise 23(a)?

24. The proof of Corollary 4.3.7(b) did not have to wait. Use mathematical induction to prove that the determinant of an upper triangular matrix is the product of the entries of its main diagonal.

25. Let $T_1, T_2 : \mathbb{R}^n \to \mathbb{R}^n$ be linear operators such that $(T_1 \circ T_2)\mathbf{u} = \mathbf{u}$ for every $\mathbf{u} \in \mathbb{R}^n$. Prove that $(T_2 \circ T_1)\mathbf{u} = \mathbf{u}$ for all $\mathbf{u} \in \mathbb{R}^n$.

26. Let E be an elementary matrix.
    (a) Prove that E^T is elementary.
    (b) Let A be a matrix the same size as E. Prove that AE affects the columns of A in the same manner as EA affects the rows.

4.4 Applications

The calculations of determinants as in Example 4.2.4 are sometimes written using terminology that is intended to name the parts of the determinant equation.

■ **Definition 4.4.1**

Let A be an $n \times n$ matrix.
    (a) The (i, j)-**minor** of A is $m_{ij} = \det A_{ij}$.
    (b) The (i, j)-**cofactor** of A is $c_{ij} = (-1)^{i+j} m_{ij}$.

Using the 3×3 matrix in Example 4.2.4, the minors of $A = \begin{bmatrix} 5 & 7 & 8 \\ 9 & 1 & 2 \\ 3 & 5 & 6 \end{bmatrix}$ using its first row are

$$m_{11} = \begin{vmatrix} 1 & 2 \\ 5 & 6 \end{vmatrix} = -4, \quad m_{12} = \begin{vmatrix} 9 & 2 \\ 3 & 6 \end{vmatrix} = 48, \quad m_{13} = \begin{vmatrix} 9 & 1 \\ 3 & 5 \end{vmatrix} = 42,$$

and the corresponding cofactors of A are

$$c_{11} = -4, \quad c_{12} = -48, \quad c_{13} = 42.$$

Thus, the determinant of A can be rewritten as

$$\det A = a_{11}c_{11} + a_{12}c_{12} + a_{13}c_{13} = 5(-4) + 7(-48) + 8(42) = -20.$$

This method of calculating the determinant is a minor modification of Definition 4.2.2 and based on Theorem 4.3.6, yet it receives its own name.

■ **Definition 4.4.2**

For any $n \times n$ matrix $A = [a_{ij}]$ and any $k = 1, 2, \ldots, n$, calculating the determinant of A by **cofactor expansion** refers to the equation,

$$\det A = \sum_{j=1}^{n} a_{kj}c_{kj} = \sum_{i=1}^{n} a_{ik}c_{ik}.$$

The Classical Adjoint

The determinant can be used to determine whether a square matrix is invertible. It can also be used to compute its inverse.

■ **Definition 4.4.3**

Let A be an $n \times n$ matrix. The **matrix of minors** of A is $M_A = [m_{ij}]$, and the **matrix of cofactors** of A is $C_A = [c_{ij}]$.

These matrices are easy but tedious to compute. Take, for instance, the matrix

$$A = \begin{bmatrix} 1 & 0 & 2 \\ 2 & -1 & 4 \\ 0 & 2 & 4 \end{bmatrix}. \tag{4.34}$$

By Definition 4.4.3, the matrix of minors is

$$M_A = \begin{bmatrix} -12 & 8 & 4 \\ -4 & 4 & 2 \\ 2 & 0 & -1 \end{bmatrix},$$

and the matrix of cofactors is

$$C_A = \begin{bmatrix} -12 & -8 & 4 \\ 4 & 4 & -2 \\ 2 & 0 & -1 \end{bmatrix}. \tag{4.35}$$

The confirmation is left to Exercise 1. However, the work is worth the trouble because it will give the inverse of A. The transpose of its cofactor matrix is

$$C_A^T = \begin{bmatrix} -12 & 4 & 2 \\ -8 & 4 & 0 \\ 4 & -2 & -1 \end{bmatrix}.$$

Compute

$$C_A^T A = \begin{bmatrix} -12 & 4 & 2 \\ -8 & 4 & 0 \\ 4 & -2 & -1 \end{bmatrix} \begin{bmatrix} 1 & 0 & 2 \\ 2 & -1 & 4 \\ 0 & 2 & 4 \end{bmatrix} = \begin{bmatrix} -4 & 0 & 0 \\ 0 & -4 & 0 \\ 0 & 0 & -4 \end{bmatrix}.$$

Notice that $\det A = -4$ and

$$-\frac{1}{4} \begin{bmatrix} -12 & 4 & 2 \\ -8 & 4 & 0 \\ 4 & -2 & -1 \end{bmatrix} \begin{bmatrix} 1 & 0 & 2 \\ 2 & -1 & 4 \\ 0 & 2 & 4 \end{bmatrix} = \begin{bmatrix} 1 & 0 & 0 \\ 0 & 1 & 0 \\ 0 & 0 & 1 \end{bmatrix},$$

so the inverse of A has been found. Finding the matrix C_A^T was the key component in calculating the inverse of A, so it receives a name. That the product of the reciprocal of the determinant of A, provided that A is invertible, and C_A^T always gives the inverse matrix is the result following the definition.

■ **Definition 4.4.4**

The **classical adjoint** of a square matrix A is the transpose of its cofactor matrix and is denoted by adj A.

■ **Theorem 4.4.5**

If A is an invertible $n \times n$ matrix, $A^{-1} = \dfrac{1}{\det A}$ adj A.

Proof

Write $A = [a_{ij}]$. Since A is invertible, $\det A \neq 0$ by Theorem 4.3.11. To complete the proof, it suffices by Theorem 4.3.16 to show that

$$A\left(\frac{1}{\det A}\text{adj } A\right) = \frac{1}{\det A}(A \text{ adj } A) = I_n. \tag{4.36}$$

To understand what happens in the calculation of A adj A in (4.36), write

$$A \text{ adj } A = \begin{bmatrix} a_{11} & a_{12} & \cdots & a_{1n} \\ a_{21} & a_{22} & \cdots & a_{2n} \\ \vdots & \vdots & & \vdots \\ a_{n1} & a_{n2} & \cdots & a_{nn} \end{bmatrix} \begin{bmatrix} c_{11} & c_{21} & \cdots & c_{n1} \\ c_{12} & c_{22} & \cdots & c_{n2} \\ \vdots & \vdots & & \vdots \\ c_{1n} & c_{2n} & \cdots & c_{nn} \end{bmatrix}.$$

First, entry i of the main diagonal of A adj A is $\sum_{j=1}^{n} a_{ij}c_{ij} = \det A$. Second, Theorem 4.3.6(a) implies that entry (i, k) that is not on the main diagonal of A adj A is of the form

$$\sum_{j=1}^{n} a_{ij}c_{kj} = \begin{vmatrix} a_{11} & a_{12} & \cdots & a_{1n} \\ \vdots & \vdots & & \vdots \\ a_{i1} & a_{i2} & \cdots & a_{in} \\ \vdots & \vdots & & \vdots \\ a_{i1} & a_{i2} & \cdots & a_{in} \\ \vdots & \vdots & & \vdots \\ a_{n1} & a_{n2} & \cdots & a_{nn} \end{vmatrix} \begin{matrix} \\ \\ \leftarrow \text{ row } i \\ \\ \leftarrow \text{ row } k \neq i. \\ \\ \end{matrix} \tag{4.37}$$

Since the computation of c_{kj} involves deleting row k of A, each entry of row k can be replaced with any value, and the resulting cofactor computation will still equal c_{kj}. For this reason row k is replaced with row i in (4.37). Hence, the determinant (4.37) equals 0 because the matrix has two identical rows (Corollary 4.2.19). These two facts imply (4.36). ■

Symmetric and Orthogonal Matrices

The transpose is formed by writing the rows of a matrix as columns. For example, the matrix

$$A = \begin{bmatrix} 1 & -1 & 5 \\ -1 & 2 & 0 \\ 5 & 0 & 3 \end{bmatrix} \tag{4.38}$$

has transpose $A^T = \begin{bmatrix} 1 & -1 & 5 \\ -1 & 2 & 0 \\ 5 & 0 & 3 \end{bmatrix}$. An examination of this particular matrix reveals that $A = A^T$, which implies that entries on opposite sides of the main diagonal are equal. This leads to a definition.

■ **Definition 4.4.6**

A square matrix A such that $A = A^T$ is **symmetric**.

■ **Example 4.4.7**

The matrix A from (4.38) is symmetric. The matrix

$$B = \begin{bmatrix} 1 & 7 & 3 \\ -2 & 5 & 1 \\ 5 & 2 & 3 \end{bmatrix} \quad \text{has transpose} \quad B^T = \begin{bmatrix} 1 & -2 & 5 \\ 7 & 5 & 2 \\ 3 & 1 & 3 \end{bmatrix}.$$

This matrix is not symmetric since $B^T \neq B$.

Consider the following. Let $A = [a_{ij}]$ be an $n \times n$ matrix. The (i, j)-entry of a matrix can be expressed as

$$a_{ij} = A\mathbf{e}_j \bullet \mathbf{e}_i.$$

For example, if $A = \begin{bmatrix} 2 & 5 \\ 7 & 3 \end{bmatrix}$, then $a_{21} = \begin{bmatrix} 2 & 5 \\ 7 & 3 \end{bmatrix}\begin{bmatrix} 1 \\ 0 \end{bmatrix} \bullet \begin{bmatrix} 0 \\ 1 \end{bmatrix} = \begin{bmatrix} 2 \\ 7 \end{bmatrix} \bullet \begin{bmatrix} 0 \\ 1 \end{bmatrix} = 7$. Because the (j, i)-entry of A^T is the (i, j)-entry of A for any $1 \leq i, j \leq n$,

$$A\mathbf{e}_j \bullet \mathbf{e}_i = \mathbf{e}_j \bullet A^T\mathbf{e}_i.$$

This interaction between the transpose of a matrix and the dot product applies more generally to any two vectors of $\mathbb{R}^n$.

■ **Theorem 4.4.8**

If A is an $n \times n$ matrix, $A\mathbf{u} \bullet \mathbf{v} = \mathbf{u} \bullet A^T\mathbf{v}$ for all $\mathbf{u}, \mathbf{v} \in \mathbb{R}^n$.

Proof

Take $\mathbf{u}, \mathbf{v} \in \mathbb{R}^n$ and write the vectors as $\mathbf{u} = u_1\mathbf{e}_1 + u_2\mathbf{e}_2 + \cdots + u_n\mathbf{e}_n$ and $\mathbf{v} = v_1\mathbf{e}_1 + v_2\mathbf{e}_2 + \cdots + v_n\mathbf{e}_n$, where $u_i, v_i \in \mathbb{R}$ for all $i = 1, 2, \ldots, n$. Then,

$$A\mathbf{u} \bullet \mathbf{v} = (u_1 A\mathbf{e}_1 + u_2 A\mathbf{e}_2 + \cdots + u_n A\mathbf{e}_n) \bullet (v_1\mathbf{e}_1 + v_2\mathbf{e}_2 + \cdots + v_n\mathbf{e}_n)$$

$$= \sum_{i=1}^{n}\sum_{j=1}^{n} u_i v_j (A\mathbf{e}_i \bullet \mathbf{e}_j)$$

$$= \sum_{i=1}^{n}\sum_{j=1}^{n} u_i v_j (\mathbf{e}_i \bullet A^T\mathbf{e}_j)$$

$$= (u_1\mathbf{e}_1 + u_2\mathbf{e}_2 + \cdots + u_n\mathbf{e}_n) \bullet (v_1 A^T\mathbf{e}_1 + v_2 A^T\mathbf{e}_2 + \cdots + v_n A^T\mathbf{e}_n)$$

$$= \mathbf{u} \bullet A^T\mathbf{v}. \quad ■$$

An immediate consequence is that symmetric matrices can be moved across the dot product without effecting the value of the dot product. Because the transpose of a symmetric matrix is the matrix itself, the next result follows immediately from Theorem 4.4.8.

■ **Corollary 4.4.9**

If A is a symmetric $n \times n$ matrix, $A\mathbf{u} \cdot \mathbf{v} = \mathbf{u} \cdot A\mathbf{v}$ for all $\mathbf{u}, \mathbf{v} \in \mathbb{R}^n$.

■ **Example 4.4.10**

The matrix $B = \begin{bmatrix} 1 & -1 & 5 \\ -1 & 2 & 0 \\ 5 & 0 & 3 \end{bmatrix}$ is clearly symmetric, so $B\mathbf{u} \cdot \mathbf{v} = \mathbf{u} \cdot B\mathbf{v}$ for

all vectors $\mathbf{u}, \mathbf{v} \in \mathbb{R}^3$. For example, taking $\mathbf{u} = \begin{bmatrix} 1 \\ 0 \\ 7 \end{bmatrix}$ and $\mathbf{v} = \begin{bmatrix} 2 \\ 1 \\ 3 \end{bmatrix}$, it is easily

checked that

$$B\mathbf{u} \cdot \mathbf{v} = \left(\begin{bmatrix} 1 & -1 & 5 \\ -1 & 2 & 0 \\ 5 & 0 & 3 \end{bmatrix} \begin{bmatrix} 1 \\ 0 \\ 7 \end{bmatrix} \right) \cdot \begin{bmatrix} 2 \\ 1 \\ 3 \end{bmatrix} = \begin{bmatrix} 36 \\ -1 \\ 26 \end{bmatrix} \cdot \begin{bmatrix} 2 \\ 1 \\ 3 \end{bmatrix} = 149$$

and

$$\mathbf{u} \cdot B\mathbf{v} = \begin{bmatrix} 1 \\ 0 \\ 7 \end{bmatrix} \cdot \left(\begin{bmatrix} 1 & -1 & 5 \\ -1 & 2 & 0 \\ 5 & 0 & 3 \end{bmatrix} \begin{bmatrix} 2 \\ 1 \\ 3 \end{bmatrix} \right) = \begin{bmatrix} 1 \\ 0 \\ 7 \end{bmatrix} \cdot \begin{bmatrix} 16 \\ 0 \\ 19 \end{bmatrix} = 149.$$

Consider the matrix $A = \begin{bmatrix} 1 & 2 & 3 \\ 6 & 5 & 4 \end{bmatrix}$ with transpose $A^T = \begin{bmatrix} 1 & 6 \\ 2 & 5 \\ 3 & 4 \end{bmatrix}$. Observe

that

$$A^T A = \begin{bmatrix} 12 & 32 & 27 \\ 32 & 29 & 26 \\ 27 & 26 & 25 \end{bmatrix},$$

which is a symmetric matrix. The computation $A^T A$ is always a symmetric matrix. This is due to two facts. The first is the obvious $(A^T)^T = A$ [Exercise 19(a)]. The second is reminiscent of the corresponding result for invertible matrices (Theorem 4.1.14).

■ **Theorem 4.4.11**

If A and B are $n \times n$ matrices, $(AB)^T = B^T A^T$.

Proof

Let $\mathbf{r}_i$ be the rows of A and $\mathbf{c}_j$ be the columns of B, where $i, j = 1, 2, \dots, n$.

Then,

$$(AB)^T = \begin{bmatrix} \mathbf{r}_1^T \cdot \mathbf{c}_1 & \mathbf{r}_2^T \cdot \mathbf{c}_1 & \cdots & \mathbf{r}_n^T \cdot \mathbf{c}_1 \\ \mathbf{r}_1^T \cdot \mathbf{c}_2 & \mathbf{r}_2^T \cdot \mathbf{c}_2 & \cdots & \mathbf{r}_n^T \cdot \mathbf{c}_2 \\ \vdots & \vdots & & \vdots \\ \mathbf{r}_1^T \cdot \mathbf{c}_n & \mathbf{r}_2^T \cdot \mathbf{c}_n & \cdots & \mathbf{r}_n^T \cdot \mathbf{c}_n \end{bmatrix}.$$

Also, the rows of B^T are $\mathbf{c}_j^T$ and the columns of A^T are $\mathbf{r}_i^T$, so

$$B^T A^T = \begin{bmatrix} (\mathbf{c}_1^T)^T \cdot \mathbf{r}_1^T & (\mathbf{c}_1^T)^T \cdot \mathbf{r}_2^T & \cdots & (\mathbf{c}_1^T)^T \cdot \mathbf{r}_n^T \\ (\mathbf{c}_2^T)^T \cdot \mathbf{r}_1^T & (\mathbf{c}_2^T)^T \cdot \mathbf{r}_2^T & \cdots & (\mathbf{c}_2^T)^T \cdot \mathbf{r}_n^T \\ \vdots & \vdots & & \vdots \\ (\mathbf{c}_n^T)^T \cdot \mathbf{r}_1^T & (\mathbf{c}_n^T)^T \cdot \mathbf{r}_2^T & \cdots & (\mathbf{c}_n^T)^T \cdot \mathbf{r}_n^T \end{bmatrix}.$$

It follows that $(AB)^T = B^T A^T$ because by Theorem 2.2.6(a),

$$(\mathbf{c}_j^T)^T \cdot \mathbf{r}_i^T = \mathbf{c}_j \cdot \mathbf{r}_i^T = \mathbf{r}_i^T \cdot \mathbf{c}_j. \blacksquare$$

Therefore, $(A^T A)^T = A^T (A^T)^T = A^T A$ by the obvious fact and Theorem 4.4.11, which proves the next result.

■ **Theorem 4.4.12**

$A^T A$ is a symmetric matrix for every matrix A.

Next, consider the matrix

$$A = \begin{bmatrix} 0 & 1 & 0 \\ \sqrt{2}/2 & 0 & \sqrt{2}/2 \\ -\sqrt{2}/2 & 0 & \sqrt{2}/2 \end{bmatrix}.$$

The inverse of A is

$$A^{-1} = \begin{bmatrix} 0 & \sqrt{2}/2 & -\sqrt{2}/2 \\ 1 & 0 & 0 \\ 0 & \sqrt{2}/2 & \sqrt{2}/2 \end{bmatrix}. \tag{4.39}$$

It is apparent that $A^{-1} = A^T$, so $A^T A = I_3$. Although $A^T A$ is always symmetric, the transpose of a square matrix is not always its inverse. This is the reason for the next definition, and the theorem that follows is an immediate consequence of Theorem 4.3.16.

■ **Definition 4.4.13**

An $n \times n$ matrix A is **orthogonal** if $A^T A = I_n$.

■ **Theorem 4.4.14**

A matrix A is orthogonal if and only if A is invertible and $A^T = A^{-1}$.

Moreover, the columns of A (4.39) are orthogonal to each other and are unit vectors, so the columns form a so-called **orthonormal set**. It is this property of A that guarantees that its inverse is simply A^T. It is also this property that makes the terminology naming the $A^T = A^{-1}$ condition make sense.

■ **Theorem 4.4.15**

A square matrix A is orthogonal if and only if the columns of A form an orthonormal set.

Proof

Let $c_1, c_2, \ldots, c_n$ be the columns of the $n \times n$ matrix A. Because the rows of A^T are $c_1^T, c_2^T, \ldots, c_n^T$,

$$A^T A = \begin{bmatrix} c_1 \cdot c_1 & c_1 \cdot c_2 & c_1 \cdot c_3 & \cdots & c_1 \cdot c_n \\ c_2 \cdot c_1 & c_2 \cdot c_2 & c_2 \cdot c_3 & \cdots & c_2 \cdot c_n \\ c_3 \cdot c_1 & c_3 \cdot c_2 & c_3 \cdot c_3 & \cdots & c_3 \cdot c_n \\ \vdots & \vdots & \vdots & & \vdots \\ c_n \cdot c_1 & c_n \cdot c_2 & c_n \cdot c_3 & \cdots & c_n \cdot c_n \end{bmatrix}.$$

Because $A^T A = I_n$ if and only if $c_i \cdot c_i = 1$ for all $i = 1, 2, \ldots n$ and $c_i \cdot c_j = 0$ when $i \neq j$, it follows that $A^T A = I_n$ if and only if the columns of A form an orthonormal set. ■

From Theorem 4.4.14 infer that if A is both symmetric and orthogonal, A is its own inverse. For example,

$$\begin{bmatrix} \sqrt{2}/2 & \sqrt{2}/2 \\ \sqrt{2}/2 & -\sqrt{2}/2 \end{bmatrix}^{-1} = \begin{bmatrix} \sqrt{2}/2 & \sqrt{2}/2 \\ \sqrt{2}/2 & -\sqrt{2}/2 \end{bmatrix}.$$

Also, notice that

$$\begin{vmatrix} \sqrt{2}/2 & \sqrt{2}/2 \\ \sqrt{2}/2 & -\sqrt{2}/2 \end{vmatrix} = -1,$$

while $\det I_n = 1$. In general, if A is orthogonal, $(\det A)^2 = 1$ because

$$\det A = \det A^T = \det A^{-1} = \frac{1}{\det A}.$$

This proves the next theorem.

■ **Theorem 4.4.16**

If A is an orthogonal matrix, $\det A = \pm 1$.

■ Example 4.4.17

Example 4.2.3 shows that the standard matrix of a rotation or a reflection on $\mathbb{R}^2$ is an orthogonal matrix with determinant ± 1. Now to show the converse. Let $A = \begin{bmatrix} a & b \\ c & d \end{bmatrix}$ be an orthogonal matrix. This implies that $a^2 + c^2 = 1$ and $b^2 + d^2 = 1$. Hence, there exists angle measures θ and ψ such that

$$a = \cos\theta, \quad c = \sin\theta, \quad b = \sin\psi, \quad d = \cos\psi.$$

Also, because the columns of A are orthogonal,

$$0 = \cos\theta \sin\psi + \sin\theta \cos\psi = \sin(\theta + \psi),$$

so there are two cases to consider. If $\psi = -\theta$, then

$$A = \begin{bmatrix} \cos\theta & \sin(-\theta) \\ \sin\theta & \cos(-\theta) \end{bmatrix} = \begin{bmatrix} \cos\theta & -\sin\theta \\ \sin\theta & \cos\theta \end{bmatrix},$$

which implies that $\det A = 1$ and A is the standard matrix of a rotation. If $\psi = \pi - \theta$, then

$$A = \begin{bmatrix} \cos\theta & \sin(\pi - \theta) \\ \sin\theta & \cos(\pi - \theta) \end{bmatrix} = \begin{bmatrix} \cos\theta & \sin\theta \\ \sin\theta & -\cos\theta \end{bmatrix},$$

which implies that $\det A = -1$ and A is the standard matrix of a reflection.

Example 4.4.17 suggests that a 2×2 orthogonal matrix is the standard matrix of an isometry by Theorems 3.3.12 and 3.3.14. That every orthogonal matrix is the standard matrix of an isometry follows from the next theorem.

■ Theorem 4.4.18

For every $n \times n$ orthogonal matrix A and $\mathbf{u}, \mathbf{v} \in \mathbb{R}^n$, $A\mathbf{u} \cdot A\mathbf{v} = \mathbf{u} \cdot \mathbf{v}$.

Proof

Let $A \in \mathbb{R}^{n \times n}$ and take $\mathbf{u}, \mathbf{v} \in \mathbb{R}^n$. Then, because the inverse of an orthogonal matrix is its transpose (Theorem 4.4.14),

$$A\mathbf{u} \cdot A\mathbf{v} = (A\mathbf{u})^T A\mathbf{v} = \mathbf{u}^T A^T A\mathbf{v} = \mathbf{u}^T \mathbf{v} = \mathbf{u} \cdot \mathbf{v}. \ ■$$

■ Corollary 4.4.19

Every orthogonal matrix is the standard matrix of an isometry.

Cramer's Rule

The next application of the determinant involves finding the solution to a system of linear equations that has an invertible coefficient matrix. To begin, some notation is needed. Let

$$A = \begin{bmatrix} a_{11} & a_{12} & \cdots & a_{1n} \\ a_{21} & a_{22} & \cdots & a_{2n} \\ \vdots & \vdots & & \vdots \\ a_{n1} & a_{n2} & \cdots & a_{nn} \end{bmatrix} \quad \text{and} \quad \mathbf{b} = \begin{bmatrix} b_1 \\ b_2 \\ \vdots \\ b_n \end{bmatrix}$$

so that $A\mathbf{x} = \mathbf{b}$ represents the system of equations,

$$\begin{array}{ccccccccc} a_{11}x_1 & + & a_{12}x_2 & + & \cdots & + & a_{1n}x_n & = & b_1, \\ a_{21}x_1 & + & a_{22}x_2 & + & \cdots & + & a_{2n}x_n & = & b_2, \\ \vdots & & \vdots & & & & \vdots & & \vdots \\ a_{n1}x_1 & + & a_{n2}x_2 & + & \cdots & + & a_{nn}x_n & = & b_n. \end{array}$$

Define $A_{j,\mathbf{b}}$ to be the matrix formed from the coefficient matrix A by replacing the jth column with $\mathbf{b}$. That is,

$$A_{j,\mathbf{b}} = \begin{bmatrix} a_{11} & \cdots & a_{1,j-1} & b_1 & a_{1,j+1} & \cdots & a_{1n} \\ a_{21} & \cdots & a_{2,j-1} & b_2 & a_{1,j+1} & \cdots & a_{2n} \\ \vdots & & \vdots & \vdots & \vdots & & \vdots \\ a_{n1} & \cdots & a_{n,j-1} & b_n & a_{n,j+1} & \cdots & a_{nn} \end{bmatrix}.$$

The system, if it has a unique solution, can now be solved with determinants.

■ Theorem 4.4.20 [Cramer's Rule]

Let A be an invertible $n \times n$ matrix. The solution set of $A\mathbf{x} = \mathbf{b}$ is

$$\left\{ \begin{bmatrix} \det A_{1,\mathbf{b}}/\det A \\ \det A_{2,\mathbf{b}}/\det A \\ \vdots \\ \det A_{n,\mathbf{b}}/\det A \end{bmatrix} \right\}.$$

Proof

First, remember that since A is invertible, $\det A \neq 0$ by Theorem 4.3.11. Next, suppose that $\mathbf{u} \in \mathbb{R}^n$ has coordinates $u_1, u_2, \ldots, u_n$ and is the solution to the system $A\mathbf{x} = \mathbf{b}$. Observe that for all $j = 1, 2, \ldots, n$,

$$\begin{aligned} A(I_n)_{j,\mathbf{u}} = A \begin{bmatrix} 1 & 0 & \cdots & u_1 & \cdots & 0 \\ 0 & 1 & \cdots & u_2 & \cdots & 0 \\ \vdots & \vdots & & \vdots & & \vdots \\ 0 & 0 & \cdots & u_n & \cdots & 1 \end{bmatrix} \\ = [A\mathbf{e}_1 \quad A\mathbf{e}_2 \quad \cdots \quad A\mathbf{u} \quad \cdots \quad A\mathbf{e}_n] \\ = [A\mathbf{e}_1 \quad A\mathbf{e}_2 \quad \cdots \quad \mathbf{b} \quad \cdots \quad A\mathbf{e}_n] = A_{j,\mathbf{b}}. \end{aligned}$$

Since $\det (I_n)_{j,\mathbf{u}} = u_j$, Theorem 4.3.19 implies that

$$\det A_{j,\mathbf{b}} = \det A(I_n)_{j,\mathbf{u}} = \det A \det (I_n)_{j,\mathbf{u}} = u_j \det A,$$

which yields $u_j = \dfrac{\det A_{j,\mathbf{b}}}{\det A}$. ∎

■ **Example 4.4.21**

Use Cramer's Rule (Theorem 4.4.20) to solve

$$
\begin{array}{rcrcrcl}
3x & - & y & + & 4z & = & 1, \\
2x & + & 8y & - & z & = & 2, \\
x & + & 5y & + & 2z & = & 4.
\end{array}
$$

Letting A be the coefficient matrix of the system and $\mathbf{b} = \begin{bmatrix} 1 \\ 2 \\ 4 \end{bmatrix}$,

$$
\det A = \begin{vmatrix} 3 & -1 & 4 \\ 2 & 8 & -1 \\ 1 & 5 & 2 \end{vmatrix} = 76, \qquad \det A_{1,\mathbf{b}} = \begin{vmatrix} 1 & -1 & 4 \\ 2 & 8 & -1 \\ 4 & 5 & 2 \end{vmatrix} = -59,
$$

$$
\det A_{2,\mathbf{b}} = \begin{vmatrix} 3 & 1 & 4 \\ 2 & 2 & -1 \\ 1 & 4 & 2 \end{vmatrix} = 43, \qquad \det A_{3,\mathbf{b}} = \begin{vmatrix} 3 & -1 & 1 \\ 2 & 8 & 2 \\ 1 & 5 & 4 \end{vmatrix} = 74.
$$

This implies that the solution set to the system is $\left\{ \begin{bmatrix} -59/76 \\ 43/76 \\ 37/38 \end{bmatrix} \right\}$.

LU Factorization

It is common in algebra to factor polynomials. The same can be done with matrices. Given a matrix A, the matrix can be **factored** if matrices B and C can be found such that $A = BC$. This is an example of a **factorization** of A. There are several factorizations in this book. The first provides a technique to quicken the process of solving systems of linear equations with matrices. Let A be an $m \times n$ matrix. Reduce A to an $m \times n$ matrix U in row echelon form using the elementary row operations without exchanging rows. Use Corollary 4.1.10 to obtain elementary $m \times m$ matrices $E_1, E_2, \dots, E_k$ such that $U = E_k \cdots E_2 E_1 A$. Because elementary matrices are invertible,

$$
A = E_1^{-1} E_2^{-1} \cdots E_k^{-1} U,
$$

so let $L = E_1^{-1} E_2^{-1} \cdots E_k^{-1}$. An examination of U and L reveals two things. First, U is upper triangular because it is in row echelon form. Second, L is lower triangular. To see this, it suffices to examine each elementary matrix in the product. For all $i = 1, 2, \dots, k$, it is the case that op $E_i \neq R_j \leftrightarrow R_l$ by assumption ($j, l = 1, 2, \dots, m$). If op $E_i = cR_j$, then E_i is a diagonal matrix, which is also lower triangular. If op $E_i = cR_j + R_l$, then E_i is lower triangular because only entries (i, j) of A with $i > j$ will be transformed to 0 in the row-reduction process from A to U. Since the product of lower triangular matrices is lower triangular (Exercise 18 of Section 4.2), L is lower triangular. This proves the factorization.

■ Theorem 4.4.22 [LU Factorization]

If A is an $m \times n$ matrix with row echelon form U that is row reduced from A without exchanging rows, there exists an $m \times m$ invertible matrix L such that L is lower triangular and $A = LU$.

■ Example 4.4.23

Find the LU factorization of $A = \begin{bmatrix} 1 & 5 & 1 & 4 \\ 2 & 11 & 5 & 0 \\ 6 & 3 & 2 & 0 \end{bmatrix}$. An echelon form of A is

$$U = \begin{bmatrix} 1 & 5 & 1 & 4 \\ 0 & 1 & 3 & -8 \\ 0 & 0 & 77 & -240 \end{bmatrix},$$

which can be obtained from A using $-2R_1 + R_2$, $-6R_1 + R_3$, and $27R_2 + R_3$. Therefore, $U = E_3E_2E_1A$, where

$$E_1 = \begin{bmatrix} 1 & 0 & 0 \\ -2 & 1 & 0 \\ 0 & 0 & 1 \end{bmatrix}, \quad E_2 = \begin{bmatrix} 1 & 0 & 0 \\ 0 & 1 & 0 \\ -6 & 0 & 1 \end{bmatrix}, \quad E_3 = \begin{bmatrix} 1 & 0 & 0 \\ 0 & 1 & 0 \\ 0 & 27 & 1 \end{bmatrix}.$$

Because

$$E_1^{-1} = \begin{bmatrix} 1 & 0 & 0 \\ 2 & 1 & 0 \\ 0 & 0 & 1 \end{bmatrix}, \quad E_2^{-1} = \begin{bmatrix} 1 & 0 & 0 \\ 0 & 1 & 0 \\ 6 & 0 & 1 \end{bmatrix}, \quad E_3^{-1} = \begin{bmatrix} 1 & 0 & 0 \\ 0 & 1 & 0 \\ 0 & -27 & 1 \end{bmatrix},$$

compute $L = E_1^{-1}E_2^{-2}E_3^{-1} = \begin{bmatrix} 1 & 0 & 0 \\ 2 & 1 & 0 \\ 6 & -27 & 1 \end{bmatrix}$. A check reveals that $A = LU$.

■ Example 4.4.24

Use LU factorization to solve the system of linear equations,

$$\begin{array}{rcrcrcrcl} 2x_1 & + & 11x_2 & + & 5x_3 & & & = & 0, \\ x_1 & + & 5x_2 & + & 1x_3 & + & 4x_4 & = & 1, \\ 6x_1 & + & 3x_2 & + & 2x_3 & & & = & 2. \end{array}$$

Since the coefficient matrix of the system equals A from Example 4.4.23, except with the first and second rows exchanged, A can be considered the coefficient matrix. The order of the equations does not matter. Write the system as $Ax = \mathbf{b}$, where

$$\mathbf{b} = \begin{bmatrix} 1 \\ 0 \\ 2 \end{bmatrix}.$$

The system can then be written as $L(U\mathbf{x}) = \mathbf{b}$. Let $\mathbf{y} = U\mathbf{x}$ and solve the system in two steps. First solve $L\mathbf{y} = \mathbf{b}$, and then solve the system $\mathbf{y} = U\mathbf{x}$. Because L is lower triangular,

$$\begin{bmatrix} 1 & 0 & 0 \\ 2 & 1 & 0 \\ 6 & -27 & 1 \end{bmatrix} \begin{bmatrix} x \\ y \\ z \end{bmatrix} = \begin{bmatrix} 1 \\ 0 \\ 2 \end{bmatrix}$$

simply requires back-substitution to solve. Its solution is

$$x = 1$$
$$y = -2$$
$$z = -58.$$

The system of linear equations,

$$\begin{bmatrix} 1 & 5 & 1 & 4 \\ 0 & 1 & 3 & -8 \\ 0 & 0 & 77 & -240 \end{bmatrix} \begin{bmatrix} x_1 \\ x_2 \\ x_3 \\ x_4 \end{bmatrix} = \begin{bmatrix} 1 \\ -2 \\ -58 \end{bmatrix},$$

is more involved because its solution requires a parameter. Letting $t = x_4$, using what might be called **forward-substitution**, find x_3, then x_2, and then x_1. The solution to the original system of linear equations is

$$x_1 = (5 - 4t)/11,$$
$$x_2 = (20 - 104t)/77,$$
$$x_3 = (-58 + 240t)/77,$$
$$x_4 = t.$$

It is evident that every system of linear equations can be solved with this method by simply arranging the equations so that no exchanges of rows are required to put the coefficient matrix into echelon form.

Area and Volume

The cross product was defined in Definition 2.3.1 using a computation reminiscent of the determinant. Specifically, if $\mathbf{u}$ and $\mathbf{v}$ are vectors in $\mathbb{R}^3$ with coordinates u_1, u_2, u_3 and v_1, v_2, v_3, respectively, using the calculus notation for the standard vectors of $\mathbb{R}^3$, the determinant can be used to calculate the cross product,

$$\begin{bmatrix} u_1 \\ u_2 \\ u_3 \end{bmatrix} \times \begin{bmatrix} v_1 \\ v_2 \\ v_3 \end{bmatrix} = \begin{vmatrix} \mathbf{i} & \mathbf{j} & \mathbf{k} \\ u_1 & u_2 & u_3 \\ v_1 & v_2 & v_3 \end{vmatrix} = \begin{vmatrix} u_2 & u_3 \\ v_2 & v_3 \end{vmatrix} \mathbf{i} - \begin{vmatrix} u_1 & u_3 \\ v_1 & v_3 \end{vmatrix} \mathbf{j} + \begin{vmatrix} u_1 & u_2 \\ v_1 & v_2 \end{vmatrix} \mathbf{k}. \qquad (4.40)$$

Moreover, in Section 2.3, the cross product was used to find the area of the parallelogram formed by two vectors (Corollary 2.3.7). It was also used to find the volume

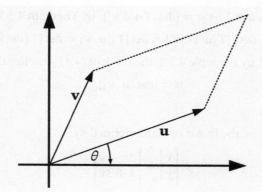

Figure 4.1 The parallelogram described by **u** and **v**.

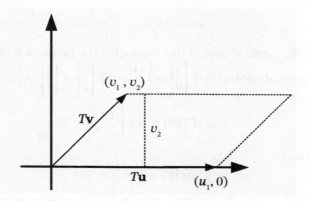

Figure 4.2 The parallelogram described by $T\mathbf{u}$ and $T\mathbf{v}$.

of the parallelepiped formed by three vectors (Example 2.3.9). This implies that these two computations should be able to be rewritten using the determinant.

Let $\mathbf{u} = \begin{bmatrix} u_1 \\ u_2 \end{bmatrix}$ and $\mathbf{v} = \begin{bmatrix} v_1 \\ v_2 \end{bmatrix}$ be two nonzero vectors in $\mathbb{R}^2$. Consider $\mathbf{u}$ to be the base of the parallelogram described by these two vectors, and let θ be the angle formed by $\mathbf{u}$ and $\mathbf{e}_1$ (Figure 4.1). Rotate the parallelogram through θ so that the image of $\mathbf{u}$ lies on the positive x-axis. This means that θ may need to be negative. Let T be this rotation (Definition 3.3.8). Theorems 3.3.11 and 3.3.14 guarantee that $T\mathbf{u}$ and $T\mathbf{v}$ describe a parallelogram that is congruent to the original parallelogram (Figure 4.2). Write

$$T\mathbf{u} = \begin{bmatrix} u_1 \\ 0 \end{bmatrix} \quad \text{and} \quad T\mathbf{v} = \begin{bmatrix} v_1 \\ v_2 \end{bmatrix}.$$

Since the area a of the parallelogram is the product of the length of its base and the length of its height, $a = |u_1 v_2| = |\det[T\mathbf{u} \ \ T\mathbf{v}]|$. By Theorem 4.3.19,

$$\det[T\mathbf{u} \ \ T\mathbf{v}] = \det[\,[T]\mathbf{u} \ \ [T]\mathbf{v}\,] = \det[T][\mathbf{u} \ \mathbf{v}] = \det[T]\det[\mathbf{u} \ \mathbf{v}]. \tag{4.41}$$

Because $\det[T] = 1$ by Example 4.2.3, the work of (4.41) implies that

$$a = |\det[\mathbf{u} \ \mathbf{v}]|. \tag{4.42}$$

■ **Example 4.4.25**

Let $T : \mathbb{R}^2 \to \mathbb{R}^2$ be the linear operator defined by

$$T\begin{bmatrix} x \\ y \end{bmatrix} = \begin{bmatrix} 2x + y \\ x + 3y \end{bmatrix}.$$

Let R be the rectangle formed by the vectors $\begin{bmatrix} 3 \\ 0 \end{bmatrix}$ and $\begin{bmatrix} 0 \\ 2 \end{bmatrix}$. By (4.42),

$$\text{area of } R = \begin{vmatrix} 3 & 0 \\ 0 & 2 \end{vmatrix} = 6,$$

which is the expected area of this rectangle. The image of R under T is the parallelogram described by $T\begin{bmatrix} 3 \\ 0 \end{bmatrix} = \begin{bmatrix} 6 \\ 3 \end{bmatrix}$ and $T\begin{bmatrix} 0 \\ 2 \end{bmatrix} = \begin{bmatrix} 2 \\ 6 \end{bmatrix}$. Again, by (4.42),

$$\text{area of } T[R] = \det\begin{bmatrix} 6 & 2 \\ 3 & 6 \end{bmatrix} = 30.$$

Observe that

$$\det[T] = \begin{vmatrix} 2 & 1 \\ 1 & 3 \end{vmatrix} = 5.$$

This work is summarized in Figure 4.3. It is not a coincidence that the area of $T[R]$ is the product of $\det[T]$ and the area of R. (4.41) shows this.

■ **Theorem 4.4.26**

If $T : \mathbb{R}^2 \to \mathbb{R}^2$ is a linear operator and $\mathbf{u}, \mathbf{v} \in \mathbb{R}^2$, the area of the parallelogram described by $T\mathbf{u}$ and $T\mathbf{v}$ equals the product of $|\det[T]|$ and the area of the parallelogram described by $\mathbf{u}$ and $\mathbf{v}$.

Using the matrices for rotations in $\mathbb{R}^3$ given by (3.26), (3.27), and (3.28), it is expected that (4.42) can be generalized to find the volume of a parallelepiped using the same technique as when finding the area of the parallelogram. However, all that is needed is an examination of the triple scalar product of Example 2.3.9. Writing $\mathbf{u} = \begin{bmatrix} u_1 \\ u_2 \\ u_3 \end{bmatrix}$, $\mathbf{v} = \begin{bmatrix} v_1 \\ v_2 \\ v_3 \end{bmatrix}$, and $\mathbf{w} = \begin{bmatrix} w_1 \\ w_2 \\ w_3 \end{bmatrix}$, it follows that

$$\mathbf{w} \cdot (\mathbf{u} \times \mathbf{v}) = w_1\begin{vmatrix} u_2 & v_2 \\ u_3 & v_3 \end{vmatrix} - w_2\begin{vmatrix} u_1 & v_1 \\ u_3 & v_3 \end{vmatrix} + w_3\begin{vmatrix} u_1 & v_1 \\ u_2 & v_2 \end{vmatrix} = \begin{vmatrix} w_1 & u_1 & v_2 \\ w_2 & u_2 & v_2 \\ w_3 & u_3 & v_3 \end{vmatrix}.$$

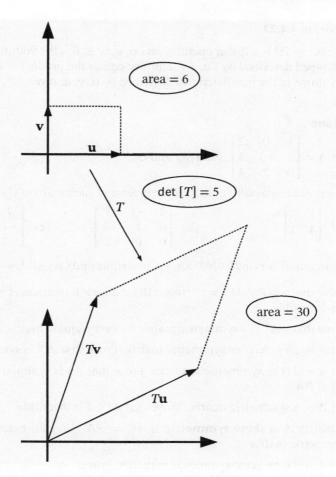

Figure 4.3 The areas of a parallelogram and its image under a linear operator and their relation to the determinant.

Thus, the volume v of the parallelepiped formed by **u**, **v**, and **w** is

$$v = |\det [\mathbf{w} \ \ \mathbf{u} \ \ \mathbf{v}]|. \tag{4.43}$$

Because of Theorem 4.2.16 and Corollary 4.3.7, the three vectors can appear as columns or rows in any order in the determinant of (4.43). The same can be said of (4.42). Lastly, because (4.41) is independent of the size of $[T]$, Theorem 4.4.26 can be generalized to $\mathbb{R}^3$, and although not stated here, the same theorem can be generalized to $\mathbb{R}^n$ using a carefully defined notion of area or volume in higher dimensions.

■ **Theorem 4.4.27**

If $T : \mathbb{R}^3 \to \mathbb{R}^3$ is a linear operator and $\mathbf{u}, \mathbf{v}, \mathbf{w} \in \mathbb{R}^3$, the volume of the parallelepiped described by $T\mathbf{u}$, $T\mathbf{v}$, and $T\mathbf{w}$ equals the product of $|\det [T]|$ and the volume of the parallelepiped described by $\mathbf{u}$, $\mathbf{v}$, and $\mathbf{w}$.

Exercises

1. Let $A = \begin{bmatrix} 1 & 0 & 2 \\ 2 & -1 & 4 \\ 0 & 2 & 4 \end{bmatrix}$. Find M_A and C_A.

2. Use the classical adjoint to find the inverse of each matrix (Theorem 4.4.5).

    (a) $\begin{bmatrix} 3 & 2 \\ 4 & 1 \end{bmatrix}$
    (b) $\begin{bmatrix} 2 & -1 & 3 \\ 0 & 1 & -1 \\ 0 & 0 & 2 \end{bmatrix}$
    (c) $\begin{bmatrix} 1 & 3 & 2 \\ -1 & 1 & 0 \\ -2 & 1 & 4 \end{bmatrix}$

3. Prove that if A is invertible, $\operatorname{adj} A$ is invertible and $(\operatorname{adj} A)^{-1} = \dfrac{1}{\det A} A$.

4. Prove that $\operatorname{adj}(P^{-1}AP) = P^{-1}(\operatorname{adj} A)P$ for all $n \times n$ matrices A and P, provided P is invertible.

5. Show that AA^T is a symmetric matrix for every square matrix A.

6. Let A be an invertible symmetric matrix. Prove that A^{-1} is symmetric.

7. Let A and B be symmetric matrices. Prove that AB is symmetric if and only if $AB = BA$.

8. Let A be a symmetric matrix. Prove that $A^2 + I$ is invertible.

9. A matrix A is **skew-symmetric** if $A^T = -A$. Give an example of a skew-symmetric matrix.

10. Let A and B be skew-symmetric matrices. Prove.

    (a) If A is invertible, A^{-1} is skew-symmetric.
    (b) $A + B$ and A^T are skew-symmetric.
    (c) rA is skew-symmetric for all $r \in \mathbb{R}$.

11. If possible, give an example of a diagonal matrix that is orthogonal.

12. If possible, give an example of an upper triangular, nondiagonal matrix that is orthogonal.

13. Demonstrate that the given matrices are orthogonal.

    (a) $\begin{bmatrix} \sqrt{3}/2 & -1/2 \\ 1/2 & \sqrt{3}/2 \end{bmatrix}$
    (b) $\begin{bmatrix} 0 & 1 \\ 1 & 0 \end{bmatrix}$
    (c) $\dfrac{\sqrt{50}}{50} \begin{bmatrix} 7 & 1 \\ -1 & 7 \end{bmatrix}$

14. Determine whether each matrix given in Exercise 13 is a rotation or a reflection. If it is a rotation, find the angle of the rotation, and if it is a reflection, find the line of the reflection. Lastly, find the image of the square described by $\mathbf{e}_1$ and $\mathbf{e}_2$ using each matrix.

15. Demonstrate that the given matrices are orthogonal.

    (a) $\begin{bmatrix} 0 & 1 & 0 \\ 1 & 0 & 0 \\ 0 & 0 & 1 \end{bmatrix}$

    (b) $\begin{bmatrix} \sqrt{5}/5 & 0 & -2\sqrt{5}/5 \\ 2\sqrt{5}/5 & 0 & \sqrt{5}/5 \\ 0 & 1 & 0 \end{bmatrix}$

16. Prove that the product of orthogonal matrices of the same size is orthogonal.

17. Prove that the transpose of an orthogonal matrix is orthogonal.

18. Prove that the inverse of an orthogonal matrix is orthogonal.

19. Use Cramer's Rule (Theorem 4.4.20) to solve each system of linear equations.

    (a)
    $$\begin{aligned} -8x &+ 3y &= -1 \\ 4x &+ 7y &= 8 \end{aligned}$$

    (b)
    $$\begin{aligned} 5x &+ 4y & &= 5 \\ 9x &+ 6y &- z &= 5 \\ 4x &+ 8y &+ 4z &= -2 \end{aligned}$$

    (c)
    $$\begin{aligned} -2x & & &- 4z &= 8 \\ -6x &+ 7y & & &= 7 \\ -x &+ y &- 4z & &= 6 \end{aligned}$$

20. Use Cramer's Rule to solve the system of linear equations.

    $$\begin{aligned} 2x_1 &- x_2 &+ 3x_3 &+ x_4 &= 2 \\ 4x_1 &+ x_2 &+ x_3 &+ 3x_4 &= -3 \\ 2x_1 &+ 4x_2 &+ 4x_3 &+ 2x_4 &= 6 \\ x_1 &- x_2 &- 6x_3 &+ 7x_4 &= -2 \end{aligned}$$

21. For each system of linear equations, find $r \in \mathbb{R}$ so that the system has a unique solution. Use Cramer's Rule.

    (a)
    $$\begin{aligned} 4x &- 5y &= 4 \\ rx &+ 4y &= 4 \end{aligned}$$

    (b)
    $$\begin{aligned} 2x &- y &- 4z &= 4 \\ 4x &+ ry & &= 3 \\ & 3y &+ z &= -1 \end{aligned}$$

22. Find an LU factorization for each matrix.

    (a) $\begin{bmatrix} 3 & 7 & 2 \\ -3 & -2 & 3 \\ 6 & 7 & 1 \end{bmatrix}$

    (b) $\begin{bmatrix} 2 & 0 & 5 & 2 \\ 4 & 1 & 6 & 1 \\ -6 & 3 & 3 & 1 \end{bmatrix}$

23. Solve each system of linear equations using an LU factorization as in Example 4.4.24.

    (a)
    $$\begin{aligned} 5y &+ 7z &= 1 \\ x - y &- z &= 0 \\ 3x + 4y &+ 2z &= 1 \end{aligned}$$

    (b)
    $$\begin{aligned} x_1 &+ 2x_2 &- x_3 & &= 3 \\ x_1 &+ x_2 & &- 2x_4 &= 2 \\ & 2x_2 &+ 3x_3 &- x_4 &= -1 \\ 2x_1 &+ x_2 &+ x_3 &+ 4x_4 &= 0 \end{aligned}$$

24. Find the area of the given polygon and then find the area of its image under the linear operator $T : \mathbb{R}^2 \to \mathbb{R}^2$ using Theorem 4.4.26. Verify the result using a direct computation of the area of the image.

(a) Parallelogram described by $\begin{bmatrix} 5 \\ 1 \end{bmatrix}$ and $\begin{bmatrix} 1 \\ 5 \end{bmatrix}$, $T\begin{bmatrix} x \\ y \end{bmatrix} = \begin{bmatrix} 2x \\ y - 6x \end{bmatrix}$.

(b) Triangle described by $\begin{bmatrix} 3 \\ 8 \end{bmatrix}$ and $\begin{bmatrix} -2 \\ 1 \end{bmatrix}$, $T\begin{bmatrix} x \\ y \end{bmatrix} = \begin{bmatrix} 5x + 3y \\ 7x - y \end{bmatrix}$.

25. Find the volume of the given solid and then find the volume of its image under the linear operator $T : \mathbb{R}^3 \rightarrow \mathbb{R}^3$ using Theorem 4.4.27. Verify the result using a direct computation of the volume of the image.

(a) Prism described by $\begin{bmatrix} 3 \\ 0 \\ 0 \end{bmatrix}$, $\begin{bmatrix} 0 \\ 2 \\ 0 \end{bmatrix}$, and $\begin{bmatrix} 0 \\ 0 \\ 5 \end{bmatrix}$; $T\begin{bmatrix} x \\ y \\ z \end{bmatrix} = \begin{bmatrix} x + y + z \\ 2x - 4z \\ 4x - 2y \end{bmatrix}$

(b) Parallelepiped by $\begin{bmatrix} 1 \\ 1 \\ 1 \end{bmatrix}$, $\begin{bmatrix} 3 \\ 2 \\ 1 \end{bmatrix}$, and $\begin{bmatrix} 2 \\ 1 \\ 5 \end{bmatrix}$; $T\begin{bmatrix} x \\ y \\ z \end{bmatrix} = \begin{bmatrix} 5x \\ 2x - 6y \\ x + 3y - 7z \end{bmatrix}$

26. Find the area of the region in $\mathbb{R} \times \mathbb{R}$ bounded by the ellipse $\dfrac{x^2}{a^2} + \dfrac{y^2}{b^2} = 1$.

27. Use Theorems 4.4.16 and 4.4.26 to explain why a linear operator with an orthogonal standard matrix is an isometry.

28. It is known that the linear operator $T : \mathbb{R}^2 \rightarrow \mathbb{R}^2$ is invertible if and only if $\det [T] = 0$. Use the notion of area to explain why this is the case.

29. If two vectors in $\mathbb{R}^2$ are scalar multiples of each other, the parallelogram described by these vectors equals 0. Explain why the area of the image of this parallelogram under any linear operator also equals 0.

30. Show that the area of the triangle in $\mathbb{R} \times \mathbb{R}$ with vertices (x_1, y_1), (x_2, y_2), and (x_3, y_3) is $\dfrac{1}{2} \begin{vmatrix} x_1 & y_1 & 1 \\ x_2 & y_2 & 1 \\ x_3 & y_3 & 1 \end{vmatrix}$.

CHAPTER 5

Abstract Vectors

5.1 Vector Spaces

The space $\mathbb{R}^n$ of Cartesian vectors is formally defined in Definition 2.1.1 as a set of $n \times 1$ matrices. Definition 2.1.4 introduced addition and scalar multiplication of these vectors. For example, in $\mathbb{R}^2$,

$$\begin{bmatrix} 1 \\ 2 \end{bmatrix} + \begin{bmatrix} 3 \\ 4 \end{bmatrix} = \begin{bmatrix} 4 \\ 6 \end{bmatrix} \quad \text{and} \quad 3 \begin{bmatrix} 1 \\ 2 \end{bmatrix} = \begin{bmatrix} 3 \\ 6 \end{bmatrix}.$$

These vectors are conveniently visualized as arrows with their initial points at the origin. Vector addition corresponds to finding the diagonal of a parallelogram, and scalar multiplication corresponds to scaling the arrow's length. In fact, the definitions of vector addition and scalar multiplication on $\mathbb{R}^n$ are motivated by their geometric interpretation as arrows in Cartesian space. The basic algebraic properties of Cartesian vectors are described in Theorem 2.1.5. In Theorem 3.2.10, it is shown that these very same algebraic properties are enjoyed by the set of all $m \times n$ matrices. As a result, working in the space of $m \times n$ matrices retains much of the flavor of working in the space of Cartesian vectors.

Linear Algebra, First Edition. Michael L. O'Leary.

245

© 2021 John Wiley & Sons, Inc. Published 2021 by John Wiley & Sons, Inc.

Companion Website: www.wiley.com/go/o'leary/linearalgebra

The scene is now set to make the concept of a vector abstract. As $\mathbb{R}^{m\times n}$ feels like $\mathbb{R}^{mn}$, this generalization will share the basic properties from Cartesian space that are found in Theorem 2.1.5. The generalization will require the operations of Definitions 1.4.34 and 1.4.41.

■ **Definition 5.1.1**

An **(abstract) vector space** is a set V, whose elements are called **vectors**, together with a binary operation $+$ called **vector addition** and a scalar operation $\cdot$ called **scalar multiplication** such that for all $\mathbf{u}, \mathbf{v}, \mathbf{w} \in V$ and $r, s \in \mathbb{R}$:

(a) $\mathbf{u} + \mathbf{v} = \mathbf{v} + \mathbf{u}$ (**Commutative Law**).

(b) $\mathbf{u} + \mathbf{v} + \mathbf{w} = \mathbf{u} + (\mathbf{v} + \mathbf{w})$ (**Associative Law**).

(c) There is an element $\mathbf{0} \in V$ such that $\mathbf{u} + \mathbf{0} = \mathbf{0} + \mathbf{u} = \mathbf{u}$ (**Additive Identity Law**).

(d) There is an element $-\mathbf{u} \in V$ such that $\mathbf{u} + (-\mathbf{u}) = -\mathbf{u} + \mathbf{u} = \mathbf{0}$ (**Additive Inverse Law**).

(e) $rs\mathbf{u} = r(s\mathbf{u})$ (**Associative Law**) .

(f) $r(\mathbf{u} + \mathbf{v}) = r\mathbf{u} + r\mathbf{v}$ (**Distributive Law**).

(g) $(r + s)\mathbf{u} = r\mathbf{u} + s\mathbf{u}$ (**Distributive Law**).

(h) $1\mathbf{u} = \mathbf{u}$ (**Multiplicative Identity**).

Denote the vector space by $(V, +, \cdot, \mathbf{0})$ and call $\mathbf{0}$ the **zero vector**. Simply identify $(V, +, \cdot, \mathbf{0})$ with V if the operations and zero vector are clear from context.

As with addition on $\mathbb{R}^n$, **vector subtraction** is defined for vectors $\mathbf{u}$ and $\mathbf{v}$ from an abstract vector space as $\mathbf{u} - \mathbf{v} = \mathbf{u} + (-\mathbf{v})$.

Definition 5.1.1 captures the important properties of the space of Cartesian vectors but is sufficiently general to apply to a variety of important mathematical systems as will be seen shortly. Many of the intuitive properties of Cartesian vectors which are not on this list can be proved as a consequence of this definition, so it would be redundant to include these properties in the definition. These include the properties enumerated in Theorem 2.1.7 for $\mathbb{R}^n$. Unlike the prior proof that relied on calculations specific to $\mathbb{R}^n$, the proof here can only rely on the specifics of Definition 5.1.1.

■ **Theorem 5.1.2**

Let V be a vector space, $\mathbf{u} \in V$, and $r \in \mathbb{R}$.

(a) $0\mathbf{u} = \mathbf{0}$,

(b) $r\mathbf{0} = \mathbf{0}$,

(c) $-\mathbf{u} = (-1)\mathbf{u}$,

(d) If $r\mathbf{u} = \mathbf{0}$, then $r = 0$ or $\mathbf{u} = \mathbf{0}$.

Proof

Because $0\mathbf{u} + 0\mathbf{u} = (0 + 0)\mathbf{u} = 0\mathbf{u}$, subtracting $0\mathbf{u}$ from both sides of this equation yields $0\mathbf{u} = \mathbf{0}$, proving (a). Similarly, $r\mathbf{0} + r\mathbf{0} = r(\mathbf{0} + \mathbf{0}) = r\mathbf{0}$ implies that $r\mathbf{0} = \mathbf{0}$, so (b) follows. To prove (c), note that

$$\mathbf{u} + (-1)\mathbf{u} = 1\mathbf{u} + (-1)\mathbf{u} = [1 + (-1)]\mathbf{u} = 0\mathbf{u} = \mathbf{0},$$

and $-\mathbf{u} + \mathbf{u} = \mathbf{0}$ follows by the commutative law [Definition 5.1.1(a)]. Lastly, to prove (d), assume $r\mathbf{u} = \mathbf{0}$ and $r \neq 0$. Then, $\mathbf{u} = \frac{1}{r}\mathbf{0} = \mathbf{0}$ by (b). ■

Examples of Vector Spaces

Before delving deeper into the study of abstract vector spaces, it is useful to develop a small catalog of familiar, concrete systems which are vector spaces. In order to prove that a system is a vector space, it must be shown that that system satisfies all of the properties of Definition 5.1.1.

■ **Example 5.1.3**

Let $+$ and $\cdot$ represent the standard vector operations on $\mathbb{R}^n$ given in Definition 2.1.4. Then,

$$\left(\mathbb{R}^n, \ +, \ \cdot, \ \begin{bmatrix} 0 \\ 0 \\ \vdots \\ 0 \end{bmatrix}\right)$$

is a vector space because vector addition on $\mathbb{R}^n$ is a binary operation and scalar multiplication on $\mathbb{R}^n$ is a scalar operation. That $\mathbb{R}^n$ satisfies the remaining properties is exactly the content of Theorem 2.1.5. This vector space is usually simply denoted by $\mathbb{R}^n$.

■ **Example 5.1.4**

Let $+$ and $\cdot$ represent the standard matrix operations on $\mathbb{R}^{m \times n}$ given in Definition 3.2.5. Then,

$$\left(\mathbb{R}^{m \times n}, \ +, \ \cdot, \ \begin{bmatrix} 0 & 0 & \cdots & 0 \\ 0 & 0 & \cdots & 0 \\ \vdots & \vdots & & \vdots \\ 0 & 0 & \cdots & 0 \end{bmatrix}\right)$$

is a vector space. The fact that $\mathbb{R}^{m \times n}$ satisfies the remaining properties is exactly the content of Theorem 3.2.10. This vector space is usually simply denoted by $\mathbb{R}^{m \times n}$.

Sets of functions provide a rich source of interesting vector spaces. The set ${}^{\mathbb{R}}\mathbb{R}$ has much in common with the Cartesian vector space $\mathbb{R}^n$. For example, two functions $f, g \in {}^{\mathbb{R}}\mathbb{R}$ are equal if they agree at every point of the domain (Theorem 1.4.6).

Compare this to the condition for equality of two Cartesian vectors in $\mathbb{R}^n$ which are equal if they agree coordinatewise (Definition 2.1.3). Similarly, addition and scalar multiplication of functions are defined in Examples 1.4.37 and 1.4.43 so that for all real numbers x,

$$
\begin{aligned}
(f + g)(x) &= f(x) + g(x), \\
(rf)(x) &= rf(x),
\end{aligned}
\tag{5.1}
$$

while addition and scalar multiplication of Cartesian vectors are defined by Definition 2.1.4 as

$$
\begin{bmatrix} u_1 \\ u_2 \\ \vdots \\ u_n \end{bmatrix} + \begin{bmatrix} v_1 \\ v_2 \\ \vdots \\ v_n \end{bmatrix} = \begin{bmatrix} u_1 + v_1 \\ u_2 + v_2 \\ \vdots \\ u_n + v_n \end{bmatrix} \quad \text{and} \quad r \begin{bmatrix} u_1 \\ u_2 \\ \vdots \\ u_n \end{bmatrix} = \begin{bmatrix} ru_1 \\ ru_1 \\ \vdots \\ ru_n \end{bmatrix}.
$$

Both of these definitions involve working with coordinates, and they differ only in the number of coordinates of their elements. Functions in $^{\mathbb{R}}\mathbb{R}$ have infinitely many coordinates, while Cartesian vectors in $\mathbb{R}^n$ have finitely many. Not surprisingly, the space of all functions on $\mathbb{R}$ with the operations of (5.1) is a vector space.

■ **Example 5.1.5**

$^{\mathbb{R}}\mathbb{R}$ is a vector space. To prove this, take $f, g, h \in {}^{\mathbb{R}}\mathbb{R}$ and $r, s \in \mathbb{R}$.

(a) It is known that function addition is a binary operation and that function scalar multiplication is a scalar operation (Examples 1.4.37 and 1.4.43).

(b) To see that $f + g = g + f$, let $x \in \mathbb{R}$. Then,

$$(f + g)(x) = f(x) + g(x) = g(x) + f(x) = (g + f)(x).$$

(c) To see that $f + g + h = f + (g + h)$, let $x \in \mathbb{R}$. Then,

$$
\begin{aligned}
(f + g + h)(x) &= (f + g)(x) + h(x) \\
&= f(x) + g(x) + h(x) \\
&= f(x) + [g(x) + h(x)] \\
&= f(x) + (g + h)(x) \\
&= (f + [g + h])(x).
\end{aligned}
$$

(d) To see that the zero function $0(x) = 0$ is the additive identity, observe that $0 + f = f$ because taking $x \in \mathbb{R}$,

$$(0 + f)(x) = 0(x) + f(x) = 0 + f(x) = f(x).$$

Because function addition is commutative, $f + 0 = f$ follows automatically.

(e) To see that $(-1)f$ is the additive inverse of f, let $x \in \mathbb{R}$ and note that

$$[f + (-1)f](x) = f(x) + (-1)f(x) = 0 = 0(x).$$

As before, $(-1)f + f = 0$ because of the commutative law.

(f) To see that $rsf = r(sf)$, let $x \in \mathbb{R}$. Then,

$$(rsf)(x) = (rs)f(x) = r[sf(x)] = r(sf)(x).$$

(g) To see that $r(f + g) = rf + rg$, take $x \in \mathbb{R}$, so

$$\begin{aligned}
[r(f + g)](x) &= r(f + g)(x) \\
&= r[f(x) + g(x)] \\
&= rf(x) + rg(x) \\
&= (rf + rg)(x).
\end{aligned}$$

(h) To see that $(r + s)f = rf + sf$, let $x \in \mathbb{R}$ and note that

$$\begin{aligned}
[(r + s)f](x) &= (r + s)f(x) \\
&= rf(x) + sf(x) \\
&= (rf)(x) + (sf)(x) \\
&= (rf + sf)(x).
\end{aligned}$$

(i) To see that the scalar 1 is the multiplicative identity, let $x \in \mathbb{R}$. Then, $1f = f$ because

$$(1f)(x) = 1f(x) = f(x).$$

The vector space $^{\mathbb{R}}\mathbb{R}$ consists of all functions defined on $\mathbb{R}$. It is often useful to consider more restrictive collections of functions. For example, calculus involves the study of differentiable functions. The set of all differentiable functions on $\mathbb{R}$ is itself a vector space which lives inside the vector space $^{\mathbb{R}}\mathbb{R}$. Before calculus, elementary algebra involves, among other things, the study of the rational and polynomial functions. The set of all rational functions is not contained in $^{\mathbb{R}}\mathbb{R}$ since, for example, $f(x) = 1/x$ is a rational function which is not defined on all of $\mathbb{R}$. While the rational functions do form a vector space (Exercise 3), this space is not a subset of $^{\mathbb{R}}\mathbb{R}$. The polynomials of degree at most n are a convenient example of a vector space that can be viewed as living inside of the space $^{\mathbb{R}}\mathbb{R}$. Compare with the definition of a polynomial in Section 3.2.

■ **Definition 5.1.6**

Let n be a natural number. Define the set of all **polynomials** of **degree** less than or equal to n to be

$$\mathbb{P}_n = \left\{ \sum_{i=0}^{n} a_i x^i : a_i \in \mathbb{R}, i = 0, 1, \dots, n \right\}.$$

Denote by $\mathbb{P}$ the set of all polynomials so that

$$\mathbb{P} = \bigcup_{n=0}^{\infty} \mathbb{P}_n.$$

The polynomials $p(x) = \sum_{i=0}^{n} a_i x^i$ and $q(x) = \sum_{i=0}^{n} b_i x^i$ in $\mathbb{P}_n$ are **equal** [written as $p(x) = q(x)$] when $a_i = b_i$ for all $i = 1, 2, \dots, n$. Furthermore, **polynomial addition** in $\mathbb{P}_n$ is defined by

$$p(x) + q(x) = \sum_{i=0}^{n} (a_i + b_i) x^i,$$

and **polynomial scalar multiplication** in $\mathbb{P}_n$ is defined so that for all $r \in \mathbb{R}$,

$$rp(x) = \sum_{i=0}^{n} r a_i x^i.$$

It should be noted that although the zero polynomial, $0 + 0x + 0x^2 + \cdots + 0x^n$, has no degree, it is nonetheless an element of $\mathbb{P}_n$ for all natural numbers n.

■ **Example 5.1.7**

Letting $p(x) = 1 + 5x + 2x^2$ and $q(x) = 7 - 2x + x^2$, then

$$p(x) + q(x) = 8 + 3x + 3x^2$$

and

$$7p(x) = 7 + 35x + 14x^2.$$

When adding polynomials in $\mathbb{P}$, let m be the maximum of the degrees of the polynomials and use the polynomial addition of $\mathbb{P}_m$. For example, letting

$$r(x) = 3 + x + 5x^2 - 7x^3 + x^4,$$

to find $p(x) + r(x)$, write

$$p(x) = 1 + 5x + 2x^2 + 0x^3 + 0x^4.$$

Because both $p(x)$ and $r(x)$ are elements of $\mathbb{P}_4$, the sum $p(x) + r(x)$ is computed as

$$(1+3) + (5+1)x + (2+5)x^2 + (0-7)x^3 + (0+1)x^4 = 4 + 6x + 7x^2 - 7x^3 + x^4.$$

This, of course, is simply combining like terms.

When computing the values of Example 5.1.7, it becomes clear that the powers of x serve as placeholders giving order to the coefficients. This resembles vectors in $\mathbb{R}^n$ and leads to the next example.

■ **Example 5.1.8**

The set $\mathbb{P}_n$ together with the addition and scalar multiplication given in Definition 5.1.6 satisfies all properties of Definition 5.1.1, so $\mathbb{P}_n$ is a vector space. The verification of these properties is an exercise in basic polynomial manipulation. Suppose that

$$p(x) = \sum_{i=0}^{n} p_i x^i, \quad q(x) = \sum_{i=0}^{n} q_i x^i, \quad r(x) = \sum_{i=0}^{n} r_i x^i$$

are elements of $\mathbb{P}_n$ and that $r, s \in \mathbb{R}$.

(a) The definition of polynomial addition and scalar multiplication make clear that $\mathbb{P}_n$ is closed under these two operations. Also, let

$$p'(x) = \sum_{i=0}^{n} p_i' x^i \quad \text{and} \quad q'(x) = \sum_{i=0}^{n} q_i' x^i.$$

Assume $p(x) = p'(x)$ and $q(x) = q'(x)$. Then, because $p_i = p_i'$ and $q_i = q_i'$ for all $i = 1, 2, \ldots, n$,

$$p(x) + q(x) = \sum_{i=0}^{n} (p_i + q_i) x^i = \sum_{i=0}^{n} (p_i' + q_i') x^i = p'(x) + q'(x).$$

This means that polynomial addition is a binary operation. Taking $r \in \mathbb{R}$,

$$rp(x) = \sum_{i=0}^{n} r p_i x^i = \sum_{i=0}^{n} r p_i' x^i = r p'(x),$$

so polynomial scalar multiplication is a scalar operation.

(b) To see that polynomial addition is commutative:

$$p(x) + q(x) = \sum_{i=0}^{n} (p_i + q_i) x^i = \sum_{i=0}^{n} (q_i + p_i) x^i = q(x) + p(x).$$

(c) To see that polynomial addition is associative:

$$\begin{aligned}
p(x) + q(x) + r(x) &= \sum_{i=0}^{n} (p_i + q_i) x^i + \sum_{i=0}^{n} r_i x^i \\
&= \sum_{i=0}^{n} (p_i + q_i + r_i) x^i \\
&= \sum_{i=0}^{n} (p_i + [q_i + r_i]) x^i \\
&= \sum_{i=0}^{n} p_i x^i + \sum_{i=0}^{n} (q_i + r_i) x^i = p(x) + [q(x) + r(x)].
\end{aligned}$$

(d) To see that the zero polynomial $0(x) = 0 + 0x + \cdots + 0x^n$ is the additive identity:

$$0(x) + p(x) = \sum_{i=0}^{n}(0 + p_i)x^i = \sum_{i=0}^{n} p_i x^i = p(x).$$

Because polynomial addition is commutative, $p(x) + 0(x) = p(x)$.

(e) The additive inverse of $p(x)$ is $(-1)p(x)$. To see this:

$$p(x) + (-1)p(x) = \sum_{i=0}^{n}(p_i + [-p_i])x^i = \sum_{i=0}^{n} 0x^i = 0(x).$$

Also, $p(x) + (-1)p(x) = 0$ by the commutative law.

(f) To see that scalar multiplication is associative:

$$rsp(x) = \sum_{i=0}^{n} rsp_i x^i = \sum_{i=0}^{n} r(sp_i)x^i = r\sum_{i=0}^{n} sp_i x^i = r[sp(x)].$$

(g) To see that the first distributive property holds:

$$r[p(x) + q(x)] = r\sum_{i=0}^{n}(p_i + q_i)x^i$$

$$= \sum_{i=0}^{n} r(p_i + q_i)x^i$$

$$= \sum_{i=0}^{n}(rp_i + rq_i)x^i$$

$$= \sum_{i=0}^{n} rp_i x^i + \sum_{i=0}^{n} rq_i x^i$$

$$= r\sum_{i=0}^{n} p_i x^i + r\sum_{i=0}^{n} q_i x^i = rp(x) + rq(x).$$

(h) To see that the second distributive property holds:

$$(r + s)p(x) = (r + s)\sum_{i=0}^{n} p_i x^i$$

$$= \sum_{i=0}^{n}(r + s)p_i x^i$$

$$= \sum_{i=0}^{n}(rp_i + sp_i)x^i$$

$$= \sum_{i=0}^{n} r p_i x^i + \sum_{i=0}^{n} s p_i x^i$$

$$= r \sum_{i=0}^{n} p_i x^i + s \sum_{i=0}^{n} p_i x^i = r p(x) + s p(x).$$

(i) To see that the scalar 1 is the multiplicative identity:

$$1 p(x) = 1 \sum_{i=0}^{n} p_i x^i = \sum_{i=0}^{n} 1 p_i x^i = \sum_{i=0}^{n} p_i x^i = p(x).$$

■ **Example 5.1.9**

$\mathbb{P}$ is a vector space. The proof is similar to that of Example 5.1.8 and is left to Exercise 1.

Linear Transformations

Recall that a linear transformation with domain $\mathbb{R}^m$ and codomain $\mathbb{R}^n$ is a function $T : \mathbb{R}^m \to \mathbb{R}^n$ that preserves addition and scalar multiplication (Definition 3.1.1). This definition generalizes to the case of a function $T : V \to W$, where V and W are vector spaces. Both vector spaces come with their own vector addition and scalar multiplication. Although the operations on V and W probably differ, the same notation will be used for both as in Definition 3.1.1. The vector space to which each operation belongs will be clear from context.

■ **Definition 5.1.10**

A function $T : V \to W$ is a **linear transformation** if V and W are vector spaces and for all $\mathbf{u}, \mathbf{v} \in V$ and $r \in \mathbb{R}$,

$$T(\mathbf{u} + \mathbf{v}) = T\mathbf{u} + T\mathbf{v} \tag{5.2}$$

and

$$T(r\mathbf{u}) = r(T\mathbf{u}). \tag{5.3}$$

T is a **linear operator** if $V = W$ and the operations on the domain are the operations on the codomain.

To clarify, in (5.2), the addition on the left is from V, and the addition on the right is from W. In (5.3), the scalar multiplication on the left is from V, and the scalar multiplication on the right is from W.

■ **Example 5.1.11**

For all $m, n \in \mathbb{Z}^+$, each function in $L(\mathbb{R}^n, \mathbb{R}^m)$ (Definition 3.2.7) continues to be a linear transformation under Definition 5.1.10.

■ **Example 5.1.12**

Prove that $T : \mathbb{P}_3 \to \mathbb{R}^{2 \times 2}$ defined by

$$T(a_0 + a_1 x + a_2 x^2 + a_3 x^3) = \begin{bmatrix} a_0 & a_1 \\ a_2 & a_3 \end{bmatrix}$$

is a linear transformation. Let $r \in \mathbb{R}$ and write $p(x) = p_0 + p_1 x + p_2 x^2 + p_3 x^3$ and $q(x) = q_0 + q_1 x + q_2 x^2 + q_3 x^3$. Then,

$$T[p(x) + q(x)] = T[(p_0 + q_0) + (p_1 + q_1)x + (p_2 + q_2)x^2 + (p_3 + q_3)x^3]$$

$$= \begin{bmatrix} p_0 + q_0 & p_1 + q_1 \\ p_2 + q_2 & p_3 + q_3 \end{bmatrix}$$

$$= \begin{bmatrix} p_0 & p_1 \\ p_2 & p_3 \end{bmatrix} + \begin{bmatrix} q_0 & q_1 \\ q_2 & q_3 \end{bmatrix}$$

$$= T(p_0 + p_1 x + p_2 x^2 + p_3 x^3) + T(q_0 + q_1 x + q_2 x^2 + q_3 x^3)$$

$$= T[p(x)] + T[q(x)],$$

and

$$T[rp(x)] = T(rp_0 + rp_1 x + rp_2 x^2 + rp_3 x^3)$$

$$= \begin{bmatrix} rp_0 & rp_1 \\ rp_2 & rp_3 \end{bmatrix}$$

$$= r \begin{bmatrix} p_0 & p_1 \\ p_2 & p_3 \end{bmatrix}$$

$$= rT(p_0 + p_1 x + p_2 x^2 + p_3 x^3) = rT[p(x)].$$

■ **Example 5.1.13**

Define the **trace** of an $n \times n$ matrix $A = [a_{ij}]$ to be the sum of the entries along its main diagonal and denote this value by tr A. That is,

$$\text{tr } A = \sum_{i=1}^{n} a_{ii}.$$

Prove that tr $: \mathbb{R}^{n \times n} \to \mathbb{R}$ is a linear transformation. Write the $n \times n$ matrix B as $[b_{ij}]$ and take $r \in \mathbb{R}$. Then,

$$\text{tr } (A + B) = \text{tr } [a_{ij} + b_{ij}] = \sum_{i=1}^{n}(a_{ii} + b_{ii}) = \sum_{i=1}^{n} a_{ii} + \sum_{i=1}^{n} b_{ii} = \text{tr } A + \text{tr } B,$$

and

$$\text{tr } (rA) = \text{tr } [ra_{ij}] = \sum_{i=1}^{n} ra_{ii} = r \sum_{i=1}^{n} a_{ii} = r \text{ tr } A.$$

The trace is a simple calculation that, nonetheless, proves very helpful. See Exercise 22 for some properties of the trace of a matrix.

Since the work is general enough, the proof of Theorem 3.1.6 suffices to prove the next theorem.

■ **Theorem 5.1.14**

Let $T : V \rightarrow W$ be a linear transformation.

(a) $T0 = 0$.

(b) $T(-\mathbf{u}) = -(T\mathbf{u})$ for all $\mathbf{u} \in \mathbb{R}^n$.

(c) $T(\mathbf{u} - \mathbf{v}) = T\mathbf{u} - T\mathbf{v}$ for all $\mathbf{u}, \mathbf{v} \in \mathbb{R}^n$.

To clarify Theorem 5.1.14, the zero vector on the left of (a) is from V, and the zero vector on the right is from W. The additive inverses on the left of (b) and (c) belong to V, and the additive inverses on the right belong to W.

■ **Example 5.1.15**

Continue Example 5.1.12 to illustrate Theorem 5.1.14. Since the zero polynomial in $\mathbb{P}_3$ is $0 + 0x + 0x^2 + 0x^3$,

$$T(0 + 0x + 0x^2 + 0x^3) = \begin{bmatrix} 0 & 0 \\ 0 & 0 \end{bmatrix}.$$

Also,

$$T[-p(x)] = T(-p_0 - p_1 x - p_2 x^2 - p_3 x^3) = \begin{bmatrix} -p_0 & -p_1 \\ -p_2 & -p_3 \end{bmatrix} = -\begin{bmatrix} p_0 & p_1 \\ p_2 & p_3 \end{bmatrix},$$

and

$$\begin{aligned} T[p(x) - q(x)] &= T[(p_0 - q_0) + (p_1 - q_1)x + (p_2 - q_2)x^2 + (p_3 - q_3)x^3] \\ &= \begin{bmatrix} p_0 - q_0 & p_1 - q_1 \\ p_2 - q_2 & p_3 - q_3 \end{bmatrix} \\ &= \begin{bmatrix} p_0 & p_1 \\ p_2 & p_3 \end{bmatrix} - \begin{bmatrix} q_0 & q_1 \\ q_2 & q_3 \end{bmatrix} = T[p(x)] - T[q(x)]. \end{aligned}$$

Many operations familiar from algebra and calculus can be viewed as linear transformations on suitably chosen vector spaces.

■ **Example 5.1.16**

Let $r \in \mathbb{R}$. Each of the given functions is a linear transformation with domain $\mathbb{P}_n$. Confirmation of this is left to Exercise 19.

- The **multiplication operator** $M_r : \mathbb{P}_n \to \mathbb{P}_n$ with

$$M_r[p(x)] = r\,p(x).$$

- The **differentiation transformation** $D : \mathbb{P}_n \to \mathbb{P}_{n-1}$ with

$$D[p(x)] = \frac{d}{dx}p(x).$$

- The **evaluation transformation** $E : \mathbb{P}_n \to \mathbb{R}$ with

$$E[p(x)] = p(0).$$

- The **shift operator** $S_r : \mathbb{P}_n \to \mathbb{P}_n$ with

$$S_r[p(x)] = p(x - r).$$

Apply each linear transformation to $p(x) = 5 + 4x + 2x^2$.

- $M_9[p(x)] = 45 + 36x + 18x^2$.
- $D[p(x)] = 4 + 4x$.
- $E[p(x)] = 5$.
- $S_8[p(x)] = 5 + 4(x - 8) + 2(x - 8)^2$.

The set of all $m \times n$ matrices $\mathbb{R}^{m \times n}$ form a vector space (Example 5.1.4) and, furthermore, $\mathbb{R}^{m \times n}$ is exactly the set of standard matrices for all linear transformations $\mathbb{R}^n \to \mathbb{R}^m$. Indeed, the set of all linear transformations between any two fixed vector spaces is itself a vector space with the following operations of addition and scalar multiplication, which are generalizations of the two operations in Definition 3.2.1. Let V and W be vector spaces. For all linear transformations $T_1, T_2 : V \to W$ and $r \in \mathbb{R}$, define $T_1 + T_2$ and rT_1 such that for all $\mathbf{u} \in V$,

$$
\begin{aligned}
(T_1 + T_2)\mathbf{u} &= T_1\mathbf{u} + T_2\mathbf{u}, \\
(rT_1)\mathbf{u} &= r(T_1\mathbf{u}).
\end{aligned}
\tag{5.4}
$$

The functions given by (5.4) are linear transformations as in Theorem 3.2.3. For this reason, it is appropriate for the next definition to be made at this time. It generalizes Definition 3.2.7. The proof of the theorem that immediately follows is left to Exercise 4.

■ **Definition 5.1.17**

Let V and W be vector spaces. Define

$$L(V, W) = \{T : T \text{ is a linear transformation } V \to W\}.$$

■ **Theorem 5.1.18**

For all vector spaces V and W, the set $L(V, W)$ with the operations of function addition and scalar multiplication is a vector space.

Exercises

1. Prove that $\mathbb{P}$ is a vector space with polynomial addition and scalar multiplication as noted in Example 5.1.9.

2. Prove that $\mathbb{R}^+$ is a vector space over $\mathbb{R}$ with addition and scalar multiplication defined by $x + y = xy$ and $kx = x^k$ for every $x, y \in \mathbb{R}^+$ and $k \in \mathbb{R}$. Is $\mathbb{R}^+$ still a vector space if we change scalar multiplication to $kx = k^x$? If it is true, prove it. Otherwise, state what conditions fail.

3. Prove that the set of **rational functions**,

$$\left\{ f(x) = \frac{p(x)}{q(x)} \; : \; p(x), q(x) \in \mathbb{P} \text{ and } q(x) \neq 0 \right\},$$

is a vector space using the function addition and scalar multiplication of (5.1).

4. Prove Theorem 5.1.18.

5. Let V be the set of invertible 2×2 matrices. For all $A, B \in V$, define $A + B = AB$, but let scalar multiplication be standard scalar multiplication of matrices. Prove or show false that V is a vector space with these operations.

6. Let $V_1, V_2, \ldots, V_k$ be vector spaces. Let

$$(\mathbf{u}_1, \mathbf{u}_2, \ldots, \mathbf{u}_k), (\mathbf{v}_1, \mathbf{v}_2, \ldots, \mathbf{v}_k) \in V_1 \times V_2 \times \cdots \times V_k$$

and $r \in \mathbb{R}$ so as to define

$$(\mathbf{u}_1, \mathbf{u}_2, \ldots, \mathbf{u}_k) + (\mathbf{v}_1, \mathbf{v}_2, \ldots, \mathbf{v}_k) = (\mathbf{u}_1 + \mathbf{v}_1, \mathbf{u}_2 + \mathbf{v}_2, \ldots, \mathbf{u}_k + \mathbf{v}_k) \quad (5.5)$$

and

$$r(\mathbf{u}_1, \mathbf{u}_2, \ldots, \mathbf{u}_k) = (r\mathbf{u}_1, r\mathbf{u}_2, \ldots, r\mathbf{u}_k), \quad (5.6)$$

where the vector addition and scalar multiplication of each coordinate comes from the corresponding vector space. Prove that $V_1 \times V_2 \times \cdots \times V_k$ is a vector space using (5.5) and (5.6).

7. Let V be a vector space and $\mathbf{u}, \mathbf{v}, \mathbf{w} \in V$.

    (a) Prove the **right cancellation law**: if $\mathbf{u} + \mathbf{w} = \mathbf{v} + \mathbf{w}$, then $\mathbf{u} = \mathbf{v}$.

    (b) State and prove the **left cancellation law** for the vector space V. Is there a proof for the left cancellation law that does not simply modify the proof of the right cancellation law?

8. Let V be a vector space with $\mathbf{u}, \mathbf{v} \in V$. Prove the for all nonzero $r \in \mathbb{R}$, if $r\mathbf{u} = r\mathbf{v}$, then $\mathbf{u} = \mathbf{v}$.

9. Let $\mathbf{u}, \mathbf{v} \in V$, where V is a vector space, and take $r, s \in \mathbb{R}$. Prove.
    (a) $(r+s)(\mathbf{u}+\mathbf{v}) = r\mathbf{u}+r\mathbf{v}+s\mathbf{u}+s\mathbf{v}$ (c) $-(\mathbf{u}+\mathbf{v}) = -\mathbf{u}-\mathbf{v}$
    (b) $-(-\mathbf{u}) = \mathbf{u}$ (d) $(-r)(-\mathbf{u}) = r\mathbf{u}$

10. Let $\mathbf{u}$ and $\mathbf{v}$ be vectors from the vector space V. Solve for $\mathbf{x}$, justifying each step as in Example 2.1.6.
    (a) $2\mathbf{u} + 4\mathbf{x} = \mathbf{v}$ (c) $7\mathbf{u} + \mathbf{x} = 2\mathbf{x} - 6\mathbf{v}$
    (b) $4(\mathbf{u} + 3\mathbf{x}) = \mathbf{0}$ (d) $\mathbf{v} - (\mathbf{u} - 7\mathbf{x}) = 5\mathbf{u} - 9\mathbf{x}$

11. Given vectors spaces V and W with their standard operations, prove that each definition of $T : V \to W$ describes a linear transformation.
    (a) $V = \mathbb{R}^2$, $W = \mathbb{R}^{2\times2}$, $T\begin{bmatrix} x \\ y \end{bmatrix} = \begin{bmatrix} x & 0 \\ 0 & y \end{bmatrix}$
    (b) $V = \mathbb{P}_2$, $W = \mathbb{P}_3$,
    $T(a_0+a_1x+a_2x^2) = (a_2-2a_1)+(a_1-4a_0)x+(a_0+a_2)x^2+(4a_1-5a_2)x^3$
    (c) $V = \mathbb{R}^{2\times3}$, $W = \mathbb{R}^{2\times2}$, $T\begin{bmatrix} a & b & c \\ d & e & f \end{bmatrix} = \begin{bmatrix} 2a & b+c \\ d+e & 4f \end{bmatrix}$
    (d) $V = \mathbb{P}_3$, $W = \mathbb{R}^2$, $T[p(x)] = \begin{bmatrix} p(0) \\ p(1) \end{bmatrix}$

12. Given vector spaces V and W with their standard operations, prove that each definition of $T : V \to W$ does not describes a linear transformation.
    (a) $V = \mathbb{P}_1$, $W = \mathbb{R}$, $T(a_0 + a_1x) = 1$
    (b) $V = \mathbb{R}^{2\times2}$, $W = \mathbb{R}^3$, $T\begin{bmatrix} a & b \\ c & d \end{bmatrix} = \begin{bmatrix} a+b \\ c^2+d^2 \\ a^3+b^3 \end{bmatrix}$
    (c) $V = {}^{\mathbb{R}}\mathbb{R}$, $W = \mathbb{R}$, $Tf = |f(0)|$

13. Define $T : V \to W$ by $T\mathbf{u} = \mathbf{0}$ for all $\mathbf{u}$, where V and W are vector spaces with $\mathbf{0}$ being the zero vector of W. Prove that T is a linear transformation. This function is called the **zero transformation** as in Example 3.1.3.

14. Prove or show false that the function $T : \mathbb{R}^{n\times n} \to \mathbb{R}$ defined by $TA = \det A$ is a linear transformation.

15. Let $T : \mathbb{R}^{4\times5} \to \mathbb{R}^{5\times4}$ be defined by $TA = A^T$. Assuming the standard matrix operations, prove that T is a linear transformation.

16. Define $T : \mathbb{P}_n \to \mathbb{P}_{n+1}$ so that for all polynomials $p(x) = a_0 + a_1x + \cdots + a_nx^n$,

$$T[p(x)] = a_0x + \frac{a_1}{2}x^2 + \cdots + \frac{a_n}{n+1}x^{n+1}.$$

Prove that T is a linear transformation.

17. The natural logarithm is a function $\mathbb{R}^+ \to \mathbb{R}$. Using the operations of Exercise 2 so that $\mathbb{R}^+$ is a vector space, prove that the natural logarithm is a linear transformation.

18. Prove that the function $T : {}^{\mathbb{R}}\mathbb{R} \to \mathbb{R}$ defined by $Tf = f(0)$, which is similar to the evaluation transformation of Example 5.1.16, is a linear transformation.

19. Prove that each function defined in Example 5.1.16 is a linear transformation.

20. Let $V_1, V_2, \ldots, V_k, W$ be vector spaces. Let $T_i : V_i \to W$ be a linear transformation for each $i = 1, 2, \ldots, k$. Using the operations defined in Exercise 6, prove that $T : V_1 \times V_2 \times \cdots \times V_k \to W$ defined by

$$T(\mathbf{u}_1, \mathbf{u}_2, \ldots, \mathbf{u}_k) = T_1\mathbf{u}_1 + T_2\mathbf{u}_2 + \cdots + T_k\mathbf{u}_k$$

is a linear transformation.

21. Let V and W be vector spaces. Prove that the function $T : V \to W$ is a linear transformation if and only if $T(r\mathbf{u}_1 + s\mathbf{u}_2) = rT\mathbf{u}_1 + sT\mathbf{u}_2$ for all $\mathbf{u}_1, \mathbf{u}_2 \in V$ and $r, s \in R$.

22. Let A and B be $n \times n$ matrices. Prove that tr $AB =$ tr BA and tr $A^T =$ tr A.

23. Prove or show false: tr $A = (\det A)(\text{tr } A^{-1})$ for all invertible 2×2 matrices A.

5.2 Subspaces

Consider the line L in $\mathbb{R}^2$ given by $y = 2x$. This line contains the zero vector, and because $\begin{bmatrix} x \\ y \end{bmatrix}$ is in this line if and only if it satisfies $y = 2x$,

$$L = \left\{ \begin{bmatrix} x \\ 2x \end{bmatrix} : x \in \mathbb{R} \right\} = S\left(\begin{bmatrix} 1 \\ 2 \end{bmatrix} \right),$$

as in Definition 2.1.21. Any two vectors $\begin{bmatrix} x_1 \\ 2x_1 \end{bmatrix}, \begin{bmatrix} x_2 \\ 2x_2 \end{bmatrix} \in L$ can be added without leaving the line L because

$$\begin{bmatrix} x_1 \\ 2x_1 \end{bmatrix} + \begin{bmatrix} x_2 \\ 2x_2 \end{bmatrix} = \begin{bmatrix} x_1 + x_2 \\ 2x_1 + 2x_2 \end{bmatrix} = \begin{bmatrix} x_1 + x_2 \\ 2(x_1 + x_2) \end{bmatrix} \in L.$$

Similarily, any vector in L can be scaled arbitrarily without leaving L because

$$r \begin{bmatrix} x_1 \\ 2x_1 \end{bmatrix} = \begin{bmatrix} rx_1 \\ 2rx_1 \end{bmatrix} \in L.$$

Stated another way, the line L is closed under vector addition (Definition 1.4.35) and closed under scalar multiplication (Definition 1.4.42).

■ **Definition 5.2.1**

A nonempty subset W of a vector space V is called a **subspace** of V if it is closed under the vector addition and scalar multiplication of V.

Examples of Subspaces

It should be noted that every vector space is a subspace of itself. This subspace is called the **improper subspace**. This means that any property of a subspace is also a property of the vector space. Also, the set that contains only the zero vector of its containing vector space is a subspace called the **trivial subspace**.

■ **Example 5.2.2**

Define

$$W = \left\{ \begin{bmatrix} x & 0 \\ y & x+y \end{bmatrix} : x, y \in \mathbb{R} \right\}.$$

Assuming the standard matrix operations, prove that W is a subspace of the vector space $\mathbb{R}^{2\times 2}$ (Example 5.1.4).

- Since

$$\begin{bmatrix} 0 & 0 \\ 0 & 0 \end{bmatrix} = \begin{bmatrix} 0 & 0 \\ 0 & 0+0 \end{bmatrix},$$

the 2×2 zero matrix is an element of W, so W is nonempty.

- Let $a, b, c, d \in \mathbb{R}$. Then,

$$\begin{bmatrix} a & 0 \\ b & a+b \end{bmatrix} + \begin{bmatrix} c & 0 \\ d & c+d \end{bmatrix} = \begin{bmatrix} a+c & 0 \\ b+d & (a+b)+(c+d) \end{bmatrix}.$$

Since $a+b, c+d \in \mathbb{R}$, conclude that W is closed under vector addition.

- Let $a, b, r \in \mathbb{R}$. Then,

$$r \begin{bmatrix} a & 0 \\ b & a+b \end{bmatrix} = \begin{bmatrix} ra & 0 \\ rb & ra+rb \end{bmatrix}.$$

Because $ra, ra, ra+rb \in \mathbb{R}$, conclude that W is closed under scalar multiplication.

■ **Example 5.2.3**

Define

$$W = \{a + bx + cx^3 : a, b, c \in \mathbb{R}\}.$$

Prove that W is a subspace of $\mathbb{P}_3$ assuming the standard operations (Example 5.1.8) by taking $p(x), q(x) \in W$ and $r \in \mathbb{R}$. Write

$$p(x) = a_1 + a_2 x + a_3 x^3 \quad \text{and} \quad q(x) = b_1 + b_2 x + b_3 x^3,$$

with $a_i, b_i \in \mathbb{R}$ for $i = 1, 2, 3$.

- $0 + 0x + 0x^2 + 0x^3 \in W$, so $W \neq \varnothing$.
- $p(x) + q(x) = a_1 + b_1 + (a_2 + b_2)x + (a_3 + b_3)x^3$. Because the sum of two real numbers is a real number, $p(x) + q(x) \in W$.
- $rp(x) = ra_1 + ra_2x + ra_3x^3$. Because the product of two real numbers is a real number, $rp(x) \in W$.

■ Example 5.2.4

Let $W_1, W_2, \ldots, W_k$ be subspaces of the vector space V. Prove that the **sum** of the subspaces,

$$W = \sum_{i=1}^{k} W_i = \{\mathbf{x}_1 + \mathbf{x}_2 + \cdots + \mathbf{x}_k : \mathbf{x}_I \in W_i \text{ for } i = 1, 2, \ldots, k\},$$

is a subspace of V.

- Since $\mathbf{0} \in W_i$ for all $i = 1, 2, \ldots, k$, the zero vector is an element of W.
- Let $\mathbf{u}, \mathbf{v} \in W$ and write $\mathbf{u} = \mathbf{u}_1 + \mathbf{u}_2 + \cdots + \mathbf{u}_k$ and $\mathbf{v} = \mathbf{v}_1 + \mathbf{v}_2 + \cdots + \mathbf{v}_k$, where $\mathbf{u}_i, \mathbf{v}_i \in W_i$ for all i. Then, $\mathbf{u} + \mathbf{v} \in W$ because

$$\mathbf{u} + \mathbf{v} = (\mathbf{u}_1 + \mathbf{v}_1) + (\mathbf{u}_2 + \mathbf{v}_2) + \cdots + (\mathbf{u}_k + \mathbf{v}_k)$$

and $\mathbf{u}_i + \mathbf{v}_i \in W_i$ for all i since each W_i is closed under vector addition.
- Take $\mathbf{u} \in W$ and $r \in \mathbb{R}$. Using the same notation as in the previous part,

$$r\mathbf{u} = r\mathbf{u}_1 + r\mathbf{u}_2 + \cdots + r\mathbf{u}_k.$$

Because each W_i is closed under scalar multiplication, $r\mathbf{u}_i \in W_i$ for all $i = 1, 2, \ldots, k$, so $r\mathbf{u} \in W$.

Properties

An alternative to Definition 5.2.1 is to define a subspace of a vector space as a subset of a vector space that is itself a vector space. A set satisfying such a definition would have the properties of Definition 5.2.1 as a result. However, Definition 5.2.1 was chosen here. The result of this choice is that any subset of a vector space that satisfies its conditions will have the property given by the alternate definition.

■ Theorem 5.2.5

A subspace of a vector space is a vector space.

Proof

Suppose that W is a subspace of a vector space V. The subspace W requires its own vector addition and scalar multiplication in order to be a vector space. Let $+$ and $\cdot$ represent the vector addition and scalar multiplication, respectively, from V. These operations are functions that can be written as

$$+ : V \times V \to V \quad \text{and} \quad \cdot : \mathbb{R} \times V \to V.$$

Restrict the domains of these two functions to W. That is, only allow inputs from $W \times W$ into $+$, and only allow inputs from $\mathbb{R} \times W$ into $\cdot$. Because W is closed under vector addition and scalar multiplication, the codomains of these restrictions are W so that vector addition and scalar multiplication can be viewed as functions

$$+ : W \times W \to W \quad \text{and} \quad \cdot : \mathbb{R} \times W \to W.$$

These two operations become the vector addition and scalar multiplication of W. This is the purpose of the closure conditions in the definition of a subspace.

Now that W has its own operations, to complete the proof that W is a vector space, confirm the properties of Definition 5.1.1.

- (a), (b), (e), (f), and (g) hold in W because $W \subseteq V$ and V satisfies these properties.

- Since W is not empty, fix some $\mathbf{w} \in W$. By closure under scalar multiplication, $\mathbf{0} = 0\mathbf{w} \in W$, so (c) holds.

- The additive inverse of any $\mathbf{w} \in W$ is $-\mathbf{w} = (-1)\mathbf{w} \in W$. Therefore, property (d) holds. ∎

■ Example 5.2.6

Let A be an $m \times n$ matrix and $\mathbf{0}$ be the zero vector of $\mathbb{R}^m$. Define

$$W = \{\mathbf{x} \in \mathbb{R}^n : A\mathbf{x} = \mathbf{0}\}.$$

This is the solution set to the homogeneous system of linear equations with coefficient matrix A [Definition 3.5.3(c)], and W is a subset of $\mathbb{R}^n$. Prove that W is a subspace of $\mathbb{R}^n$.

- W is not empty because $A\mathbf{0} = \mathbf{0}$.

- Let $\mathbf{u}, \mathbf{v} \in W$. This means that $A\mathbf{u} = \mathbf{0}$ and $A\mathbf{v} = \mathbf{0}$. Then, $\mathbf{u} + \mathbf{v} \in W$ because

$$A(\mathbf{u} + \mathbf{v}) = A\mathbf{u} + A\mathbf{v} = \mathbf{0} + \mathbf{0} = \mathbf{0}.$$

- Let $\mathbf{u} \in W$ and $r \in \mathbb{R}$. It follows that $r\mathbf{u} \in W$ because

$$A(r\mathbf{u}) = rA\mathbf{u} = r\mathbf{0} = \mathbf{0}.$$

Therefore, W is a subspace of $\mathbb{R}^n$ and is called a **solution space**. In addition, by Theorem 5.2.5, W can be considered a vector space on its own. This means that every theorem proved about vector spaces also applies to W.

■ Example 5.2.7

Continuing Example 5.2.6, it is important that the system of linear equations is homogeneous because, otherwise, the solution set could not be shown to be closed under vector addition or scalar multiplication. For example, take

$$\begin{bmatrix} 1 & 2 \\ 2 & 4 \end{bmatrix} \begin{bmatrix} x \\ y \end{bmatrix} = \begin{bmatrix} 5 \\ 10 \end{bmatrix}.$$

Let W still represent the solution set. Both $\begin{bmatrix} 3 \\ 1 \end{bmatrix}$ and $\begin{bmatrix} -3 \\ 4 \end{bmatrix}$ are elements of W. However,

$$\begin{bmatrix} 1 & 2 \\ 2 & 4 \end{bmatrix} \begin{bmatrix} 0 \\ 5 \end{bmatrix} = \begin{bmatrix} 10 \\ 20 \end{bmatrix},$$

showing that W is not closed under vector addition, so in this case W is not a subspace of $\mathbb{R}^2$. Also,

$$\begin{bmatrix} 1 & 2 \\ 2 & 4 \end{bmatrix} \begin{bmatrix} 9 \\ 3 \end{bmatrix} = \begin{bmatrix} 15 \\ 30 \end{bmatrix},$$

showing that W in this case is not closed under scalar multiplication. There are other issues with W. For instance, the zero vector is not an element of W, and although $\begin{bmatrix} 3 \\ 1 \end{bmatrix} \in W$, its additive inverse is not in W because

$$\begin{bmatrix} 1 & 2 \\ 2 & 4 \end{bmatrix} \begin{bmatrix} -3 \\ -1 \end{bmatrix} = \begin{bmatrix} -5 \\ -10 \end{bmatrix}.$$

Although the existence of the zero vector and additive inverses in a set is not a guarantee that the subset will be a subspace, these are requirements of any subspace. This is seen in the proof of Theorem 5.2.5 but stated here for easy reference.

■ Theorem 5.2.8

Let W be a subspace of a vector space V.

(a) The zero vector of V is an element of W.

(b) If $\mathbf{u} \in W$, then $-\mathbf{u} \in W$.

Continuity and differentiability of functions play an important role in calculus. Functions exhibiting varying degrees of differentiability can be categorized using vector spaces. Recall that a function f is **continuous** on $\mathbb{R}$ means that f is continuous for each $x \in \mathbb{R}$.

■ **Definition 5.2.9**

For any positive integer n,

$C^0(\mathbb{R}) = \{f : f$ is a continuous function $\mathbb{R} \to \mathbb{R}\}$,
$C^n(\mathbb{R}) = \{f : f$ is a function $\mathbb{R} \to \mathbb{R}$ with n continuous derivatives$\}$,
$C^\infty(\mathbb{R}) = \{f : f$ is a function $\mathbb{R} \to \mathbb{R}$ with derivatives of all orders$\}$.

■ **Theorem 5.2.10**

For all $n \in \mathbb{Z}^+$, the sets $C^0(\mathbb{R})$, $C^n(\mathbb{R})$, and $C^\infty(\mathbb{R})$ are subspaces of $^\mathbb{R}\mathbb{R}$.

Proof

Prove the theorem for $C^1(\mathbb{R})$, leaving the rest of the proof to Exercise 10. First, $C^1(\mathbb{R}) \neq \varnothing$ because the zero map has a continuous derivative with domain $\mathbb{R}$. Next, let $f, g \in C^1(\mathbb{R})$ and $r \in \mathbb{R}$. This means that f' and g' exist and are continuous with domains equal to $\mathbb{R}$. Therefore, the derivatives of $f + g$ and rf exist and are continuous with $\mathbb{R}$ as their domains, which implies that $f + g$ and rf are elements of $C^1(\mathbb{R})$. ■

■ **Example 5.2.11**

Viewing the elements of $\mathbb{P}_n$ as functions, as opposed to simply polynomials, leads to the sequence of subspaces,

$$\mathbb{P}_0 \subset \mathbb{P}_1 \subset \mathbb{P}_2 \subset \cdots \subset \mathbb{P} \subset C^\infty(\mathbb{R}) \subset \cdots \subset C^2(\mathbb{R}) \subset C^1(\mathbb{R}) \subset C^0(\mathbb{R}) \subset {}^\mathbb{R}\mathbb{R}.$$

Every vector space in the sequence is a subspace of every vector space to its right in the sequence. It is also worth noting that these vector spaces are a natural context for the derivative transformation D. For each $n \in \mathbb{Z}^+$, the surjection D can be restricted to other surjections of the form

$$D : C^n(\mathbb{R}) \to C^{n-1}(\mathbb{R}) \quad \text{and} \quad D : \mathbb{P}_n \to \mathbb{P}_{n-1}.$$

Spanning Sets

Recall that the span of a finite collection of Cartesian vectors is defined to be the set of all linear combinations of said vectors (Definitions 2.1.18 and 2.1.20). A linear combination and the span of a (possibly infinite) set of vectors from any vector space are defined analogously.

■ **Definition 5.2.12**

Let V be a vector space and $\mathcal{S} \subseteq V$. The vector $\mathbf{u} \in V$ is a **linear combination** of $\mathcal{S}$ if there exist $\mathbf{v}_1, \mathbf{v}_2, \dots, \mathbf{v}_k \in \mathcal{S}$ and $r_1, r_2, \dots, r_k \in \mathbb{R}$ such that

$$\mathbf{u} = r_1\mathbf{v}_1 + r_2\mathbf{v}_2 + \cdots + r_k\mathbf{v}_k.$$

■ **Definition 5.2.13**

Let V be a vector space and $\mathcal{S} \subseteq V$. The **span** of $\mathcal{S}$ is

$$\mathsf{S}(\mathcal{S}) = \{r_1\mathbf{v}_1 + r_2\mathbf{v}_2 + \cdots + r_k\mathbf{v}_k : \mathbf{v}_i \in \mathcal{S} \text{ and } r_i \in \mathbb{R} \text{ for } i = 1, 2, \ldots, k\}.$$

If $\mathcal{S} = \{\mathbf{v}_1, \mathbf{v}_2, \ldots, \mathbf{v}_n\}$, write $\mathsf{S}(\mathcal{S}) = \mathsf{S}(\mathbf{v}_1, \mathbf{v}_2, \ldots, \mathbf{v}_n)$, and if $\mathsf{S}(\mathcal{S}) = V$, then $\mathcal{S}$ is called a **spanning set** of V.

■ **Example 5.2.14**

Let $\mathcal{S} = \left\{ \begin{bmatrix} 1 & 0 \\ 0 & 0 \end{bmatrix}, \begin{bmatrix} 0 & 1 \\ 0 & 0 \end{bmatrix}, \begin{bmatrix} 0 & 0 \\ 1 & 0 \end{bmatrix}, \begin{bmatrix} 0 & 0 \\ 0 & 1 \end{bmatrix} \right\}$. The set $\mathcal{S}$ is a spanning set of $\mathbb{R}^{2\times 2}$ because $\begin{bmatrix} a & b \\ c & d \end{bmatrix}$ is a linear combination of $\mathcal{S}$ for all 2×2 matrices $\begin{bmatrix} a & b \\ c & d \end{bmatrix}$. This is because

$$\begin{bmatrix} a & b \\ c & d \end{bmatrix} = a\begin{bmatrix} 1 & 0 \\ 0 & 0 \end{bmatrix} + b\begin{bmatrix} 0 & 1 \\ 0 & 0 \end{bmatrix} + c\begin{bmatrix} 0 & 0 \\ 1 & 0 \end{bmatrix} + d\begin{bmatrix} 0 & 0 \\ 0 & 1 \end{bmatrix}.$$

Also, a spanning set of the subspace from Example 5.2.2,

$$W = \left\{ \begin{bmatrix} x & 0 \\ y & x+y \end{bmatrix} : x, y \in \mathbb{R} \right\},$$

is

$$\left\{ \begin{bmatrix} 1 & 0 \\ 0 & 1 \end{bmatrix}, \begin{bmatrix} 0 & 0 \\ 1 & 1 \end{bmatrix} \right\}.$$

Other spanning sets include W itself,

$$\left\{ \begin{bmatrix} 4 & 0 \\ 0 & 4 \end{bmatrix}, \begin{bmatrix} 0 & 0 \\ 2 & 2 \end{bmatrix} \right\} \quad \text{and} \quad \left\{ \begin{bmatrix} 1 & 0 \\ 0 & 1 \end{bmatrix}, \begin{bmatrix} 0 & 0 \\ 1 & 1 \end{bmatrix}, \begin{bmatrix} 1 & 0 \\ 1 & 2 \end{bmatrix} \right\}.$$

■ **Theorem 5.2.15**

The span of a subset of a vector space V is a subspace of V.

Proof

Let $\mathcal{S} \subseteq V$. Because $\mathbf{0} \in \mathsf{S}(\mathcal{S})$, this set is nonempty. Suppose r is a real number and $\mathbf{u}, \mathbf{v} \in \mathsf{S}(\mathcal{S})$. This means that there exist scalars $a_1, a_2, \ldots, a_k$ and $b_1, b_1, \ldots, b_k$ and vectors $\mathbf{w}_1, \mathbf{w}_2, \ldots, \mathbf{w}_k \in \mathcal{S}$ such that

$$\mathbf{u} = a_1\mathbf{w}_1 + a_2\mathbf{w}_2 + \cdots + a_k\mathbf{w}_k$$

and

$$\mathbf{v} = b_1\mathbf{w}_1 + b_2\mathbf{w}_2 + \cdots + b_k\mathbf{w}_k.$$

Then, $\mathbf{u} + \mathbf{v} \in S(\mathcal{S})$ since

$$\mathbf{u} + \mathbf{v} = (a_1 + b_1)\mathbf{w}_1 + (a_2 + b_2)\mathbf{w}_2 + \cdots + (a_k + b_k)\mathbf{w}_k,$$

and $r\mathbf{u} \in S(\mathcal{S})$ because

$$r\mathbf{u} = r(a_1\mathbf{w}_1 + a_2\mathbf{w}_2 + \cdots + a_k\mathbf{w}_k) = (ra_1)\mathbf{w}_1 + (ra_2)\mathbf{w}_2 + \cdots + (ra_k)\mathbf{w}_k.$$

Hence, $S(\mathcal{S})$ is a subspace of V. ∎

■ Example 5.2.16

Because each coset of $\mathbb{R}^n$ (Definition 2.1.20) that contains the zero vector is a subspace of $\mathbb{R}^n$ by Theorem 5.2.15, lines (Definition 2.1.21) and planes (Definition 2.1.25) that contain the zero vector are subspaces of $\mathbb{R}^n$.

■ Example 5.2.17

By Example 5.1.8, the space $\mathbb{P}_n$ can be viewed as a subspace of $^{\mathbb{R}}\mathbb{R}$ for all $n \in \mathbb{N}$. In fact, the tedious verification in Example 5.1.8 can be entirely supplanted by observing that $\mathbb{P}_n = S(1, x, \ldots, x^n)$. Furthermore, $\{1, x, x^2, \ldots\}$ is a spanning set for $\mathbb{P}$, which does not have a finite spanning set.

Kernel and Range

Every linear transformation between vector spaces is associated with two important sets. The first is an extension of Definition 3.4.1.

■ Definition 5.2.18

The **kernel** of a linear transformation $T : V \to W$ is $\{\mathbf{u} \in V : T\mathbf{u} = \mathbf{0}\}$ and denoted by ker T.

The proof that the kernel of a linear transformation is a subspace of the domain is essentially a copy of the proof of Theorem 3.4.2. Likewise, that a linear transformation being one-to-one is equivalent with a trivial kernel is also true in the general case, and its proof is nearly identical to that of the earlier proof involving only $\mathbb{R}^n$ in Theorem 3.4.3. These general results are stated here.

■ Theorem 5.2.19

The kernel of a linear transformation $T : V \to W$ is a subspace of V.

■ Theorem 5.2.20

The linear transformation $T : V \to W$ is one-to-one if and only if ker $T = \{\mathbf{0}\}$.

■ **Example 5.2.21**

Consider the linear transformation $T : \mathbb{R}^{2\times 2} \to \mathbb{P}_2$ defined by

$$T\begin{bmatrix} a & b \\ c & d \end{bmatrix} = (a + 2b + c + 3d) + (3a + 2b + 2c - d)x + (9a - 6b + 3c + 2d)x^2.$$

The kernel of T is the subspace of $\mathbb{R}^{2\times 2}$ consisting of all 2×2 matrices A such that $T(A) = 0 + 0x + 0x^2$. In other words, the kernel of T is

$$\left\{ \begin{bmatrix} a & b \\ c & d \end{bmatrix} \in \mathbb{R}^{2\times 2} : \begin{array}{rrrrrrrrr} a & + & 2b & + & c & + & 3d & = & 0, \\ 3a & + & 2b & + & 2c & - & d & = & 0, \\ 9a & - & 6b & + & 3c & + & 2d & = & 0 \end{array} \right\}. \tag{5.7}$$

Use Gauss–Jordan elimination to take the coefficient matrix that defines the set of (5.7),

$$\begin{bmatrix} 1 & 2 & 1 & 3 \\ 3 & 2 & 2 & -1 \\ 9 & -6 & 3 & 2 \end{bmatrix}, \tag{5.8}$$

to its reduced row echelon form,

$$\begin{bmatrix} 1 & 0 & 1/2 & 0 \\ 0 & 1 & 1/4 & 0 \\ 0 & 0 & 0 & 1 \end{bmatrix}.$$

The parametric equations for the solution set to this system are

$$a = -\frac{1}{2}t,$$

$$b = -\frac{1}{4}t,$$

$$c = t,$$

$$d = 0,$$

which implies that T is not one-to-one by Theorem 5.2.20 because

$$\ker T = \mathsf{S}\left(\begin{bmatrix} -1/2 & -1/4 \\ 1 & 0 \end{bmatrix}\right) = \mathsf{S}\left(\begin{bmatrix} 2 & 1 \\ -4 & 0 \end{bmatrix}\right).$$

■ **Example 5.2.22**

The set of all functions satisfying $f(0) = 0$ is a subspace of $\mathbb{R}^{\mathbb{R}}$. To see this, note that $\{f \in {}^{\mathbb{R}}\mathbb{R} : f(0) = 0\}$ is exactly the kernel of the linear transformation $T : {}^{\mathbb{R}}\mathbb{R} \to \mathbb{R}$ defined by $Tf = f(0)$. However, the set of all functions satisfying $f(0) = 1$ is a not subspace of $\mathbb{R}^{\mathbb{R}}$ since it is not closed under scalar multiplication or addition. For example, the constant function defined by $g(x) = 1$ belongs to this set, but $g + g$ and $5g$ do not.

The other important set related to an abstract linear transformation is its range. It is defined in Definition 1.4.1, but its definition specific to linear transformations is given again here for convenience. That the range is a subspace of the codomain is proved immediately after the definition.

■ Definition 5.2.23

The **range** of a linear transformation $T : V \to W$ is $\{T\mathbf{u} \; : \; \mathbf{u} \in V\}$ and is denoted by ran T.

■ Theorem 5.2.24

The range of a linear transformation $T : V \to W$ is a subspace of W.

Proof

The range of T is not empty because $\mathbf{0} \in$ ran T by Theorem 5.1.14(a). Now, suppose that $\mathbf{v}_1, \mathbf{v}_2 \in$ ran T and $r \in \mathbb{R}$. This means that there exist $\mathbf{u}_1, \mathbf{u}_2 \in V$ such that $T\mathbf{u}_1 = \mathbf{v}_1$ and $T\mathbf{u}_2 = \mathbf{v}_2$. Then, $\mathbf{v}_1 + \mathbf{v}_2 \in$ ran T since

$$T(\mathbf{u}_1 + \mathbf{u}_2) = T\mathbf{u}_1 + T\mathbf{u}_2 = \mathbf{v}_1 + \mathbf{v}_2,$$

and $r\mathbf{v}_1 \in$ ran T because $T(r\mathbf{u}_1) = r(T\mathbf{u}_1) = r\mathbf{v}_1$. ■

■ Example 5.2.25

Consider computing a spanning set for the range of T from Example 5.2.21. Since an arbitrary polynomial in ran T is of the form

$$p(x) = (a + 2b + c + 3d) + (3a + 2b + 2c - d)x + (9a - 6b + 3c + 2d)x^2,$$

it follows that

$$p(x) = a(1 + 3x + 9x^2) + b(2 + 2x - 6x^2) + c(1 + 2x + 3x^2) + d(3 - x + 2x^2),$$

so

$$\text{ran } T = \mathsf{S}\left(1 + 3x + 9x^2, 2 + 2x - 6x^2, 1 + 2x + 3x^2, 3 - x + 2x^2\right).$$

It should be noted that the coefficients of the polynomials in the spanning set of ran T are the columns of the matrix given in (5.8). This is reminiscent of Theorem 3.4.5.

Exercises

1. Determine whether each set is a subspace of $\mathbb{R}^2$ and prove the result.

    (a) $\left\{ \begin{bmatrix} x \\ y \end{bmatrix} \in \mathbb{R}^2 : x + y = 0 \right\}$ (b) $\left\{ \begin{bmatrix} x \\ y \end{bmatrix} \in \mathbb{R}^2 : x - y = 1 \right\}$

(c) $\left\{ \begin{bmatrix} x \\ y \end{bmatrix} \in \mathbb{R}^2 : \dfrac{x}{y} = 1 \right\}$ 　　　　　(d) $\left\{ \begin{bmatrix} x \\ y \end{bmatrix} \in \mathbb{R}^2 : x^2 + y^2 = 0 \right\}$

2. Is $\left\{ \begin{bmatrix} a & b \\ c & d \end{bmatrix} \in \mathbb{R}^{2\times2} : a^2 = d^2 \right\}$ a subspace of $\mathbb{R}^{2\times2}$? Explain.

3. Show that the set of all vectors in $\mathbb{R}^3$ orthogonal to $\begin{bmatrix} 1 \\ -1 \\ 1 \end{bmatrix}$ form a subspace of $\mathbb{R}^3$. What does this subspace look like?

4. Show that the set of all vectors in $\mathbb{R}^3$ orthogonal to both $\begin{bmatrix} 1 \\ 0 \\ 1 \end{bmatrix}$ and $\begin{bmatrix} 0 \\ 1 \\ 1 \end{bmatrix}$ form a subspace of $\mathbb{R}^3$. What does this subspace look like?

5. Prove that the set of diagonal $n \times n$ matrices is a subspace of $\mathbb{R}^{n\times n}$.

6. Prove that the set of upper triangular $n \times n$ matrices is a subspace of $\mathbb{R}^{n\times n}$. In addition, prove that the same result holds true for lower triangular matrices. Is the set of triangular $n \times n$ matrices a subspace of $\mathbb{R}^{n\times n}$?

7. Let $W_1, W_2, \ldots, W_k$ be subspaces of the vector space V. Prove that the intersection $W_1 \cap W_2 \cap \cdots \cap W_k$ is a subspace of V.

8. Let W_1 and W_2 be subspaces of V. What must be true so that $W_1 \cup W_2$ is a subspace of V?

9. Let $W_1, W_2, \ldots, W_k$ be subspaces of the vector spaces $V_1, V_2, \ldots, V_k$, respectively. Prove that $W_1 \times W_2 \times \cdots \times W_k$ is a subspace of $V_1 \times V_2 \times \cdots \times V_k$ using the operations of Exercise 6 from Section 5.1.

10. Complete the proof of Theorem 5.2.10.

11. Determine whether each set is a subspace of $C[-1, 1]$, the vector space of all continuous functions defined on $[-1, 1]$.

　　(a) $\{ f \in C[-1, 1] : f(-1) = f(1) \}$

　　(b) $\{ f \in C[-1, 1] : f(-1) = f(1) = 0 \}$

　　(c) $\{ f \in C[-1, 1] : f(-1) \cdot f(1) = 0 \}$

12. Determine whether each set is a subspace of $C[-1, 1]$. See Exercise 11

　　(a) $\left\{ f \in C[-1, 1] : \displaystyle\int_{-1}^{1} f(x)dx = 1 \right\}$

　　(b) $\left\{ f \in C[-1, 1] : \displaystyle\int_{-1}^{1} f(x)dx = 0 \right\}$

　　(c) $\left\{ f \in C[-1, 1] : \displaystyle\int_{-1}^{1} |f(x)|^2 dx = 0 \right\}$

13. Determine whether each set is a subspace of $^\mathbb{R}\mathbb{R}$. See Exercises 14 and 15 of Section 1.4 for the definition of increasing and decreasing functions.

   (a) The set of increasing functions

   (b) The set of functions which are either nonincreasing on all of $[-1,1]$ or nondecreasing on all of $[-1,1]$

   (c) The set of functions which are equal to 0 at all but finitely many points

14. Use the strategy of Example 5.2.22 to prove that the set of all functions f satisfying $f(-1) = 2f(1)$ is a subspace of $^\mathbb{R}\mathbb{R}$.

15. Find spanning sets for the given subspaces.

   (a) $\{r + (r + s)x + (2r + 3s)x^2 : r, s \in \mathbb{R}\}$

   (b) $\left\{ \begin{bmatrix} 0 & r \\ s + 2t & t \end{bmatrix} : r, s, t \in \mathbb{R} \right\}$

16. True or false. Explain.

   (a) $4 + 2x + 3x^2 \in \mathsf{S}\left(2 + 2x + x^2, 2 + x + 4x^2, 4 + 3x^2\right)$

   (b) $\begin{bmatrix} 4 & 2 \\ 1 & 3 \end{bmatrix} \in \mathsf{S}\left(\begin{bmatrix} 3 & 3 \\ 0 & 2 \end{bmatrix}, \begin{bmatrix} 4 & 3 \\ 1 & 2 \end{bmatrix} \right)$

17. Prove that $\mathbb{P}_3 = \mathsf{S}\left(1 + x, 1 + x^2, 1 + x^3, 1 + x + x^2 + x^3\right)$.

18. Prove that $\mathbb{R}^{2\times 2} = \mathsf{S}\left(\begin{bmatrix} 3 & 2 \\ -1 & 3 \end{bmatrix}, \begin{bmatrix} 2 & 0 \\ 1 & -1 \end{bmatrix}, \begin{bmatrix} -1 & 1 \\ 0 & -1 \end{bmatrix}, \begin{bmatrix} 3 & 1 \\ 2 & 1 \end{bmatrix} \right)$.

19. Prove that a line (Definition 2.1.21) is a subspace of $\mathbb{R}^n$ if and only if it contains the zero vector. Prove that the same holds for all cosets [Definition 2.1.20(b)].

20. Assume that $\mathcal{S}$ is a subset of the vector space V. Prove that if $V = \mathsf{S}(\mathcal{S})$, then $V = \mathsf{S}(\mathcal{S} \cup \{\mathbf{u}\})$ for all $\mathbf{u} \in V$.

21. Let V be a vector space. Take $\mathbf{u}_1, \mathbf{u}_2, \mathbf{u}_3 \in V$ such that $r_1\mathbf{u}_1 + r_2\mathbf{u}_2 + r_3\mathbf{u}_3 = \mathbf{0}$ for nonzero scalars r_1, r_2, and r_3. Prove that $\mathsf{S}(\mathbf{u}_1, \mathbf{u}_2) = \mathsf{S}(\mathbf{u}_2, \mathbf{u}_3)$.

22. Let $T : V \to W$ be a linear transformation. Let $\{\mathbf{u}_1, \mathbf{u}_2, \dots, \mathbf{u}_n\} \subset V$.

   (a) Prove false: If $\{\mathbf{u}_1, \mathbf{u}_2, \dots \mathbf{u}_n\}$ is a spanning set for V, $\{T\mathbf{u}_1, T\mathbf{u}_2, \dots, T\mathbf{u}_n\}$ is a spanning set for W.

   (b) Find an additional condition on T that makes the implication of Exercise 22(a) true and prove the result.

   (c) Prove false: If $\{T\mathbf{u}_1, T\mathbf{u}_2, \dots, T\mathbf{u}_n\}$ is a spanning set for W, $\{\mathbf{u}_1, \mathbf{u}_2, \dots, \mathbf{u}_n\}$ is a spanning set for V.

   (d) Find an additional condition on T that makes the implication of Exercise 22(c) true and prove the result.

23. Let $\mathcal{S}$ be a subset of the vector space V. Prove that $\mathsf{S}(\mathcal{S})$ is the smallest subspace of V that contains $\mathcal{S}$. In other words, prove that for any subspace W of V, if $\mathcal{S} \subseteq W$, then $\mathsf{S}(\mathcal{S}) \subseteq W$.

24. Find the kernel and range of each linear transformation in Exercise 11 of Section 5.1. Which of these functions are one-to-one and which are onto?

25. Assume that V_1 and V_2 are subspaces of the vector space V. Let $T_1 : V_1 \to W$ and $T_2 : V_2 \to W$ be linear transformations. Prove.

    (a) $T_1 \cap T_2$ is a linear transformation.
    (b) $\ker(T_1 \cap T_2) = \ker T_1 \cap \ker T_2$ and $\operatorname{ran}(T_1 \cap T_2) = \operatorname{ran} T_1 \cap \operatorname{ran} T_2$

26. Let $T : V \to V$ be a linear operator. Prove that the following are equivalent.

    • $[T]$ is invertible.   • T is one-to-one.   • T is onto.

27. Let $T : U \to V$ be a linear transformation. Prove.

    (a) If W is a subspace of U, then $T[W]$ is a subspace of V.
    (b) If W is a subspace of V, then $T^{-1}[W]$ is a subspace of U.

28. Use Exercise 27(a) to prove that the range of a linear transformation is a subspace of its codomain.

29. Prove that if W_1 is a subspace of W_2 and W_2 is a subspace of the vector space V, then W_1 is a subspace of V.

30. Let W be a subspace of the vector space V. Define the set of cosets (Definition 2.1.20), called the **quotient vector space**,

$$V/W = \{\mathbf{u} + W : \mathbf{u} \in V\}.$$

Also, for all $\mathbf{u}, \mathbf{v} \in V$ and $r \in \mathbb{R}$, define the binary operation,

$$(\mathbf{u} + W) + (\mathbf{v} + W) = (\mathbf{u} + \mathbf{v}) + W, \tag{5.9}$$

and the scalar operation,

$$r(\mathbf{u} + W) = (r\mathbf{u}) + W. \tag{5.10}$$

Prove that V/W is a vector space with (5.9) as its vector addition and (5.10) as its scalar multiplication.

31. Let W be a subspace of the vector space V. Prove that $\mathbf{u} + W = \mathbf{v} + W$ if and only if $\mathbf{u} - \mathbf{v} \in W$ for all $\mathbf{u}, \mathbf{v} \in V$.

32. Find two nonzero vectors in each quotient vector space V/W, multiply both by 5, and add the results. Also, find the zero vector of each quotient vector space and prove that it is the additive identity.

    (a) $V = \mathbb{R}^3$, $W = S(\mathbf{e}_1, \mathbf{e}_2)$
    (b) $V = \mathbb{P}_3$, $W = S(1, x, x^2)$
    (c) $V = \mathbb{R}^{2\times 2}$, $W = S\left(\begin{bmatrix} 1 & 0 \\ 0 & 1 \end{bmatrix}, \begin{bmatrix} 0 & 1 \\ 1 & 0 \end{bmatrix}\right)$

5.3 Linear Independence

The subspace of $\mathbb{R}^3$ defined by the span of the vectors $\mathbf{e}_1$ and $\mathbf{e}_2$ is the set of all linear combinations of $\mathbf{e}_1$ and $\mathbf{e}_2$. This space is easy to describe because it is the set of all vectors in $\mathbb{R}^3$ with third component equal to 0:

$$S(\mathbf{e}_1, \mathbf{e}_2) = \left\{ r \begin{bmatrix} 1 \\ 0 \\ 0 \end{bmatrix} + s \begin{bmatrix} 0 \\ 1 \\ 0 \end{bmatrix} : r, s \in \mathbb{R} \right\} = \left\{ \begin{bmatrix} r \\ s \\ 0 \end{bmatrix} : r, s \in \mathbb{R} \right\}.$$

Imagine the various ways to augment this subspace by adding a new vector $\mathbf{u}$ to the spanning set to form $S(\mathbf{e}_1, \mathbf{e}_2, \mathbf{u})$. If the new vector is taken as $\mathbf{u} = \mathbf{e}_3$, the subspace is enlarged to all of $\mathbb{R}^3$ because

$$S(\mathbf{e}_1, \mathbf{e}_2, \mathbf{e}_3) = \left\{ r \begin{bmatrix} 1 \\ 0 \\ 0 \end{bmatrix} + s \begin{bmatrix} 0 \\ 1 \\ 0 \end{bmatrix} + t \begin{bmatrix} 0 \\ 0 \\ 1 \end{bmatrix} : r, s, t \in \mathbb{R} \right\} = \mathbb{R}^3. \tag{5.11}$$

However, if the new vector is taken as $\mathbf{u} = \mathbf{e}_1 + \mathbf{e}_2$, the space is not enlarged at all because

$$S(\mathbf{e}_1, \mathbf{e}_2, \mathbf{e}_1 + \mathbf{e}_2) = \left\{ r \begin{bmatrix} 1 \\ 0 \\ 0 \end{bmatrix} + s \begin{bmatrix} 0 \\ 1 \\ 0 \end{bmatrix} + t \begin{bmatrix} 1 \\ 1 \\ 0 \end{bmatrix} : r, s, t \in \mathbb{R} \right\}$$

$$= \left\{ \begin{bmatrix} r+t \\ s+t \\ 0 \end{bmatrix} : r, s, t \in \mathbb{R} \right\} = S(\mathbf{e}_1, \mathbf{e}_2). \tag{5.12}$$

The difference between these two cases can be easily explained. The reason that $S(\mathbf{e}_1, \mathbf{e}_2)$ does not grow under the addition of $\mathbf{e}_1 + \mathbf{e}_2$ is that $\mathbf{e}_1 + \mathbf{e}_2 \in S(\mathbf{e}_1, \mathbf{e}_2)$. Its addition to $S(\mathbf{e}_1, \mathbf{e}_2)$ creates no new vectors in the subspace. On the other hand, $\mathbf{e}_3$ is not present in $S(\mathbf{e}_1, \mathbf{e}_2)$, so its addition to $S(\mathbf{e}_1, \mathbf{e}_2)$ adds new vectors to the subspace. Specifically, it adds the vectors with a nonzero third coordinate.

Viewed another way, the vectors of the spanning set $\{\mathbf{e}_1, \mathbf{e}_2, \mathbf{e}_3\}$ in (5.11) are not redundant. It is impossible to write any of these vectors as a linear combination of the other two. That is, none of these vectors is in the spanning set of the other two. The vectors of the spanning set $\{\mathbf{e}_1, \mathbf{e}_2, \mathbf{e}_1 + \mathbf{e}_2\}$ of (5.12) are redundant. Each of the vectors can be written as a linear combination of the other two:

$$\mathbf{e}_1 = -1\mathbf{e}_2 + 1(\mathbf{e}_1 + \mathbf{e}_2),$$
$$\mathbf{e}_2 = -1\mathbf{e}_1 + 1(\mathbf{e}_1 + \mathbf{e}_2), \tag{5.13}$$
$$\mathbf{e}_1 + \mathbf{e}_2 = 1\mathbf{e}_1 + 1\mathbf{e}_2.$$

Each of the three relationships of (5.13) reduce to the single equation,

$$1\mathbf{e}_1 + 1\mathbf{e}_2 - 1(\mathbf{e}_1 + \mathbf{e}_2) = \mathbf{0},$$

which expresses the redundancy as a dependence among the three vectors $\mathbf{e}_1$, $\mathbf{e}_2$, and $\mathbf{e}_1 + \mathbf{e}_2$ with coefficients 1, 1, and -1, respectively. Any collection of vectors that exhibit such a redundancy with some of the coefficients nonzero is identified by the next definition.

■ Definition 5.3.1

A set of vectors $\mathcal{S}$ is **linearly dependent** means that there exists distinct vectors $\mathbf{u}_1, \mathbf{u}_2, \ldots, \mathbf{u}_n \in \mathcal{S}$ and scalars $r_1, r_2, \ldots, r_n \in \mathbb{R}$ not all 0 such that

$$r_1\mathbf{u}_1 + r_2\mathbf{u}_2 + \cdots + r_n\mathbf{u}_n = \mathbf{0}. \tag{5.14}$$

A set of vectors that is not linearly dependent is **linearly independent**.

To show that a set of vectors is linearly dependent, it is enough to find specific scalars $r_1, r_2, \ldots, r_n$ not all 0 which satisfy (5.14). Linear independence is a bit more subtle. To show that a set of vectors is linearly independent, it must be shown that the vectors are not linearly dependent. This means that it must be shown that for all distinct $\mathbf{u}_1, \mathbf{u}_2, \ldots, \mathbf{u}_n \in \mathcal{S}$ and $r_1, r_2, \ldots, r_n \in \mathbb{R}$,

$$\text{if } r_1\mathbf{u}_1 + r_2\mathbf{u}_2 + \cdots + r_n\mathbf{u}_n = \mathbf{0}, \text{ then } r_1 = 0, \ r_2 = 0, \ \ldots, \ r_n = 0.$$

In other words, the only scalars $r_1, r_2, \ldots, r_n$ that satisfy the equation of (5.14) are the trivial $r_1 = r_2 = \cdots = r_n = 0$, and property (5.14) fails to hold with any other selection of $r_1, r_2, \ldots, r_n$. With this in mind, when are small sets of vectors linearly independent?

■ Theorem 5.3.2

Let V be a vector space and $\mathbf{u}, \mathbf{v} \in V$.

(a) $\varnothing$ is linearly independent.

(b) $\mathbf{v} \neq \mathbf{0}$ if and only if $\{\mathbf{v}\}$ is linearly independent.

(c) $\mathbf{u}, \mathbf{v}$ are not multiples of each other if and only if $\{\mathbf{u}, \mathbf{v}\}$ is linearly independent.

Proof

The statement that $\varnothing$ is linearly independent starts with a universal quantifier with an empty domain. Such a sentence is true by Example 1.2.2. This proves (a).

To prove (b), note that $5\mathbf{0} = \mathbf{0}$ proves that $\{\mathbf{0}\}$ is linearly dependent. Next suppose that $\mathbf{v} \neq \mathbf{0}$ and assume that $r\mathbf{v} = \mathbf{0}$. By Lemma 5.1.2(b), $r = 0$, showing that $\{\mathbf{v}\}$ is linearly independent.

Lastly, to prove (c), let $\mathbf{u} = r\mathbf{v}$ for some $r \in \mathbb{R}$. Then, $\mathbf{u} - r\mathbf{v} = \mathbf{0}$, showing that $\{\mathbf{u}, \mathbf{v}\}$ is linearly dependent. Conversely, if the set is linearly dependent, then $r\mathbf{u} + s\mathbf{v} = \mathbf{0}$, where it can be assumed that $r \neq 0$. This implies that $\mathbf{u} = (s/r)\mathbf{v}$. ■

Euclidean Examples

The first examples of linear dependence and linear independence are from the familiar space of $\mathbb{R}^n$. The techniques used are from Section 3.5.

■ **Example 5.3.3**

In $\mathbb{R}^3$ the set formed by the vectors,

$$\mathbf{u}_1 = \begin{bmatrix} 1 \\ 2 \\ 0 \end{bmatrix}, \quad \mathbf{u}_2 = \begin{bmatrix} 0 \\ 2 \\ 1 \end{bmatrix}, \quad \mathbf{u}_3 = \begin{bmatrix} -1 \\ 0 \\ 1 \end{bmatrix},$$

is linearly dependent. To see this, write

$$1 \begin{bmatrix} 1 \\ 2 \\ 0 \end{bmatrix} - 1 \begin{bmatrix} 0 \\ 2 \\ 1 \end{bmatrix} + 1 \begin{bmatrix} -1 \\ 0 \\ 1 \end{bmatrix} = \begin{bmatrix} 0 \\ 0 \\ 0 \end{bmatrix}.$$

However, $\{\mathbf{u}_1, \mathbf{u}_2\}$ and $\{\mathbf{u}_1\}$ are linearly independent by Theorem 5.3.2.

The constants in the previous example were easy to guess. This is not always the case. Consider a slight modification of the previous example.

■ **Example 5.3.4**

Do the vectors, $\mathbf{v}_1 = \begin{bmatrix} 1 \\ 2 \\ 3 \end{bmatrix}$, $\mathbf{v}_2 = \begin{bmatrix} 4 \\ 5 \\ 6 \end{bmatrix}$, and $\mathbf{v}_3 = \begin{bmatrix} 7 \\ 8 \\ 9 \end{bmatrix}$, form a linearly dependent or linearly independent set? It is not obvious how to find the scalars r_1, r_2, r_3 not all 0 required of Definition 5.3.1, or whether such constants even exist. Suppose, by some method yet unknown, that these constants were found. This would mean that

$$r_1 \begin{bmatrix} 1 \\ 2 \\ 3 \end{bmatrix} + r_2 \begin{bmatrix} 4 \\ 5 \\ 6 \end{bmatrix} + r_3 \begin{bmatrix} 7 \\ 8 \\ 9 \end{bmatrix} = \begin{bmatrix} 0 \\ 0 \\ 0 \end{bmatrix}.$$

This can be written as a homogeneous system of linear equations,

$$\begin{aligned} r_1 &+ 4r_2 &+ 7r_3 &= 0, \\ 2r_1 &+ 5r_2 &+ 8r_3 &= 0, \\ 3r_1 &+ 6r_2 &+ 9r_3 &= 0. \end{aligned} \tag{5.15}$$

Use Gauss–Jordan elimination to take the coefficient matrix of this system,

$$\begin{bmatrix} 1 & 4 & 7 \\ 2 & 5 & 8 \\ 3 & 6 & 9 \end{bmatrix},$$

to its reduced row echelon form,

$$R = \begin{bmatrix} 1 & 0 & -1 \\ 0 & 1 & 2 \\ 0 & 0 & 0 \end{bmatrix}.$$

The matrix R shows that the system of linear equations (5.15) has a nontrivial solution. Hence, $\{\mathbf{v}_1, \mathbf{v}_2, \mathbf{v}_3\}$ is linearly dependent, and because

$$r_1 = t,$$
$$r_2 = -2t,$$
$$r_3 = t,$$

$\mathbf{v}_3 = -\mathbf{v}_1 + 2\mathbf{v}_2$. However, the standard vectors in the first two columns of R show that $\{\mathbf{v}_1, \mathbf{v}_2\}$ is linearly independent.

■ **Example 5.3.5**

Do the vectors in $\mathbb{R}^4$, $\mathbf{u}_1 = \begin{bmatrix} 1 \\ 0 \\ 3 \\ 1 \end{bmatrix}$, $\mathbf{u}_2 = \begin{bmatrix} 3 \\ 2 \\ -5 \\ 1 \end{bmatrix}$, and $\mathbf{u}_3 = \begin{bmatrix} 2 \\ 1 \\ 0 \\ -3 \end{bmatrix}$, form a set that

is linearly dependent or linearly independent? To find all $r_1, r_2, r_3 \in \mathbb{R}$ such that

$$r_1 \begin{bmatrix} 1 \\ 0 \\ 3 \\ 1 \end{bmatrix} + r_2 \begin{bmatrix} 3 \\ 2 \\ -5 \\ 1 \end{bmatrix} + r_3 \begin{bmatrix} 2 \\ 1 \\ 0 \\ -3 \end{bmatrix} = \begin{bmatrix} 0 \\ 0 \\ 0 \\ 0 \end{bmatrix}, \tag{5.16}$$

the system is recast into a homogeneous system of linear equations with coefficient matrix

$$\begin{bmatrix} 1 & 3 & 2 \\ 0 & 2 & 1 \\ 3 & -5 & 0 \\ 1 & 1 & -3 \end{bmatrix}.$$

The reduced row echelon form of this matrix is

$$\begin{bmatrix} 1 & 0 & 0 \\ 0 & 1 & 0 \\ 0 & 0 & 1 \\ 0 & 0 & 0 \end{bmatrix},$$

which implies that (5.16) has only the trivial solution. Therefore, $\{\mathbf{u}_1, \mathbf{u}_2, \mathbf{u}_3\}$ is linearly independent.

In Examples 5.3.4 and 5.3.5, linear dependence and linear independence were determined by finding the solution set of a particular homogeneous system of linear equations. Since every $m \times n$ matrix is the standard matrix of a linear transformation $\mathbb{R}^n \to \mathbb{R}^m$, conditions for linear dependence and linear independence can be written using linear transformations. This suggests the following theorem.

■ **Theorem 5.3.6**

Let $T : \mathbb{R}^n \to \mathbb{R}^m$ be the linear transformation so that $[T] = [\mathbf{u}_1 \ \mathbf{u}_2 \ \cdots \ \mathbf{u}_n]$ has n distinct columns. The set $\{\mathbf{u}_1, \mathbf{u}_2, \dots, \mathbf{u}_n\}$ is linearly independent if and only if $\ker T = \{\mathbf{0}\}$.

Proof

Let $\mathcal{S} = \{\mathbf{u}_1, \mathbf{u}_2, \dots, \mathbf{u}_n\}$. The set $\mathcal{S}$ is linearly independent if and only if

$$r_1\mathbf{u}_1 + r_2\mathbf{u}_2 + \cdots + r_n\mathbf{u}_n = \mathbf{0}$$

has only the trivial solution $r_1 = r_2 = \cdots = r_n = 0$. This is the case if and only if the linear system,

$$[\mathbf{u}_1 \ \mathbf{u}_2 \ \cdots \ \mathbf{u}_n]\begin{bmatrix} r_1 \\ r_2 \\ \vdots \\ r_n \end{bmatrix} = \begin{bmatrix} 0 \\ 0 \\ \vdots \\ 0 \end{bmatrix},$$

has only the trivial solution, which is the same as $\ker T = \{\mathbf{0}\}$. ■

Abstract Vector Space Examples

The next collection of examples of linear dependence and linear independence lie in vector spaces other than $\mathbb{R}^n$. Nonetheless, the work will quickly repeat the techniques used in Examples 5.3.4 and 5.3.5.

■ **Example 5.3.7**

Do $p_1(x) = 1 + x + x^2 + x^3$, $p_2(x) = 1 + x^2$, and $p_3(x) = 1 - x^3$ form a linearly dependent or linearly independent set in $\mathbb{P}_3$? In other words, is it possible to find $r_1, r_2, r_3 \in \mathbb{R}$, not all zero, such that

$$r_1(1 + x + x^2 + x^3) + r_2(1 + x^2) + r_3(1 - x^3) = 0.$$

Collecting like terms leads to the equation,

$$(r_1 + r_2 + r_3) + r_1 x + (r_1 + r_2)x^2 + (r_1 - r_3)x^3 = 0,$$

which is equivalent to the system,

$$\begin{aligned} r_1 + r_2 + r_3 &= 0, \\ r_1 &= 0, \\ r_1 + r_2 &= 0, \\ r_1 - r_3 &= 0. \end{aligned}$$

The coefficient matrix of this system of linear equations,

$$\begin{bmatrix} 1 & 1 & 1 \\ 1 & 0 & 0 \\ 1 & 1 & 0 \\ 1 & 0 & -1 \end{bmatrix},$$

reduces to the row echelon form,

$$\begin{bmatrix} 1 & 1 & 1 \\ 0 & 1 & 1 \\ 0 & 0 & 1 \\ 0 & 0 & 0 \end{bmatrix}.$$

Back substitution shows that this system has only the trivial solution, which proves that $\{p_1(x), p_2(x), p_3(x)\}$ is linearly independent. Of course, the reduced row echelon form of the coefficient matrix yields the same result.

■ Example 5.3.8

The functions $f(x) = \cos 2x$, $g(x) = \cos^2 x$, and $h(x) = 1$ form a linearly dependent set in the vector space $^{\mathbb{R}}\mathbb{R}$ since $f - 2g + h = 0$. The reason for this is that $\cos 2x = 2\cos^2 x - 1$ for all $x \in \mathbb{R}$.

■ Example 5.3.9

Let n be a positive integer. Assume that $\{f_1, f_2, \dots, f_n\}$ is a linearly dependent subset of $C^{n-1}(\mathbb{R})$ (Definition 5.2.9). Take $r_1, r_2, \dots, r_n \in \mathbb{R}$, not all zero, such that

$$r_1 f_1 + r_2 f_2 + \cdots + r_n f_n = 0.$$

This implies that for all $x \in \mathbb{R}$,

$$r_1 f_1(x) + r_2 f_2(x) + \cdots + r_n f_n(x) = 0. \tag{5.17}$$

Since the functions have their first $n - 1$ derivatives, continually differentiate (5.17) to find that for all $x \in \mathbb{R}$,

$$\begin{array}{ccccccc} r_1 f_1(x) & + & r_2 f_2(x) & + & \cdots & + & r_n f_n(x) & = & 0, \\ r_1 f_1'(x) & + & r_2 f_2'(x) & + & \cdots & + & r_n f_n'(x) & = & 0, \\ \vdots & & \vdots & & \vdots & & \vdots & & \\ r_1 f_1^{(n-1)}(x) & + & r_2 f_2^{(n-1)}(x) & + & \cdots & + & r_n f_n^{(n-1)}(x) & = & 0. \end{array} \tag{5.18}$$

Hence, the system of (5.18) has a nontrivial solution for all $x \in \mathbb{R}$, which implies by Theorem 4.3.11 that for all $x \in \mathbb{R}$,

$$w(x) = \begin{vmatrix} f_1(x) & f_2(x) & \cdots & f_n(x) \\ f_1'(x) & f_2'(x) & \cdots & f_n'(x) \\ \vdots & \vdots & & \vdots \\ f_1^{(n-1)}(x) & f_2^{(n-1)}(x) & \cdots & f_n^{(n-1)}(x) \end{vmatrix} = 0.$$

The function w is called the **Wronskian**, and it was just shown that

$$\begin{array}{l} \text{if } w(x) \neq 0 \text{ for some } x \in \mathbb{R}, \\ \{f_1, f_2, \dots, f_n\} \text{ is linearly independent.} \end{array} \tag{5.19}$$

To illustrate, $\{\sin, \cos\}$ is linearly independent because

$$\begin{vmatrix} \sin 0 & \cos 0 \\ \cos 0 & -\sin 0 \end{vmatrix} = \begin{vmatrix} 0 & 1 \\ 1 & 0 \end{vmatrix} = -1.$$

However, the converse of (5.19) is false. Consider the functions $f(x) = x^2$ and $g(x) = x|x|$. The set $\{f, g\}$ is linearly independent (Exercise 17). However, clearly $w(x) = 0$ for all $x \geq 0$, and for $x < 0$,

$$w(x) = \begin{vmatrix} x^2 & -x^2 \\ 2x & -2x \end{vmatrix} = 0.$$

Suppose that $T : V \to W$ is a linear transformation. It is not true that the image of a linearly independent collection of vectors is necessarily linearly independent. A trivial example is the zero transformation on $\mathbb{R}^2$ which maps the linearly independent $\{\mathbf{e}_1, \mathbf{e}_2\}$ to the linearly dependent $\{\mathbf{0}\}$. Linear dependence, however, is preserved under linear transformations.

■ **Theorem 5.3.10**

Suppose that $T : V \to W$ is a linear transformation.

   (a) If $\{\mathbf{u}_1, \mathbf{u}_2, \dots, \mathbf{u}_n\} \subset V$ is linearly dependent, $\{T\mathbf{u}_1, T\mathbf{u}_2, \dots, T\mathbf{u}_n\}$ is linearly dependent.

   (b) If $\{T\mathbf{u}_1, T\mathbf{u}_2, \dots, T\mathbf{u}_n\} \subset W$ is linearly independent, $\{\mathbf{u}_1, \mathbf{u}_2, \dots, \mathbf{u}_n\}$ is linearly independent.

Proof

For (a), suppose that $\{\mathbf{u}_1, \mathbf{u}_2, \dots, \mathbf{u}_n\}$ is linearly dependent so that

$$r_1\mathbf{u}_1 + r_2\mathbf{u}_2 + \cdots + r_n\mathbf{u}_n = \mathbf{0}$$

for some $r_1, r_2, \dots, r_n \in \mathbb{R}$ with not all r_i equal to 0. Because T is a linear transformation, applying T to both sides yields

$$r_1 T\mathbf{u}_1 + r_2 T\mathbf{u}_2 + \cdots + r_n T\mathbf{u}_n = T\mathbf{0} = \mathbf{0}$$

with, again, not all r_i equal to 0. Hence, the set of images $\{T\mathbf{u}_1, T\mathbf{u}_2, \dots, T\mathbf{u}_n\}$ is linearly dependent. Observe that (b) follows immediately as the contrapositive of (a). ■

Exercises

1. Prove that the given sets of vectors are linearly dependent. Write one of the vectors in each set as a linear combination of the other vectors in the set.

   (a) $\left\{ \begin{bmatrix} 1 \\ 2 \\ 5 \end{bmatrix}, \begin{bmatrix} 0 \\ 0 \\ 0 \end{bmatrix}, \begin{bmatrix} 2 \\ 0 \\ 1 \end{bmatrix} \right\}$

(b) $\left\{ \begin{bmatrix} 0 & -1 \\ 0 & -1 \end{bmatrix}, \begin{bmatrix} 3 & -3 \\ -2 & -2 \end{bmatrix}, \begin{bmatrix} 1 & -2 \\ 0 & 3 \end{bmatrix}, \begin{bmatrix} 3 & -4 \\ 0 & -1 \end{bmatrix}, \begin{bmatrix} 0 & 1 \\ 3 & 3 \end{bmatrix} \right\}$

(c) $\{5 + 3x + 6x^2 + 3x^3, 9 + 5x + 8x^2 + 5x^3, 1 + x + 4x^2 + x^3\}$

2. Prove that the given sets of vectors are linearly independent.

(a) $\left\{ \begin{bmatrix} 2 \\ 5 \\ 1 \end{bmatrix}, \begin{bmatrix} 3 \\ 1 \\ 3 \end{bmatrix}, \begin{bmatrix} 0 \\ 5 \\ 4 \end{bmatrix} \right\}$

(b) $\left\{ \begin{bmatrix} 6 & 3 \\ 3 & 6 \end{bmatrix}, \begin{bmatrix} 5 & 1 \\ 5 & 3 \end{bmatrix}, \begin{bmatrix} 0 & 2 \\ 0 & 1 \end{bmatrix}, \begin{bmatrix} 4 & 1 \\ 1 & 1 \end{bmatrix} \right\}$

(c) $\{4 + 3x, 2 + 3x^2, 4 + x\}$

3. Verify Theorem 5.3.2 by directly proving that the given sets are linearly independent or linearly dependent.

(a) $\left\{ \begin{bmatrix} 0 \\ 0 \end{bmatrix} \right\}$

(b) $\left\{ \begin{bmatrix} 6 \\ -2 \end{bmatrix}, \begin{bmatrix} -3 \\ 1 \end{bmatrix} \right\}$

(c) $\left\{ \begin{bmatrix} 0 & 0 \\ 0 & 0 \end{bmatrix}, \begin{bmatrix} 1 & 2 \\ 3 & 4 \end{bmatrix} \right\}$

(d) $\left\{ \begin{bmatrix} 1 \\ 5 \end{bmatrix} \right\}$

(e) $\left\{ \begin{bmatrix} 6 \\ -2 \end{bmatrix}, \begin{bmatrix} 3 \\ 1 \end{bmatrix} \right\}$

(f) $\{1 + x + x^3, 3 - x + 5x^2\}$

4. Determine whether the given sets of vectors are linearly dependent or linearly independent. If the set is linearly dependent, write one of the vectors in the set as a linear combination of the other vectors in the set.

(a) $\left\{ \begin{bmatrix} 1 \\ 2 \end{bmatrix}, \begin{bmatrix} 3 \\ 4 \end{bmatrix}, \begin{bmatrix} 4 \\ 5 \end{bmatrix}, \begin{bmatrix} 6 \\ 7 \end{bmatrix} \right\}$

(b) $\left\{ \begin{bmatrix} 3 & 3 & 2 \\ 5 & 2 & 4 \end{bmatrix}, \begin{bmatrix} 0 & 6 & 6 \\ 4 & 0 & 6 \end{bmatrix}, \begin{bmatrix} 0 & 2 & 6 \\ 4 & 1 & 3 \end{bmatrix} \right\}$

(c) $\{5 + 2x + 4x^2 + 3x^3 + 6x^4, 2x + 3x^2 + 3x^3 + 2x^4, 5x + 4x^2 + 3x^3 + 3x^4\}$

5. Given each linear transformation T, verify Theorem 5.3.6 by directly showing one of the following:

- The columns of $[T]$ form a linearly independent set and $\ker T = \{0\}$.

- The columns of $[T]$ form a linearly dependent set and $\ker T \neq \{0\}$.

(a) $T : \mathbb{R}^2 \to \mathbb{R}^3$, $T \begin{bmatrix} x \\ y \end{bmatrix} = \begin{bmatrix} 3x - 2y \\ x + y \\ 5x - 5y \end{bmatrix}$

(b) $T : \mathbb{R}^2 \to \mathbb{R}^3$, $T \begin{bmatrix} x \\ y \end{bmatrix} = \begin{bmatrix} x - 3y \\ 3x - 9y \\ -2x + 6y \end{bmatrix}$

(c) $T : \mathbb{R}^3 \to \mathbb{R}^2$, $T \begin{bmatrix} x \\ y \\ z \end{bmatrix} = \begin{bmatrix} 3x - y + z \\ x + y - 2z \\ x - z \end{bmatrix}$

(d) $T : \mathbb{R}^3 \to \mathbb{R}^3$, $T \begin{bmatrix} x \\ y \\ z \end{bmatrix} = \begin{bmatrix} 4x - 3y + 7z \\ 7x + 2y + 5z \\ 8x + y + 7z \end{bmatrix}$.

(e) $T : \mathbb{R}^2 \to \mathbb{R}^2$, $T \begin{bmatrix} x \\ y \end{bmatrix} = \begin{bmatrix} 2x - y \\ x - 2y \end{bmatrix}$.

6. Prove that a set $\mathcal{S}$ of at least two vectors is linearly independent if and only if no vector in $\mathcal{S}$ can be written as a linear combination of the other vectors in $\mathcal{S}$.

7. Let V be a vector space and $\mathbf{u}, \mathbf{v}, \mathbf{w} \in V$. If $\{\mathbf{u}, \mathbf{v}, \mathbf{w}\}$ is linearly independent, show that $\{\mathbf{u} + \mathbf{v}, \mathbf{u} - \mathbf{v}, \mathbf{u} - 2\mathbf{v} + \mathbf{w}\}$ is also linearly independent.

8. Find $r \in \mathbb{R}$ such that the set $\{\mathbf{u}_2 - \mathbf{u}_1, r\mathbf{u}_3 - \mathbf{u}_2, \mathbf{u}_1 - \mathbf{u}_3\}$ is linearly independent given that $\{\mathbf{u}_1, \mathbf{u}_2, \mathbf{u}_3\}$ is a linearly independent subset of the vector space V.

9. Let V be a vector space and $\{\mathbf{u}_1, \mathbf{u}_2, \mathbf{u}_3, \mathbf{u}_4\} \subseteq V$ be linearly independent. Prove that the given sets are linearly independent or find a counterexample.

   (a) $\{\mathbf{u}_1 + \mathbf{u}_2, \mathbf{u}_2 + \mathbf{u}_3, \mathbf{u}_3 + \mathbf{u}_4, \mathbf{u}_4 + \mathbf{u}_1\}$

   (b) $\{\mathbf{u}_1 + \mathbf{u}_2, \mathbf{u}_2 + \mathbf{u}_3, \mathbf{u}_3 + \mathbf{u}_4, \mathbf{u}_4 - \mathbf{u}_1\}$

10. Let $T : \mathbb{R}^n \to \mathbb{R}^n$ be a linear operator. Prove that the following are equivalent.

    - $[T]$ is invertible.
    - The columns of $[T]$ form a linearly independent set.
    - The columns of the reduced row echelon form of $[T]$ are standard vectors.
    - $\ker T = \{\mathbf{0}\}$
    - $\operatorname{ran} T = \mathbb{R}^n$.

11. Show that the converse of Theorem 5.3.10(b) is false. What additional assumption is required for T so that $\{\mathbf{u}_1, \mathbf{u}_2, \dots, \mathbf{u}_n\}$ being linearly independent in V implies that $\{T\mathbf{u}_1, T\mathbf{u}_2, \dots, T\mathbf{u}_n\}$ is independent in W? Prove the result.

12. Let $\mathcal{S}_1 \subseteq \mathcal{S}_2$ be subsets of the vector space V. Prove the following.

    (a) If $\mathcal{S}_1$ is linearly dependent, $\mathcal{S}_2$ is linearly dependent.

    (b) If $\mathcal{S}_2$ is linearly independent, $\mathcal{S}_1$ is linearly independent.

13. Let $\mathcal{S} = \{\mathbf{u}_1, \mathbf{u}_2, \dots, \mathbf{u}_n\}$ be a linearly independent subset of the vector space V. Prove that if $\mathbf{u} \in V \setminus \mathsf{S}(\mathcal{S})$, then $\mathcal{S} \cup \{\mathbf{u}\}$ is linearly independent.

14. Let a, b, c be distinct real numbers, prove the linear independence of

$$\left\{ \begin{bmatrix} 1 \\ 1 \\ 1 \end{bmatrix}, \begin{bmatrix} a \\ b \\ c \end{bmatrix}, \begin{bmatrix} a^2 \\ b^2 \\ c^2 \end{bmatrix} \right\}.$$

15. Let A be an $m \times n$ matrix and E an $m \times m$ elementary matrix. Prove that if the columns of A form a linearly independent set and E does not have a row of zeros, the columns of EA form a linearly independent set.

16. Prove that the nonzero rows of a matrix in reduced row echelon form give a linearly independent set.

17. Let $f(x) = x^2$ and $g(x) = x|x|$. Prove that $\{f, g\}$ is linearly independent.

18. Use the Wronskian of Example 5.3.9 to determine whether $\{\sin, \sin^2, \cos^2, 1\}$ and $\{\sin, \sin^2\}$ are linearly independent or linearly dependent.

5.4 Basis and Dimension

The vectors $\mathbf{e}_1, \mathbf{e}_2, \mathbf{e}_3 \in \mathbb{R}^3$ can be used to describe every vector in $\mathbb{R}^3$ as a linear combination. Indeed, any vector $\begin{bmatrix} r_1 \\ r_2 \\ r_3 \end{bmatrix} \in \mathbb{R}^3$ is described as

$$\begin{bmatrix} r_1 \\ r_2 \\ r_3 \end{bmatrix} = r_1 \mathbf{e}_1 + r_2 \mathbf{e}_2 + r_3 \mathbf{e}_3. \tag{5.20}$$

In other words, $\{\mathbf{e}_1, \mathbf{e}_2, \mathbf{e}_3\}$ is a spanning set for $\mathbb{R}^3$, so write $\mathbb{R}^3 = \mathrm{S}(\mathbf{e}_1, \mathbf{e}_2, \mathbf{e}_3)$. Furthermore, any vector in $\mathbb{R}^3$ has exactly one such representation because the system of equations derived from (5.20) can be represented by

$$\begin{bmatrix} 1 & 0 & 0 & r_1 \\ 0 & 1 & 0 & r_2 \\ 0 & 0 & 1 & r_3 \end{bmatrix}.$$

These properties of $\mathbb{R}^3$ are found in all vector spaces.

Basis

In summary, any vector in $\mathbb{R}^3$ can be written in exactly one way as a linear combination of the vectors $\mathbf{e}_1, \mathbf{e}_2, \mathbf{e}_3$. The argument supporting this conclusion relied on two key facts. That $\{\mathbf{e}_1, \mathbf{e}_2, \mathbf{e}_3\}$ is linearly independent and that $\{\mathbf{e}_1, \mathbf{e}_2, \mathbf{e}_3\}$ is a spanning set for $\mathbb{R}^3$.

■ **Definition 5.4.1**

A linearly independent set of vectors that spans a vector space V is called a **basis** for V.

■ **Example 5.4.2**

The vector space $\{\mathbf{0}\}$ has only two subsets: $\emptyset$, which is not a spanning set of $\{\mathbf{0}\}$ yet is linearly independent by Theorem 5.3.2(a), and $\{\mathbf{0}\}$, which is linearly dependent by Theorem 5.3.2(b) but is a spanning set. This means that $\{\mathbf{0}\}$ has no subset that is both linearly independent and a spanning set, but for reasons that will be made clear, $\emptyset$ is defined to be the basis of $\{\mathbf{0}\}$.

■ **Example 5.4.3**

Each of the following sets of vectors are considered a **standard basis** for their respective vector spaces.

- $\{\mathbf{e}_1, \mathbf{e}_2, \dots, \mathbf{e}_n\}$ is the standard basis for $\mathbb{R}^n$.
- $\{1, x, x^2, \dots, x^n\}$ is the standard basis for $\mathbb{P}_n$.
- $\{1, x, x^2, x^3 \dots\}$ is the standard basis for $\mathbb{P}$.
- $\left\{ \begin{bmatrix} 1 & 0 \\ 0 & 0 \end{bmatrix}, \begin{bmatrix} 0 & 1 \\ 0 & 0 \end{bmatrix}, \begin{bmatrix} 0 & 0 \\ 1 & 0 \end{bmatrix}, \begin{bmatrix} 0 & 0 \\ 0 & 1 \end{bmatrix} \right\}$ is the standard basis for $\mathbb{R}^{2 \times 2}$.

■ **Example 5.4.4**

The set of vectors $\mathscr{B} = \left\{ \begin{bmatrix} 1 \\ 1 \\ 0 \end{bmatrix}, \begin{bmatrix} 0 \\ 1 \\ 1 \end{bmatrix}, \begin{bmatrix} 1 \\ 0 \\ 1 \end{bmatrix} \right\}$ is a basis for $\mathbb{R}^3$. First, $\mathscr{B}$ is linearly independent because if $r_1, r_2, r_3 \in \mathbb{R}$ such that

$$r_1 \begin{bmatrix} 1 \\ 1 \\ 0 \end{bmatrix} + r_2 \begin{bmatrix} 0 \\ 1 \\ 1 \end{bmatrix} + r_3 \begin{bmatrix} 1 \\ 0 \\ 1 \end{bmatrix} = \begin{bmatrix} 0 \\ 0 \\ 0 \end{bmatrix},$$

it must be the case that $r_1 = r_2 = r_3 = 0$ as in Example 5.3.4. Second, $\mathscr{B}$ is a spanning set because the equation,

$$r_1 \begin{bmatrix} 1 \\ 1 \\ 0 \end{bmatrix} + r_2 \begin{bmatrix} 0 \\ 1 \\ 1 \end{bmatrix} + r_3 \begin{bmatrix} 1 \\ 0 \\ 1 \end{bmatrix} = \begin{bmatrix} b_1 \\ b_2 \\ b_3 \end{bmatrix},$$

which is written in matrix form as

$$\begin{bmatrix} 1 & 0 & 1 & | & b_1 \\ 1 & 1 & 0 & | & b_2 \\ 0 & 1 & 1 & | & b_3 \end{bmatrix},$$

always has a solution because it reduces to

$$\begin{bmatrix} 1 & 0 & 0 & | & (b_1 + b_2 - b_3)/2 \\ 0 & 1 & 0 & | & (-b_1 + b_2 + b_3)/2 \\ 0 & 0 & 1 & | & (b_1 - b_2 + b_3)/2 \end{bmatrix}.$$

This shows that $\mathscr{B}$ is a basis of $\mathbb{R}^3$.

■ **Example 5.4.5**

Let V be a vector space with basis $\mathcal{B} = \{\mathbf{b}_1, \mathbf{b}_2, \dots, \mathbf{b}_n\}$. Define the **dual** of V as

$$V^* = \{T : T \text{ is a linear transformation } V \rightarrow \mathbb{R}\}.$$

Each element of V^* is called a **linear functional**. Let $\mathcal{B}^* = \{b_1^*, b_2^*, \dots, b_n^*\}$ be the subset of the dual of V such that for all $i, j = 1, 2, \dots, n$,

$$b_i^* \mathbf{b}_j = \delta_{ij}, \tag{5.21}$$

where δ_{ij} (the **Kronecker delta**) is defined by

$$\delta_{ij} = \begin{cases} 0 & \text{if } i \neq j, \\ 1 & \text{if } i = j. \end{cases}$$

- Let $r_1, r_2, \dots, r_n \in \mathbb{R}$ such that

$$r_1 b_1^* + r_2 b_2^* + \cdots + r_n b_n^* = 0,$$

where 0 is the zero transformation $V \rightarrow \mathbb{R}$. Then, for $i = 1, 2, \dots, n$,

$$(r_1 b_1^* + r_2 b_2^* + \cdots + r_n b_n^*)\mathbf{b}_i = 0\mathbf{b}_i = 0.$$

However,

$$(r_1 b_1^* + r_2 b_2^* + \cdots + r_n b_n^*)\mathbf{b}_i = r_1 b_1^* \mathbf{b}_i + r_2 b_2^* \mathbf{b}_i \cdots + r_n b_n^* \mathbf{b}_i = r_i b_i^* \mathbf{b}_i$$

because $b_j^* \mathbf{b}_i = 0$ when $i \neq j$. Since $b_i^* \mathbf{b}_i = 1$, it follows that $r_i = 0$, proving that $\mathcal{B}^*$ is linearly independent.

- Let $T : V \rightarrow \mathbb{R}$ be a linear transformation. Define $s_i = T\mathbf{b}_i$ for every $i = 1, 2, \dots, n$. Take $\mathbf{u} \in V$ and write

$$\mathbf{u} = r_1 \mathbf{b}_1 + r_2 \mathbf{b}_2 + \cdots + r_n \mathbf{b}_n$$

for some $r_1, r_2, \dots, r_n \in \mathbb{R}$. Because of (5.21),

$$\begin{aligned} T\mathbf{u} &= T(r_1 \mathbf{b}_1 + r_2 \mathbf{b}_2 + \cdots + r_n \mathbf{b}_n) \\ &= r_1 T\mathbf{b}_1 + r_2 T\mathbf{b}_2 + \cdots + r_n T\mathbf{b}_n \\ &= r_1 s_1 + r_2 s_2 + \cdots + r_n s_n \\ &= r_1 s_1 b_1^* \mathbf{b}_1 + r_2 s_2 b_2^* \mathbf{b}_2 + \cdots + r_n s_n b_n^* \mathbf{b}_n \\ &= s_1 b_1^* \mathbf{u} + s_2 b_2^* \mathbf{u} + \cdots + s_n b_n^* \mathbf{u} \\ &= (s_1 b_1^* + s_2 b_2^* + \cdots + s_n b_n^*)\mathbf{u}. \end{aligned}$$

This implies that $T = s_1 b_1^* + s_2 b_2^* + \cdots + s_n b_n^*$. Hence, $\mathcal{B}^*$ spans V^*. Therefore, $\mathcal{B}^*$ is a basis of V^*. Call $\mathcal{B}^*$ the **dual basis** for V.

A basis for a vector space V is defined as a linearly independent spanning set. If a vector is removed from a basis, the resulting set of vectors is no longer a basis since it is no longer a spanning set. If any vector is added to the basis, the resulting set of vectors is no longer a basis since it is not linearly independent. Loosely speaking, a basis has exactly the right number of linearly independent elements to span the vector space. The next two results make this idea precise.

■ Lemma 5.4.6

Suppose that V is a vector space. If $\{\mathbf{u}_1, \mathbf{u}_2, \ldots, \mathbf{u}_m\}$ is linearly independent and $\{\mathbf{v}_1, \mathbf{v}_2, \ldots, \mathbf{v}_n\}$ is a spanning set of V, then $m \leq n$.

Proof

Towards contradiction, suppose that $m > n$. Since $\{\mathbf{u}_1, \mathbf{u}_2, \ldots, \mathbf{u}_m\}$ is a linearly independent set and $m > n$, it must be true that the set of the first $n+1$ vectors $\{\mathbf{u}_1, \mathbf{u}_2, \ldots, \mathbf{u}_{n+1}\}$ is linearly independent [Exercise 12(b) of Section 5.3]. Since $\{\mathbf{v}_1, \mathbf{v}_2, \ldots, \mathbf{v}_n\}$ is a spanning set of V, expand each of these $\mathbf{u}_i$ with the scalars a_{ij} ($i = 1, 2, \ldots, n$ and $j = 1, 2, \ldots, n+1$):

$$\mathbf{u}_1 = a_{11}\mathbf{v}_1 + a_{21}\mathbf{v}_2 + \cdots + a_{n1}\mathbf{v}_n,$$
$$\mathbf{u}_2 = a_{12}\mathbf{v}_1 + a_{22}\mathbf{v}_2 + \cdots + a_{n2}\mathbf{v}_n,$$
$$\vdots$$
$$\mathbf{u}_n = a_{1,n}\mathbf{v}_1 + a_{2,n}\mathbf{v}_2 + \cdots + a_{n,n}\mathbf{v}_n,$$
$$\mathbf{u}_{n+1} = a_{1,n+1}\mathbf{v}_1 + a_{2,n+1}\mathbf{v}_2 + \cdots + a_{n,n+1}\mathbf{v}_n.$$

Because $\{\mathbf{u}_1, \mathbf{u}_2, \ldots, \mathbf{u}_{n+1}\}$ is linearly independent,

$$r_1\mathbf{u}_1 + r_2\mathbf{u}_2 + \cdots + r_{n+1}\mathbf{u}_{n+1} = \mathbf{0} \tag{5.22}$$

has only the trivial solution $r_1 = r_2 = \cdots = r_{n+1} = 0$. Upon substituting for the $\mathbf{u}_i$, this equation becomes

$$r_1 \sum_{i=1}^{n} a_{i1}\mathbf{v}_i + r_2 \sum_{i=1}^{n} a_{i2}\mathbf{v}_i + \cdots + r_{n+1} \sum_{i=1}^{n} a_{i,n+1}\mathbf{v}_i = \mathbf{0},$$

which simplifies to

$$\left(\sum_{i=1}^{n+1} r_i a_{1i} \right)\mathbf{v}_1 + \left(\sum_{i=1}^{n+1} r_i a_{2i} \right)\mathbf{v}_2 + \cdots + \left(\sum_{i=1}^{n+1} r_i a_{ni} \right)\mathbf{v}_n = \mathbf{0}.$$

This system is satisfied by any $r_1, r_2, \ldots, r_n$ which satisfy the related system of equations written in matrix form as

$$
\begin{bmatrix}
a_{11} & a_{12} & \cdots & a_{1n} & a_{1,n+1} \\
a_{21} & a_{22} & \cdots & a_{2n} & a_{2,n+1} \\
\vdots & \vdots & & \vdots & \vdots \\
a_{n1} & a_{n2} & \cdots & a_{nn} & a_{n,n+1}
\end{bmatrix}
\begin{bmatrix}
r_1 \\
r_2 \\
\vdots \\
r_n \\
r_{n+1}
\end{bmatrix}
=
\begin{bmatrix}
0 \\
0 \\
\vdots \\
0 \\
0
\end{bmatrix}.
$$

However, this is an underdetermined homogeneous linear system of n equations in $n + 1$ unknowns. It must, therefore, have a nontrivial solution for $r_1, r_2, \ldots, r_n$. These scalars also satisfy (5.22), which contradicts the fact that $\{\mathbf{u}_1, \mathbf{u}_2, \ldots, \mathbf{u}_n\}$ is linearly independent. ■

■ Theorem 5.4.7

If the vector space V has a basis consisting of n vectors, every basis of V contains exactly n vectors.

Proof

Suppose that $\mathscr{B}_1 = \{\mathbf{u}_1, \mathbf{u}_2, \ldots, \mathbf{u}_m\}$ and $\mathscr{B}_2 = \{\mathbf{v}_1, \mathbf{v}_2, \ldots, \mathbf{v}_n\}$ are each a basis for V. Because $\mathscr{B}_1$ is linearly independent and $\mathscr{B}_2$ spans V, $m \leq n$ by Lemma 5.4.6. On the other hand, $\mathscr{B}_2$ is linearly independent and $\mathscr{B}_1$ spans V, so $n \leq m$ by another application of Lemma 5.4.6. Hence, $m = n$. ■

Zorn's Lemma

From Example 5.4.3 and other work in this section, it appears that every vector space has a basis. The proof of this fact requires some technical results from set theory, beginning with the following definition.

■ Definition 5.4.8

A family of sets $\mathscr{C}$ is a **chain** if $A \subseteq B$ or $B \subseteq A$ for all $A, B \in \mathscr{C}$.

■ Example 5.4.9

Define a set of open intervals $\mathscr{C} = \{(-n, n) : n \in \mathbb{Z}^+\}$. Let $m_1, m_2 \in \mathbb{Z}^+$. If $m_1 \leq m_2$, then $(-m_1, m_1) \subseteq (-m_2, m_2)$, else $(-m_2, m_2) \subseteq (-m_1, m_1)$, proving that $\mathscr{C}$ is a chain. In addition, it should be noted that $\bigcup \mathscr{C} = \mathbb{R}$ and $\bigcap \mathscr{C} = (-1, 1)$, the proof of which is left to Exercise 20.

■ Example 5.4.10

The family of sets $\mathscr{P} = \{\mathbb{P}_n : n \in \mathbb{N}\}$ is a chain in $\mathbb{P}$ and

$$\bigcup \mathscr{P} = \mathbb{P}.$$

Because $\{1, x, x^2, \ldots, x^n\}$ is a basis of $\mathbb{P}_n$, the chain $\mathscr{P}$ can be written as

$$\mathscr{P} = \{S(1), S(1, x), S(1, x, x^2), \ldots, S(1, x, x^2, \ldots, x^n), \ldots\}.$$

The chains of Examples 5.4.9 and 5.4.10 are infinite and have no element that contains all of the other elements of the chain. That is, the chains do not possess the following set.

■ **Definition 5.4.11**

Let $\mathscr{F}$ be a family of sets. The set $M \in \mathscr{F}$ is a **maximal element** means that for all $A \in \mathscr{F}$, if $M \subseteq A$, then $M = A$.

Think of a maximal element as a set that is not a proper subset of any other set in the family. Notice that a maximal element may not be a **maximum**, a set $G \in \mathscr{F}$ with the property that $B \subseteq G$ for all $B \in \mathscr{F}$. In fact, $\mathscr{F}$ can have many maximal elements, but $\mathscr{F}$ can have at most one maximum. Furthermore, a maximum is a maximal element, but a maximal element may not be a maximum.

■ **Example 5.4.12**

Let A be a set. Define the **power set** of A by

$$P(A) = \{B : B \subseteq A\}.$$

The maximum of $P(A)$ is A because $A \in P(A)$ and all elements of $P(A)$ are subsets of A. In addition, if A has at least two elements, $P(A)$ is not a chain. For example,

$$P(\{1, 2\}) = \{\varnothing, \{1\}, \{2\}, \{1, 2\}\},$$

but $P(\{1, 2\})$ does contain chains. Two examples are $\{\{1\}, \{1, 2\}\}$ and $\{\varnothing, \{2\}\}$. Notice that the union of both of these chains belongs to $P(A)$. Specifically,

$$\bigcup\{\{1\}, \{1, 2\}\} = \{1, 2\} \quad \text{and} \quad \bigcup\{\varnothing, \{2\}\} = \{2\}.$$

In fact, the union of every chain in $P(A)$, or the union of every subset for that matter, is an element of $P(A)$. To prove this, let $\mathscr{C} \subseteq P(A)$. Take $a \in \bigcup \mathscr{C}$. This implies that $a \in C$ for some $C \in \mathscr{C}$. Because $C \subseteq A$, it follows that $a \in A$. Thus, $\bigcup \mathscr{C} \subseteq A$, so $\bigcup \mathscr{C} \in P(A)$.

■ **Example 5.4.13**

Consider the family of sets $\mathscr{F} = P(A) \setminus \{A\}$. Unlike $P(A)$, this family has no maximum, but if A contains at least two elements, $\mathscr{F}$ does have maximal elements. To illustrate, suppose $A = \{1, 2, 3\}$. The following chains in $\mathscr{F}$ have three elements:

$$\varnothing \subset \{1\} \subset \{1, 2\}, \quad \varnothing \subset \{1\} \subset \{1, 3\}, \quad \varnothing \subset \{2\} \subset \{1, 2\},$$
$$\varnothing \subset \{2\} \subset \{2, 3\}, \quad \varnothing \subset \{3\} \subset \{1, 3\}, \quad \varnothing \subset \{3\} \subset \{2, 3\}.$$

The maximal elements of $\mathscr{F}$ are $\{1, 2\}$, $\{1, 3\}$, and $\{2, 3\}$ because there are no elements of $\mathscr{F}$ that properly contain any of these three sets.

The families of sets found in Examples 5.4.12 and 5.4.13 are easy to examine and find maximal elements because they are finite. If the family is infinite and a maximal element needs to be found, it is common to rely on the following axiom. In this context, its title appears to be a misnomer. The axiom is equivalent to other statements that are sometimes listed among the axioms of set theory, so although it is an assumption here, in other texts it follows as a theorem. For this reason, its traditional name is reasonable and will be used here.

■ **Axiom 5.4.14 [Zorn's Lemma]**

If $\mathcal{S}$ is a nonempty family of sets such that $\bigcup \mathcal{C} \in \mathcal{S}$ for all chains $\mathcal{C} \subseteq \mathcal{S}$, then $\mathcal{S}$ contains a maximal element.

The purpose of stating Zorn's Lemma is that it is required to prove the expected result about vector spaces. It does not matter the size of the basis of the vector space. The result holds for both finite bases and infinite bases, but it is the infinite case that requires the technical Zorn's Lemma. In fact, Zorn's Lemma and the next theorem are equivalent.

■ **Theorem 5.4.15**

Every vector space has a basis.

Proof

Let V be a vector space. If $V = \{\mathbf{0}\}$, its basis is $\varnothing$, so suppose that $V \neq \{\mathbf{0}\}$ and let $\mathcal{S}$ be the set of all linearly independent subsets of V. Notice that $\mathcal{S}$ is not empty because $\{\mathbf{v}\} \in \mathcal{S}$ for all nonzero $\mathbf{v} \in V$ [Theorem 5.3.2(b)]. Suppose that $\mathcal{C} \subseteq \mathcal{S}$ is a chain. To show that $\bigcup \mathcal{C}$ is linearly independent, take $\mathbf{u}_1, \mathbf{u}_2, \ldots, \mathbf{u}_k \in \bigcup \mathcal{C}$. Since $\mathcal{C}$ is a chain, there exists $C \in \mathcal{C}$ such that $\mathbf{u}_1, \mathbf{u}_2, \ldots, \mathbf{u}_k \in C$. Therefore, $\{\mathbf{u}_1, \mathbf{u}_2, \ldots, \mathbf{u}_k\}$ is linearly independent because it is a subset of a linearly independent set of vectors. This implies that $\bigcup \mathcal{C}$ is an element of $\mathcal{S}$. By Zorn's Lemma (5.4.14), there exists a maximal element $\mathcal{B} \in \mathcal{S}$. It should be noted that $\mathcal{B}$ is nonempty. To show that $\mathcal{B}$ spans V, let $\mathbf{v} \in V \backslash S(\mathcal{B})$ and $\mathbf{v}_1, \mathbf{v}_2, \ldots, \mathbf{v}_k \in \mathcal{B}$. Assume that there exists $r, r_1, \ldots, r_k \in \mathbb{R}$ such that $r\mathbf{v} + r_1\mathbf{v}_1 + \cdots + r_k\mathbf{v}_k = \mathbf{0}$. If $r \neq 0$, then

$$\mathbf{v} = -\frac{r_1}{r}\mathbf{v}_1 - \frac{r_2}{r}\mathbf{v}_2 - \cdots - \frac{r_k}{r}\mathbf{v}_k,$$

which implies that $\mathbf{v} \in S(\mathcal{B})$, a contradiction. It follows that $r = 0$, so because $\mathcal{B}$ is linearly independent, $r_1 = r_2 = \cdots = r_k = 0$. Thus, $\mathcal{B} \cup \{\mathbf{v}\}$ is linearly independent, which contradicts the maximality if $\mathcal{B}$. Therefore, $V \backslash S(\mathcal{B})$ is empty, so $\mathcal{B}$ is a spanning set of V, and the conclusion that $\mathcal{B}$ is a basis for V is reached. ■

Dimension

It is understood that $\mathbb{R}^n$ has n dimensions. The question is how is this generalized to an arbitrary vector space? Since the size of a basis is invariant (Theorem 5.4.7), the answer is to use the basis.

■ **Definition 5.4.16**

Let V be a vector space. The **dimension** of V, denoted by $\dim V$, is the number of vectors in a basis of V. If a basis of V consists of a finite number of vectors, V is called **finite dimensional**, otherwise V is **infinite dimensional**. If the dimension of V is n, then V is called **n-dimensional**.

Therefore, $\mathbb{R}^n$ has dimension n because the standard vectors, $\mathbf{e}_1, \mathbf{e}_2, \ldots, \mathbf{e}_n$, form a basis for $\mathbb{R}^n$. Also, $\dim\{\mathbf{0}\} = 0$ because $\varnothing$ is the basis of $\{\mathbf{0}\}$. Also,

$$\dim \mathbb{P}_n = n + 1,$$

because $\{1, x, x^2, \ldots, x^n\}$ is its basis, and

$$\dim \mathbb{R}^{m \times n} = mn$$

because there are mn distinct $m \times n$ matrices consisting of a 1 in an entry and a 0 in all other entries. Since every vector space has a basis by Theorem 5.4.15, every vector space has a dimension.

■ **Example 5.4.17**

The subspace of $\mathbb{R}^{2 \times 2}$ from Example 5.2.2, $W = \left\{ \begin{bmatrix} x & 0 \\ y & x + y \end{bmatrix} : x, y \in \mathbb{R} \right\}$,

has a basis of $\mathscr{B} = \left\{ \begin{bmatrix} 1 & 0 \\ 0 & 1 \end{bmatrix}, \begin{bmatrix} 0 & 0 \\ 1 & 1 \end{bmatrix} \right\}$. To prove this, $\mathscr{B}$ is linearly independent because of Theorem 5.3.2(c), and $\mathscr{B}$ is a spanning set of W because for all $r, s \in \mathbb{R}$,

$$\begin{bmatrix} r & 0 \\ s & r + s \end{bmatrix} = r \begin{bmatrix} 1 & 0 \\ 0 & 1 \end{bmatrix} + s \begin{bmatrix} 0 & 0 \\ 1 & 1 \end{bmatrix}.$$

All of this implies that $\dim W = 2$.

■ **Example 5.4.18**

Consider the hyperplane (Definition 2.2.16) in $\mathbb{R}^4$,

$$H = \left\{ \begin{bmatrix} x_1 \\ x_2 \\ x_3 \\ x_4 \end{bmatrix} \in \mathbb{R}^4 : x_1 + 8x_2 - x_3 + 7x_4 = 0 \right\}.$$

By Theorem 5.2.15, H is a subspace of $\mathbb{R}^4$. Because H can be described by the parametric equations,

$$x_1 = -8r + s - 7t,$$
$$x_2 = r,$$
$$x_3 = s,$$
$$x_4 = t,$$

$\dim H = 3$ because H has a basis of

$$\left\{ \begin{bmatrix} -8 \\ 1 \\ 0 \\ 0 \end{bmatrix}, \begin{bmatrix} 1 \\ 0 \\ 1 \\ 0 \end{bmatrix}, \begin{bmatrix} -7 \\ 0 \\ 0 \\ 1 \end{bmatrix} \right\}.$$

Because the hyperplane of Example 5.4.18 is a three-dimensional subspace of $\mathbb{R}^4$, it is expected that any three linearly independent vectors in this hyperplane would span the hyperplane. These vectors could not span a subspace of dimension only 2, and certainly the vectors could not span all of $\mathbb{R}^4$, which has dimension 4. Hence, the vectors must span a three-dimensional subspace of $\mathbb{R}^4$, which, since the vectors lie in the hyperplane, must be the hyperplane itself. This discussion can be generalized to any vector space and leads to the next result.

■ **Theorem 5.4.19**

If V is an n-dimensional vector space, every set of n linearly independent vectors in V is a basis of V.

Proof

Let $\mathscr{B} = \{\mathbf{u}_1, \mathbf{u}_2, \dots, \mathbf{u}_n\}$ be a linearly independent set. In order to obtain a contradiction, suppose that $\mathscr{B}$ is not a spanning set of V. Then, there exists a vector $\mathbf{w} \in V \backslash S(\mathbf{u}_1, \mathbf{u}_2, \dots, \mathbf{u}_n)$. As in the proof of Zorn's Lemma (5.4.14) or by Exercise 13 of Section 5.3, this implies that $\{\mathbf{u}_1, \mathbf{u}_2, \dots, \mathbf{u}_n, \mathbf{w}\}$ is a linearly independent set consisting of $n + 1 > \dim V$ vectors. This contradicts Lemma 5.4.6. ■

■ **Example 5.4.20**

The polynomials,

$$p_1(x) = 1 + x, \quad p_2(x) = 1 + x^2, \quad p_3(x) = 1 + x^3, \quad p_4(x) = 1 + x + x^2 + x^3,$$

form a basis for the vector space $\mathbb{P}_3$. To see this, first note that $\dim \mathbb{P}_3 = 4$ since $\{1, x, x^2, x^3\}$ is a known basis. Thus, by Theorem 5.4.19 it suffices to show that $\mathscr{B} = \{p_1(x), p_2(x), p_3(x), p_4(x)\}$ is a linearly independent set. The method to show this is similar to that used in Example 5.3.7. Expanding the equation

$$r_1 p_1(x) + r_2 p_2(x) + r_3 p_3(x) + r_4 p_4(x) = 0$$

and collecting like terms reveals that it is equivalent to the matrix equation

$$\begin{bmatrix} 1 & 1 & 1 & 1 \\ 1 & 0 & 0 & 1 \\ 0 & 1 & 0 & 1 \\ 0 & 0 & 1 & 1 \end{bmatrix} \begin{bmatrix} r_1 \\ r_2 \\ r_3 \\ r_4 \end{bmatrix} = \begin{bmatrix} 0 \\ 0 \\ 0 \\ 0 \end{bmatrix}.$$

Notice that the columns of this matrix match the coefficients of the polynomials. This will always be the case. The coefficient matrix of this system is row equivalent to the matrix in reduced row echelon form,

$$\begin{bmatrix} 1 & 0 & 0 & 0 \\ 0 & 1 & 0 & 0 \\ 0 & 0 & 1 & 0 \\ 0 & 0 & 0 & 1 \end{bmatrix},$$

which has only the trivial solution $r_1 = r_2 = r_3 = r_4 = 0$. Hence, $\mathcal{B}$ is a set of linearly independent vectors in a four-dimensional vector space, so it follows that $\mathcal{B}$ is a basis for $\mathbb{P}_3$.

Expansions and Reductions

Consider the linearly independent subset of $\mathbb{R}^{2\times2}$,

$$\left\{ \begin{bmatrix} 1 & 0 \\ 0 & 2 \end{bmatrix}, \begin{bmatrix} 0 & 3 \\ 4 & 0 \end{bmatrix} \right\}.$$

This set can be expanded to a basis of $\mathbb{R}^{2\times2}$,

$$\left\{ \begin{bmatrix} 1 & 0 \\ 0 & 2 \end{bmatrix}, \begin{bmatrix} 0 & 3 \\ 4 & 0 \end{bmatrix}, \begin{bmatrix} 1 & 0 \\ 1 & 0 \end{bmatrix}, \begin{bmatrix} 0 & 1 \\ 0 & 1 \end{bmatrix} \right\}.$$

Turning this around, consider the spanning set of $\mathbb{P}_2$,

$$\{1, 1 + x, x + x^2, 1 + x + x^2\}.$$

This set is not linearly independent because $\dim \mathbb{P}_2 = 3$ (Theorem 5.4.19). However, it can be reduced to the basis,

$$\{1, 1 + x, x + x^2\}.$$

Expansions and **reductions** as these can be done in all finite-dimensional vector spaces, and this is the result of the next sequence of theorems.

■ **Theorem 5.4.21**

If V is a finite-dimensional vector space, any linearly independent collection of $k < \dim V$ vectors in V can be completed to a basis by introducing exactly $(\dim V) - k$ additional vectors to the collection.

Proof

Let $n = \dim V$ and suppose that $\{\mathbf{u}_1, \mathbf{u}_2, \dots, \mathbf{u}_k\}$ is linearly independent with $k < n$. Then, $\mathrm{S}(\mathbf{u}_1, \mathbf{u}_2, \dots, \mathbf{u}_k)$ cannot equal V since in that case $\{\mathbf{u}_1, \mathbf{u}_2, \dots, \mathbf{u}_k\}$ would be a basis of V, and any basis of V must have exactly n vectors by Theorem 5.4.7. Take $\mathbf{v}_{k+1} \in V \backslash \mathrm{S}(\mathbf{u}_1, \mathbf{u}_2, \dots, \mathbf{u}_k)$. The set $\{\mathbf{u}_1, \mathbf{u}_2, \dots, \mathbf{u}_k, \mathbf{v}_{k+1}\}$ must be linearly independent because otherwise $\mathbf{v}_{k+1} \in \mathrm{S}(\mathbf{u}_1, \mathbf{u}_2, \dots, \mathbf{u}_k)$. Repeat this process for a total of $n - k$ times to obtain a set of n linearly independent vectors $\{\mathbf{u}_1, \mathbf{u}_2, \dots, \mathbf{u}_k, \mathbf{v}_{k+1}, \mathbf{v}_{k+2}, \dots, \mathbf{v}_n\}$. By Theorem 5.4.19 this set is a basis for V. ■

■ **Theorem 5.4.22**

Let V be a finite-dimensional vector space with subspace W.

(a) $\dim W \leq \dim V$.

(b) $\dim W = \dim V$ if and only if $V = W$.

Proof

Clearly, if $V = W$, dim W = dim V, so suppose $V \neq W$ and let $\mathcal{B}$ be a basis of W. Because $\mathcal{B}$ is not a spanning set of V, Theorem 5.4.21 implies that $\mathcal{B}$ can be extended to a basis $\mathcal{B}'$ of V where $\mathcal{B} \subset \mathcal{B}'$. It follows that dim W < dim V, proving (a). To prove (b), it remains to assume that dim W < dim V. Since any basis of W will not be a spanning set of V, there must be a vector in V not in W. ∎

■ **Example 5.4.23**

Use Theorem 5.4.22 to show that $\mathbb{P}$ is not finite dimensional. Suppose that there exists a positive integer n such that dim $\mathbb{P} = n$. Since $\mathbb{P}_{n-1}$ is a subspace of $\mathbb{P}$ and is known to have dimension n by Example 5.4.3, this implies that $\mathbb{P} = \mathbb{P}_{n-1}$, which is false.

■ **Theorem 5.4.24**

Every spanning set of a finite-dimensional vector space V has a finite subset that is a basis for V.

Proof

If $V = \{\mathbf{0}\}$, its basis is $\varnothing$, which is a subset of every set, so let dim $V \in \mathbb{Z}^+$. Suppose that S is a spanning set of V. Then, S has at least dim V vectors by Theorem 5.4.22. Take any nonzero $\mathbf{u}_1 \in S$. This implies that $\{\mathbf{u}_1\}$ is linearly independent by Theorem 5.3.2(b). If dim $V = 1$, then $\{\mathbf{u}_1\}$ is a basis for V by Theorem 5.4.22. Otherwise, choose $\mathbf{u}_2 \in S \setminus S(\mathbf{u}_1)$, which implies that $\{\mathbf{u}_1, \mathbf{u}_2\}$ is linearly independent and is a basis for V if dim $V = 2$. Continue this process until a linearly independent subset of S of size dim V is found. This set will be a basis of V. ∎

■ **Corollary 5.4.25**

Let V be an n-dimensional vector space. Every spanning set of V of n vectors is a basis for V.

■ **Example 5.4.26**

Consider the linearly independent set of vectors,

$$S = \left\{ \begin{bmatrix} 1 \\ 0 \\ 2 \\ 0 \end{bmatrix}, \begin{bmatrix} 1 \\ 1 \\ 1 \\ 1 \end{bmatrix} \right\}.$$

Since S is a subset of the four-dimensional vector space $\mathbb{R}^4$, Theorem 5.4.21 implies that there exists two vectors in $\mathbb{R}^4$ that will extend S to a basis. The technique for finding these vectors involves Theorem 5.4.24. Add the standard vectors to S forming the spanning set of $\mathbb{R}^4$,

$$S' = \left\{ \begin{bmatrix} 1 \\ 0 \\ 2 \\ 0 \end{bmatrix}, \begin{bmatrix} 1 \\ 1 \\ 1 \\ 1 \end{bmatrix}, \begin{bmatrix} 1 \\ 0 \\ 0 \\ 0 \end{bmatrix}, \begin{bmatrix} 0 \\ 1 \\ 0 \\ 0 \end{bmatrix}, \begin{bmatrix} 0 \\ 0 \\ 1 \\ 0 \end{bmatrix}, \begin{bmatrix} 0 \\ 0 \\ 0 \\ 1 \end{bmatrix} \right\}.$$

Theorem 5.4.24 guarantees that S' has a subset that is a basis for $\mathbb{R}^4$. To find it, consider the vector equation,

$$r_1 \begin{bmatrix} 1 \\ 0 \\ 2 \\ 0 \end{bmatrix} + r_2 \begin{bmatrix} 1 \\ 1 \\ 1 \\ 1 \end{bmatrix} + r_3 \begin{bmatrix} 1 \\ 0 \\ 0 \\ 0 \end{bmatrix} + r_4 \begin{bmatrix} 0 \\ 1 \\ 0 \\ 0 \end{bmatrix} + r_5 \begin{bmatrix} 0 \\ 0 \\ 1 \\ 0 \end{bmatrix} + r_6 \begin{bmatrix} 0 \\ 0 \\ 0 \\ 1 \end{bmatrix} = \begin{bmatrix} 0 \\ 0 \\ 0 \\ 0 \end{bmatrix}.$$

This equation corresponds to a system of linear equations with coefficient matrix consisting of the vectors of S' as columns,

$$\begin{bmatrix} 1 & 1 & 1 & 0 & 0 & 0 \\ 0 & 1 & 0 & 1 & 0 & 0 \\ 2 & 1 & 0 & 0 & 1 & 0 \\ 0 & 1 & 0 & 0 & 0 & 1 \end{bmatrix},$$

which has the reduced row echelon form of

$$R = \begin{bmatrix} 1 & 0 & 0 & 0 & 1/2 & -1/2 \\ 0 & 1 & 0 & 0 & 0 & 1 \\ 0 & 0 & 1 & 0 & -1/2 & -1/2 \\ 0 & 0 & 0 & 1 & 0 & -1 \end{bmatrix}.$$

The first four columns of R show that the first four vectors of S' are linearly independent and, thus, a basis of $\mathbb{R}^4$ by Theorem 5.4.19, Call this basis $\mathcal{B}$ and note that it is an extension of S. The last two columns of R reveal the scalars needed to write $\mathbf{e}_3$ and $\mathbf{e}_4$ as linear combinations of the vectors of $\mathcal{B}$. Specifically,

$$\begin{bmatrix} 0 \\ 0 \\ 1 \\ 0 \end{bmatrix} = \frac{1}{2} \begin{bmatrix} 1 \\ 0 \\ 2 \\ 0 \end{bmatrix} - \frac{1}{2} \begin{bmatrix} 1 \\ 0 \\ 0 \\ 0 \end{bmatrix},$$

and

$$\begin{bmatrix} 0 \\ 0 \\ 0 \\ 1 \end{bmatrix} = -\frac{1}{2} \begin{bmatrix} 1 \\ 0 \\ 2 \\ 0 \end{bmatrix} + \begin{bmatrix} 1 \\ 1 \\ 1 \\ 1 \end{bmatrix} - \frac{1}{2} \begin{bmatrix} 1 \\ 0 \\ 0 \\ 0 \end{bmatrix} - \begin{bmatrix} 0 \\ 1 \\ 0 \\ 0 \end{bmatrix}.$$

Consider the vector space $\mathbb{R}^3$. The set,

$$\left\{ \begin{bmatrix} 1 \\ 0 \\ 0 \end{bmatrix}, \begin{bmatrix} 0 \\ 1 \\ 0 \end{bmatrix}, \begin{bmatrix} 0 \\ 0 \\ 1 \end{bmatrix}, \begin{bmatrix} 2 \\ 4 \\ 5 \end{bmatrix} \right\},$$

is not linearly independent because it contains too many vectors. If fact, any set containing more than three vectors in $\mathbb{R}^3$ will be linearly dependent. The set,

$$\left\{ \begin{bmatrix} 4 \\ 2 \\ 1 \end{bmatrix}, \begin{bmatrix} 1 \\ 1 \\ 1 \end{bmatrix} \right\},$$

does not span $\mathbb{R}^3$ because it has too few vectors. In fact, any set containing fewer than three vectors will not span $\mathbb{R}^3$. This result applies to arbitrary vector spaces.

■ **Theorem 5.4.27**

Let $S = \{\mathbf{u}_1, \mathbf{u}_2, \ldots, \mathbf{u}_k\}$ be a subset of the n-dimensional vector space V.

(a) If $k > n$, then S is linearly dependent.

(b) If $k < n$, then S does not span V.

Proof

First, suppose that $k > n$. If S is linearly independent, then it is a basis of a subspace of V, contradicting Theorem 5.4.22(a). Second, suppose that $k < n$. If S spans V, Theorem 5.4.24 implies that S contains a basis of V, which contradicts Theorem 5.4.7. ■

Exercises

1. Find two nonstandard bases of the vector space $\mathbb{R}$.

2. Find two nonstandard bases of the vector space $\mathbb{R}^3$.

3. Find a basis and the dimension of the given subspaces of V.

(a) $\left\{ \begin{bmatrix} x_1 \\ x_2 \\ x_1 + x_2 \\ x_1 - x_2 \end{bmatrix} : x_1, x_2, x_3, x_4 \in \mathbb{R} \right\}, \quad V = \mathbb{R}^4$

(b) $\left\{ \begin{bmatrix} a+b & a+c \\ b+c & 2a-3b \\ 4a-5c & 6b-7c \end{bmatrix} : a, b, c \in \mathbb{R} \right\}, \quad V = \mathbb{R}^{3 \times 2}$

(c) $\{a + (2a+b)x + (3a+2b+c)x^2 : a, b, c \in \mathbb{R}\}, \quad V = \mathbb{R}^3$

4. Prove that $\{1, 1+x, x+x^2, x^2+x^3\}$ is a basis for $\mathbb{P}_3$.

5. Let $W = \{p(x) \in \mathbb{P}_3 : p(1) = 0\}$.

   (a) Prove that W is a subspace of $\mathbb{P}_3$.

   (b) Find a basis and the dimension of W.

6. Let W be the set of symmetric 2×2 matrices.

   (a) Prove that W is a subspace of $\mathbb{R}^{2\times 2}$.

   (b) Find a basis and the dimension of W.

7. Prove or show false: $\{1 + 2x + 2x^2, 3 + 2x\}$ is a basis for $S\left(2x + x^2, 2 + x + x^2\right)$ in $\mathbb{P}_2$.

8. Prove or show false: $\left\{ \begin{bmatrix} 2 \\ 1 \\ 0 \\ 3 \end{bmatrix}, \begin{bmatrix} 1 \\ -1 \\ 1 \\ 0 \end{bmatrix} \right\}$ is a basis for $S\left(\begin{bmatrix} 3 \\ 0 \\ 1 \\ 3 \end{bmatrix}, \begin{bmatrix} -1 \\ -2 \\ 1 \\ -3 \end{bmatrix}, \begin{bmatrix} 5 \\ -2 \\ 3 \\ 3 \end{bmatrix} \right)$.

9. Find a basis and the dimension of the solution space of each system of linear equations in Exercise 2 in Section 3.5.

10. Find a basis and the dimension of the solution space of the given system of linear equations,

$$
\begin{array}{ccccccccc}
x_1 & + & x_2 & + & x_3 & & & & = & 0, \\
 & & x_2 & + & x_3 & + & x_4 & & = & 0, \\
 & & & & x_3 & + & x_4 & + & x_5 & & = & 0, \\
 & & & & & & x_4 & + & x_5 & + & x_6 & = & 0.
\end{array}
$$

11. Recall that a function $f : \mathbb{R} \to \mathbb{R}$ is **even** means that $f(x) = f(-x)$ for all $x \in \mathbb{R}$ and f is **odd** means that $f(x) = -f(-x)$ for all $x \in \mathbb{R}$.

   (a) Prove that the set F_e of even functions is a subspace of $^\mathbb{R}\mathbb{R}$ and that the set F_o of odd functions is also a subspace of $^\mathbb{R}\mathbb{R}$.

   (b) Prove that the set P_e of even polynomials is a subspace of $\mathbb{P}$ and that the set P_o of odd polynomials is also a subspace of $\mathbb{P}$.

   (c) Prove that P_e and P_o are infinite dimensional.

   (d) Prove that F_e and F_o are infinite dimensional.

   (e) Let $S = \{p(x) \in \mathbb{P}_3 : p(x)$ is an even polynomial$\}$. Find a basis and the dimension of S.

12. If the reduced row echelon form of an $n \times n$ matrix A is the identity matrix, prove that the columns of A form a basis of $\mathbb{R}^n$.

13. Let $\{\mathbf{u}_1, \mathbf{u}_2, \mathbf{u}_3, \mathbf{u}_4\}$ be a basis of a vector space V. Show that

$$\{\mathbf{u}_1, \mathbf{u}_1 + \mathbf{u}_2, \mathbf{u}_1 + \mathbf{u}_2 + \mathbf{u}_3, \mathbf{u}_1 + \mathbf{u}_2 + \mathbf{u}_3 + \mathbf{u}_4\}$$

   is also a basis of V.

14. Let $\{\mathbf{u}_1, \mathbf{u}_2, \ldots, \mathbf{u}_n\}$ be a basis for $\mathbb{R}^n$. Prove that if A is an invertible $n \times n$ matrix, $\{A\mathbf{u}_1, A\mathbf{u}_2, \ldots, A\mathbf{u}_n\}$ is also a basis for $\mathbb{R}^n$.

15. Suppose that $\mathscr{B} = \{\mathbf{u}_1, \mathbf{u}_2, \ldots, \mathbf{u}_n\}$ is a basis for $\mathbb{R}^n$. Prove that if A is an $n \times n$ matrix such that $\mathscr{B}' = \{A\mathbf{u}_1, A\mathbf{u}_2, \ldots, A\mathbf{u}_n\}$ contains n vectors, then $\mathscr{B}'$ is a basis for $\mathbb{R}^n$.

16. Let $\mathcal{S}$ be a set of vectors from a finite-dimensional vector space V. Prove that if every vector in V can be written as a unique linear combination of vectors from $\mathcal{S}$, then $\mathcal{S}$ is a basis.

17. Without using Zorn's Lemma (5.4.14), prove that every finite-dimensional vector space has a basis.

18. Prove that a basis is a minimal spanning set. In other words, prove that for any vector space V with basis $\mathscr{B}$, if $\mathcal{S} \subset \mathscr{B}$, then $\mathcal{S}$ is not a spanning set of V.

19. Prove that a basis is a maximal independent set. In other words, prove that for any vector space V with basis $\mathscr{B}$ and any $\mathcal{S} \subseteq V$, if $\mathscr{B} \subset \mathcal{S}$, then $\mathcal{S}$ is not linearly independent.

20. Let $\mathscr{C} = \{(-n, n) : n \in \mathbb{Z}^+\}$. Prove that $\bigcup \mathscr{C} = \mathbb{R}$ and $\bigcap \mathscr{C} = (-1, 1)$.

21. Let $\mathscr{C}$ be a chain of subspaces of the vector space V. Prove that $\bigcup \mathscr{C}$ is a subspace of V.

22. Prove that a family of sets can have at most one maximum.

23. Let V_1 and V_2 be finite-dimensional vector spaces. Using the definition in Exercise 6 of Section 5.1, find a basis for $V_1 \times V_2$ and give an expression for its dimension.

24. Let W_1 and W_2 be subspaces of the finite-dimensional vector space V. Using the notation of Example 5.2.4, demonstrate that

$$\dim(W_1 + W_2) = \dim W_1 + \dim W_2 - \dim(W_1 \cap W_2).$$

25. Prove that each set is linearly independent and then expand the set to a basis of the given vector space V.

(a) $\left\{ \begin{bmatrix} 1 \\ 2 \\ 3 \\ 4 \end{bmatrix}, \begin{bmatrix} 1 \\ 1 \\ 1 \\ 1 \end{bmatrix} \right\}$, $V = \mathbb{R}^4$.

(b) $\left\{ \begin{bmatrix} 1 & 2 & 0 \\ 0 & 0 & 0 \end{bmatrix}, \begin{bmatrix} 0 & 0 & 3 \\ 0 & 0 & 4 \end{bmatrix}, \begin{bmatrix} 0 & 0 & 0 \\ 5 & 6 & 0 \end{bmatrix} \right\}$, $V = \mathbb{R}^{2\times3}$

26. Prove that each set is a spanning set of the vector space V and then reduce the spanning set to a basis of V.

(a) $\left\{ \begin{bmatrix} 0 \\ 0 \\ 0 \end{bmatrix}, \begin{bmatrix} -4 \\ 2 \\ -3 \end{bmatrix}, \begin{bmatrix} -3 \\ 1 \\ -1 \end{bmatrix}, \begin{bmatrix} 4 \\ 1 \\ -6 \end{bmatrix}, \begin{bmatrix} 2 \\ 2 \\ 1 \end{bmatrix} \right\}$, $V = \mathbb{R}^3$

(b) $\{\sin^2, \cos^2, 1, 0\}$, $V = \mathsf{S}\left(\sin^2, \cos^2, 1, 0\right)$

27. Let $S = \{1 + 3x, 1 + 4x + x^2, 2x - 5x^2\}$

    (a) Expand to a basis of $\mathbb{P}_4$.
    (b) Reduce to a basis of $\mathbb{P}_1$.

5.5   Rank and Nullity

Define the linear transformation $T : \mathbb{R}^{2\times 4} \to \mathbb{P}_3$ by

$$T\begin{bmatrix} a & b & c & d \\ e & f & g & h \end{bmatrix} = (a + f) + (g + d)x + (b - c)x^3. \tag{5.23}$$

The kernel of T is

$$\ker T = \left\{ \begin{bmatrix} r & t & t & -s \\ u & -r & s & v \end{bmatrix} : r, s, t, u, v \in \mathbb{R} \right\},$$

which implies that a basis for $\ker T$ is

$$\left\{ \begin{bmatrix} 1 & 0 & 0 & 0 \\ 0 & -1 & 0 & 0 \end{bmatrix}, \begin{bmatrix} 0 & 0 & 0 & -1 \\ 0 & 0 & 1 & 0 \end{bmatrix}, \begin{bmatrix} 0 & 1 & 1 & 0 \\ 0 & 0 & 0 & 0 \end{bmatrix}, \right.$$
$$\left. \begin{bmatrix} 0 & 0 & 0 & 0 \\ 1 & 0 & 0 & 0 \end{bmatrix}, \begin{bmatrix} 0 & 0 & 0 & 0 \\ 0 & 0 & 0 & 1 \end{bmatrix} \right\}, \tag{5.24}$$

so

$$\dim(\ker T) = 5. \tag{5.25}$$

Also, the range of T is

$$\operatorname{ran} T = \{a + bx + cx^3 : a, b, c \in \mathbb{R}\},$$

which implies that a basis for $\operatorname{ran} T$ is $\{1, x, x^3\}$, so

$$\dim(\operatorname{ran} T) = 3. \tag{5.26}$$

The kernel and range of a linear transformation are important subspaces. Their dimensions are also important and enable much to be known about the transformation simply by knowing their values. For this reason, these dimensions are named by the next definition.

■ **Definition 5.5.1**

Let $T : V \to W$ be a linear transformation. The **rank** of T, rank T, is the dimension of ran T, and the **nullity** of T, null T, is the dimension of ker T.

Rank-Nullity Theorem

Because a basis for the domain of the linear transformation T defined by (5.23) contains eight matrices, $\dim(\text{dom } T) = 8$. Also, because of (5.26), $\text{rank } T = 3$, and because of (5.25), $\text{null } T = 5$. In this case, the sum of the rank of the linear transformation and its nullity equals the dimension of the domain of the transformation. This holds for all linear transformations.

■ **Theorem 5.5.2 [Rank-Nullity Theorem]**

Suppose that $T : V \to W$ is a linear transformation. Then,

$$\text{rank } T + \text{null } T = \dim V.$$

Proof

Let $\dim V = n$ and $\text{rank}(T) = k$. By the definition of rank, there exists vectors $\mathbf{u}_1, \mathbf{u}_2, \dots, \mathbf{u}_k \in V$ such that $\{T\mathbf{u}_1, T\mathbf{u}_2, \dots, T\mathbf{u}_k\}$ is a basis for $\text{ran } T$. By Theorem 5.3.10, the set of vectors $\{\mathbf{u}_1, \mathbf{u}_2, \dots, \mathbf{u}_k\}$ is also linearly independent, and by Theorem 5.4.21 it can be completed to a basis of V by adding $n - k$ new vectors. Write this basis as

$$\{\mathbf{u}_1, \mathbf{u}_2, \dots, \mathbf{u}_k, \mathbf{v}_{k+1}, \mathbf{v}_{k+2}, \dots, \mathbf{v}_n\}.$$

Expand each of $T\mathbf{v}_{k+1}, T\mathbf{v}_{k+2}, \dots, T\mathbf{v}_n$ by writing

$$T\mathbf{v}_{k+1} = r_{1,k+1}T\mathbf{u}_1 + r_{2,k+1}T\mathbf{u}_2 + \cdots + r_{k,k+1}T\mathbf{u}_k,$$
$$T\mathbf{v}_{k+2} = r_{1,k+2}T\mathbf{u}_1 + r_{2,k+2}T\mathbf{u}_2 + \cdots + r_{k,k+2}T\mathbf{u}_k,$$
$$\vdots$$
$$T\mathbf{v}_n = r_{1,n}T\mathbf{u}_1 + r_{2,n}T\mathbf{u}_2 + \cdots + r_{k,n}T\mathbf{u}_k.$$

Therefore,

$$T(\mathbf{v}_{k+1} - r_{1,k+1}\mathbf{u}_1 - r_{2,k+1}\mathbf{u}_2 - \cdots - r_{k,k+1}\mathbf{u}_k) = \mathbf{0},$$
$$T(\mathbf{v}_{k+2} - r_{1,k+2}\mathbf{u}_1 - r_{2,k+2}\mathbf{u}_2 - \cdots - r_{k,k+2}\mathbf{u}_k) = \mathbf{0},$$
$$\vdots$$
$$T(\mathbf{v}_n - r_{1,n}\mathbf{u}_1 - r_{2,n}\mathbf{u}_2 - \cdots - r_{k,n}\mathbf{u}_k) = \mathbf{0}.$$

It follows that

$$\mathbf{v}'_{k+1} = \mathbf{v}_{k+1} - r_{1,k+1}\mathbf{u}_1 - r_{2,k+1}\mathbf{u}_2 - \cdots - r_{k,k+1}\mathbf{u}_k \in \ker T,$$
$$\mathbf{v}'_{k+2} = \mathbf{v}_{k+2} - r_{1,k+2}\mathbf{u}_1 - r_{2,k+2}\mathbf{u}_2 - \cdots - r_{k,k+2}\mathbf{u}_k \in \ker T,$$
$$\vdots$$
$$\mathbf{v}'_n = \mathbf{v}_n - r_{1n}\mathbf{u}_1 - r_{2,n}\mathbf{u}_2 - \cdots - r_{k,n}\mathbf{u}_k \in \ker T,$$

where the linear independence of $\{\mathbf{v}_{k+1}, \mathbf{v}_{k+2}, \dots, \mathbf{v}_n\}$ implies that the vectors $\mathbf{v}'_{k+1}, \mathbf{v}'_{k+2}, \dots, \mathbf{v}'_n$ form a linearly independent set (Exercise 6).

To complete the proof it will be shown that

$$\mathcal{B} = \{\mathbf{v}'_{k+1}, \mathbf{v}'_{k+2}, \dots, \mathbf{v}'_n\}$$

is a basis for ker T so that null $T = n - k$. Toward this end, observe that

$$\{\mathbf{u}_1, \mathbf{u}_2, \dots, \mathbf{u}_k, \mathbf{v}'_{k+1}, \mathbf{v}'_{k+2}, \dots, \mathbf{v}'_n\}$$

is a spanning set for V. This is because

$$\mathbf{v}_{k+1}, \mathbf{v}_{k+2}, \dots, \mathbf{v}_{k+1} \in S\left(\mathbf{u}_1, \mathbf{u}_2, \dots, \mathbf{u}_k, \mathbf{v}'_{k+1}, \mathbf{v}'_{k+2}, \dots, \mathbf{v}'_n\right),$$

which is true because

$$\mathbf{v}_{k+1} = \mathbf{v}'_{k+1} + r_{1,k+1}\mathbf{u}_1 + r_{2,k+1}\mathbf{u}_2 + \cdots + r_{k,k+1}\mathbf{u}_k,$$
$$\mathbf{v}_{k+2} = \mathbf{v}'_{k+2} + r_{1,k+2}\mathbf{u}_1 + r_{2,k+2}\mathbf{u}_2 + \cdots + r_{k,k+2}\mathbf{u}_k,$$
$$\vdots$$
$$\mathbf{v}_n = \mathbf{v}'_n + r_{1,n}\mathbf{u}_1 + r_{2,n}\mathbf{u}_2 + \cdots + r_{k,n}\mathbf{u}_k.$$

To prove that $\mathcal{B}$ is a spanning set of ker T, suppose there exists a vector $\mathbf{v} \in \ker T$ that is not in $S\left(\mathbf{v}'_{k+1}, \mathbf{v}'_{k+2}, \dots, \mathbf{v}'_n\right)$. Expand this vector as

$$\mathbf{v} = r_1\mathbf{u}_1 + r_2\mathbf{u}_2 + \cdots + r_k\mathbf{u}_k + r_{k+1}\mathbf{v}'_{k+1} + r_{k+2}T\mathbf{v}'_{k+2} + \cdots + r_n\mathbf{v}'_n.$$

Hence, because $\mathcal{B} \subset \ker T$,

$$\mathbf{0} = r_1T\mathbf{u}_1 + r_2T\mathbf{u}_2 + \cdots + r_kT\mathbf{u}_k + r_{k+1}T\mathbf{v}'_{k+1} + r_{k+2}T\mathbf{v}'_{k+2} + \cdots + r_nT\mathbf{v}'_n$$
$$= r_1T\mathbf{u}_1 + r_2T\mathbf{u}_2 + \cdots + r_kT\mathbf{u}_k.$$

But, $\mathbf{v} \notin S\left(\mathbf{v}'_{k+1}, \mathbf{v}'_{k+2}, \dots, \mathbf{v}'_n\right)$ implies that not all of $r_1, r_2, \dots, r_k$ are 0, which violates the linear independence of $\{T\mathbf{u}_1, T\mathbf{u}_2, \dots, T\mathbf{u}_k\}$. ∎

Before any direct applications of the Rank-Nullity Theorem (5.5.2), a concrete illustration of its important proof would be beneficial. Continue working with the linear transformation $T : \mathbb{R}^{2\times 4} \to \mathbb{P}_3$ as defined in (5.23). First, find pre-images of the basis vectors of ran T and write

$$\mathbf{u}_1 = \begin{bmatrix} 1 & 0 & 0 & 0 \\ 0 & 0 & 0 & 0 \end{bmatrix}, \quad \mathbf{u}_2 = \begin{bmatrix} 0 & 0 & 0 & 1 \\ 0 & 0 & 0 & 0 \end{bmatrix}, \quad \mathbf{u}_3 = \begin{bmatrix} 0 & 1 & 0 & 0 \\ 0 & 0 & 0 & 0 \end{bmatrix},$$

so $T\mathbf{u}_1 = 1$, $T\mathbf{u}_2 = x$, and $T\mathbf{u}_3 = x^3$. Notice that these pre-images are not unique because T is not one-to-one. Following the procedure of the proof of the Rank-Nullity Theorem, expand $\{\mathbf{u}_1, \mathbf{u}_2, \mathbf{u}_3\}$ to a basis of $\mathbb{R}^{2\times 4}$ by adding $\mathbf{v}_4, \mathbf{v}_5, \mathbf{v}_6, \mathbf{v}_7, \mathbf{v}_8$. The obvious choice for these new vectors are the other standard vectors of $\mathbb{R}^{2\times 4}$, so choose

$$\mathbf{v}_4 = \begin{bmatrix} 0 & 0 & 1 & 0 \\ 0 & 0 & 0 & 0 \end{bmatrix}, \quad \mathbf{v}_5 = \begin{bmatrix} 0 & 0 & 0 & 0 \\ 1 & 0 & 0 & 0 \end{bmatrix}, \quad \mathbf{v}_6 = \begin{bmatrix} 0 & 0 & 0 & 0 \\ 0 & 1 & 0 & 0 \end{bmatrix},$$

$$\mathbf{v}_7 = \begin{bmatrix} 0 & 0 & 0 & 0 \\ 0 & 0 & 1 & 0 \end{bmatrix}, \quad \mathbf{v}_8 = \begin{bmatrix} 0 & 0 & 0 & 0 \\ 0 & 0 & 0 & 1 \end{bmatrix}.$$

However, these new vectors do not form a basis for ker T. For example, the matrix $\begin{bmatrix} 0 & 1 & 1 & 0 \\ 0 & 0 & 0 & 0 \end{bmatrix}$ is not an element of $S(\mathbf{v}_4, \mathbf{v}_5, \mathbf{v}_6, \mathbf{v}_7, \mathbf{v}_8)$. To find a basis for ker T, compute the vectors $\mathbf{v}_4', \mathbf{v}_5', \mathbf{v}_6', \mathbf{v}_7', \mathbf{v}_8'$ as in the proof. To do this, first note that

$$T\mathbf{v}_4 = -x^3, \quad T\mathbf{v}_5 = 0, \quad T\mathbf{v}_6 = 1, \quad T\mathbf{v}_7 = x, \quad \mathbf{v}_8 = 0.$$

Then,

$$\mathbf{v}_4' = \mathbf{v}_4 - 0\mathbf{u}_1 - 0\mathbf{u}_2 + 1\mathbf{u}_3 = \begin{bmatrix} 0 & 1 & 1 & 0 \\ 0 & 0 & 0 & 0 \end{bmatrix},$$

$$\mathbf{v}_5' = \mathbf{v}_5 - 0\mathbf{u}_1 - 0\mathbf{u}_2 - 0\mathbf{u}_3 = \begin{bmatrix} 0 & 0 & 0 & 0 \\ 1 & 0 & 0 & 0 \end{bmatrix},$$

$$\mathbf{v}_6' = \mathbf{v}_6 - 1\mathbf{u}_1 - 0\mathbf{u}_2 - 0\mathbf{u}_3 = \begin{bmatrix} -1 & 0 & 0 & 0 \\ 0 & 1 & 0 & 0 \end{bmatrix},$$

$$\mathbf{v}_7' = \mathbf{v}_7 - 0\mathbf{u}_1 - 1\mathbf{u}_2 - 0\mathbf{u}_3 = \begin{bmatrix} 0 & 0 & 0 & -1 \\ 0 & 0 & 1 & 0 \end{bmatrix},$$

$$\mathbf{v}_8' = \mathbf{v}_8 - 0\mathbf{u}_1 - 0\mathbf{u}_2 - 0\mathbf{u}_3 = \begin{bmatrix} 0 & 0 & 0 & 0 \\ 0 & 0 & 0 & 1 \end{bmatrix}.$$

$\mathcal{B} = \{\mathbf{v}_4', \mathbf{v}_5', \mathbf{v}_6', \mathbf{v}_7', \mathbf{v}_8'\}$ is clearly a basis for ker T considering (5.24), and the union $\{\mathbf{u}_1, \mathbf{u}_2, \mathbf{u}_3\} \cup \mathcal{B}$ is a basis for $\mathbb{R}^{2 \times 4}$. These sets of vectors are illustrated in Figure 5.1. Counting vectors in the bases confirms the Rank-Nullity Theorem.

The decomposition of the basis for V in the proof of the Rank-Nullity Theorem provides information concerning the action of T on vectors of V. For example, take $A = \begin{bmatrix} 1 & 2 & 3 & 4 \\ 5 & 6 & 7 & 8 \end{bmatrix} \in \mathbb{R}^{2 \times 4}$. It can be shown that

$$A = 7\mathbf{u}_1 + 11\mathbf{u}_2 - \mathbf{u}_3 + 3\mathbf{v}_4' + 5\mathbf{v}_5' + 6\mathbf{v}_6' + 7\mathbf{v}_7' + 8\mathbf{v}_8'.$$

Since $3\mathbf{v}_4' + 5\mathbf{v}_5' + 6\mathbf{v}_6' + 7\mathbf{v}_7' + 8\mathbf{v}_8' \in \ker T$,

$$TA = T(7\mathbf{u}_1 + 11\mathbf{u}_2 - \mathbf{u}_3 + 3\mathbf{v}_4' + 5\mathbf{v}_5' + 6\mathbf{v}_6' + 7\mathbf{v}_7' + 8\mathbf{v}_8')$$

$$= T(7\mathbf{u}_1 + 11\mathbf{u}_2 - \mathbf{u}_3) + T(3\mathbf{v}_4' + 5\mathbf{v}_5' + 6\mathbf{v}_6' + 7\mathbf{v}_7' + 8\mathbf{v}_8')$$

$$= T(7\mathbf{u}_1 + 11\mathbf{u}_2 - \mathbf{u}_3).$$

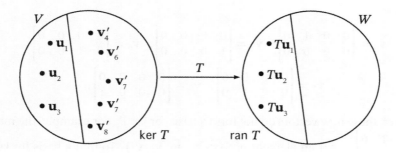

Figure 5.1 Illustrating the Rank-Nullity Theorem.

In fact, $T(7\mathbf{u}_1 + 11\mathbf{u}_2 - \mathbf{u}_3 + \mathbf{v}) = T(7\mathbf{u}_1 + 11\mathbf{u}_2 - \mathbf{u}_3)$ for all $\mathbf{w} \in \ker T$. Since the kernel of T is nontrivial, there are infinitely many vectors in $\ker T$, so there are infinitely many pre-images of $T(7\mathbf{u}_1 + 11\mathbf{u}_2 - \mathbf{u}_3)$. If T had been one-to-one, the only vector in $\ker T$ would be the zero vector, so in this case there would be only one pre-image of $T(7\mathbf{u}_1 + 11\mathbf{u}_2 - \mathbf{u}_3)$.

■ **Example 5.5.3**

Let $T : \mathbb{R}^n \to \mathbb{R}^m$ be a linear transformation.

- Suppose that $n > m$. Since $\operatorname{ran} T$ is a subspace of $\mathbb{R}^m$ (Theorem 5.2.24), Theorem 5.4.22(a) and the Rank-Nullity Theorem (5.5.2) imply

$$\operatorname{rank} T \leq m < n,$$
$$\operatorname{rank} T < \operatorname{rank} T + \operatorname{null} T.$$

Therefore, $\operatorname{null} T > 0$, and although T might be onto, it is not one-to-one.

- Assume that $n < m$. This implies that

$$\operatorname{rank} T + \operatorname{null} T < m,$$

so $\operatorname{rank} T < m$. Hence, although T might be one-to-one, it is not onto.

The case when $n = m$ is the next corollary.

■ **Corollary 5.5.4**

Let V and W be n-dimensional vector spaces and $T : V \to W$ be a linear transformation. Then, T is one-to-one if and only if T is onto.

Proof

Suppose that T is one-to-one. This implies that $\ker T = \{\mathbf{0}\}$, so $\operatorname{null} T = 0$. Since $\dim V = n$, by the Rank-Nullity Theorem (5.5.2), $\operatorname{rank} T = n$, which implies that T is onto. Conversely, let T be onto. Then, $\operatorname{rank} T = n$, and this gives $\operatorname{null} T = 0$, which means that T is one-to-one. ■

■ **Example 5.5.5**

Consider the linear operator $T : \mathbb{R}^3 \to \mathbb{R}^3$ defined in Example 5.2.21 by the matrix

$$A = \begin{bmatrix} 1 & 2 & 1 \\ 3 & 2 & 2 \\ 9 & -6 & 3 \end{bmatrix}.$$

It was shown that the kernel of T is the one-dimensional subspace,

$$\ker T = \mathsf{S}\left(\begin{bmatrix} 2 \\ 1 \\ -4 \end{bmatrix} \right).$$

By Theorem 5.5.2, the range of T must have dimension

$$\dim \mathbb{R}^3 - \text{null } T = 3 - 1 = 2. \tag{5.27}$$

Now, in Example 5.2.25 it was shown that $\text{ran } T = \mathsf{S}\left(\begin{bmatrix} 1 \\ 3 \\ 9 \end{bmatrix}, \begin{bmatrix} 2 \\ 2 \\ -6 \end{bmatrix}, \begin{bmatrix} 1 \\ 2 \\ 3 \end{bmatrix} \right)$. By (5.27) a basis for $\text{ran } T$ should have two vectors not three, so it should be possible to remove one of the vectors from the above spanning set by writing one of the vectors as a linear combination of the others. Using the fact that $\begin{bmatrix} 2 \\ 1 \\ -4 \end{bmatrix} \in \ker T$,

$$\begin{bmatrix} 1 & 2 & 1 \\ 3 & 2 & 2 \\ 9 & -6 & 3 \end{bmatrix} \begin{bmatrix} 2 \\ 1 \\ -4 \end{bmatrix} = \begin{bmatrix} 0 \\ 0 \\ 0 \end{bmatrix}$$

can be rewritten as

$$2 \begin{bmatrix} 1 \\ 3 \\ 9 \end{bmatrix} + 1 \begin{bmatrix} 2 \\ 2 \\ -6 \end{bmatrix} - 4 \begin{bmatrix} 1 \\ 2 \\ 3 \end{bmatrix} = \begin{bmatrix} 0 \\ 0 \\ 0 \end{bmatrix}$$

so that

$$\begin{bmatrix} 2 \\ 2 \\ -6 \end{bmatrix} = -2 \begin{bmatrix} 1 \\ 3 \\ 9 \end{bmatrix} + 4 \begin{bmatrix} 1 \\ 2 \\ 3 \end{bmatrix}.$$

Hence, the spanning set for $\text{ran } T$ can be reduced to $\text{ran } T = \mathsf{S}\left(\begin{bmatrix} 1 \\ 3 \\ 9 \end{bmatrix}, \begin{bmatrix} 1 \\ 2 \\ 3 \end{bmatrix} \right)$.

■ Example 5.5.6

Take the derivative transformation $D : \mathbb{P}_n \to \mathbb{P}_{n-1}$. Using facts from basic calculus, $\ker D = \mathbb{P}_0$ and $\operatorname{ran} D = \mathbb{P}_{n-1}$ so that

$$\operatorname{rank} D = n \quad \text{and} \quad \operatorname{null} D = 1.$$

The Rank-Nullity Theorem (5.5.2) expresses the fact that

$$\operatorname{rank} D + \operatorname{null} D = \dim \mathbb{P}_n$$
$$\dim \mathbb{P}_{n-1} + \dim \mathbb{P}_0 = \dim \mathbb{P}_n$$
$$n + 1 = n + 1.$$

More generally, in the context of the kth derivative $D^k : \mathbb{P}_n \to \mathbb{P}_{n-k}$ the Rank-Nullity Theorem expresses the fact that

$$\operatorname{rank} D^k + \operatorname{null} D^k = \dim \mathbb{P}_n$$
$$\dim \mathbb{P}_{n-k} + \dim \mathbb{P}_{k-1} = \dim \mathbb{P}_n$$
$$(n - k + 1) + (k - 1 + 1) = n + 1.$$

Fundamental Subspaces

Consider the linear transformation $T : \mathbb{R}^5 \to \mathbb{R}^3$ with standard matrix

$$A = \begin{bmatrix} 1 & 2 & -1 & 4 & 1 \\ 2 & -3 & 0 & -1 & 5 \\ 5 & -4 & -1 & 2 & 11 \end{bmatrix}. \tag{5.28}$$

It can be shown that the reduced row echelon form of A is

$$R = \begin{bmatrix} 1 & 0 & -3/7 & 10/7 & 13/7 \\ 0 & 1 & -2/7 & 9/7 & -3/7 \\ 0 & 0 & 0 & 0 & 0 \end{bmatrix}. \tag{5.29}$$

Viewing A as the coefficient matrix of a homogeneous system of equations, R can be used to show that the solution space of the system has basis

$$\left\{ \begin{bmatrix} 3 \\ 2 \\ 7 \\ 0 \\ 0 \end{bmatrix}, \begin{bmatrix} -10 \\ -9 \\ 0 \\ 7 \\ 0 \end{bmatrix}, \begin{bmatrix} -13 \\ 3 \\ 0 \\ 0 \\ 7 \end{bmatrix} \right\}. \tag{5.30}$$

Next, look at the column and rows of A. Every spanning set of a finite-dimensional vector space can be reduced to a basis by Theorem 5.4.24. A basis of the vector space spanned by the columns of A can be found by examining R. Because the first two columns of R are standard vectors and the other columns are not,

$$\left\{ \begin{bmatrix} 1 \\ 2 \\ 5 \end{bmatrix}, \begin{bmatrix} 2 \\ -3 \\ -4 \end{bmatrix} \right\} \tag{5.31}$$

is a basis of the subspace of $\mathbb{R}^3$ spanned by the columns of A. To find a basis for the vector space spanned by the rows of A, examine

$$A^T = \begin{bmatrix} 1 & 2 & 5 \\ 2 & -3 & -4 \\ -1 & 0 & -1 \\ 4 & -1 & 2 \\ 1 & 5 & 11 \end{bmatrix},$$

and compute its reduced row echelon form, which is

$$\begin{bmatrix} 1 & 0 & 1 \\ 0 & 1 & 2 \\ 0 & 0 & 0 \\ 0 & 0 & 0 \\ 0 & 0 & 0 \end{bmatrix}.$$

As with the columns, this implies that

$$\left\{ \begin{bmatrix} 1 \\ 2 \\ -1 \\ 4 \\ 1 \end{bmatrix}, \begin{bmatrix} 2 \\ -3 \\ 0 \\ -1 \\ 5 \end{bmatrix} \right\} \tag{5.32}$$

is a basis for the vector space spanned by the transposes of the rows of A. This analysis of A can be done to any $m \times n$ matrix A.

■ **Definition 5.5.7**

The $m \times n$ matrix A has associated with it 3 **fundamental subspaces**.

(a) The **column space** of A is the subspace of $\mathbb{R}^m$ that is spanned by the columns of A and is denoted by CS A.

(b) The **row space** of A is the subspace of $\mathbb{R}^n$ that is spanned by the columns of A^T and is denoted by RS A.

(c) The **null space** of A is the solution space of $A\mathbf{x} = \mathbf{0}$ and is denoted by NS A.

Using the matrix A given in (5.28), because of (5.31),

$$\text{CS}\,A = \text{S}\left(\begin{bmatrix} 1 \\ 2 \\ 5 \end{bmatrix}, \begin{bmatrix} 2 \\ -3 \\ -4 \end{bmatrix} \right),$$

because of (5.32),

$$\text{RS}\,A = \text{S}\left(\begin{bmatrix} 1 \\ 2 \\ -1 \\ 4 \\ 1 \end{bmatrix}, \begin{bmatrix} 2 \\ -3 \\ 0 \\ -1 \\ 5 \end{bmatrix} \right),$$

and because of (5.30),

$$
\text{NS}\, A = \text{S}\left(\begin{bmatrix} 3 \\ 2 \\ 7 \\ 0 \\ 0 \end{bmatrix}, \begin{bmatrix} -10 \\ -9 \\ 0 \\ 7 \\ 0 \end{bmatrix}, \begin{bmatrix} -13 \\ 3 \\ 0 \\ 0 \\ 7 \end{bmatrix}\right).
$$

It should be noted that the elementary row operations **preserve** the row space and null space of a matrix. More precisely, the following theorem holds.

■ Theorem 5.5.8

Let A and A' be $m \times n$ matrices. If A' is obtained from A by an elementary row operation, $\text{RS}\, A = \text{RS}\, A'$ and $\text{NS}\, A = \text{NS}\, A'$.

Proof

$\text{RS}\, A = \text{RS}\, A'$ follows from Exercise 6 of Section 3.5 and $\text{NS}\, A = \text{NS}\, A'$ follows from Theorem 3.5.7. ■

A result of Theorem 5.5.8 is that the row space of a matrix equals the row space of its reduced row echelon form. For example, again using A from (5.28),

$$
\text{RS}\, A = \text{S}\left(\begin{bmatrix} 7 \\ 0 \\ -3 \\ 10 \\ 13 \end{bmatrix}, \begin{bmatrix} 0 \\ 7 \\ -2 \\ 9 \\ -3 \end{bmatrix}\right)
$$

because of (5.29). On the other hand, it should be evident that the elementary row operations do not preserve the column space. This is illustrated by the fact that the column space of A and the column space of its reduced row echelon form (5.29) are different vector spaces.

Rank and Nullity of a Matrix

Let $T : \mathbb{R}^n \to \mathbb{R}^m$ be a linear transformation. The standard matrix $[T]$ of this transformation is an $m \times n$ matrix. Let $\mathbf{c}_1, \mathbf{c}_2, \ldots, \mathbf{c}_n \in \mathbb{R}^m$ be the columns of $[T]$. The rank of $[T]$ is found by examining the range of T. Specifically,

$$
\begin{aligned}
\text{ran}\, T &= \{T\mathbf{x} : \mathbf{x} \in \mathbb{R}^n\} \\
&= \{[T]\mathbf{x} : \mathbf{x} \in \mathbb{R}^n\} \\
&= \{x_1\mathbf{c}_1 + x_2\mathbf{c}_2 + \cdots + x_n\mathbf{c}_n : x_1, x_2, \ldots, x_n \in \mathbb{R}\} = \text{CS}\,[T].
\end{aligned}
$$

Therefore,

$$
\dim \text{CS}\,[T] = \dim(\text{ran}\, T) = \text{rank}\, T,
$$

and this motivates the next definition.

■ **Definition 5.5.9**

The **rank** of a matrix A is the dimension of its column space and is denoted by rank A.

Given a matrix A, let R be the reduced row echelon form of A. The number of standard vectors among the columns of R equals the dimension of the column space of A. Each 1 in those columns is a leading 1, and all other entries in columns with a leading 1 are 0. This means an examination of the rows of R will have a pattern of 1s and 0s so that the transpose of the rows of R form a linearly independent set of vectors.

For example, consider

$$A = \begin{bmatrix} 2 & 1 & 9 \\ -8 & 9 & 8 \\ 5 & -10 & -6 \\ 3 & 7 & 7 \end{bmatrix}.$$

The reduced row echelon form of A is

$$R = \begin{bmatrix} 1 & 0 & 0 \\ 0 & 1 & 0 \\ 0 & 0 & 1 \\ 0 & 0 & 0 \end{bmatrix}.$$

This shows that

$$\left\{ \begin{bmatrix} 2 \\ -8 \\ 5 \\ 3 \end{bmatrix}, \begin{bmatrix} 1 \\ 9 \\ -10 \\ 7 \end{bmatrix}, \begin{bmatrix} 9 \\ 8 \\ -6 \\ 7 \end{bmatrix} \right\}$$

is a linearly independent set and, moreover, a basis for the column space of A, so $\dim \mathrm{CS}\, A = 3$. Since the row space is preserved under the elementary row operations,

$$S\left(\begin{bmatrix} 2 \\ 1 \\ 9 \end{bmatrix}, \begin{bmatrix} -8 \\ 9 \\ 8 \end{bmatrix}, \begin{bmatrix} 5 \\ -10 \\ -6 \end{bmatrix}, \begin{bmatrix} 3 \\ 7 \\ 7 \end{bmatrix} \right) = S\left(\begin{bmatrix} 1 \\ 0 \\ 0 \end{bmatrix}, \begin{bmatrix} 0 \\ 1 \\ 0 \end{bmatrix}, \begin{bmatrix} 0 \\ 0 \\ 1 \end{bmatrix} \right),$$

showing that the dimension of the row space of A is 3. Notice how each leading 1 of R identify both a vector in a basis for the column space and a vector in a basis for the row space. All of this explains the next theorem.

■ **Theorem 5.5.10**

For any matrix A, rank $A = \dim \mathrm{CS}\, A = \dim \mathrm{RS}\, A$.

Returning to the linear transformation $T : \mathbb{R}^n \to \mathbb{R}^m$, the nullity of T is found by examining the kernel of T. Specifically,

$$\ker T = \{ \mathbf{x} \in \mathbb{R}^n : T\mathbf{x} = \mathbf{0} \} = \{ \mathbf{x} \in \mathbb{R}^n : [T]\mathbf{x} = \mathbf{0} \} = \mathrm{NS}\,[T].$$

Therefore, $\dim \mathrm{NS}\,[T] = \dim (\ker T) = \mathrm{null}\, T$, and this motivates the next definition.

■ **Definition 5.5.11**

The **nullity** of a matrix A is the dimension of the solution space of $A\mathbf{x} = \mathbf{0}$ and is denoted by null A.

■ **Example 5.5.12**

The reduced row echelon form of the matrix $A = \begin{bmatrix} -2 & 2 & 6 \\ -1 & 7 & 9 \\ 3 & 4 & -2 \\ -1 & -3 & -1 \end{bmatrix}$ is the ma-

trix $\begin{bmatrix} 1 & 0 & -2 \\ 0 & 1 & 1 \\ 0 & 0 & 0 \\ 0 & 0 & 0 \end{bmatrix}$. Therefore, $\left\{ \begin{bmatrix} -2 \\ -1 \\ 3 \\ -1 \end{bmatrix}, \begin{bmatrix} 2 \\ 7 \\ 4 \\ -3 \end{bmatrix} \right\}$ is a basis for the column space

of A, and both $\left\{ \begin{bmatrix} 1 \\ 0 \\ -2 \end{bmatrix}, \begin{bmatrix} 0 \\ 1 \\ 1 \end{bmatrix} \right\}$ and $\left\{ \begin{bmatrix} -2 \\ 2 \\ 6 \end{bmatrix}, \begin{bmatrix} -1 \\ 7 \\ 9 \end{bmatrix} \right\}$ are a bases for the row

space of A, so rank $A = 2$. Also, $\left\{ \begin{bmatrix} 2 \\ -1 \\ 1 \end{bmatrix} \right\}$ is a basis for the null space of A, so

null $A = 1$. It is not a coincidence that the sum of the rank of A and its nullity equals the number of columns of A.

■ **Theorem 5.5.13 [Rank-Nullity Theorem for Matrices]**

For every $m \times n$ matrix A, rank A + null $A = n$.

Proof

Let $T : \mathbb{R}^n \to \mathbb{R}^m$ be the linear transformation such that $[T] = A$. By the Rank-Nullity Theorem (5.5.2),

$$\begin{aligned} \text{rank } A + \text{null } A &= \dim \text{CS } A + \dim \text{NS } A \\ &= \dim (\text{ran } T) + \dim (\ker T) \\ &= \text{rank } T + \text{null } T = \dim \mathbb{R}^n = n. \blacksquare \end{aligned}$$

Let A be an invertible $n \times n$ matrix. Theorem 4.3.15(b) implies that $A\mathbf{x} = \mathbf{0}$ has only the trivial solution. Hence, null $A = 0$, and by the Rank-Nullity Theorem for Matrices (5.5.13), rank $A = n$. On the other hand, suppose that A is not invertible. Then, $A\mathbf{x} = \mathbf{0}$ has a nontrivial solution, so null A is positive, which implies that rank $A < n$. Since a linear operator $T : \mathbb{R}^n \to \mathbb{R}^n$ is invertible if and only if $[T]$ is invertible, the same argument can be used to show that T is invertible if and only if rank $T = n$ and null $T = 0$. These results are summarized in parts (a)–(c) of the following theorem and its corollary. The proof of the equivalence of the remaining parts is left to Exercise 16.

■ **Theorem 5.5.14**

Let A be an $n \times n$ matrix. The following are equivalent.

(a) A is invertible.

(b) rank $A = n$.

(c) null $A = 0$.

(d) The columns of A are linearly independent.

(e) The transpose of the rows of A are linearly independent.

■ **Corollary 5.5.15**

Let $T : \mathbb{R}^n \to \mathbb{R}^n$ be a linear operator. The following are equivalent.

(a) T is invertible.

(b) rank $T = n$.

(c) null $T = 0$.

(d) The columns of $[T]$ are linearly independent.

(e) The transpose of the rows of $[T]$ are linearly independent.

Exercises

1. Find the rank and nullity of the given linear transformations.

(a) $T : \mathbb{R}^{2\times2} \to \mathbb{R}^2$, $T \begin{bmatrix} a & b \\ c & d \end{bmatrix} = \begin{bmatrix} -6b + c \\ -5a - 3b + 5c \end{bmatrix}$

(b) $T : \mathbb{R}^2 \to \mathbb{P}_2$, $T \begin{bmatrix} a \\ b \end{bmatrix} = (6a - 6b) + (9a - b)x + (6x - 4y)x^2$

(c) $T : \mathbb{R}^3 \to \mathbb{R}^3$, $T \begin{bmatrix} x \\ y \\ z \end{bmatrix} = \begin{bmatrix} 7x - 3y - 2z \\ 3x + 7y - 5z \\ 3x - 5y + z \end{bmatrix}$

(d) $T : \mathbb{P}_3 \to \mathbb{R}^{2\times2}$, $T(a + bx + cx^2 + dx^3) = \begin{bmatrix} 3b + 4d & -3a - 5c \\ -5c + 4d & -3a + 3b \end{bmatrix}$

2. Use the given information to determine whether the linear transformation T is one-to-one or onto.

(a) $T : \mathbb{R}^3 \to \mathbb{P}_2$, rank $T = 2$ (c) $T : \mathbb{P}_4 \to \mathbb{R}^5$, rank $T = 5$

(b) $T : \mathbb{R}^{2\times2} \to \mathbb{R}^3$, rank $T = 3$ (d) $T : \mathbb{R}^{2\times3} \to \mathbb{R}^{4\times5}$, rank $T = 6$

3. The trace function tr $: \mathbb{R}^{n\times n} \to \mathbb{R}$ is known to be a linear transformation by Example 5.1.13. Find the rank and nullity of tr .

4. Find the rank and nullity of the linear transformations of Example 5.1.16. Use these values to determine whether the transformations are one-to-one or onto.

5. If the rank of the linear transformation $T : V \to W$ is 0, what does this say about T?

6. Prove that the set $\{v'_{k+1}, v'_{k+2}, \ldots, v'_n\}$ from the proof of the Rank-Nullity Theorem (5.5.2) is linearly independent.

7. Let $T_1 : \mathbb{R}^3 \to \mathbb{R}^2$ be defined by $T_1 \begin{bmatrix} x \\ y \\ z \end{bmatrix} = \begin{bmatrix} 3x + 5y - z \\ 4x - 2y - 6z \end{bmatrix}$.

    (a) Find the rank and nullity of T_1 and use this information to determine whether T_1 is one-to-one or onto.
    (b) Define $T_2 : \mathbb{R}^2 \to \mathbb{R}^3$ so that $[T_2] = [T_1]^T$. Find a vector equation for T_2.
    (c) Find the rank and nullity of T_2 and use this information to determine whether T_2 is one-to-one or onto.
    (d) Is it possible to predict whether T_2 is one-to-one or onto by simply examining T_1 without looking at the transpose of its standard matrix? Explain.

8. Find a basis and the dimension for the column space, row space, and null space of given matrices.

    (a) $A = \begin{bmatrix} 7 & 1 \\ 8 & 8 \end{bmatrix}$

    (b) $B = \begin{bmatrix} 1 & 6 & -2 \\ 0 & -2 & 2 \\ 5 & 6 & 8 \\ -5 & 5 & -3 \end{bmatrix}$

    (c) $C = \begin{bmatrix} -2 & 0 & -2 & -6 \\ -2 & 9 & 7 & 3 \\ -1 & 5 & 4 & 2 \end{bmatrix}$

    (d) $D = \begin{bmatrix} -1 & -6 & 5 & 5 \\ -2 & 6 & -2 & 4 \end{bmatrix}$

9. Find the rank and nullity of the matrices in Exercise 8.

10. Answer true or false, and explain using the matrices of Exercise 8.

    (a) $\begin{bmatrix} 3 \\ 7 \end{bmatrix} \in RS\,A$

    (b) $\begin{bmatrix} 1 \\ 1 \\ 1 \\ 1 \end{bmatrix} \in CS\,B$

    (c) $\begin{bmatrix} -4 \\ -2 \\ 1 \\ 1 \end{bmatrix} \in NS\,C$

    (d) $\begin{bmatrix} 6 \\ -6 \end{bmatrix} \in NS\,D$

11. Let A be a 5×5 matrix and $\mathbf{u}$ be a nonzero vector in $\mathbb{R}^5$. Answer true, false, or possible, and explain.

    (a) $\mathbf{u} \in CS\,A$ if rank $A = 1$
    (b) $\mathbf{u} \in NS\,A$ if null $A = 3$
    (c) $\mathbf{u} \in RS\,A$ if rank $A = 0$
    (d) $\mathbf{u} \in NS\,A$ if null $A = 0$
    (e) $\mathbf{u} \in RS\,A$ if rank $A = 5$
    (f) $\mathbf{u} \in CS\,A$ if null $A = 5$

12. Use the given information to determine the rank of the given matrix A.

    (a) The nullity of the 4×8 matrix A is 3.
    (b) A is an 5×7 matrix with dim NS $A = 2$.
    (c) The 9×6 matrix A is the coefficient matrix of the system of linear equations $Ax = \mathbf{0}$ with a four-dimensional solution space.
    (d) A is a 4×4 matrix with linearly independent columns.

13. Use the given information to determine if the 5×5 matrix A is invertible. Explain.

    (a) rank $A = 5$ (c) null $A = 4$
    (b) rank $A = 3$ (d) null $A = 0$

14. Let $T : \mathbb{R}^n \to \mathbb{R}^m$ be a linear transformation. Prove.

    (a) T is one-to-one if and only if NS $[T] = \{\mathbf{0}\}$.
    (b) T is onto if and only if CS $[T] = \mathbb{R}^m$.

15. Let A be an $m \times n$ matrix. Prove.

    (a) The rank of A equals the number of leading 1s in the reduced row echelon form of A.

    (b) The nullity of A equals the number of parameters required to describe the solution of $A\mathbf{x} = \mathbf{0}$.

16. Complete the proof of Theorem 5.5.14.

17. Consider the block diagonal matrix with A and B along its diagonal. Prove that
$$\text{rank} \begin{bmatrix} A & 0 \\ 0 & B \end{bmatrix} = \text{rank}\, A + \text{rank}\, B.$$

18. Let A be an $m \times n$ matrix and $\mathbf{b} \in \mathbb{R}^n$. Denote the matrix A augmented by $\mathbf{b}$ by $[A \mid \mathbf{b}]$. Let $p = \text{rank}\, A$ and $q = \text{rank}\, [A \mid \mathbf{b}]$. Determine the number of solutions of $A\mathbf{x} = \mathbf{b}$ under the given conditions.

    (a) $p < q$
    (b) $p = q$ and $p < n$
    (c) $p = q = n$

19. Let $A \in M_n(\mathbb{R})$ and $\mathbf{u}_1, \mathbf{u}_2, \ldots, \mathbf{u}_n \in \mathbb{R}^n$ form a linearly independent set. Show that rank $A = n$ if and only if $\{A\mathbf{u}_1, A\mathbf{u}_2, \ldots, A\mathbf{u}_n\}$ is linearly independent.

20. Let A be an $m \times n$ matrix. Prove that if $\mathbf{u} \in \text{CS}\, A$, then $\mathbf{u} = A\mathbf{v}$ for some $\mathbf{v} \in \mathbb{R}^n$.

21. Prove that rank $A = \text{rank}\, A^T$ for all matrices A.

22. Let A be an $m \times n$ matrix. Prove.

    (a) dim RS $A + \dim$ NS $A = n$
    (b) dim CS $A + \dim$ NS $A^T = m$

23. Assume that A is an $m \times n$ matrix and $\mathbf{b} \in \mathbb{R}^m$. Prove that $A\mathbf{x} = \mathbf{b}$ is consistent if and only if $A^T\mathbf{x} = \mathbf{0}$ has only the trivial solution.

24. Let A be an $m \times n$ matrix and B be an $n \times p$ matrix. Prove.

    (a) rank $AB \leq \text{rank}\, A$
    (b) rank $AB \leq \text{rank}\, B$

5.6 Isomorphism

It is often required to determine whether two given vector spaces are identical in structure. The examples of vector spaces in Section 5.1 suggest that this is not uncommon. For example, $\mathbb{R}^6$ appears to resemble $\mathbb{R}^{3\times2}$ because a 3×2 matrix has ordered entries like a six-dimensional vector has ordered coordinates, matrices add like vectors, and scalar multiplication works essentially the same way on both sets. Maybe $\mathbb{R}^{3\times2}$ is simply $\mathbb{R}^6$ written differently. If this is the case, there is a bijection that pairs every element of $\mathbb{R}^6$ with exactly one matrix of $\mathbb{R}^{3\times2}$, thus renaming the elements of $\mathbb{R}^6$. But, this is not enough. If $\mathbb{R}^6$ and $\mathbb{R}^{3\times2}$ are truly copies of each other as vector spaces, the bijection should also be a linear transformation so that the operations are also seen as copies of each other.

■ **Definition 5.6.1**

An **isomorphism** is a linear transformation that is also a bijection.

Here is an isomorphism $\mathbb{R}^6 \rightarrow \mathbb{R}^{3\times2}$. Define T by

$$T \begin{bmatrix} a_1 \\ a_2 \\ a_3 \\ a_4 \\ a_5 \\ a_6 \end{bmatrix} = \begin{bmatrix} a_1 & a_2 \\ a_3 & a_4 \\ a_5 & a_6 \end{bmatrix}. \tag{5.33}$$

Notice that the function uniquely renames the vectors of $\mathbb{R}^6$, for example,

$$T \begin{bmatrix} 1 \\ 2 \\ 4 \\ 6 \\ 8 \\ 0 \end{bmatrix} = \begin{bmatrix} 1 & 2 \\ 4 & 6 \\ 8 & 0 \end{bmatrix} \text{ and } T \begin{bmatrix} 3 \\ 1 \\ 5 \\ 3 \\ 7 \\ 5 \end{bmatrix} = \begin{bmatrix} 3 & 1 \\ 5 & 3 \\ 7 & 5 \end{bmatrix}.$$

Since T is a linear transformation, T also renames vector addition, as with

$$T \left(\begin{bmatrix} 1 \\ 2 \\ 4 \\ 6 \\ 8 \\ 0 \end{bmatrix} + \begin{bmatrix} 3 \\ 1 \\ 5 \\ 3 \\ 7 \\ 5 \end{bmatrix} \right) = \begin{bmatrix} 1 & 2 \\ 4 & 6 \\ 8 & 0 \end{bmatrix} + \begin{bmatrix} 3 & 1 \\ 5 & 3 \\ 7 & 5 \end{bmatrix} = T \begin{bmatrix} 1 \\ 2 \\ 4 \\ 6 \\ 8 \\ 0 \end{bmatrix} + T \begin{bmatrix} 3 \\ 1 \\ 5 \\ 3 \\ 7 \\ 5 \end{bmatrix},$$

and T renames scalar multiplication, as with

$$T\left(7\begin{bmatrix}1\\2\\4\\6\\8\\0\end{bmatrix}\right) = 7\begin{bmatrix}1 & 2\\4 & 6\\8 & 0\end{bmatrix} = 7T\begin{bmatrix}1\\2\\4\\6\\8\\0\end{bmatrix}.$$

Generalize this argument in Exercise 2 to prove that T is an isomorphism. Moreover, there are other functions that rename each element of $\mathbb{R}^6$ with exactly one element of $\mathbb{R}^{3\times2}$ and copies the operations of $\mathbb{R}^6$. Two other such isomorphisms are

$$T_1\begin{bmatrix}a_1\\a_2\\a_3\\a_4\\a_5\\a_6\end{bmatrix} = \begin{bmatrix}a_1 & a_4\\a_2 & a_5\\a_3 & a_6\end{bmatrix} \quad \text{and} \quad T_2\begin{bmatrix}a_1\\a_2\\a_3\\a_4\\a_5\\a_6\end{bmatrix} = \begin{bmatrix}2a_3 & 7a_4\\4a_1 & 5a_2\\3a_5 & 4a_6\end{bmatrix}, \tag{5.34}$$

yet only one such function is needed to show that one vector space is a copy of another.

■ **Definition 5.6.2**

Vector spaces V and W are **isomorphic** if there is an isomorphism $V \to W$.

This means that (5.33) shows that $\mathbb{R}^6$ is isomorphic to $\mathbb{R}^{3\times2}$. Taken individually, the linear transformations of (5.34) also show this.

■ **Example 5.6.3**

$\mathbb{P}_3$ is isomorphic to $\mathbb{R}^{2\times2}$. Define $T : \mathbb{P}_3 \to \mathbb{R}^{2\times2}$ by

$$T(a_0 + a_1x + a_2x^2 + a_3x^3) = \begin{bmatrix}a_0 & a_1\\a_2 & a_3\end{bmatrix}.$$

That T is a linear transformation is proved in Example 5.1.12. Also, T is one-to-one by Theorem 5.2.20 because $\ker T = \{0 + 0x + 0x^2 + 0x^3\}$. Therefore, since $\dim \mathbb{P}_3 = \dim \mathbb{R}^{2\times2} = 4$, conclude that T is a bijection by Corollary 5.5.4 and, thus, is an isomorphism.

■ **Example 5.6.4**

$V = \left\{\begin{bmatrix}a & 0\\0 & b\end{bmatrix} : a, b \in \mathbb{R}\right\}$ is isomorphic to $\mathbb{R}^2$. This is shown by defining $T : V \to \mathbb{R}^2$ by $T\begin{bmatrix}a & 0\\0 & b\end{bmatrix} = \begin{bmatrix}a\\b\end{bmatrix}$. That T is an isomorphism is Exercise 3.

Since $\mathbb{P}_3$ is isomorphic to $\mathbb{R}^{2\times 2}$, it follows that $\mathbb{R}^{2\times 2}$ is isomorphic to $\mathbb{P}_3$. Furthermore, since $\mathbb{R}^{2\times 2}$ is isomorphic to $\mathbb{R}^4$, it is the case that $\mathbb{P}_3$ is also isomorphic to $\mathbb{R}^4$. There are similar results for real numbers. For example, $a = a$, if $a = b$, then $b = a$, and if $a = b$ and $b = c$, then $a = c$. Putting this together, this means that isomorphic vector spaces are equal in some sense.

■ **Lemma 5.6.5**

If $T_1 : U \to V$ and $T_2 : V \to W$ are isomorphisms, then T_1^{-1} and $T_2 \circ T_1$ are isomorphisms.

Proof

The proof that the composition of linear transformations is a linear transformation follows the proof of Theorem 3.2.12, and the proof that the inverse of a linear transformation is a linear transformation follows the proof of Theorem 4.1.1. That the composition of bijections and the inverse of a bijection are both bijections is Theorem 1.4.30. ■

■ **Theorem 5.6.6**

Let V_1, V_2, and V_3 be vector spaces.

   (a) V_1 is isomorphic to itself.

   (b) If V_1 is isomorphic to V_2, then V_2 is isomorphic to V_1.

   (c) If V_1 is isomorphic to V_2 and V_2 is isomorphic to V_3, then V_1 is isomorphic to W.

Proof

Since the identity map is an isomorphism, (a) follows. Parts (b) and (c) are implied by Lemma 5.6.5. ■

According to Theorem 5.1.14, a linear transformation will map the zero vector of its domain to the zero vector of its range. It will also map additive inverses to additive inverses. This means that linear transformations preserve certain aspects of a vector space. There are other properties, such as being a linearly independent set or even a basis, that are not preserved by linear transformations in general. For example, the image of a basis under the zero transformation will not be linearly independent, and thus not a basis. The linear transformation needs to be an isomorphism to preserve some of these stronger properties. Consider the following which includes the converse of Theorem 5.3.10.

■ **Lemma 5.6.7**

Let V and W be vector spaces with $\mathcal{S} \subset V$. Let $T : V \to W$ be a linear transformation.

   (a) If T is an injection and $\mathcal{S}$ is linearly independent, then $T[\mathcal{S}]$ is linearly independent.

(b) If T is a surjection and S is a spanning set of V, then $T[S]$ is a spanning set for W.

Proof

First, assume that $T : V \to W$ is one-to-one. Let S be linearly independent. Take $\mathbf{u}_1, \mathbf{u}_2, \dots, \mathbf{u}_k \in S$ and $r_1, r_2, \dots, r_k \in \mathbb{R}$ and assume that

$$r_1(T\mathbf{u}_1) + r_2(T\mathbf{u}_2) + \cdots + r_k(T\mathbf{u}_k) = \mathbf{0}.$$

Because T is a linear transformation,

$$T(r_1\mathbf{u}_1 + r_2\mathbf{u}_2 + \cdots + r_k\mathbf{u}_k) = \mathbf{0},$$

so by Theorem 5.2.20,

$$r_1\mathbf{u}_1 + r_2\mathbf{u}_2 + \cdots + r_k\mathbf{u}_k = \mathbf{0}.$$

Therefore, because S is linearly independent, $r_1 = r_2 = \cdots = r_n = 0$.

Now, suppose that T is onto and that S is a spanning set of V. Let $\mathbf{w} \in W$. Since T is a surjection, there exists $\mathbf{v} \in V$ such that $T\mathbf{v} = \mathbf{w}$. Write for some $\mathbf{u}_1, \mathbf{u}_2, \dots, \mathbf{u}_k \in S$ and $r_1, r_2, \dots, r_n \in \mathbb{R}$,

$$\mathbf{v} = r_1\mathbf{u}_1 + r_2\mathbf{u}_2 + \cdots + r_k\mathbf{u}_k,$$

so since T is a linear transformation,

$$\mathbf{w} = T\mathbf{v} = T(r_1\mathbf{u}_1 + r_2\mathbf{u}_2 + \cdots + r_k\mathbf{u}_k) = r_1T\mathbf{u}_1 + r_2T\mathbf{u}_2 + \cdots + r_kT\mathbf{u}_k. \ \blacksquare$$

■ **Theorem 5.6.8**

Let V and W be vector spaces with $\mathcal{B}$ being a basis for V. If $T : V \to W$ is an isomorphism, then $T[\mathcal{B}]$ is a basis for W.

Proof

Let $T : V \to W$ be an isomorphism and apply Lemma 5.6.7. The image $T[\mathcal{B}]$ is linearly independent because T is one-to-one and $\mathcal{B}$ is linearly independent, and $T[\mathcal{B}]$ is a spanning set of W because T is onto and $\mathcal{B}$ is a spanning set of V. ■

The corollary follows immediately from Theorem 5.6.8 because $\{\mathbf{b}_1, \mathbf{b}_2, \dots, \mathbf{b}_n\}$ and $\{T\mathbf{b}_1, T\mathbf{b}_2, \dots, T\mathbf{b}_n\}$ have the same number of elements if T is a bijection.

■ **Corollary 5.6.9**

If V and W are isomorphic finite dimensional vector spaces, $\dim V = \dim W$.

■ **Example 5.6.10**

Define $T : \mathbb{R}^3 \to \mathbb{P}_2$ by $T \begin{bmatrix} a \\ b \\ c \end{bmatrix} = (a + 3c) + 2cx + (a - b)x^2$. This is an iso-

morphism. To prove this, take $\begin{bmatrix} u_1 \\ u_2 \\ u_3 \end{bmatrix}, \begin{bmatrix} v_1 \\ v_2 \\ v_3 \end{bmatrix} \in \mathbb{R}^3$ and $r \in \mathbb{R}$. Then,

$$T \begin{bmatrix} u_1 + v_1 \\ u_2 + v_2 \\ u_3 + v_3 \end{bmatrix} = T \begin{bmatrix} u_1 \\ u_2 \\ u_3 \end{bmatrix} + T \begin{bmatrix} v_1 \\ v_2 \\ v_3 \end{bmatrix}$$

because

$$u_1 + v_1 + 3(u_3 + v_3) + 2(u_3 + v_3)x + (u_1 + v_1 - [u_2 + v_2])x^2$$
$$= u_1 + 3u_3 + 2u_3x + (u_1 - u_2)x^2 + v_1 + 3v_3 + 2v_3x + (v_1 - v_2)x^2$$

and

$$T \begin{bmatrix} ru_1 \\ ru_2 \\ ru_3 \end{bmatrix} = r \left(T \begin{bmatrix} u_1 \\ u_2 \\ u_3 \end{bmatrix} \right)$$

because

$$ru_1 + 3ru_3 + 2ru_3x + (ru_1 - ru_2)x^2 = r[u_1 + 3u_3 + 2u_3x + (u_1 - u_2)x^2]$$

Next, solve $T \begin{bmatrix} a \\ b \\ c \end{bmatrix} = 0 + 0x + 0x^2$ to check the kernel of T. That is, solve the

system,

$$
\begin{array}{rcrcl}
a & & + & 3c & = & 0, \\
& & & 2c & = & 0, \\
a & - & b & & = & 0,
\end{array}
$$

finding $a = b = c = 0$, which implies that ker $(T) = \{\mathbf{0}\}$. Therefore, T is one-to-one, and by Corollary 5.5.4, T is also onto. Since T is an isomorphism, and

$$\left\{ \begin{bmatrix} 1 \\ 1 \\ 0 \end{bmatrix}, \begin{bmatrix} 0 \\ 1 \\ 1 \end{bmatrix}, \begin{bmatrix} 1 \\ 0 \\ 1 \end{bmatrix} \right\}$$

is a basis for $\mathbb{R}^3$, Theorem 5.6.8 implies that

$$\{1, 3 + 2x - x^2, 4 + 2x + x^2\}$$

is a basis for $\mathbb{P}_2$. Moreover, it should be noted that dim $\mathbb{R}^3 = $ dim $\mathbb{P}_2$.

■ Example 5.6.11

As a result of Corollary 5.6.9, $\mathbb{R}^7$ is not isomorphic to $\mathbb{R}^5$. A more interesting example is the following. For every positive integer n, the vector space $^R\mathbb{R}$ is not isomorphic to $\mathbb{R}^n$ because if they were isomorphic, Corollary 5.6.9 would imply that the dimension of $^R\mathbb{R}$ would be n. This is impossible because $\mathbb{P}$ is a subspace of $^R\mathbb{R}$ (Example 5.2.11) and $\mathbb{P}$ is not finite dimensional (Example 5.4.23).

Coordinates

Since an isomorphism can be regarded as a renaming of the elements of one vector space with the elements of another, it is natural to consider the possibility of uniform names for vectors from an abstract vector space. This will be accomplished by relying on the following theorem.

■ Theorem 5.6.12

Let V be a finite-dimensional vector space with basis $\mathcal{B}$. Every vector in V can be written uniquely as a linear combination of the vectors of $\mathcal{B}$.

Proof

Write $\mathcal{B} = \{\mathbf{b}_1, \mathbf{b}_2, \dots, \mathbf{b}_n\}$. Take $\mathbf{u} \in V$ and scalars $r_i, s_i \in \mathbb{R}$ for $i = 1, 2, \dots, n$ such that

$$\mathbf{u} = r_1\mathbf{b}_1 + r_2\mathbf{b}_2 + \cdots + r_n\mathbf{b}_n = s_1\mathbf{b}_1 + s_2\mathbf{b}_2 + \cdots + s_n\mathbf{b}_n.$$

Then,

$$(r_1\mathbf{b}_1 + r_2\mathbf{b}_2 + \cdots + r_n\mathbf{b}_n) - (s_1\mathbf{b}_1 + s_2\mathbf{b}_2 + \cdots + s_n\mathbf{b}_n) = \mathbf{0},$$

so

$$(r_1 - s_1)\mathbf{b}_1 + (r_2 - s_2)\mathbf{b}_2 + \cdots + (r_n - s_n)\mathbf{b}_n = \mathbf{0}.$$

But, $\mathcal{B}$ is linearly independent, so

$$r_1 - s_1 = r_2 - s_2 = \cdots = r_n - s_n = 0$$

or, in other words, $r_1 = s_1, \quad r_2 = s_2, \quad \dots, \quad r_n = s_n.$ ■

Let V be a finite-dimensional vector space. An **ordered basis** of V is a basis $\{\mathbf{b}_1, \mathbf{b}_2, \dots, \mathbf{b}_n\}$ such that the order in which the vectors are written matters. For example, although as sets

$$\left\{ \begin{bmatrix} 1 \\ 0 \end{bmatrix}, \begin{bmatrix} 0 \\ 1 \end{bmatrix} \right\} = \left\{ \begin{bmatrix} 0 \\ 1 \end{bmatrix}, \begin{bmatrix} 1 \\ 0 \end{bmatrix} \right\},$$

these sets are considered different when viewed as ordered bases.

Consider the vector space $\mathbb{P}_2$ with ordered basis

$$\mathscr{B} = \{1 + x + x^2, 1 + x, 1\}. \tag{5.35}$$

For every polynomial $p(x)$ of degree at most 2, Theorem 5.6.12 implies that there are unique scalars r_1, r_2, r_3, in that order, that can be paired with the polynomials of $\mathscr{B}$ so that

$$p(x) = r_1(1 + x + x^2) + r_2(1 + x) + r_3. \tag{5.36}$$

Conversely, given $\begin{bmatrix} r_1 \\ r_2 \\ r_3 \end{bmatrix} \in \mathbb{R}^3$, there exists a unique $p(x) \in \mathbb{P}_2$ such that (5.36) holds. For example, given the polynomial $8 + x + 3x^2$,

$$8 + x + 3x^2 = 3(1 + x + x^2) - 2(1 + x) + 7,$$

so the vector $\begin{bmatrix} 3 \\ -2 \\ 7 \end{bmatrix}$ uniquely identifies $8 + x + 3x^2$. Observe, that if the basis was written as

$$\mathscr{B}' = \{x + 1, 1 + x + x^2, 1\}, \tag{5.37}$$

$\begin{bmatrix} -2 \\ 3 \\ 7 \end{bmatrix}$ would represent $8 + x + 3x^2$, instead. It appears, then, that all of this describes a function relating elements of an n-dimensional vector space with vectors in $\mathbb{R}^n$ that is dependent on the choice of the ordered basis.

To generalize, take a finite-dimensional vector space V with the ordered basis $\mathscr{B} = \{\mathbf{b}_1, \mathbf{b}_2, \dots, \mathbf{b}_n\}$. For every $\mathbf{u} \in V$, Theorem 5.6.12 implies that there exists unique $r_1, r_2, \dots, r_n \in \mathbb{R}$ such that $\mathbf{u} = r_1 \mathbf{b}_1 + r_2 \mathbf{b}_2 + \cdots + r_n \mathbf{b}_n$. These unique scalars can be used to uniquely identify or name the vector $\mathbf{u}$. This is the purpose of the next definition.

■ **Definition 5.6.13**

Let $\mathscr{B} = \{\mathbf{b}_1, \mathbf{b}_2, \dots, \mathbf{b}_n\}$ be an ordered basis for $\mathbb{R}^n$. For all $\mathbf{u} \in \mathbb{R}^n$, let $r_1, r_2, \dots, r_n$ be the unique real numbers such that

$$\mathbf{u} = r_1 \mathbf{b}_1 + r_2 \mathbf{b}_2 + \cdots + r_n \mathbf{b}_n.$$

Define the $\mathscr{B}$-**coordinate vector** of $\mathbf{u}$ to be $\mathbf{u}_{\mathscr{B}} = \begin{bmatrix} r_1 \\ r_2 \\ \vdots \\ r_n \end{bmatrix}$.

To understand what is happening in Definition 5.6.13, take a vector $\mathbf{w} = \begin{bmatrix} a \\ b \end{bmatrix}$ in $\mathbb{R}^2$ as illustrated in Figure 5.2. Let $\mathscr{E} = \{\mathbf{e}_1, \mathbf{e}_2\}$ be the standard basis of $\mathbb{R}^2$.

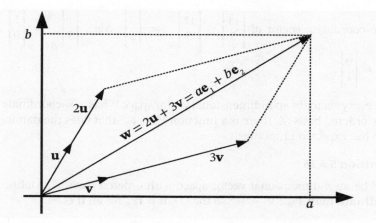

Figure 5.2   Writing **w** as a linear combination of two bases.

Viewing $\mathcal{E}$ as an ordered basis, $\mathbf{w}_{\mathcal{E}} = \begin{bmatrix} a \\ b \end{bmatrix}$. If the order of $\mathcal{E}$ is reversed resulting in $\mathcal{E}' = \{\mathbf{e}_2, \mathbf{e}_1\}$, the $\mathcal{E}'$-coordinate vector of **w** is $\mathbf{w}_{\mathcal{E}'} = \begin{bmatrix} b \\ a \end{bmatrix}$. Lastly, let $\mathcal{B} = \{\mathbf{u}, \mathbf{v}\}$ be a basis of $\mathbb{R}^2$ so that $\mathbf{w} = 2\mathbf{u} + 3\mathbf{v}$ as illustrated in Figure 5.2. This implies $\mathbf{w}_{\mathcal{B}} = \begin{bmatrix} 2 \\ 3 \end{bmatrix}$. This illustrates how each vector in a vector space has a unique name, which is its coordinate vector, for every ordered basis of the vector space.

■ Example 5.6.14

Take $\mathbf{u} \in \mathbb{R}^n$ with coordinates $u_1, u_2, \ldots, u_n \in \mathbb{R}$. Letting the standard basis $\mathcal{E} = \{\mathbf{e}_1, \mathbf{e}_2, \ldots, \mathbf{e}_n\}$ be considered an ordered basis, the $\mathcal{E}$-coordinate vector of **u** is

$$\mathbf{u}_{\mathcal{E}} = \begin{bmatrix} u_1 \\ u_2 \\ \vdots \\ u_n \end{bmatrix}_{\mathcal{E}} = \begin{bmatrix} u_1 \\ u_2 \\ \vdots \\ u_n \end{bmatrix}$$

because $\mathbf{u} = u_1\mathbf{e}_1 + u_2\mathbf{e}_2 + \cdots + u_n\mathbf{e}_n$.

■ Example 5.6.15

Let $\mathcal{B} = \left\{ \begin{bmatrix} 1 \\ 1 \end{bmatrix}, \begin{bmatrix} 4 \\ 1 \end{bmatrix} \right\}$. Because

$$\begin{bmatrix} 11 \\ 5 \end{bmatrix} = 3\begin{bmatrix} 1 \\ 1 \end{bmatrix} + 2\begin{bmatrix} 4 \\ 1 \end{bmatrix},$$

the $\mathscr{B}$-coordinate vector of $\begin{bmatrix} 11 \\ 5 \end{bmatrix}$ is $\begin{bmatrix} 11 \\ 5 \end{bmatrix}_{\mathscr{B}} = \begin{bmatrix} 3 \\ 2 \end{bmatrix}$. Also, $\begin{bmatrix} 1 \\ 1 \end{bmatrix}_{\mathscr{B}} = \begin{bmatrix} 1 \\ 0 \end{bmatrix}$ and $\begin{bmatrix} 4 \\ 1 \end{bmatrix}_{\mathscr{B}} = \begin{bmatrix} 0 \\ 1 \end{bmatrix}$.

Since every vector in an n-dimensional vector space V has a $\mathscr{B}$-coordinate vector for every ordered basis $\mathscr{B}$, there is a function $V \to \mathbb{R}^n$ that does the naming. This function has expected properties.

■ **Definition 5.6.16**

Let V be an n-dimensional vector space with ordered basis $\mathscr{B}$. Define the **coordinate map** $T_{\mathscr{B}} : V \to \mathbb{R}^n$ so that $T_{\mathscr{B}}\mathbf{u} = \mathbf{u}_{\mathscr{B}}$ for all $\mathbf{u} \in V$.

■ **Example 5.6.17**

Using the coordinate map $T_{\mathscr{B}} : \mathbb{P}_2 \to \mathbb{R}^n$ with $\mathscr{B}$ as in (5.35),

$$T_{\mathscr{B}}(8 + x + 3x^2) = \begin{bmatrix} 3 \\ -2 \\ 7 \end{bmatrix},$$

and using $T_{\mathscr{B}'} : \mathbb{P}_2 \to \mathbb{R}^n$ with $\mathscr{B}'$ as in (5.37),

$$T'_{\mathscr{B}}(8 + x + 3x^2) = \begin{bmatrix} -2 \\ 3 \\ 7 \end{bmatrix}.$$

Of course, writing $\mathscr{E} = \{x^2, x, 1\}$ yields

$$T_{\mathscr{E}}(8 + x + 3x^2) = \begin{bmatrix} 3 \\ 1 \\ 8 \end{bmatrix}.$$

■ **Theorem 5.6.18**

A coordinate map is an isomorphism.

Proof

Take an ordered basis $\mathscr{B} = \{\mathbf{b}_1, \mathbf{b}_2, \dots, \mathbf{b}_n\}$ for the vector space V. Let $\mathbf{u}, \mathbf{v} \in V$ with $r \in \mathbb{R}$ and write for some $u_i, v_i \in \mathbb{R}$ with $i = 1, 2, \dots, n$,

$$\mathbf{u} = u_1\mathbf{b}_1 + u_2\mathbf{b}_2 + \cdots + u_n\mathbf{b}_n \quad \text{and} \quad \mathbf{v} = v_1\mathbf{b}_1 + v_2\mathbf{b}_2 + \cdots + v_n\mathbf{b}_n.$$

This implies that

$$T_{\mathscr{B}}\mathbf{u} = \begin{bmatrix} u_1 \\ u_2 \\ \vdots \\ u_n \end{bmatrix} \quad \text{and} \quad T_{\mathscr{B}}\mathbf{v} = \begin{bmatrix} v_1 \\ v_2 \\ \vdots \\ v_n \end{bmatrix}.$$

The coordinate map $T_{\mathcal{B}} : V \to \mathbb{R}^n$ is seen to be a linear transformation because

$$\mathbf{u} + \mathbf{v} = (u_1 + v_1)\mathbf{b}_1 + (u_2 + v_2)\mathbf{b}_2 + \cdots + (u_n + v_n)\mathbf{b}_n,$$

it follows that

$$T_{\mathcal{B}}(\mathbf{u} + \mathbf{v}) = \begin{bmatrix} u_1 + v_1 \\ u_2 + v_2 \\ \vdots \\ u_n + v_n \end{bmatrix} = \begin{bmatrix} u_1 \\ u_2 \\ \vdots \\ u_n \end{bmatrix} + \begin{bmatrix} v_1 \\ v_2 \\ \vdots \\ v_n \end{bmatrix} = T_{\mathcal{B}}\mathbf{u} + T_{\mathcal{B}}\mathbf{v},$$

and because

$$r\mathbf{u} = ru_1\mathbf{b}_1 + ru_2\mathbf{b}_2 + \cdots + ru_n\mathbf{b}_n,$$

it follows that

$$T_{\mathcal{B}}(r\mathbf{u}) = \begin{bmatrix} ru_1 \\ ru_2 \\ \vdots \\ ru_n \end{bmatrix} = r\begin{bmatrix} u_1 \\ u_2 \\ \vdots \\ u_n \end{bmatrix} = r(T_{\mathcal{B}}\mathbf{u}).$$

Furthermore, $T_{\mathcal{B}}$ is a surjection because for any $\begin{bmatrix} a_1 \\ a_2 \\ \vdots \\ a_n \end{bmatrix} \in \mathbb{R}^n$,

$$T_{\mathcal{B}}(a_1\mathbf{b}_1 + a_2\mathbf{b}_2 + \cdots + a_n\mathbf{b}_n) = \begin{bmatrix} a_1 \\ a_2 \\ \vdots \\ a_n \end{bmatrix},$$

and by Corollary 5.5.4, it follows that $T_{\mathcal{B}}$ is also an injection. ∎

■ Corollary 5.6.19

If V is an n-dimensional vector space, V is isomorphic to $\mathbb{R}^n$.

For example, if

$$\mathcal{B} = \left\{ \begin{bmatrix} 1 & 0 \\ 0 & 0 \end{bmatrix}, \begin{bmatrix} 0 & 1 \\ 0 & 0 \end{bmatrix}, \begin{bmatrix} 0 & 0 \\ 1 & 0 \end{bmatrix}, \begin{bmatrix} 0 & 0 \\ 0 & 1 \end{bmatrix} \right\},$$

the coordinate map $\mathbb{R}^{2\times 2} \to \mathbb{R}^4$ with respect to $\mathcal{B}$ shows that $\mathbb{R}^{2\times 2}$ is isomorphic to $\mathbb{R}^4$.

■ Corollary 5.6.20

For all V and W that are finite-dimensional vector spaces, V is isomorphic to W if and only if $\dim V = \dim W$.

Proof

Suppose that $\dim V = \dim W = n$. Appealing to Corollary 5.6.19, it follows that both V and W are isomorphic to $\mathbb{R}^n$. Thus, V is isomorphic to W by Theorem 5.6.6. The converse is Corollary 5.6.9. ■

For example, let V be a finite-dimensional vector space. Example 5.4.5 shows that $\dim V = \dim V^*$, which implies by Corollary 5.6.20 that V is isomorphic to V^*.

Change of Basis

Every vector in a vector space has a unique $\mathcal{B}$-coordinate vector for every basis $\mathcal{B}$ of the vector space. Furthermore, every vector in $\mathbb{R}^n$ is a coordinate vector for a vector in a vector space with respect to a given basis. Since each coordinate vector provides a name for the vector, it is natural to ask whether there is a simple relation between coordinate vectors with respect to different bases.

Take the ordered basis in $\mathbb{R}^2$,

$$\mathcal{B} = \left\{ \begin{bmatrix} 1 \\ 2 \end{bmatrix}, \begin{bmatrix} 6 \\ 3 \end{bmatrix} \right\}. \tag{5.38}$$

Let $\mathbf{u} = \begin{bmatrix} 20 \\ 13 \end{bmatrix}$. Notice that

$$\begin{bmatrix} 20 \\ 13 \end{bmatrix} = 2 \begin{bmatrix} 1 \\ 2 \end{bmatrix} + 3 \begin{bmatrix} 6 \\ 3 \end{bmatrix} = \begin{bmatrix} 1 & 6 \\ 2 & 3 \end{bmatrix} \begin{bmatrix} 2 \\ 3 \end{bmatrix}.$$

This implies that the coordinate map $T_{\mathcal{B}} : \mathbb{R}^2 \to \mathbb{R}^2$ gives

$$T_{\mathcal{B}} \begin{bmatrix} 20 \\ 13 \end{bmatrix} = \begin{bmatrix} 2 \\ 3 \end{bmatrix} \quad \text{and} \quad T_{\mathcal{B}}^{-1} \begin{bmatrix} 2 \\ 3 \end{bmatrix} = \begin{bmatrix} 1 & 6 \\ 2 & 3 \end{bmatrix} \begin{bmatrix} 2 \\ 3 \end{bmatrix} = \begin{bmatrix} 20 \\ 13 \end{bmatrix}.$$

Since the inverse of an isomorphism is an isomorphism (Lemma 5.6.5) and the standard matrix of a linear transformation is unique (Theorem 3.1.22),

$$[T_{\mathcal{B}}^{-1}] = \begin{bmatrix} 1 & 6 \\ 2 & 3 \end{bmatrix}.$$

■ **Theorem 5.6.21**

If $\mathcal{B} = \{\mathbf{b}_1, \mathbf{b}_2, \dots, \mathbf{b}_n\}$ is an ordered basis for $\mathbb{R}^n$ and $T_{\mathcal{B}}$ is the coordinate map $\mathbb{R}^n \to \mathbb{R}^n$,

$$[T_{\mathcal{B}}^{-1}] = \begin{bmatrix} \mathbf{b}_1 & \mathbf{b}_2 & \dots & \mathbf{b}_n \end{bmatrix}.$$

Proof

Let $\mathbf{u} \in \mathbb{R}^n$ and write $\mathbf{u} = r_1\mathbf{b}_1 + r_2\mathbf{b}_2 + \cdots + r_n\mathbf{b}_n$ for some $r_1, r_2, \ldots, r_n \in \mathbb{R}$. Then,

$$T_{\mathscr{B}}\mathbf{u} = \begin{bmatrix} r_1 \\ r_2 \\ \vdots \\ r_n \end{bmatrix}.$$

Since $T_{\mathscr{B}}$ is invertible (Theorem 5.6.18),

$$T_{\mathscr{B}}^{-1} \begin{bmatrix} r_1 \\ r_2 \\ \vdots \\ r_n \end{bmatrix} = \mathbf{u} = \begin{bmatrix} \mathbf{b}_1 & \mathbf{b}_2 & \cdots & \mathbf{b}_n \end{bmatrix} \begin{bmatrix} r_1 \\ r_2 \\ \vdots \\ r_n \end{bmatrix}. \quad \blacksquare$$

Continuing to use basis $\mathscr{B}$ of (5.38), the conversion from a vector in $\mathbb{R}^2$ to the $\mathscr{B}$-coordinate vector is achieved by reversing the process. Find the $\mathscr{B}$-coordinate vector for $\begin{bmatrix} 8 \\ 7 \end{bmatrix}$. By Theorem 5.6.21, if

$$\begin{bmatrix} 8 \\ 7 \end{bmatrix}_{\mathscr{B}} = \begin{bmatrix} a_1 \\ a_2 \end{bmatrix},$$

then

$$\begin{bmatrix} 1 & 6 \\ 2 & 3 \end{bmatrix} \begin{bmatrix} a_1 \\ a_2 \end{bmatrix} = \begin{bmatrix} 8 \\ 7 \end{bmatrix},$$

and since $\begin{bmatrix} 1 & 6 \\ 2 & 3 \end{bmatrix}$ is invertible (Theorem 5.5.14),

$$\begin{bmatrix} a_1 \\ a_2 \end{bmatrix} = \begin{bmatrix} 1 & 6 \\ 2 & 3 \end{bmatrix}^{-1} \begin{bmatrix} 8 \\ 7 \end{bmatrix}.$$

Specifically,

$$\begin{bmatrix} 1 & 6 \\ 2 & 3 \end{bmatrix}^{-1} = \begin{bmatrix} -1/3 & 2/3 \\ 2/9 & -1/9 \end{bmatrix}.$$

Hence, the $\mathscr{B}$-coordinate vector for $\begin{bmatrix} 8 \\ 7 \end{bmatrix}$ is $\begin{bmatrix} 8 \\ 7 \end{bmatrix}_{\mathscr{B}} = \begin{bmatrix} 2 \\ 1 \end{bmatrix}$ because

$$\begin{bmatrix} -1/3 & 2/3 \\ 2/9 & -1/9 \end{bmatrix} \begin{bmatrix} 8 \\ 7 \end{bmatrix} = \begin{bmatrix} 2 \\ 1 \end{bmatrix}.$$

The generalization of this argument is the next lemma. It follows directly from Theorem 5.6.21, so the details of its proof are left to Exercise 26.

■ **Theorem 5.6.22**

If $\mathscr{B} = \{\mathbf{b}_1, \mathbf{b}_2, \dots, \mathbf{b}_n\}$ is an ordered basis for $\mathbb{R}^n$ and $T_{\mathscr{B}}$ is the coordinate map $\mathbb{R}^n \to \mathbb{R}^n$,

$$[T_{\mathscr{B}}] = \begin{bmatrix} \mathbf{b}_1 & \mathbf{b}_2 & \cdots & \mathbf{b}_n \end{bmatrix}^{-1}.$$

To determine the connection between two coordinate vectors of $\mathbf{u} \in \mathbb{R}^n$, let $\mathscr{B}$ and $\mathscr{B}'$ be ordered bases of $\mathbb{R}^n$. Consider $\mathbf{v} \in \mathbb{R}^n$ to be the $\mathscr{B}$-coordinate vector of $\mathbf{u}$. This means that to **find u** given $\mathbf{v}$, apply $T_{\mathscr{B}}^{-1}$ to $\mathbf{v}$. That is, compute

$$\mathbf{u} = T_{\mathscr{B}}^{-1}\mathbf{v}.$$

To **convert u** to its $\mathscr{B}'$-coordinate vector, apply $T_{\mathscr{B}'}$ to obtain the equation

$$T_{\mathscr{B}'}\mathbf{u} = (T_{\mathscr{B}'} \circ T_{\mathscr{B}}^{-1})\mathbf{v}.$$

Therefore, directly applying the composition $T_{\mathscr{B}'} \circ T_{\mathscr{B}}^{-1}$ to $\mathbf{v}$ will **change** the $\mathscr{B}$-coordinate vector of $\mathbf{u}$ to the $\mathscr{B}'$-coordinate vector of $\mathbf{u}$. That is, the vector $\mathbf{v}$ and $(T_{\mathscr{B}'} \circ T_{\mathscr{B}}^{-1})\mathbf{v}$ both name $\mathbf{u}$ but using different ordered bases. This process is illustrated in Figure 5.3. The linear transformation $T_{\mathscr{B}'} \circ T_{\mathscr{B}}^{-1}$ is called the **change of basis transformation** $\mathscr{B} \to \mathscr{B}'$. A **function diagram** is a generalization of the function arrow notation of Definition 1.4.1. The purpose of a function diagram is to display the domains and codomains of related functions. The diagram for the change of basis transformation is found in Figure 5.4. The function diagram is a **commutative diagram** because following arrows along different paths will, nonetheless, describe the same function.

■ **Example 5.6.23**

Let $\mathscr{B} = \left\{ \begin{bmatrix} 1 \\ 1 \end{bmatrix}, \begin{bmatrix} 1 \\ -1 \end{bmatrix} \right\}$ and $\mathscr{B}' = \left\{ \begin{bmatrix} 1 \\ 2 \end{bmatrix}, \begin{bmatrix} 3 \\ 4 \end{bmatrix} \right\}$. View these bases for $\mathbb{R}^2$ as ordered. Let $\mathbf{u} \in \mathbb{R}^2$ such that

$$\mathbf{u}_{\mathscr{B}} = \begin{bmatrix} 3 \\ 4 \end{bmatrix}.$$

To find the $\mathscr{B}'$-coordinate vector of $\mathbf{u}$, first find $\mathbf{u}$ by applying Theorem 5.6.21:

$$[T_{\mathscr{B}}^{-1}]\begin{bmatrix} 3 \\ 4 \end{bmatrix} = \begin{bmatrix} 1 & 1 \\ 1 & -1 \end{bmatrix}\begin{bmatrix} 3 \\ 4 \end{bmatrix} = \begin{bmatrix} 7 \\ -1 \end{bmatrix}.$$

Now, to convert this vector to its $\mathscr{B}'$-coordinate vector, use $[T_{\mathscr{B}'}^{-1}] = \begin{bmatrix} 1 & 3 \\ 2 & 4 \end{bmatrix}$ to find

$$[T_{\mathscr{B}'}] = \begin{bmatrix} 1 & 3 \\ 2 & 4 \end{bmatrix}^{-1} = \begin{bmatrix} -2 & 3/2 \\ 1 & -1/2 \end{bmatrix},$$

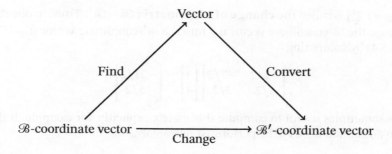

Figure 5.3 Change of basis $\mathscr{B} \to \mathscr{B}'$.

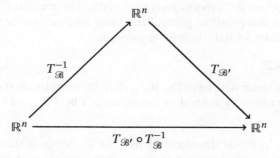

Figure 5.4 Diagram for the change of basis transformation $\mathscr{B} \to \mathscr{B}'$.

and then use Theorem 5.6.22 to convert the vector $\begin{bmatrix} 7 \\ -1 \end{bmatrix}$ to the $\mathscr{B}'$-coordinate vector:

$$T_{\mathscr{B}'} \begin{bmatrix} 7 \\ -1 \end{bmatrix} = \begin{bmatrix} -2 & 3/2 \\ 1 & -1/2 \end{bmatrix} \begin{bmatrix} 7 \\ -1 \end{bmatrix} = \begin{bmatrix} -31/2 \\ 15/2 \end{bmatrix}.$$

All of this can be summarized by stating that the vector **u** with $\mathscr{B}$-coordinate vector $\mathbf{u}_{\mathscr{B}} = \begin{bmatrix} 3 \\ 4 \end{bmatrix}$ also has $\mathscr{B}'$-coordinate vector $\mathbf{u}_{\mathscr{B}'} = \begin{bmatrix} -31/2 \\ 15/2 \end{bmatrix}$. Again, this procedure is illustrated in Figure 5.3.

■ **Example 5.6.24**

Although it is advisable to understand and follow the process found in Example 5.6.23 to perform change of basis calculations, the same outcome can be achieved in one step using

$$[T_{\mathscr{B}'} \circ T_{\mathscr{B}}^{-1}] = \begin{bmatrix} -2 & 3/2 \\ 1 & -1/2 \end{bmatrix} \begin{bmatrix} 1 & 1 \\ 1 & -1 \end{bmatrix} = \begin{bmatrix} -1/2 & -7/2 \\ 1/2 & 3/2 \end{bmatrix}.$$

$[T_{\mathscr{B}'} \circ T_{\mathscr{B}}^{-1}]$ is called the **change of basis matrix** $\mathscr{B} \rightarrow \mathscr{B}'$. Thus, in one step change the $\mathscr{B}$-coordinate vector $\mathbf{u}_{\mathscr{B}}$ into the $\mathscr{B}'$-coordinate vector $\mathbf{u}_{\mathscr{B}'}$ (Figure 5.4) by computing

$$\begin{bmatrix} -1/2 & -7/2 \\ 1/2 & 3/2 \end{bmatrix} \begin{bmatrix} 3 \\ 4 \end{bmatrix} = \begin{bmatrix} -31/2 \\ 15/2 \end{bmatrix}.$$

It is sometimes useful to compute this matrix explicitly, for example, if this transformation needs to be performed repeatedly.

Matrix of a Linear Transformation

A problem similar to the change of basis problem is the following. Every linear transformation $\mathbb{R}^n \rightarrow \mathbb{R}^m$ can uniquely be described by a matrix (Theorems 3.1.19 and 3.1.22). Can this result be generalized to an arbitrary linear transformation, and if so, how? Start with the following theorem.

■ **Theorem 5.6.25**

If V is a vector space with basis $\{\mathbf{b}_1, \mathbf{b}_2, \dots, \mathbf{b}_n\}$, then any linear transformation $T : V \rightarrow W$ is uniquely defined by the values of $T\mathbf{b}_1, T\mathbf{b}_2, \dots, T\mathbf{b}_n$.

Proof

Let T_1 and T_2 be linear transformations $V \rightarrow W$. Suppose that $T_1 \mathbf{b}_i = T_2 \mathbf{b}_i$ for all $i = 1, 2, \dots, n$. Taking $\mathbf{u} \in V$, there exists $r_1, r_2, \dots, r_n \in \mathbb{R}$ such that

$$\mathbf{u} = r_1 \mathbf{b}_1 + r_2 \mathbf{b}_2 + \cdots + r_n \mathbf{b}_n,$$

so Theorem 1.4.6 implies $T_1 = T_2$ because

$$\begin{aligned} T_1 \mathbf{u} &= T_1(r_1 \mathbf{b}_1 + r_2 \mathbf{b}_2 + \cdots + r_n \mathbf{b}_n) \\ &= r_1 T_1 \mathbf{b}_1 + r_2 T_1 \mathbf{b}_2 + \cdots + r_n T_1 \mathbf{b}_n \\ &= r_1 T_2 \mathbf{b}_1 + r_2 T_2 \mathbf{b}_2 + \cdots + r_n T_2 \mathbf{b}_n \\ &= T_2(r_1 \mathbf{b}_1 + r_2 \mathbf{b}_2 + \cdots + r_n \mathbf{b}_n) = T_2 \mathbf{u}. \ ■ \end{aligned}$$

■ **Example 5.6.26**

Define the linear operators $T_1, T_2, T_3 : \mathbb{R}^2 \rightarrow \mathbb{R}^2$ by

$$T_1 \begin{bmatrix} x \\ y \end{bmatrix} = \begin{bmatrix} 2x - y \\ -x \end{bmatrix}, \quad T_2 \begin{bmatrix} x \\ y \end{bmatrix} = \begin{bmatrix} y \\ x + 5y \end{bmatrix}, \quad T_3 \begin{bmatrix} x \\ y \end{bmatrix} = \begin{bmatrix} -x - 3y \\ -y \end{bmatrix}.$$

To confirm that $T_1 \circ T_2 = T_3$, use Theorem 5.6.25 by choosing any basis of $\mathbb{R}^2$. The easy choice is the standard basis, so apply $T_1 \circ T_2$ and T_3 to the standard vectors and compare the results:

$$(T_1 \circ T_2) \begin{bmatrix} 1 \\ 0 \end{bmatrix} = T_1 \begin{bmatrix} 0 \\ 1 \end{bmatrix} = \begin{bmatrix} -1 \\ 0 \end{bmatrix} = T_3 \begin{bmatrix} 1 \\ 0 \end{bmatrix},$$

and

$$(T_1 \circ T_2) \begin{bmatrix} 0 \\ 1 \end{bmatrix} = T_1 \begin{bmatrix} 1 \\ 5 \end{bmatrix} = \begin{bmatrix} -3 \\ -1 \end{bmatrix} = T_3 \begin{bmatrix} 0 \\ 1 \end{bmatrix}.$$

Hence, $T_1 \circ T_2 = T_3$ is proved.

■ **Example 5.6.27**

A matrix such as $A = \begin{bmatrix} 1 & 2 & 3 \\ 0 & 1 & 5 \\ 2 & 4 & 0 \end{bmatrix}$ defines a linear operator T on $\mathbb{R}^3$ which

maps $\mathbf{u} = \begin{bmatrix} u_1 \\ u_2 \\ u_3 \end{bmatrix}$ to $A\mathbf{u}$. Theorem 5.6.25 implies that it is enough to define T

to know that

$$A\mathbf{e}_1 = \begin{bmatrix} 1 \\ 0 \\ 2 \end{bmatrix}, \quad A\mathbf{e}_2 = \begin{bmatrix} 2 \\ 1 \\ 4 \end{bmatrix}, \quad A\mathbf{e}_3 = \begin{bmatrix} 3 \\ 5 \\ 0 \end{bmatrix}.$$

The calculation goes like this:

$$T\mathbf{u} = u_1 A\mathbf{e}_1 + u_2 A\mathbf{e}_2 + u_3 A\mathbf{e}_3 = u_1 \begin{bmatrix} 1 \\ 0 \\ 2 \end{bmatrix} + u_2 \begin{bmatrix} 2 \\ 1 \\ 4 \end{bmatrix} + u_3 \begin{bmatrix} 3 \\ 5 \\ 0 \end{bmatrix}.$$

Consider the linear transformation $T : \mathbb{R}^{2\times 2} \to \mathbb{P}_2$ defined by

$$T \begin{bmatrix} a & b \\ c & d \end{bmatrix} = 4c + (3d + b)x + 2ax^2. \tag{5.39}$$

This function behaves essentially the same as $T_0 : \mathbb{R}^4 \to \mathbb{R}^3$ defined by

$$T_0 \begin{bmatrix} a \\ b \\ c \\ d \end{bmatrix} = \begin{bmatrix} 4c \\ 3d + b \\ 2a \end{bmatrix},$$

which is the linear transformation with standard matrix

$$[T_0] = \begin{bmatrix} 0 & 0 & 4 & 0 \\ 0 & 1 & 0 & 3 \\ 2 & 0 & 0 & 0 \end{bmatrix}. \tag{5.40}$$

So, how can $[T_0]$ be associated with T so that it can be said that the matrix of T is $[T_0]$? The key to making this work is to use coordinate maps.

Let V be an n-dimensional vector space and W be an m-dimensional vector space. Assume that $\mathscr{B}$ is an ordered basis of V and $\mathscr{B}'$ is an ordered basis of W. Take a linear transformation $T : V \to W$. To find a matrix that could represent T, modify

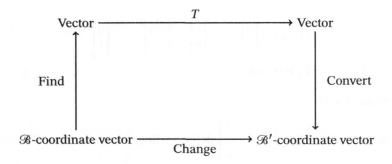

Figure 5.5   Change of basis $\mathscr{B} \to \mathscr{B}'$ with linear transformation T.

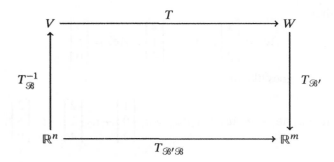

Figure 5.6   Diagram for the standard transformation $\mathscr{B} \to \mathscr{B}'$.

the routine of Figure 5.3 by mapping the vector at the top of the diagram to another vector using T. The result is the procedure illustrated in Figure 5.5. To find the function that represents the change in coordinate vectors, modify the change of basis transformation by using the coordinate maps $T_{\mathscr{B}} : V \to \mathbb{R}^n$ and $T_{\mathscr{B}'} : W \to \mathbb{R}^m$ along with the linear transformation T to define the **standard transformation** $\mathscr{B} \to \mathscr{B}'$ as the function $T_{\mathscr{B}'\mathscr{B}} : \mathbb{R}^n \to \mathbb{R}^m$ such that

$$T_{\mathscr{B}'\mathscr{B}} = T_{\mathscr{B}'} \circ T \circ T_{\mathscr{B}}^{-1}. \tag{5.41}$$

The commutative diagram for the standard transformation is in Figure 5.6. As the composition of linear transformations, the standard transformation is indeed a linear transformation. Hence, $T_{\mathscr{B}'\mathscr{B}}$ has a standard matrix, which will be used to represent T.

■ **Definition 5.6.28**

Suppose that V is an n-dimensional vector space with ordered basis $\mathscr{B}$ and W is an m-dimensional vector space with ordered basis $\mathscr{B}'$. The **standard matrix** $\mathscr{B} \to \mathscr{B}'$ of the linear transformation $T : V \to W$ is

$$[T]_{\mathscr{B}'\mathscr{B}} = [T_{\mathscr{B}'\mathscr{B}}].$$

It is relatively easy to find the standard matrix of a general linear transformation. Write $\mathscr{B} = \{\mathbf{b}_1, \mathbf{b}_2, \dots, \mathbf{b}_n\}$. Because $T_{\mathscr{B}'\mathscr{B}}$ is the composition given in (5.41),

$$[T_{\mathscr{B}'\mathscr{B}}] = [T_{\mathscr{B}'}][T][T_{\mathscr{B}}^{-1}]. \tag{5.42}$$

By Theorem 5.6.21,

$$[T_{\mathscr{B}}^{-1}] = [\mathbf{b}_1 \ \ \mathbf{b}_2 \ \ \dots \ \ \mathbf{b}_n],$$

so $[T_{\mathscr{B}'\mathscr{B}}] = [T_{\mathscr{B}'} \circ T][\mathbf{b}_1 \ \ \mathbf{b}_2 \ \ \dots \ \ \mathbf{b}_n]$, which implies that

$$[T_{\mathscr{B}'\mathscr{B}}] = \left[(T_{\mathscr{B}'} \circ T)\mathbf{b}_1 \ \ (T_{\mathscr{B}'} \circ T)\mathbf{b}_2 \ \ \cdots \ \ (T_{\mathscr{B}'} \circ T)\mathbf{b}_n\right]. \tag{5.43}$$

■ **Example 5.6.29**

Given the linear transformation $T : \mathbb{R}^3 \to \mathbb{R}^2$ defined by

$$T\begin{bmatrix} x \\ y \\ z \end{bmatrix} = \begin{bmatrix} -x + 2y \\ y - 2z \\ 3x - 4y + 5z \end{bmatrix},$$

the standard matrix of T using Definition 3.1.20, is

$$[T] = \begin{bmatrix} T\mathbf{e}_1 & T\mathbf{e}_2 & T\mathbf{e}_3 \end{bmatrix} = \begin{bmatrix} -1 & 2 & 0 \\ 0 & 1 & -2 \\ 3 & -4 & 5 \end{bmatrix}.$$

Let $\mathscr{E} = \{\mathbf{e}_1, \mathbf{e}_2, \mathbf{e}_3\}$. Because

$$(T_{\mathscr{E}} \circ T)\mathbf{e}_1 = \begin{bmatrix} -1 \\ 0 \\ 3 \end{bmatrix}, \quad (T_{\mathscr{E}} \circ T)\mathbf{e}_2 = \begin{bmatrix} 2 \\ 1 \\ -4 \end{bmatrix}, \quad (T_{\mathscr{E}} \circ T)\mathbf{e}_3 = \begin{bmatrix} 3 \\ -4 \\ 5 \end{bmatrix},$$

the standard matrix $\mathscr{E} \to \mathscr{E}$ of T is

$$[T]_{\mathscr{E}\mathscr{E}} = \left[(T_{\mathscr{E}} \circ T)\mathbf{e}_1 \ \ (T_{\mathscr{E}} \circ T)\mathbf{e}_2 \ \ (T_{\mathscr{E}} \circ T)\mathbf{e}_3\right] = \begin{bmatrix} -1 & 2 & 0 \\ 0 & 1 & -2 \\ 3 & -4 & 5 \end{bmatrix},$$

illustrating that Definition 5.6.28 is a generalization of Definition 3.1.20.

■ **Example 5.6.30**

Let $T : \mathbb{R}^{2 \times 2} \to \mathbb{P}_2$ be the function defined by (5.39),

$$T\begin{bmatrix} a & b \\ c & d \end{bmatrix} = 4c + (3d + b)x + 2ax^2$$

Let $\mathscr{B}$ be the standard basis of $\mathbb{R}^{2 \times 2}$ and $\mathscr{B}'$ be the standard basis of $\mathbb{P}_2$. View both as ordered bases. Then, $[T]_{\mathscr{B}'\mathscr{B}} = [T_0]$ as found in (5.40).

■ **Example 5.6.31**

Redo Example 5.6.30 using

$$\mathscr{B} = \left\{ \begin{bmatrix} 1 & 1 \\ 0 & 0 \end{bmatrix}, \begin{bmatrix} 0 & 1 \\ 0 & 1 \end{bmatrix}, \begin{bmatrix} 0 & 0 \\ 1 & 1 \end{bmatrix}, \begin{bmatrix} 0 & 0 \\ 0 & 1 \end{bmatrix} \right\}$$

as the ordered basis for $\mathbb{R}^{2\times 2}$ and $\mathscr{B}' = \{x^2, -x + x^2, -1 + x\}$ as the ordered basis for $\mathbb{P}_2$. Note that

$$T\begin{bmatrix} 1 & 2 \\ 3 & 4 \end{bmatrix} = 12 + 14x + 2x^2.$$

To find the standard matrix $[T]_{\mathscr{B}'\mathscr{B}}$, compute

$$T\begin{bmatrix} 1 & 1 \\ 0 & 0 \end{bmatrix} = x + 2x^2, \quad T\begin{bmatrix} 0 & 1 \\ 0 & 1 \end{bmatrix} = 4x,$$

$$T\begin{bmatrix} 0 & 0 \\ 1 & 1 \end{bmatrix} = 4 + 3x, \quad T\begin{bmatrix} 0 & 0 \\ 0 & 1 \end{bmatrix} = 3x,$$

and

$$T_{\mathscr{B}'}(x + 2x^2) = \begin{bmatrix} 3 \\ -1 \\ 0 \end{bmatrix}, \quad T_{\mathscr{B}'}(4x) = \begin{bmatrix} 4 \\ -4 \\ 0 \end{bmatrix},$$

$$T_{\mathscr{B}'}(4 + 3x) = \begin{bmatrix} 7 \\ -7 \\ -4 \end{bmatrix}, \quad T_{\mathscr{B}'}(3x) = \begin{bmatrix} 3 \\ -3 \\ 0 \end{bmatrix}.$$

Therefore, the standard matrix $\mathscr{B} \to \mathscr{B}'$ for T is

$$[T]_{\mathscr{B}'\mathscr{B}} = \begin{bmatrix} 3 & 4 & 7 & 3 \\ -1 & -4 & -7 & -3 \\ 0 & 0 & -4 & 0 \end{bmatrix}.$$

To illustrate, observe that

$$T_{\mathscr{B}}\begin{bmatrix} 1 & 2 \\ 3 & 4 \end{bmatrix} = \begin{bmatrix} 1 \\ 1 \\ 3 \\ 0 \end{bmatrix}$$

and

$$(T_{\mathscr{B}'} \circ T)\begin{bmatrix} 1 & 2 \\ 3 & 4 \end{bmatrix} = T_{\mathscr{B}'}(12 + 14x + 2x^2) = \begin{bmatrix} 28 \\ -26 \\ -12 \end{bmatrix}.$$

This is confirmed by multiplying by $[T]_{\mathcal{B}'\mathcal{B}}$,

$$\begin{bmatrix} 3 & 4 & 7 & 3 \\ -1 & -4 & -7 & -3 \\ 0 & 0 & -4 & 0 \end{bmatrix} \begin{bmatrix} 1 \\ 1 \\ 3 \\ 0 \end{bmatrix} = \begin{bmatrix} 28 \\ -26 \\ -12 \end{bmatrix}.$$

Exercises

1. Prove that each function T is an isomorphism.

   (a) $T : \mathbb{R}^3 \to \mathbb{R}^3$, $T\begin{bmatrix} x \\ y \\ z \end{bmatrix} = \begin{bmatrix} 3x + 2y - 5z \\ y - 4z \\ 6z \end{bmatrix}$

   (b) $T : \mathbb{P}_1 \to \mathbb{R}^2$, $T(a + bx) = \begin{bmatrix} 4x + 7y \\ 6x - 2y \end{bmatrix}$

   (c) $T : \mathbb{R}^4 \to \mathbb{R}^{2\times 2}$, $T\begin{bmatrix} x_1 \\ x_2 \\ x_3 \\ x_4 \end{bmatrix} = \begin{bmatrix} x_1 + 2x_2 - x_3 & 3x_2 - 2x_3 + 6x_4 \\ 3x_1 + x_2 - 4x_3 + 6x_4 & 9x_1 + 3x_2 - 2x_3 \end{bmatrix}$

   (d) $T : \mathbb{R}^{3\times 2} \to \mathbb{P}_5$,

   $T\begin{bmatrix} a & b \\ c & d \\ e & f \end{bmatrix} = (a+b)+(c+d)x+(e+f)x^2+(a-c)x^3+(d-f)x^4+(a-f)x^5$

2. Prove that T defined by (5.33) is an isomorphism.

3. Prove that the function given in Example 5.6.4 is an isomorphism.

4. Prove that each function T is not an isomorphism.

   (a) $T : \mathbb{R}^3 \to \mathbb{R}^3$, $T\begin{bmatrix} x \\ y \\ z \end{bmatrix} = \begin{bmatrix} x + y + z \\ 1 \\ 0 \end{bmatrix}$

   (b) $T : \mathbb{P}_3 \to \mathbb{R}^{4\times 2}$, $T(a + bx + cx^2 + dx^3) = \begin{bmatrix} a & b & c \\ d & 0 & 0 \end{bmatrix}$

   (c) $T : \mathbb{R}^3 \to \mathbb{P}_2$, $T\begin{bmatrix} a \\ b \\ c \end{bmatrix} = (a + 3b + c) + (a - b + c)x + (2a + 4b - 8c)x^2$

5. Prove that $\mathbb{P}$ is not isomorphic to $\mathbb{P}_n$ for all positive integers n.

6. Given the isomorphism T, find the image of the subspace W under T.

   (a) $T : \mathbb{R}^2 \to \mathbb{R}^2$, $T\begin{bmatrix} x \\ y \end{bmatrix} = \begin{bmatrix} 4x - 4y \\ x + 3y \end{bmatrix}$, $W = S\left(\begin{bmatrix} 5 \\ -6 \end{bmatrix} \right)$

   (b) $T : \mathbb{R}^3 \to \mathbb{R}^3$, $T\begin{bmatrix} x \\ y \\ z \end{bmatrix} = \begin{bmatrix} 5x - y + 2z \\ 3x + 5y - 3z \\ 2x - 4z \end{bmatrix}$, $W = S\left(\begin{bmatrix} 5 \\ 1 \\ -6 \end{bmatrix}, \begin{bmatrix} 1 \\ 0 \\ 3 \end{bmatrix} \right)$

(c) $T : \mathbb{P}_3 \rightarrow \mathbb{R}^{2\times2}, \quad T(a + bx + bx^2 + cx^3) = \begin{bmatrix} b & c \\ d & a \end{bmatrix}, \quad W = \mathsf{S}\left(x, x^2, x^3\right)$

7. Verify that isomorphisms preserve dimension by confirming for each T, W pair in Exercise 6 that $\dim W = \dim T[W]$.

8. Let $T : \mathbb{R}^n \rightarrow \mathbb{R}^n$ be a linear operator. Prove that T is an isomorphism if and only if $\det[T] \neq 0$.

9. Which of the linear operators of Section 3.3 are isomorphisms and which are not? Explain.

10. Which of the linear transformations of Example 5.1.16 are isomorphisms and which are not? Explain.

11. Must an isometry be isomorphism? Prove or give a counterexample.

12. Must an isomorphism $\mathbb{R}^n \rightarrow \mathbb{R}^n$ be isometry? Prove or give a counterexample.

13. Let $T : \mathbb{R}^{3\times3} \rightarrow \mathbb{R}^{3\times3}$ be an isomorphism.
    (a) Find the rank and nullity of $[T]$.
    (b) Prove or show false: $\det A = 0$ if and only if $\det TA = 0$ for all 3×3 matrices A.
    (c) Prove or show false: $\det A = \det TA$ for all 3×3 matrices A.
    (d) Prove or show false: $\operatorname{rank} A = \operatorname{rank} TA$ for all 3×3 matrices A.
    (e) Prove or show false: $\operatorname{null} A = \operatorname{null} TA$ for all 3×3 matrices A.

14. Let $T_1 : \mathbb{R}^n \rightarrow \mathbb{R}^n$ be an isomorphism. Let $T_2 : \mathbb{R}^n \rightarrow \mathbb{R}^n$ be a linear operator such that $[T_2] = [T_1]^T$. Prove that T_2 is an isomorphism.

15. Suppose that $T_1 : \mathbb{R}^n \rightarrow \mathbb{R}^n$ is a linear operator so that $[T_1]$ is a diagonal matrix. Prove that T_1 is an isomorphism if and only if there are no 0 entries on the main diagonal of $[T_1]$.

16. If the standard matrix of a linear operator $T : \mathbb{R}^n \rightarrow \mathbb{R}^n$ is an orthogonal matrix, is T an isomorphism? Explain.

17. Let $T : V_1 \rightarrow V_2$ be an isomorphism. Prove.
    (a) If W is a subspace of V_1, then W is isomorphic to $T[W]$.
    (b) If W is a subspace of V_2, then W is isomorphic to $T^{-1}[W]$.

18. Assume that $T_1, T_2 : V \rightarrow V$ are linear operators such that $T_1 \circ T_2$ is an isomorphism. Prove or show false.
    (a) If T_1 is an isomorphism, T_2 is an isomorphism.
    (b) If T_2 is an isomorphism, T_1 is an isomorphism.

19. Let V be an n-dimensional vector space and W be an m-dimensional vector space. Prove.
    (a) $L(V, W)$ is isomorphic to $\mathbb{R}^{m\times n}$.   (c) $\dim L(V, W) = (\dim V)(\dim W)$.
    (b) $L(V, W)$ is isomorphic to $\mathbb{R}^{n\times m}$.

20. Let $T : V \rightarrow W$ be a linear transformation. Using the quotient vector space defined in Problem 30 of Section 5.1, prove that ran T is isomorphic to $V/\ker T$. This is called the **First Isomorphism Theorem**.

21. Let W_1 and W_2 be subspaces of the vector space V. Prove that $(W_1 + W_2)/W_2$ is isomorphic to $W_1/(W_1 \cap W_2)$. This is called the **Second Isomorphism Theorem**.

22. Suppose that W_1 is a subspace of W_2 and W_2 is a subspace of the vector space V. Prove that $(V/W_2)/(W_1/W_2)$ is isomorphic to V/W_1. This is called the **Third Isomorphism Theorem**.

23. Show that $\mathscr{B} = \left\{ \begin{bmatrix} 1 \\ 1 \\ 0 \end{bmatrix}, \begin{bmatrix} 1 \\ 0 \\ 1 \end{bmatrix}, \begin{bmatrix} 0 \\ 1 \\ 1 \end{bmatrix} \right\}$ is a basis for $\mathbb{R}^3$. Viewing $\mathscr{B}$ as an ordered

basis, find the $\mathscr{B}$-coordinate vectors of $\mathbf{u}_1 = \begin{bmatrix} 2 \\ 0 \\ 0 \end{bmatrix}, \mathbf{u}_2 = \begin{bmatrix} 1 \\ 0 \\ 0 \end{bmatrix}$, and $\mathbf{u}_3 = \begin{bmatrix} 1 \\ 1 \\ 1 \end{bmatrix}$.

24. Let V be a vector space with ordered basis $\mathscr{B}$. Find $\mathbf{u} \in V$ given $\mathbf{u}_{\mathscr{B}}$.

   (a) $V = \mathbb{R}^2$, $\mathscr{B} = \left\{ \begin{bmatrix} 3 \\ 6 \end{bmatrix}, \begin{bmatrix} -5 \\ 1 \end{bmatrix} \right\}$, $\mathbf{u}_{\mathscr{B}} = \begin{bmatrix} 5 \\ -2 \end{bmatrix}$

   (b) $V = \mathbb{R}^3$, $\mathscr{B} = \left\{ \begin{bmatrix} 5 \\ -5 \\ 6 \end{bmatrix}, \begin{bmatrix} 7 \\ -6 \\ 6 \end{bmatrix}, \begin{bmatrix} -3 \\ 6 \\ 5 \end{bmatrix} \right\}$, $\mathbf{u}_{\mathscr{B}} = \begin{bmatrix} 2 \\ -4 \\ 0 \end{bmatrix}$

   (c) $V = \mathbb{P}_3$, $\mathscr{B} = \{1, x - x^2, x^2 - x^3, 1 - x + x^2 - x^3\}$, $\mathbf{u}_{\mathscr{B}} = \begin{bmatrix} 1 \\ 2 \\ 3 \\ 4 \end{bmatrix}$

   (d) $V = \mathbb{R}^{2\times2}$, $\mathscr{B} = \left\{ \begin{bmatrix} 5 & 6 \\ 1 & 5 \end{bmatrix}, \begin{bmatrix} 6 & 2 \\ 6 & 4 \end{bmatrix}, \begin{bmatrix} 4 & 5 \\ 4 & 4 \end{bmatrix}, \begin{bmatrix} 4 & 2 \\ 9 & 8 \end{bmatrix} \right\}$, $\mathbf{u}_{\mathscr{B}} = \begin{bmatrix} 4 \\ 1 \\ -2 \\ 7 \end{bmatrix}$

25. Let V be a vector space with ordered basis $\mathscr{B}$. Find $\mathbf{u}_{\mathscr{B}}$ given $\mathbf{u} \in V$.

   (a) $V = \mathbb{R}^2$, $\mathscr{B} = \left\{ \begin{bmatrix} -2 \\ 3 \end{bmatrix}, \begin{bmatrix} 6 \\ 9 \end{bmatrix} \right\}$, $\mathbf{u} = \begin{bmatrix} 2 \\ 2 \end{bmatrix}$

   (b) $V = \mathbb{R}^3$, $\mathscr{B} = \left\{ \begin{bmatrix} 3 \\ 2 \\ -1 \end{bmatrix}, \begin{bmatrix} 1 \\ 1 \\ 0 \end{bmatrix}, \begin{bmatrix} -3 \\ 1 \\ 3 \end{bmatrix} \right\}$, $\mathbf{u} = \begin{bmatrix} 2 \\ 2 \\ -1 \end{bmatrix}$

   (c) $V = \mathbb{P}_4$, $\mathscr{B} = \{x^4, 1, x^2, x, x^3\}$, $\mathbf{u} = 5 + 8x - 6x^2 + 3x^3 + 9x^4$

   (d) $V = \mathbb{R}^{2\times2}$, $\mathscr{B} = \left\{ \begin{bmatrix} 2 & 1 \\ 0 & 1 \end{bmatrix}, \begin{bmatrix} 0 & 3 \\ 2 & -1 \end{bmatrix}, \begin{bmatrix} 1 & -1 \\ 2 & 0 \end{bmatrix}, \begin{bmatrix} 1 & 1 \\ 0 & -3 \end{bmatrix} \right\}$, $\mathbf{u} = \begin{bmatrix} -2 & 0 \\ 3 & 2 \end{bmatrix}$

26. Prove Theorem 5.6.22.

27. Let $\mathcal{S}$ be a spanning set of the vector space V. Demonstrate that if every vector of V can be written as a linear combination of vectors in $\mathcal{S}$ in more than one way (not counting order), $\mathcal{S}$ is linearly dependent.

28. Let V be a vector space with ordered bases $\mathcal{B}$ and $\mathcal{B}'$. Prove that $\mathcal{B} = \mathcal{B}'$ if and only if $\mathbf{u}_{\mathcal{B}} = \mathbf{u}_{\mathcal{B}'}$ for all $\mathbf{u} \in V$.

29. Suppose that V is a vector space with basis $\mathcal{B}$. Let $\mathbf{u}_1, \mathbf{u}_2, \ldots, \mathbf{u}_n \in V$. Prove that $\{\mathbf{u}_1, \mathbf{u}_2, \ldots, \mathbf{u}_n\}$ is a basis for V if and only if $\{(\mathbf{u}_1)_{\mathcal{B}}, (\mathbf{u}_2)_{\mathcal{B}}, \ldots, (\mathbf{u}_n)_{\mathcal{B}}\}$ is a basis for $\mathbb{R}^n$.

30. Using Exercise 29, use $\mathcal{B}$-coordinate vectors to demonstrate that the given sets of vectors are linearly independent or linearly dependent. If the set is linearly dependent, write one vector as a linear combination of the others.

    (a) $V = \mathbb{R}^3$, $\left\{ \begin{bmatrix} 1 \\ 2 \\ 5 \end{bmatrix}, \begin{bmatrix} 5 \\ 2 \\ 1 \end{bmatrix}, \begin{bmatrix} 2 \\ 0 \\ 1 \end{bmatrix} \right\}$, $\mathcal{B} = \{\mathbf{e}_2, \mathbf{e}_1, \mathbf{e}_3\}$

    (b) $V = \mathbb{R}^3$, $\left\{ \begin{bmatrix} 2 \\ 5 \\ 1 \end{bmatrix}, \begin{bmatrix} 3 \\ 1 \\ 3 \end{bmatrix}, \begin{bmatrix} 0 \\ 5 \\ 4 \end{bmatrix} \right\}$, $\mathcal{B} = \left\{ \begin{bmatrix} 1 \\ 0 \\ 1 \end{bmatrix}, \begin{bmatrix} 1 \\ 1 \\ -1 \end{bmatrix}, \begin{bmatrix} 1 \\ -2 \\ 3 \end{bmatrix} \right\}$

    (c) $V = \mathbb{R}^{2\times2}$, $\left\{ \begin{bmatrix} 2 & 1 \\ 1 & 2 \end{bmatrix}, \begin{bmatrix} 5 & 1 \\ 5 & 3 \end{bmatrix}, \begin{bmatrix} 1 & 2 \\ -2 & 3 \end{bmatrix} \right\}$,

    $\mathcal{B} = \left\{ \begin{bmatrix} 0 & 1 \\ 0 & 0 \end{bmatrix}, \begin{bmatrix} 0 & 0 \\ 1 & 0 \end{bmatrix}, \begin{bmatrix} 1 & 0 \\ 0 & 0 \end{bmatrix}, \begin{bmatrix} 0 & 0 \\ 0 & 1 \end{bmatrix} \right\}$

    (d) $V = \mathbb{P}_1$, $\{4 + 3x, 2 + 3x^2, 4 + x\}$, $\mathcal{B} = \{3 + x, 1 - 6x\}$

31. Using Exercise 29, use $\mathcal{B}$-coordinate vectors to demonstrate that the given sets of vectors are or are not spanning sets of the vector space V. If the set is not a spanning set, find a vector in V not in the span of the given set.

    (a) $V = \mathbb{R}^2$, $\left\{ \begin{bmatrix} 4 \\ 2 \end{bmatrix}, \begin{bmatrix} 3 \\ 6 \end{bmatrix} \right\}$, $\mathcal{B} = \left\{ \begin{bmatrix} 0 \\ 3 \end{bmatrix}, \begin{bmatrix} 1 \\ 2 \end{bmatrix} \right\}$

    (b) $V = \mathbb{R}^3$, $\left\{ \begin{bmatrix} 8 \\ -4 \\ 0 \end{bmatrix}, \begin{bmatrix} 1 \\ 0 \\ 4 \end{bmatrix}, \begin{bmatrix} 2 \\ -2 \\ -8 \end{bmatrix} \right\}$, $\mathcal{B} = \{3\mathbf{e}_3, 2\mathbf{e}_2, \mathbf{e}_1\}$

    (c) $V = \mathbb{R}^{2\times2}$, $\left\{ \begin{bmatrix} 3 & 2 \\ -1 & 3 \end{bmatrix}, \begin{bmatrix} 2 & 0 \\ 1 & -1 \end{bmatrix}, \begin{bmatrix} -1 & 1 \\ 0 & -1 \end{bmatrix}, \begin{bmatrix} 3 & 1 \\ 2 & 1 \end{bmatrix} \right\}$

    $\mathcal{B} = \left\{ \begin{bmatrix} 2 & -1 \\ 0 & 5 \end{bmatrix}, \begin{bmatrix} 1 & 0 \\ 2 & -2 \end{bmatrix}, \begin{bmatrix} 1 & 1 \\ -1 & 0 \end{bmatrix}, \begin{bmatrix} 2 & 0 \\ -2 & 6 \end{bmatrix} \right\}$

    (d) $V = \mathbb{P}_3$, $\{1+x, 1+x^2, 1+x^3, 1+x+x^2+x^3\}$, $\mathcal{B} = \{1-x, x-x^2, x^2-x^3, x^3\}$

32. Viewing $\mathscr{B}$ and $\mathscr{B}'$ as ordered bases, find the change of basis matrix $\mathscr{B} \to \mathscr{B}'$ and apply the matrix to $\mathbf{u}_{\mathscr{B}}$ to find $\mathbf{u}_{\mathscr{B}'}$. Finally, find $\mathbf{u}_{\mathscr{B}'}$ directly to confirm the result.

(a) $\mathscr{B} = \left\{ \begin{bmatrix} 0 \\ 1 \end{bmatrix}, \begin{bmatrix} 1 \\ 0 \end{bmatrix} \right\}$, $\mathscr{B}' = \left\{ \begin{bmatrix} 3 \\ 6 \end{bmatrix}, \begin{bmatrix} 6 \\ 2 \end{bmatrix} \right\}$, $\mathbf{u}_{\mathscr{B}} = \begin{bmatrix} 1 \\ 1 \end{bmatrix}$

(b) $\mathscr{B} = \left\{ \begin{bmatrix} -1 \\ 5 \\ -1 \end{bmatrix}, \begin{bmatrix} -4 \\ 3 \\ 0 \end{bmatrix}, \begin{bmatrix} 7 \\ 0 \\ 7 \end{bmatrix} \right\}$, $\mathscr{B}' = \left\{ \begin{bmatrix} 1 \\ -4 \\ 1 \end{bmatrix}, \begin{bmatrix} 0 \\ -2 \\ 0 \end{bmatrix}, \begin{bmatrix} 2 \\ -1 \\ -1 \end{bmatrix} \right\}$, $\mathbf{u}_{\mathscr{B}} = \begin{bmatrix} 2 \\ -5 \\ 1 \end{bmatrix}$

(c) $\mathscr{B} = \{\mathbf{e}_1, \mathbf{e}_2, \mathbf{e}_3, \mathbf{e}_4\}$, $\mathscr{B}' = \{\mathbf{e}_2, \mathbf{e}_1, \mathbf{e}_4, \mathbf{e}_3\}$, $\mathbf{u}_{\mathscr{B}} = 3\mathbf{e}_1 - 2\mathbf{e}_2 + 4\mathbf{e}_3 - \mathbf{e}_4$

33. Repeat Exercise 32 except that the change of basis matrix $\mathscr{B}' \to \mathscr{B}$ should be found instead.

34. Suppose that $\mathscr{B}$ and $\mathscr{B}'$ are ordered bases for $\mathbb{R}^n$. Let A be the change of basis matrix $\mathscr{B} \to \mathscr{B}'$. Prove that A^{-1} is the change of basis matrix $\mathscr{B}' \to \mathscr{B}$.

35. Suppose that $\mathscr{B}$, $\mathscr{B}'$, and $\mathscr{B}''$ are ordered bases for $\mathbb{R}^n$. Let A be the change of basis matrix $\mathscr{B} \to \mathscr{B}'$ and B be the change of basis matrix $\mathscr{B}' \to \mathscr{B}''$. Prove that BA is the change of basis matrix $\mathscr{B} \to \mathscr{B}''$.

36. Viewing $\mathscr{B}$ and $\mathscr{B}'$ as ordered bases, find the standard matrix $\mathscr{B} \to \mathscr{B}'$ for each linear transformation T.

(a) $T : \mathbb{R}^2 \to \mathbb{R}^3$, $T \begin{bmatrix} x \\ y \end{bmatrix} = \begin{bmatrix} 2x + 5y \\ x - 3y \\ 3x + y \end{bmatrix}$,

$\mathscr{B} = \left\{ \begin{bmatrix} 1 \\ 1 \end{bmatrix}, \begin{bmatrix} 2 \\ 0 \end{bmatrix} \right\}$, $\mathscr{B}' = \left\{ \begin{bmatrix} 1 \\ 1 \\ 0 \end{bmatrix}, \begin{bmatrix} 1 \\ 0 \\ 1 \end{bmatrix}, \begin{bmatrix} 0 \\ 1 \\ 1 \end{bmatrix} \right\}$

(b) $T : \mathbb{P}_3 \to \mathbb{R}$, $T(p(x)) = p(1)$, $\mathscr{B} = \{1 + x^2, x - x^2, x^3 - x^2, x^2\}$, $\mathscr{B}' = \{5\}$

(c) $T : \mathbb{P}_2 \to \mathbb{P}_2$, T, $\mathscr{B} = \{1 + x + x^2, 1 + 2x + 2x^2, 1 + 3x + 4x^2\}$, $\mathscr{B}' = \{x^2, x, 1\}$

(d) $T : \mathsf{S}(\cos, \sin) \to \mathbb{P}_1$, $T \cos = 3$ and $T \sin = 2 + x$, $\mathscr{B} = \{\sin, \cos\}$, $\mathscr{B}' = \{1, x\}$

37. Let $T : V \to W$ be a linear transformation. Prove that T is an isomorphism if and only if every standard matrix of T is invertible. Can T have an invertible standard matrix at the same time that it has a noninvertible standard matrix? Explain.

38. Assume that $T : V \to W$ is an invertible linear transformation. Let $\mathscr{B}$ be an ordered basis of V and $\mathscr{B}'$ be an ordered basis of W. Let A be the standard matrix $\mathscr{B} \to \mathscr{B}'$ of T. Prove that A is invertible and $[T^{-1}]_{\mathscr{B}'\mathscr{B}} = A^{-1}$.

39. Let the standard matrix $\mathscr{B} \to \mathscr{B}'$ of the linear transformation $T_1 : V_1 \to V_2$ be A and the standard matrix $\mathscr{B}' \to \mathscr{B}''$ of the linear transformation $T_2 : V_2 \to V_3$ be B, where $\mathscr{B}$ is an ordered matrix of V_1, $\mathscr{B}'$ is an ordered matrix of V_2, and $\mathscr{B}''$ is an ordered matrix of V_3. Prove that the standard matrix $\mathscr{B} \to \mathscr{B}''$ of $T_2 \circ T_1$ is BA.

40. Assume that $\mathscr{B}$ is an ordered basis of the vector space V and $\mathscr{B}'$ is an ordered basis of the vector space W. Let $T : V \to W$ be a linear transformation. Prove.

   (a) $\mathrm{CS}\,[T]_{\mathscr{B}'\mathscr{B}}$ is isomorphic to ran T.

   (b) $\mathrm{NS}\,[T]_{\mathscr{B}'\mathscr{B}}$ is isomorphic to ker T.

CHAPTER 6

Inner Product Spaces

6.1 Inner Products

There is no geometry on an vector space. It is simply a set with two operations, one binary (Definition 1.4.34) and the other scalar (Definition 1.4.41). This is true even for $\mathbb{R}^n$. The geometry of Chapter 2 is natural, yet it all follows from the assumption that $\mathbb{R}^2$ can be viewed as the Euclidean plane, $\mathbb{R}^3$ can be viewed as Euclidean space, and $\mathbb{R}^n$ can be interpreted as a generalization to higher dimensions. This natural interpretation is missing when the vector space consists of objects such as matrices, polynomials, or functions. An appropriate strategy to view these types of vector spaces geometrically is to attempt to model the geometry given to $\mathbb{R}^n$. To accomplish this, the source of its geometry needs to be identified. This source is known, and it is the dot product. The dot product is used to find the angle between two nonzero vectors with Definition 2.2.2 and to define what it means for vectors to be orthogonal using Definition 2.2.7. It is also used to define the length of a vector and the distance between two vectors using Theorem 2.2.17 and to describe geometric objects in $\mathbb{R}^n$, among which are lines and planes as in Examples 2.2.14 and 2.2.15. As a vector space was defined in Definition 5.1.1 using the key proper-

Linear Algebra, First Edition. Michael L. O'Leary.
© 2021 John Wiley & Sons, Inc. Published 2021 by John Wiley & Sons, Inc.
Companion Website: www.wiley.com/go/o'leary/linearalgebra

ties of $\mathbb{R}^n$ as listed in Theorem 2.1.5, give geometry to an arbitrary vector space by generalizing the results of Theorem 2.2.6 to define the analog of the dot product.

■ **Definition 6.1.1**

Let V be a vector space. A function $V \times V \to \mathbb{R}$, where the image of $(\mathbf{v}_1, \mathbf{v}_2)$ is denoted by $\langle \mathbf{v}_1, \mathbf{v}_2 \rangle$, is an **inner product** if for all $\mathbf{u}, \mathbf{v}, \mathbf{w} \in V$ and $r \in \mathbb{R}$,

(a) $\langle \mathbf{u}, \mathbf{v} \rangle = \langle \mathbf{v}, \mathbf{u} \rangle$,

(b) $\langle \mathbf{u}, \mathbf{v} + \mathbf{w} \rangle = \langle \mathbf{u}, \mathbf{v} \rangle + \langle \mathbf{u}, \mathbf{w} \rangle$,

(c) $\langle r\mathbf{u}, \mathbf{v} \rangle = r \langle \mathbf{u}, \mathbf{v} \rangle$,

(d) $\langle \mathbf{u}, \mathbf{u} \rangle > 0$ if $\mathbf{u} \neq \mathbf{0}$.

A vector space paired with an inner product is an **inner product space**.

It should be noted that Definition 6.1.1 only includes what is needed to obtain the key properties of the dot product. The other properties are immediate results to Definition 6.1.1 which complete the generalization of Theorem 2.2.6. The proof is left to Exercise 8.

■ **Theorem 6.1.2**

Let V be an inner product space. For all $\mathbf{u}, \mathbf{v}, \mathbf{w} \in V$ and $r \in \mathbb{R}$,

(a) $\langle \mathbf{u} + \mathbf{v}, \mathbf{w} \rangle = \langle \mathbf{u}, \mathbf{w} \rangle + \langle \mathbf{v}, \mathbf{w} \rangle$,

(b) $\langle \mathbf{u}, r\mathbf{v} \rangle = r \langle \mathbf{u}, \mathbf{v} \rangle$,

(c) $\langle \mathbf{u}, \mathbf{u} \rangle = 0$ if and only if $\mathbf{u} = \mathbf{0}$.

Theorem 2.2.6 shows that the dot product is an inner product on $\mathbb{R}^n$. This is not the only inner product on this vector space. It should be noted that replacing the dot product with another inner product on $\mathbb{R}^n$ will change the geometry of $\mathbb{R}^n$.

■ **Example 6.1.3**

Choose positive real numbers $t_1, t_2, \dots, t_n$. Call each of these real numbers a **weight** and define the **weighted inner product** on $\mathbb{R}^n$ as

$$\left\langle \begin{bmatrix} x_1 \\ x_2 \\ \vdots \\ x_n \end{bmatrix}, \begin{bmatrix} y_1 \\ y_2 \\ \vdots \\ y_n \end{bmatrix} \right\rangle = \sum_{i=1}^{n} t_i x_i y_i. \tag{6.1}$$

Confirm that this is an inner product by letting

$$\mathbf{u} = \begin{bmatrix} u_1 \\ u_2 \\ \vdots \\ u_n \end{bmatrix}, \quad \mathbf{v} = \begin{bmatrix} v_1 \\ v_2 \\ \vdots \\ v_n \end{bmatrix}, \quad \mathbf{w} = \begin{bmatrix} w_1 \\ w_2 \\ \vdots \\ w_n \end{bmatrix},$$

and $r \in \mathbb{R}$, and then check that the four conditions of Definition 6.1.1 hold.

- $\langle \mathbf{u}, \mathbf{v} \rangle = \sum_{i=1}^{n} t_i u_i v_i = \sum_{i=1}^{n} t_i v_i u_i = \langle \mathbf{v}, \mathbf{u} \rangle.$

- $\langle \mathbf{u}, \mathbf{v} + \mathbf{w} \rangle = \langle \mathbf{u}, \mathbf{v} \rangle + \langle \mathbf{u}, \mathbf{w} \rangle$ because

$$\sum_{i=1}^{n} t_i u_i (v_i + w_i) = \sum_{i=1}^{n} (t_i u_i v_i + t_i u_i w_i) = \sum_{i=1}^{n} t_i u_i v_i + \sum_{i=1}^{n} t_i u_i w_i.$$

- Since $\sum_{i=1}^{n} t_i (r u_i) v_i = r \sum_{i=1}^{n} t_i u_i v_i$, it follows that $\langle r\mathbf{u}, \mathbf{v} \rangle = r \langle \mathbf{u}, \mathbf{v} \rangle.$

- Suppose that $\mathbf{u} \neq \mathbf{0}$. This implies that $u_j \neq 0$ for some $j = 1, 2, \ldots, n$. Hence, for all $i = 1, 2, \ldots, n$, it follows that $t_i u_i^2 \geq 0$ and $t_j u_j^2 > 0$ because the weights are positive. This implies that

$$\langle \mathbf{u}, \mathbf{u} \rangle = \sum_{i=1}^{n} t_i u_i^2 > 0.$$

To illustrate this inner product, define

$$\left\langle \begin{bmatrix} u_1 \\ u_2 \\ u_3 \end{bmatrix}, \begin{bmatrix} v_1 \\ v_2 \\ v_3 \end{bmatrix} \right\rangle = 2u_1 v_1 + 3u_2 v_2 + 4u_3 v_3, \tag{6.2}$$

so although $\begin{bmatrix} 1 \\ 5 \\ -2 \end{bmatrix} \cdot \begin{bmatrix} 1 \\ 2 \\ 4 \end{bmatrix} = 3$, with the inner product of (6.2),

$$\left\langle \begin{bmatrix} 1 \\ 5 \\ -2 \end{bmatrix}, \begin{bmatrix} 1 \\ 2 \\ 4 \end{bmatrix} \right\rangle = 0,$$

which should make clear that the understanding of the geometry of $\mathbb{R}^3$ is different under (6.2) than under the dot product.

■ **Example 6.1.4**

For all $A, B \in \mathbb{R}^{m \times n}$, use the trace of a matrix given in Example 5.1.13 to define

$$\langle A, B \rangle = \operatorname{tr} AB^T. \tag{6.3}$$

This is a function $\mathbb{R}^{m \times n} \times \mathbb{R}^{m \times n} \to \mathbb{R}$. To illustrate the computation, if

$$A = \begin{bmatrix} 1 & 2 & 3 \\ 4 & 5 & 6 \end{bmatrix} \quad \text{and} \quad B = \begin{bmatrix} 7 & 8 & 9 \\ 0 & 1 & 2 \end{bmatrix},$$

then

$$\langle A,B\rangle = \text{tr}\begin{bmatrix}1 & 2 & 3\\4 & 5 & 6\end{bmatrix}\begin{bmatrix}7 & 0\\8 & 1\\9 & 2\end{bmatrix} = \text{tr}\begin{bmatrix}50 & 8\\122 & 17\end{bmatrix} = 67.$$

Note that (6.3) represents the sum of the products of all corresponding pairs of entries from the $m\times n$ matrices A and B. To see this, let $\mathbf{r}_i$ be the rows of A and $\mathbf{s}_i$ be the rows of B, where $i = 1, 2, \ldots, m$. Because $B^T = \begin{bmatrix}\mathbf{s}_1^T & \mathbf{s}_2^T & \cdots & \mathbf{s}_m^T\end{bmatrix}$, it follows that

$$AB^T = \begin{bmatrix}\mathbf{r}_1^T \bullet \mathbf{s}_1^T & \mathbf{r}_1^T \bullet \mathbf{s}_2^T & \cdots & \mathbf{r}_1^T \bullet \mathbf{s}_m^T\\\mathbf{r}_2^T \bullet \mathbf{s}_1^T & \mathbf{r}_2^T \bullet \mathbf{s}_2^T & \cdots & \mathbf{r}_2^T \bullet \mathbf{s}_m^T\\\vdots & \vdots & & \vdots\\\mathbf{r}_m^T \bullet \mathbf{s}_1^T & \mathbf{r}_m^T \bullet \mathbf{s}_2^T & \cdots & \mathbf{r}_m^T \bullet \mathbf{s}_m^T\end{bmatrix},$$

which is an $m\times m$ matrix. Hence, writing for $i = 1, 2, \ldots, m$,

$$\mathbf{r}_i = \begin{bmatrix}a_{i1} & a_{i2} & \cdots & a_{in}\end{bmatrix} \quad \text{and} \quad \mathbf{s}_i = \begin{bmatrix}b_{i1} & b_{i2} & \cdots & b_{in}\end{bmatrix},$$

the inner product is calculated as

$$\text{tr}\, AB^T = \sum_{i=1}^{m}\mathbf{r}_i^T \bullet \mathbf{s}_i^T = \sum_{i=1}^{m}(a_{i1}b_{i1} + a_{i2}b_{i2} + \cdots + a_{in}b_{in}),$$

so

$$\langle A,B\rangle = \sum_{i=1}^{m}\sum_{j=1}^{n}a_{ij}b_{ij}. \tag{6.4}$$

Use (6.4) to confirm that (6.3) is an inner product. Let $A, B, C \in \mathbb{R}^{m\times n}$ and write $A = [a_{ij}]$, $B = [b_{ij}]$ and $C = [c_{ij}]$.

- $\langle A,B\rangle = \displaystyle\sum_{i=1}^{m}\sum_{j=1}^{n}a_{ij}b_{ij} = \sum_{i=1}^{m}\sum_{j=1}^{n}b_{ij}a_{ij} = \langle B,A\rangle.$

- $\langle A,B+C\rangle = \langle A,B\rangle + \langle A,C\rangle$ because

$$\sum_{i=1}^{m}\sum_{j=1}^{n}a_{ij}(b_{ij}+c_{ij}) = \sum_{i=1}^{m}\sum_{j=1}^{n}(a_{ij}b_{ij} + a_{ij}c_{ij})$$

$$= \sum_{i=1}^{m}\sum_{j=1}^{n}a_{ij}b_{ij} + \sum_{i=1}^{m}\sum_{j=1}^{n}a_{ij}c_{ij}.$$

- Letting $r \in \mathbb{R}$, $\langle rA,B\rangle = \displaystyle\sum_{i=1}^{m}\sum_{j=1}^{n}(ra_{ij})b_{ij} = r\sum_{i=1}^{m}\sum_{j=1}^{n}a_{ij}b_{ij} = r\langle A,B\rangle.$

- Suppose that entry a_{ij} is a nonzero entry of A. Then, $\langle A,A\rangle \neq 0$ since

$$a_{i1}^2 + \cdots + a_{ij}^2 + \cdots + a_{in}^2 \neq 0.$$

Thus, (6.3) is an inner product, which is called a **Frobenius inner product**.

■ **Example 6.1.5**

The vector space $\mathbb{P}_{n-1}$ becomes an inner product space by choosing scalars $r_1, r_2, \ldots, r_n \in \mathbb{R}$ and defining for all $p(x), q(x) \in \mathbb{P}_{n-1}$,

$$\langle p(x), q(x) \rangle = \sum_{i=1}^{n} p(r_i)q(r_i). \qquad (6.5)$$

For example, taking $r_1 = 1$, $r_2 = 2$, and $r_3 = 4$, the inner product of the polynomials $p(x) = x + x^2$ and $q(x) = -1 + 2x^2$ in $\mathbb{P}_2$ is

$$\langle p(x), q(x) \rangle = p(1)q(1) + p(2)q(2) + p(4)q(4) = 664.$$

(6.5) is proved to be an inner product in Exercise 2. It is called an **evaluation inner product**.

■ **Example 6.1.6**

Generalize the definition given in Exercise 11 from Section 5.2 and denote the vector space consisting of all continuous functions $[a, b] \to \mathbb{R}$ by $C[a, b]$. Define

$$\langle f, g \rangle = \int_a^b f(x)g(x)\,dx$$

defines an inner product, called an **integral inner product**. To confirm this, let $f, g, h \in C[a, b]$ and $r \in \mathbb{R}$.

- $\langle f, g \rangle = \int_a^b f(x)g(x)\,dx = \int_a^b g(x)f(x)\,dx = \langle g, f \rangle$.

- $\langle f, g + h \rangle = \langle f, g \rangle + \langle f, h \rangle$ because

$$\int_a^b f(x)[g(x) + h(x)]\,dx = \int_a^b f(x)g(x)\,dx + \int_a^b f(x)h(x)\,dx.$$

- $\langle rf, g \rangle = r\langle f, g \rangle$ since $\int_a^b [rf(x)]g(x)\,dx = r\int_a^b f(x)g(x)\,dx$.

- Assume that f is not the zero function on $[a, b]$ (Example 1.4.40). This implies that there exists $x_0 \in [a, b]$ such that $f(x_0) \neq 0$. Since f is continuous, there exists $[c, d] \subseteq [a, b]$ with $c \neq d$ so that $x_0 \in [c, d]$ and $f(x) \neq 0$ for all $x \in [c, d]$. Hence, $\langle f, f \rangle > 0$ because

$$\int_a^b [f(x)]^2\,dx \geq \int_c^d [f(x)]^2\,dx > 0.$$

As computations of length, distance, and angle measure in $\mathbb{R}^n$ can be made with the dot product, the inner product is used to compute these magnitudes in a general inner product space. This can be done because every inner product satisfies an important inequality (compare Theorem 2.1.11).

■ **Theorem 6.1.7 [Cauchy–Schwartz Inequality]**

Let V be an inner product space. For all $\mathbf{u}, \mathbf{v} \in V$,

$$\langle \mathbf{u}, \mathbf{v} \rangle \le \sqrt{\langle \mathbf{u}, \mathbf{u} \rangle} \sqrt{\langle \mathbf{v}, \mathbf{v} \rangle}.$$

Proof

Take $\mathbf{u}, \mathbf{v} \in V$ and $x \in \mathbb{R}$. Then, by various parts of Definition 6.1.1 and Theorem 6.1.2,

$$
\begin{aligned}
0 &\le \langle \mathbf{u} + x\mathbf{v}, \mathbf{u} + x\mathbf{v} \rangle \\
&= \langle \mathbf{u} + x\mathbf{v}, \mathbf{u} \rangle + \langle \mathbf{u} + x\mathbf{v}, x\mathbf{v} \rangle \\
&= \langle \mathbf{u}, \mathbf{u} \rangle + \langle x\mathbf{v}, \mathbf{u} \rangle + \langle \mathbf{u}, x\mathbf{v} \rangle + \langle x\mathbf{v}, x\mathbf{v} \rangle \\
&= \langle \mathbf{u}, \mathbf{u} \rangle + 2x \langle \mathbf{u}, \mathbf{v} \rangle + x^2 \langle \mathbf{v}, \mathbf{v} \rangle.
\end{aligned}
$$

Therefore, the graph of the quadratic equation in $\mathbb{R} \times \mathbb{R}$,

$$y = \langle \mathbf{u}, \mathbf{u} \rangle + 2 \langle \mathbf{u}, \mathbf{v} \rangle x + \langle \mathbf{v}, \mathbf{v} \rangle x^2,$$

does not drop below the x-axis, so its discriminant is not positive. In other words,

$$(2 \langle \mathbf{u}, \mathbf{v} \rangle)^2 - 4 \langle \mathbf{u}, \mathbf{u} \rangle \langle \mathbf{v}, \mathbf{v} \rangle \le 0.$$

Therefore, $\langle \mathbf{u}, \mathbf{v} \rangle^2 \le \langle \mathbf{u}, \mathbf{u} \rangle \langle \mathbf{v}, \mathbf{v} \rangle$, and taking square roots gives the result. ■

To illustrate, take $\begin{bmatrix} 1 \\ 2 \\ -3 \end{bmatrix}, \begin{bmatrix} 4 \\ -1 \\ 2 \end{bmatrix} \in \mathbb{R}^3$. Using the dot product,

$$\begin{bmatrix} 1 \\ 2 \\ -3 \end{bmatrix} \cdot \begin{bmatrix} 4 \\ -1 \\ 2 \end{bmatrix} = -4, \quad \begin{bmatrix} 1 \\ 2 \\ -3 \end{bmatrix} \cdot \begin{bmatrix} 1 \\ 2 \\ -3 \end{bmatrix} = 14, \quad \begin{bmatrix} 4 \\ -1 \\ 2 \end{bmatrix} \cdot \begin{bmatrix} 4 \\ -1 \\ 2 \end{bmatrix} = 21.$$

Hence,

$$\left(\begin{bmatrix} 1 \\ 2 \\ -3 \end{bmatrix} \cdot \begin{bmatrix} 4 \\ -1 \\ 2 \end{bmatrix} \right)^2 \le \left(\begin{bmatrix} 1 \\ 2 \\ -3 \end{bmatrix} \cdot \begin{bmatrix} 1 \\ 2 \\ -3 \end{bmatrix} \right) \left(\begin{bmatrix} 4 \\ -1 \\ 2 \end{bmatrix} \cdot \begin{bmatrix} 4 \\ -1 \\ 2 \end{bmatrix} \right).$$

Using the weighted inner product with weights $t_1 = 2$, $t_2 = 3$, and $t_3 = 4$ (6.2),

$$\left\langle \begin{bmatrix} 1 \\ 2 \\ -3 \end{bmatrix}, \begin{bmatrix} 4 \\ -1 \\ 2 \end{bmatrix} \right\rangle = -22, \quad \left\langle \begin{bmatrix} 1 \\ 2 \\ -3 \end{bmatrix}, \begin{bmatrix} 1 \\ 2 \\ -3 \end{bmatrix} \right\rangle = 50, \quad \left\langle \begin{bmatrix} 4 \\ -1 \\ 2 \end{bmatrix}, \begin{bmatrix} 4 \\ -1 \\ 2 \end{bmatrix} \right\rangle = 51.$$

Hence,

$$\left\langle \begin{bmatrix} 1 \\ 2 \\ -3 \end{bmatrix}, \begin{bmatrix} 4 \\ -1 \\ 2 \end{bmatrix} \right\rangle^2 \le \left\langle \begin{bmatrix} 1 \\ 2 \\ -3 \end{bmatrix}, \begin{bmatrix} 1 \\ 2 \\ -3 \end{bmatrix} \right\rangle \left\langle \begin{bmatrix} 4 \\ -1 \\ 2 \end{bmatrix}, \begin{bmatrix} 4 \\ -1 \\ 2 \end{bmatrix} \right\rangle.$$

Norms

The effect of each weight k_i is to distort distances parallel to $\mathbf{e}_i$ relative to the other standard vectors. To see this, it is required to be able to measure length. This is done with a certain type of function. Since the function is to measure length, its range should be a subset of the nonnegative real numbers. The function should also have the property that if the vector doubles in size, the measure of the double should be twice the measure of the original vector. The same goes for any other scalar factor of a vector. Lastly, the function should satisfy the triangle inequality because the measure of the sum of two vectors should never be greater than the sum of the measures of the vectors taken individually.

■ **Definition 6.1.8**

Let V be a vector space. A function $\nu : V \rightarrow \mathbb{R}$ is a **norm** if for all $\mathbf{u}, \mathbf{v} \in V$ and $r \in \mathbb{R}$,

(a) $\nu(\mathbf{u}) > 0$ if $\mathbf{u} \neq \mathbf{0}$,

(b) $\nu(r\mathbf{u}) = |r|\nu(\mathbf{u})$,

(c) $\nu(\mathbf{u} + \mathbf{v}) \leq \nu(\mathbf{u}) + \nu(\mathbf{v})$ (**Triangle Inequality**).

Let ν be a norm on a vector space V. No matter how ν measures vectors of positive length, it should be the case that the zero vector has measure 0. To prove that this is indeed the case, simply compute using (b) of Definition 6.1.8,

$$\nu(\mathbf{0}) = \nu(0\mathbf{0}) = 0\nu(\mathbf{0}) = 0.$$

Theorem 2.1.15 shows that the magnitude of a vector in $\mathbb{R}^n$ given by Definition 2.1.13 is a norm. In fact, Definition 2.1.13 can serve as a model to follow when defining a norm on any inner product space. Let V be an inner product space. For all $\mathbf{u}, \mathbf{v} \in V$, define

$$\nu(\mathbf{u}) = \sqrt{\langle \mathbf{u}, \mathbf{u} \rangle}. \tag{6.6}$$

Confirm that this satisfies the properties of a norm given in Definition 6.1.8.

• Suppose that $\mathbf{u} \neq \mathbf{0}$. Then, $\langle \mathbf{u}, \mathbf{u} \rangle > 0$ by Definition 6.1.1(d), so $\sqrt{\langle \mathbf{u}, \mathbf{u} \rangle} > 0$.

• Let $r \in \mathbb{R}$. Parts (a) and (c) of Definition 6.1.1 imply that

$$\sqrt{\langle r\mathbf{u}, r\mathbf{u} \rangle} = \sqrt{r^2 \langle \mathbf{u}, \mathbf{u} \rangle} = |r| \sqrt{\langle \mathbf{u}, \mathbf{u} \rangle}.$$

• The last part follows by applying the Cauchy–Schwartz Inequality (Theorem 6.1.7) to find

$$\langle \mathbf{u} + \mathbf{v}, \mathbf{u} + \mathbf{v} \rangle = \langle \mathbf{u}, \mathbf{u} \rangle + 2 \langle \mathbf{u}, \mathbf{v} \rangle + \langle \mathbf{v}, \mathbf{v} \rangle$$
$$\leq \langle \mathbf{u}, \mathbf{u} \rangle + 2\sqrt{\langle \mathbf{u}, \mathbf{u} \rangle} \sqrt{\langle \mathbf{v}, \mathbf{v} \rangle} + \langle \mathbf{v}, \mathbf{v} \rangle$$
$$= \left(\sqrt{\langle \mathbf{u}, \mathbf{u} \rangle} + \sqrt{\langle \mathbf{v}, \mathbf{v} \rangle} \right)^2.$$

Since $\langle \mathbf{u} + \mathbf{v}, \mathbf{u} + \mathbf{v} \rangle \geq 0$, it is legal to take the square roots of both sides to conclude that

$$\sqrt{\langle \mathbf{u} + \mathbf{v}, \mathbf{u} + \mathbf{v} \rangle} \leq \sqrt{\langle \mathbf{u}, \mathbf{u} \rangle} + \sqrt{\langle \mathbf{v}, \mathbf{v} \rangle}.$$

Therefore, (6.6) is a norm. Since it is an important norm that is defined using an inner product, it receives a name and uses the same notation as in Definition 2.1.13.

■ **Definition 6.1.9**

Let V be an inner product space. The **(induced) norm** on V is the norm defined by $\|\mathbf{u}\| = \sqrt{\langle \mathbf{u}, \mathbf{u} \rangle}$ for all $\mathbf{u} \in V$.

As before, the norm is viewed as the length of a vector. Using the weighted inner product on $\mathbb{R}^2$ defined by

$$\left\langle \begin{bmatrix} x_1 \\ y_1 \end{bmatrix}, \begin{bmatrix} x_2 \\ y_2 \end{bmatrix} \right\rangle = \frac{1}{9} x_1 x_2 + \frac{1}{25} y_1 y_2, \tag{6.7}$$

observe that

$$\left\| \begin{bmatrix} 1 \\ 0 \end{bmatrix} \right\| = \frac{1}{3} \quad \text{and} \quad \left\| \begin{bmatrix} 0 \\ 1 \end{bmatrix} \right\| = \frac{1}{5}.$$

However, a **unit vector**, a vector that has length 1 using the induced norm of the inner product space, can still be found as in Theorem 2.1.17. For example,

$$3 \begin{bmatrix} 1 \\ 0 \end{bmatrix} = \begin{bmatrix} 3 \\ 0 \end{bmatrix} \quad \text{and} \quad 5 \begin{bmatrix} 0 \\ 1 \end{bmatrix} = \begin{bmatrix} 0 \\ 5 \end{bmatrix}.$$

are unit vectors. A calculation with (6.7) will confirm this. Furthermore, for any inner product space V, the set of unit vectors is

$$\{\mathbf{u} : \|\mathbf{u}\| = 1 \text{ and } \mathbf{u} \in V\},$$

which is called the **unit circle**, especially when $V = \mathbb{R}^2$. The equation of the unit circle using the inner product (6.7) is $\sqrt{\frac{1}{9} x^2 + \frac{1}{25} y^2} = 1$. That is, the equation for the unit circle in this inner product space is

$$\frac{x^2}{9} + \frac{y^2}{25} = 1,$$

which is an ellipse when lengths are computed with the dot product. This illustrates how the weights distort the space differently depending on the direction.

Metrics

Once vectors can be measured, the next step is to define the distance between them. The method used will be a generalization of the properties of the Euclidean distance (Definition 2.1.9) given in Theorem 2.1.12.

■ **Definition 6.1.10**

Let V be a vector space. A **metric** or a **distance function** on V is a function $\delta : V \times V \to \mathbb{R}$ such that for all $\mathbf{u}, \mathbf{v}, \mathbf{w} \in V$,

(a) $\delta(\mathbf{u}, \mathbf{v}) \geq 0$,

(b) $\delta(\mathbf{u}, \mathbf{v}) = 0$ if and only if $\mathbf{u} = \mathbf{v}$,

(c) $\delta(\mathbf{u}, \mathbf{v}) = \delta(\mathbf{v}, \mathbf{u})$,

(d) $\delta(\mathbf{u}, \mathbf{v}) \leq \delta(\mathbf{u}, \mathbf{w}) + \delta(\mathbf{w}, \mathbf{v})$ **(Triangle Inequality)**.

An inner product spaces paired with a metric is called an **metric space**.

Definition 6.1.10 incorporates the basic properties of what it means for a function to represent distance. The function should only return nonnegative values (a). The distance between vectors should be 0 when the vectors are equal (b). A distance function should be commutative (c). The distance between two vectors should not be greater than the sum of the distances between the vectors and a third vector (d). Put another way, the direct distance between two points should not be greater than the path between the points that includes a detour to a third point. Such a function can always be found in an inner product space as with Theorem 2.1.14.

■ **Definition 6.1.11**

Let V be an inner product space and take $\mathbf{u}, \mathbf{v} \in V$. The **(induced) distance** between $\mathbf{u}$ and $\mathbf{v}$ is $d(\mathbf{u}, \mathbf{v}) = \|\mathbf{u} - \mathbf{v}\|$.

Observe that the distance of Definition 6.1.11 makes any inner product space a metric space. This is the next theorem. Its proof is left to Exercise 18.

■ **Theorem 6.1.12**

The induced distance on an inner product space is a metric on the inner product space.

Using the inner product of (6.7), the induced distance on $\mathbb{R}^2$ is

$$d\left(\begin{bmatrix} x_1 \\ y_1 \end{bmatrix}, \begin{bmatrix} x_2 \\ y_2 \end{bmatrix} \right) = \left\| \begin{bmatrix} x_1 - x_2 \\ y_1 - y_2 \end{bmatrix} \right\| = \sqrt{\frac{(x_1 - x_2)^2}{9} + \frac{(y_1 - y_2)^2}{25}}.$$

For example,

$$d\left(\begin{bmatrix} 1 \\ 0 \end{bmatrix}, \begin{bmatrix} 0 \\ 1 \end{bmatrix} \right) = \frac{\sqrt{34}}{15} \approx 0.389,$$

while the distance between $\begin{bmatrix} 1 \\ 0 \end{bmatrix}$ and $\begin{bmatrix} 0 \\ 1 \end{bmatrix}$ is $\sqrt{2}$ when the inner product is the dot product.

■ **Example 6.1.13**

The vector form of the equation of a line in $\mathbb{R}^n$ (2.23) can be generalized to this setting. Pair $\mathbb{P}_2$ with the evaluation inner product with $r_1 = 1$, $r_2 = 2$, and $r_3 = 4$ of Example 6.1.4 to define an inner product space. Since two vectors are enough to determine a line, find the line joining $1 + 3x^2$ and $x + 6x^2$. The direction vector for this line is

$$(1 + 3x^2) - (x + 6x^2) = 1 - x - 3x^2,$$

so one possible equation is

$$f(x) = (1 + 3x^2) + t(1 - x - 3x^2).$$

Observe that $p(x) = 3 - 2x - 3x^2 = (1 + 3x^2) + 2(1 - x - 3x^2)$ is in the line. Furthermore, using the coordinate map $T_{\mathscr{E}} : \mathbb{P}_2 \to \mathbb{R}^3$ of Definition 5.6.16, where $\mathscr{E} = \{1, x, x^2\}$ is the standard basis of $\mathbb{P}_2$,

$$T_{\mathscr{E}}[f(x)] = \begin{bmatrix} 1 \\ 0 \\ 3 \end{bmatrix} + t \begin{bmatrix} 1 \\ -1 \\ -3 \end{bmatrix}$$

because $T_{\mathscr{E}}$ is an isomorphism (Theorem 5.6.18). This means that the isomorphism $T_{\mathscr{E}}$ maps a line consisting of degree at most two polynomials to a line in $\mathbb{R}^3$.

■ **Example 6.1.14**

Continuing Example 6.1.14, write $q(x) = 1 + 3x^2$, so both $p(x)$ and $q(x)$ are in the line. Compute the distance,

$$d(p(x), q(x)) = \|p(x) - q(x)\| = \sqrt{\langle p(x) - q(x), p(x) - q(x) \rangle} = 105.4.$$

It is also worth noting that because $\|q(x)\| = \sqrt{\langle q(x), q(x) \rangle} = \sqrt{2586}$, it follows that $q(x)/\sqrt{2586}$ is a unit vector in the direction of $q(x)$.

Angles

The definition of the angle between two vectors in $\mathbb{R}^n$ (Definition 2.2.2) is based on the the law of cosines, but it is unclear how to apply the law in a general inner product space V. However, the Cauchy–Schwartz Inequality (Theorem 6.1.7) does hold in V. Because of this, if $\mathbf{u}, \mathbf{v} \in V$,

$$-1 \le \frac{\langle \mathbf{u}, \mathbf{v} \rangle}{\sqrt{\langle \mathbf{u}, \mathbf{u} \rangle \langle \mathbf{v}, \mathbf{v} \rangle}} \le 1,$$

and then using Definition 6.1.9,

$$-1 \le \frac{\langle \mathbf{u}, \mathbf{v} \rangle}{\|\mathbf{u}\| \, \|\mathbf{v}\|} \le 1,$$

This makes the next concept well-defined.

■ **Definition 6.1.15**

Let **u** and **v** be nonzero vectors from an inner product space. Define the **angle between u** and **v** to have measure θ where $0 \leq \theta \leq \pi$ and

$$\cos\theta = \frac{\langle \mathbf{u}, \mathbf{v} \rangle}{\|\mathbf{u}\|\|\mathbf{v}\|}.$$

■ **Example 6.1.16**

Consider the vector space $C[0, 1]$. Using the integral inner product of Example 6.1.6, the measure θ of the angle between sin and cos requires the following computations:

$$\langle \cos, \sin \rangle = \int_0^1 \cos x \sin x \, dx \approx 0.3540,$$

$$\langle \cos, \cos \rangle = \int_0^1 \cos^2 x \, dx \approx 0.7273,$$

and

$$\langle \sin, \sin \rangle = \int_0^1 \sin^2 x \, dx \approx 0.2727.$$

Therefore,

$$\| \cos \| = \sqrt{\langle \cos, \cos \rangle} \approx 0.8528 \quad \text{and} \quad \| \sin \| = \sqrt{\langle \sin, \sin \rangle} \approx 0.5222,$$

so using Definition 6.1.15, $\cos\theta \approx 0.7951$. This means that the angle formed by sin and cos in the inner product space $C[0, 1]$ with the integral inner product is about 37°.

Because of the close association between $\mathbb{R}^n$ and Euclidean space, the definition of *perpendicular* in $\mathbb{R}^n$ made sense with the definition of *orthogonal* being a special case. In this general setting, the concept orthogonality is more natural, so the generalization of Definition 2.2.7 is better when it starts with orthogonal vectors.

■ **Definition 6.1.17**

Let **u**, **v** be vectors from an inner product space.

(a) **u** and **v** are **orthogonal** if $\langle \mathbf{u}, \mathbf{v} \rangle = 0$.

(b) **u** is **perpendicular** to **v** if **u** and **v** are nonzero and orthogonal.

■ **Example 6.1.18**

As with Theorem 2.2.8, for all nonzero **u** and **v** from an inner product space forming an angle of measure θ,

$$\langle \mathbf{u}, \mathbf{v} \rangle = 0 \text{ if and only if } \theta = 0.$$

For example, using the weighted inner product of (6.7) with $t_1 = 1/9$ and $t_2 = 1/25$,

$$\left\langle \begin{bmatrix} 1 \\ 0 \end{bmatrix}, \begin{bmatrix} 0 \\ 1 \end{bmatrix} \right\rangle = 0,$$

so $\begin{bmatrix} 1 \\ 0 \end{bmatrix}$ is orthogonal (and perpendicular) to $\begin{bmatrix} 0 \\ 1 \end{bmatrix}$ in this inner product space. However, because

$$\left\langle \begin{bmatrix} 3 \\ 4 \end{bmatrix}, \begin{bmatrix} -4 \\ 3 \end{bmatrix} \right\rangle = -\frac{64}{75},$$

$\begin{bmatrix} 3 \\ 4 \end{bmatrix}$ and $\begin{bmatrix} -4 \\ 3 \end{bmatrix}$ are not orthogonal in this inner product space. Furthermore, since

$$\left\langle \begin{bmatrix} 3 \\ 4 \end{bmatrix}, \begin{bmatrix} 3 \\ 4 \end{bmatrix} \right\rangle = \frac{41}{25} \quad \text{and} \quad \left\langle \begin{bmatrix} -4 \\ 3 \end{bmatrix}, \begin{bmatrix} -4 \\ 3 \end{bmatrix} \right\rangle = \frac{481}{225},$$

the angle θ between $\begin{bmatrix} 3 \\ 4 \end{bmatrix}$ and $\begin{bmatrix} -4 \\ 3 \end{bmatrix}$ is found by solving

$$\cos \theta = \frac{\left\langle \begin{bmatrix} 3 \\ 4 \end{bmatrix}, \begin{bmatrix} -4 \\ 3 \end{bmatrix} \right\rangle}{\left\| \begin{bmatrix} 3 \\ 4 \end{bmatrix} \right\| \left\| \begin{bmatrix} -4 \\ 3 \end{bmatrix} \right\|} = -\frac{64}{\sqrt{19721}},$$

which implies that the angle formed by these two vectors in this inner product space is about 2.04 radians (117°).

Although the inner product controls the geometry of the inner product space, the geometry is still Euclidean in that a key theorem proved by Euclid still holds.

■ **Theorem 6.1.19 [Pythagoras]**

Let V be an inner product space. The vectors $\mathbf{u}, \mathbf{v} \in V$ are orthogonal if and only if $\|\mathbf{u}\|^2 + \|\mathbf{v}\|^2 = \|\mathbf{u} - \mathbf{v}\|^2$.

Proof

Take $\mathbf{u}, \mathbf{v} \in V$ such that $\langle \mathbf{u}, \mathbf{v} \rangle = 0$. Then,

$$\begin{aligned} \|\mathbf{u} - \mathbf{v}\|^2 &= \langle \mathbf{u} - \mathbf{v}, \mathbf{u} - \mathbf{v} \rangle \\ &= \langle \mathbf{u}, \mathbf{u} \rangle - 2 \langle \mathbf{u}, \mathbf{v} \rangle + \langle \mathbf{v}, \mathbf{v} \rangle \\ &= \langle \mathbf{u}, \mathbf{u} \rangle + \langle \mathbf{v}, \mathbf{v} \rangle = \|\mathbf{u}\|^2 + \|\mathbf{v}\|^2. \end{aligned}$$

The converse follows because $\langle \mathbf{u}, \mathbf{u} \rangle - 2 \langle \mathbf{u}, \mathbf{v} \rangle + \langle \mathbf{v}, \mathbf{v} \rangle = \langle \mathbf{u}, \mathbf{u} \rangle + \langle \mathbf{v}, \mathbf{v} \rangle$ implies that $\langle \mathbf{u}, \mathbf{v} \rangle = 0$. ■

Orthogonal Projection

An important application of the dot product is the orthogonal projection. Unlike in Section 2.2 where the projection was defined after a discussion that showed that its construction is orthogonal to the vector component, here the discussion begins with the generalization of Definition 2.2.18. This is followed by a proof that the vector defined as the generalization of the projection has the same orthogonality property as before.

■ **Definition 6.1.20**

Let $\mathbf{u}$ and $\mathbf{v}$ be vectors from an inner product space and assume that $\mathbf{v} \neq \mathbf{0}$. The **orthogonal projection** (or simply the **projection**) of $\mathbf{u}$ in the direction of $\mathbf{v}$ is

$$\text{proj}_{\mathbf{v}}\mathbf{u} = \frac{\langle \mathbf{u}, \mathbf{v} \rangle}{\langle \mathbf{v}, \mathbf{v} \rangle}\mathbf{v}.$$

The vector $\mathbf{u} - \text{proj}_{\mathbf{v}}\mathbf{u}$ is the **vector component of u orthogonal to v**.

■ **Theorem 6.1.21**

Let V be an inner project space. If $\mathbf{u}, \mathbf{v} \in V$ and $\mathbf{v} \neq \mathbf{0}$, then $\text{proj}_{\mathbf{v}}\mathbf{u}$ is orthogonal to $\mathbf{u} - \text{proj}_{\mathbf{v}}\mathbf{u}$.

Proof

Simply compute using Definition 6.1.1,

$$\langle \mathbf{u} - \text{proj}_{\mathbf{v}}\mathbf{u}, \text{proj}_{\mathbf{v}}\mathbf{u} \rangle = \left\langle \mathbf{u}, \frac{\langle \mathbf{u}, \mathbf{v} \rangle}{\langle \mathbf{v}, \mathbf{v} \rangle}\mathbf{v} \right\rangle - \left\langle \frac{\langle \mathbf{u}, \mathbf{v} \rangle}{\langle \mathbf{v}, \mathbf{v} \rangle}\mathbf{v}, \frac{\langle \mathbf{u}, \mathbf{v} \rangle}{\langle \mathbf{v}, \mathbf{v} \rangle}\mathbf{v} \right\rangle$$

$$= \frac{\langle \mathbf{u}, \mathbf{v} \rangle}{\langle \mathbf{v}, \mathbf{v} \rangle}\langle \mathbf{u}, \mathbf{v} \rangle - \frac{\langle \mathbf{u}, \mathbf{v} \rangle^2}{\langle \mathbf{v}, \mathbf{v} \rangle^2}\langle \mathbf{v}, \mathbf{v} \rangle$$

$$= \frac{\langle \mathbf{u}, \mathbf{v} \rangle^2}{\langle \mathbf{v}, \mathbf{v} \rangle} - \frac{\langle \mathbf{u}, \mathbf{v} \rangle^2}{\langle \mathbf{v}, \mathbf{v} \rangle} = 0. \blacksquare$$

Although the calculations will be different, it is worth noting that the projection using an abstract inner product has the same basic properties as the orthogonal projection defined using the dot product. To see an example of this, let V be an inner product space. Let $\mathbf{u}, \mathbf{v} \in V$ with $\mathbf{v} \neq \mathbf{0}$ and $r \in \mathbb{R}$. Because $\|r\mathbf{u}\| = |r| \|\mathbf{u}\|$ as in Theorem 2.1.15(c), the length of the projection is found by calculating

$$\|\text{proj}_{\mathbf{v}}\mathbf{u}\| = \left\| \frac{\langle \mathbf{u}, \mathbf{v} \rangle}{\langle \mathbf{v}, \mathbf{v} \rangle}\mathbf{v} \right\| = \left\| \frac{\langle \mathbf{u}, \mathbf{v} \rangle}{\sqrt{\langle \mathbf{v}, \mathbf{v} \rangle}} \frac{1}{\sqrt{\langle \mathbf{v}, \mathbf{v} \rangle}}\mathbf{v} \right\| = \frac{|\langle \mathbf{u}, \mathbf{v} \rangle|}{\|\mathbf{v}\|}.$$

The last equation follows because $\left\| \dfrac{1}{\sqrt{\langle \mathbf{v}, \mathbf{v} \rangle}}\mathbf{v} \right\| = \left\| \dfrac{1}{\|\mathbf{v}\|}\mathbf{v} \right\| = \dfrac{1}{\|\mathbf{v}\|}\|\mathbf{v}\| = 1.$

■ **Example 6.1.22**

Basic geometry dictates that three vectors will determine a unique plane, so take $1 + 2x + 3x^2$, $4 - x^2$, and $x + x^2$ in $\mathbb{P}_2$ paired with the evaluation inner product with $r_1 = 1$, $r_2 = 2$, and $r_3 = 3$ (Example 6.1.13) and find the plane P described by these polynomials. Subtract two pairs of polynomials to find direction vectors of the plane. They can be taken to be

$$m_1(x) = -3 + 2x + 4x^2 \quad \text{and} \quad m_2(x) = 1 + x + 2x^2.$$

The cross product can only be used to find a normal vector to a plane in $\mathbb{R}^3$ when the inner product is the dot product. Since in this case the evaluation inner product is being used, let $n(x) = a + bx + cx^2$ be a normal to the plane and adapt the technique discussed at the start of Section 2.3 to find $n(x)$ by solving the system of linear equations given by

$$\langle n(x), m_1(x) \rangle = \langle n(x), m_2(x) \rangle = 0.$$

The system is

$$
\begin{aligned}
59a &+ 154b &+ 422c &= 0, \\
37a &+ 92b &+ 246c &= 0,
\end{aligned}
$$

and its solution is

$$a = \frac{94}{27}t,$$
$$b = -\frac{110}{27}t,$$
$$c = t.$$

Therefore, $n(x)$ can be taken to be $94 - 110x + 27x^2$. Now, solving

$$\langle 94 - 110x + 27x^2, (u_1 - 1) + (u_2 - 2)x + (u_3 - 3)x^2 \rangle = 0$$

will give parametric equations,

$$u_1 = s,$$
$$u_2 = t,$$
$$u_3 = 2t - 1.$$

Therefore,
$$P = \{s + tx + (2t - 1)x^2 : s, t \in \mathbb{R}\}, \tag{6.8}$$

and following (2.31) and (2.33), if $u_1 + u_2 x + u_3 x^2 \in P$, a vector-normal equation for P is

$$\langle 94 - 110x + 27x^2, u_1 + u_2 x + u_3 x^2 \rangle = \langle 94 - 110x + 27x^2, 1 + 2x + 3x^2 \rangle.$$

■ **Example 6.1.23**

Continuing Example 6.1.22, to illustrate the projection with this inner product, find the distance from $p(x) = 4 + 3x^2$ to the plane P. Follow the strategy of Example 2.2.21 using Figure 2.11 as assistance. Let $q(x) = -x^2$, which is in the plane, and subtract $p(x) - q(x) = 4 + 4x^2$. Project this polynomial onto a normal of the plane and calculate the length of the resulting vector,

$$\|\text{proj}_{n(x)} 4 + 4x^2\| = \frac{|\langle n(x), 4 + 4x^2 \rangle|}{\sqrt{\langle n(x), n(x) \rangle}} = \frac{8}{\sqrt{494}}.$$

Exercises

1. Find the inner product of each pair of vectors using the given inner product.

    (a) $\begin{bmatrix} 3 \\ 5 \\ -2 \end{bmatrix}$ and $\begin{bmatrix} 1 \\ 0 \\ -5 \end{bmatrix}$, weighted inner product with weights $k_1 = 4$, $k_2 = 5$, and $k_3 = 2$.

    (b) $4 + x - 2x^2$ and $7 + 3x + x^2$, evaluation inner product with $r_1 = 4$, $r_2 = 2$ and $r_3 = 1$.

    (c) $\begin{bmatrix} 1 & 2 & 3 \\ 4 & 5 & 6 \end{bmatrix}$ and $\begin{bmatrix} 7 & 8 & 9 \\ 0 & 1 & 2 \end{bmatrix}$, Frobenius inner product.

    (d) $f, g : [0, 1] \to \mathbb{R}$, where $f(x) = 9 + 4x$ and $g(x) = 6 - 7x$; integral inner product.

2. Prove that the function $\mathbb{P}_n \times \mathbb{P}_n \to \mathbb{R}$ from Example 6.1.5 is an inner product.

3. Define a function $\mathbb{P}_n \times \mathbb{P}_n \to \mathbb{R}$ such that for all polynomials $p(x), q(x) \in \mathbb{P}_n$ where $p(x) = a_0 + a_1 x + \cdots + a_n x^n$ and $q(x) = b_0 + b_1 x + \cdots + b_n x^n$,

    $$\langle p(x), q(x) \rangle = a_0 b_0 + a_1 b_1 + \cdots + a_n b_n.$$

    Prove that this defines an inner product.

4. Evaluate $\langle 3 + x + 5x^2, 1 - 7x + 2x^2 \rangle$ using the inner product of Exercise 3.

5. Let $f, g \in C[-1, 1]$. Show that

    $$\langle f, g \rangle = \int_{-1}^{1} x^2 f(x) g(x) \, dx$$

    defines an inner product on $C[-1, 1]$.

6. Evaluate $\langle f, g \rangle$, where $f(x) = x^3$ and $g(x) = x$, using the inner product of Exercise 5.

7. Let $T : V \to \mathbb{R}^n$ be one-to-one. Prove that the function $V \times V \to \mathbb{R}$ such that $\langle \mathbf{u}, \mathbf{v} \rangle = T\mathbf{u} \cdot T\mathbf{v}$ is an inner product.

8. Prove Theorem 6.1.2.

9. Use the Cauchy–Schwartz Inequality to prove that $(x + y)^2 \le 2(x^2 + y^2)$ for all $x, y \in \mathbb{R}$.

10. Let A be an $n{\times}n$ matrix. View $\mathbb{R}^n$ as an inner product space and take $\mathbf{u}, \mathbf{v} \in \mathbb{R}^n$. Prove these generalizations of Theorem 4.4.8 and Corollary 4.4.9.

    (a) $\langle A\mathbf{u}, \mathbf{v} \rangle = \langle \mathbf{u}, A^T\mathbf{v} \rangle$

    (b) If A is symmetric, $\langle A\mathbf{u}, \mathbf{v} \rangle = \langle \mathbf{u}, A\mathbf{v} \rangle$.

11. Compute the norm for each vector using the given inner product.

    (a) $\begin{bmatrix} 1 \\ 0 \end{bmatrix}$, weighted inner product with weights $k_1 = 4$ and $k_2 = 7$.

    (b) $\begin{bmatrix} 5 & -3 \\ 0 & 9 \end{bmatrix}$, Frobenius inner product.

    (c) $3 + x - 7x^2 + 3x^3$, evaluation inner product with $r_1 = 4$ and $r_2 = 7$.

    (d) $f : [0, 1] \to \mathbb{R}$, where $f(x) = 3x + e^x$; integral inner product.

12. Find the unit circle in each inner product space with the given inner product.

    (a) $\mathbb{R}^2$, weighted inner product with weights $k_1 = 16$ and $k_2 = 9$.

    (b) $\mathbb{R}^3$, weighted inner product with $k_1 = 1/25$, $k_2 = 1/9$, and $k_3 = 1/25$. (The unit circle in $\mathbb{R}^3$ is usually called the **unit sphere**.)

    (c) $\mathbb{R}^{2\times 2}$, Frobenius inner product.

    (d) $\mathbb{P}_3$, evaluation inner product with $r_1 = 2$ and $r_2 = 4$.

13. Let $\mathbf{u}$ and $\mathbf{v}$ be vectors in an inner product space. Prove the **Parallelogram Law**:
$$\|\mathbf{u} + \mathbf{v}\|^2 + \|\mathbf{u} - \mathbf{v}\|^2 = 2\|\mathbf{u}\|^2 + 2\|\mathbf{v}\|^2.$$

14. Let V be an inner product space. Prove that for all $\mathbf{u}, \mathbf{v} \in V$,
$$\langle \mathbf{u}, \mathbf{v} \rangle = \frac{1}{4}(\|\mathbf{u} + \mathbf{v}\|^2 - \|\mathbf{u} - \mathbf{v}\|^2).$$

15. Suppose that V is an inner product space. For every $\mathbf{u}, \mathbf{v} \in V$, prove that $\langle \mathbf{u} + \mathbf{v}, \mathbf{u} - \mathbf{v} \rangle = \|\mathbf{u}\|^2 - \|\mathbf{v}\|^2$.

16. Let $\mathbf{u}$ and $\mathbf{v}$ be vectors from an inner product space. Prove that if $\mathbf{u}$ and $\mathbf{v}$ are orthogonal, $\|\mathbf{u} + \mathbf{v}\| = \|\mathbf{u} - \mathbf{v}\|$. Is the converse true?

17. Prove that $\langle \mathbf{u}, \mathbf{v} \rangle = \frac{1}{2}(\|\mathbf{u}\|^2 + \|\mathbf{v}\|^2 - \|\mathbf{u} - \mathbf{v}\|^2)$ for all $\mathbf{u}, \mathbf{v}$ in an inner product space.

18. Let V be an inner product space. Complete the proof of Theorem 6.1.12 by proving the following for all $\mathbf{u}, \mathbf{v} \in V$.

    (a) $d(\mathbf{u}, \mathbf{v}) = 0$ if and only if $\mathbf{u} = \mathbf{v}$.   (b) $d(\mathbf{u}, \mathbf{v}) = d(\mathbf{v}, \mathbf{u})$.

19. Find the distance between each pair of vectors using the given inner product.

    (a) $\begin{bmatrix} 1 \\ 0 \end{bmatrix}$ and $\begin{bmatrix} 0 \\ 1 \end{bmatrix}$, weighted inner product with weights $k_1 = 4$ and $k_2 = 7$.

    (b) $\begin{bmatrix} 5 & -3 \\ 0 & 9 \end{bmatrix}$ and $\begin{bmatrix} 1 & 5 \\ 0 & -3 \end{bmatrix}$, Frobenius inner product.

    (c) $3 + x - 7x^2 + 3x^3$ and $4x - x^2 + x^3$, evaluation inner product with $r_1 = 4$ and $r_2 = 7$.

    (d) $f, \cos \in C[0, \pi]$, where $f(x) = e^x$; integral inner product.

20. Find the angle described by each pair of vectors using the given inner product.

    (a) $\begin{bmatrix} 1 \\ 0 \end{bmatrix}$ and $\begin{bmatrix} 0 \\ 1 \end{bmatrix}$, weighted inner product with weights $k_1 = 4$ and $k_2 = 7$.

    (b) $\begin{bmatrix} 5 & -3 \\ 0 & 9 \end{bmatrix}$ and $\begin{bmatrix} 1 & 5 \\ 0 & -3 \end{bmatrix}$, Frobenius inner product.

    (c) $3 + x - 7x^2 + 3x^3$ and $4x - x^2 + x^3$, evaluation inner product with $r_1 = 4$ and $r_2 = 7$.

    (d) $f, \cos \in C[0, \pi]$, where $f(x) = e^x$; integral inner product.

21. Determine whether each pair of vectors is orthogonal using the given inner product.

    (a) $\begin{bmatrix} 1 \\ 0 \end{bmatrix}$ and $\begin{bmatrix} 0 \\ 1 \end{bmatrix}$, weighted inner product with weights $k_1 = 4$ and $k_2 = 7$.

    (b) $\begin{bmatrix} 3 & 1 \\ 0 & 5 \end{bmatrix}$ and $\begin{bmatrix} -2 & 1 \\ 7 & 1 \end{bmatrix}$, Frobenius inner product.

    (c) $5x - 2x^2$ and $4 - x + 3x^2$, evaluation inner product with $r_1 = 0$, $r_2 = 1$, and $r_3 = 2$.

    (d) $f, g : [0, 1] \to \mathbb{R}$, where $f(x) = e^x$ and $g(x) = \ln(x + 1)$; integral inner product.

22. Let $\mathbf{u}, \mathbf{v}_1, \mathbf{v}_2, \dots, \mathbf{v}_k$ be vectors from an inner product space. Prove that if $\mathbf{u}$ is orthogonal to $\mathbf{v}_i$ for every $i = 1, 2, \dots, k$, then $\mathbf{u}$ is orthogonal to $\mathbf{w}$ for all $\mathbf{w} \in S(\mathbf{v}_1, \mathbf{v}_2, \dots, \mathbf{v}_k)$.

23. Find the the plane containing the given vectors using the given inner product. Write the equation in vector-normal form if possible. See Example 6.1.22.

    (a) $\begin{bmatrix} 1 \\ 0 \\ 2 \end{bmatrix}, \begin{bmatrix} 2 \\ 1 \\ -1 \end{bmatrix}, \begin{bmatrix} 0 \\ 0 \\ 1 \end{bmatrix}$; weighted inner product with weights $k_1 = 1$, $k_2 = 2$, and $k_3 = 3$.

    (b) $1 + x, 1 + x^2, 2 + x^2$; evaluation inner product with $r_1 = 0$, $r_2 = 2$, and $r_3 = 4$.

    (c) $\begin{bmatrix} 1 & 3 \\ 0 & 2 \end{bmatrix}, \begin{bmatrix} 0 & 1 \\ 1 & 0 \end{bmatrix}, \begin{bmatrix} 0 & 0 \\ 0 & 0 \end{bmatrix}$; Frobenius inner product.

24. Modify the work of Example 6.1.22 to find the line containing the given vectors using the given inner product. Write the equation in vector-normal form if possible.

    (a) $\begin{bmatrix} 1 \\ 0 \end{bmatrix}$ and $\begin{bmatrix} 2 \\ 1 \end{bmatrix}$, weighted inner product with weights $k_1 = 2$, and $k_2 = 3$.

    (b) $\begin{bmatrix} 1 & 3 \\ 0 & 2 \end{bmatrix}$ and $\begin{bmatrix} 0 & 1 \\ 1 & 0 \end{bmatrix}$, Frobenius inner product.

    (c) $\sin, \cos \in C[0, \pi]$, integral inner product.

25. For each pair of vectors, find the orthogonal projection of the first vector onto the second vector using the given inner product.

    (a) $\begin{bmatrix} 3 \\ 5 \\ -2 \end{bmatrix}$ and $\begin{bmatrix} 1 \\ 0 \\ -5 \end{bmatrix}$, weighted inner product with weights $k_1 = 4$, $k_2 = 5$, and $k_3 = 2$.

    (b) $4 + x - 2x^2$ and $7 + 3x + x^2$, evaluation inner product with $r_1 = 4, r_2 = 2$ and $r_3 = 1$.

    (c) $\begin{bmatrix} 1 & 2 & 3 \\ 4 & 5 & 6 \end{bmatrix}$ and $\begin{bmatrix} 7 & 8 & 9 \\ 0 & 1 & 2 \end{bmatrix}$, Frobenius inner product.

    (d) $f, g : [0, 1] \to \mathbb{R}$, where $f(x) = 9 + 4x$ and $g(x) = 6 - 7x$; integral inner product.

26. Let V be an inner product space and $\mathbf{v} \in V \setminus \{\mathbf{0}\}$. Prove that $\text{proj}_{\mathbf{v}}$ is a linear operator $V \to V$.

6.2 Orthonormal Bases

Consider the set of vectors from $\mathbb{R}^3$,

$$\left\{ \begin{bmatrix} 1 \\ 2 \\ -1 \end{bmatrix}, \begin{bmatrix} 3 \\ -1 \\ 1 \end{bmatrix}, \begin{bmatrix} 1 \\ -4 \\ -7 \end{bmatrix} \right\}. \tag{6.9}$$

Each pair of distinct vectors from this set is orthogonal using the dot product because

$$\begin{bmatrix} 1 \\ 2 \\ -1 \end{bmatrix} \cdot \begin{bmatrix} 3 \\ -1 \\ 1 \end{bmatrix} = 0, \quad \begin{bmatrix} 1 \\ 2 \\ -1 \end{bmatrix} \cdot \begin{bmatrix} 1 \\ -4 \\ -7 \end{bmatrix} = 0, \quad \begin{bmatrix} 3 \\ -1 \\ 1 \end{bmatrix} \cdot \begin{bmatrix} 1 \\ -4 \\ -7 \end{bmatrix} = 0. \tag{6.10}$$

If a set like that given in (6.9) can be found in $\mathbb{R}^3$, such a set should be able to be found in an arbitrary inner product space. The definition that follows is a generalization of that found in Section 4.4.

■ **Definition 6.2.1**

Let V be an inner product space. If $\mathcal{S}$ is a nonempty subset of V, then $\mathcal{S}$ is an **orthogonal set** means that for all $\mathbf{u}, \mathbf{v} \in \mathcal{S}$, if $\mathbf{u} \neq \mathbf{v}$, then $\mathbf{u}$ is orthogonal to $\mathbf{v}$. An orthogonal set consisting only of unit vectors is an **orthonormal set**.

■ **Example 6.2.2**

Define a function $\mathbb{P}_n \times \mathbb{P}_n \to \mathbb{R}$ such that for all $p(x), q(x) \in \mathbb{P}_n$, where $p(x) = a_0 + a_1 x + \cdots + a_n x^n$ and $q(x) = b_0 + b_1 x + \cdots + b_n x^n$,

$$\langle p(x), q(x) \rangle = a_0 b_0 + a_1 b_1 + \cdots + a_n b_n. \tag{6.11}$$

That this function is an inner product on $\mathbb{P}_n$ is due to Exercise 3 of Section 6.1. Since (6.11) mimics the dot product on $\mathbb{R}^{n+1}$, the calculations of (6.10) can be modified to show that

$$\{1 + 2x - x^2, 3 - x + x^2, 1 - 4x - 7x^2\}$$

is an orthogonal set in $\mathbb{P}_n$ if $n \geq 2$, but it is not orthonormal because

$$\|1 + 2x - x^2\| = \sqrt{6}, \quad \|3 - x + x^2\| = \sqrt{11}, \quad \|1 - 4x - 7x^2\| = \sqrt{66}.$$

Use these lengths to find the orthonormal set,

$$\left\{ \frac{1 + 2x - x^2}{\sqrt{6}}, \frac{3 - x + x^2}{\sqrt{11}}, \frac{1 - 4x - 7x^2}{\sqrt{66}} \right\}.$$

Another orthonormal set in $\mathbb{P}_n$, which is easier to find, is $\{1, x, x^2, \ldots, x^n\}$.

■ **Example 6.2.3**

Consider $C[0, 1]$ as an inner product space using the integral inner product of Example 6.1.6. Let $f, g \in C[0, 1]$ be defined as

$$f(x) = \left| x - \frac{1}{2} \right| - \frac{1}{4} \quad \text{and} \quad g(x) = 4.$$

Then,

$$\int_0^1 f(x) g(x)\, dx = \int_0^1 \left(4 \left| x - \frac{1}{2} \right| - 1 \right) dx = 0.$$

Therefore, $\{f, g\}$ is an orthogonal set in $C[0, 1]$. Also,

$$\|f\| = \frac{\sqrt{3}}{12} \quad \text{and} \quad \|g\| = 4$$

because

$$\langle f, f \rangle = \int_0^1 \left(\left| x - \frac{1}{2} \right| - \frac{1}{4} \right)^2 dx = \frac{1}{48} \quad \text{and} \quad \langle g, g \rangle = \int_0^1 16\, dx = 16.$$

This implies that $\left\{ \left(4\sqrt{3} \right) f, g/4 \right\}$ is an orthonormal.

The set of standard vectors $\{\mathbf{e}_1, \mathbf{e}_2, \dots, \mathbf{e}_n\}$ is a basis for $\mathbb{R}^n$ that is also an orthonormal set with respect to the dot product. The set of vectors $\{1, x, x^2, \dots, x^n\}$ is a basis for $\mathbb{P}_n$ and is also an orthonormal set with respect to the inner product of (6.11). This leads to a natural definition.

■ **Definition 6.2.4**

Let V be an inner product space.

(a) An **orthogonal basis** of V is a basis that is an orthogonal set.

(b) An **orthonormal basis** of V is a basis that is an orthonormal set.

■ **Example 6.2.5**

View $\mathbb{R}^{2\times2}$ as an inner product space using the Frobenius inner product for 2×2 matrices (Example 6.1.4). Then,

$$\left\{\begin{bmatrix} 1 & 0 \\ 0 & 0 \end{bmatrix}, \begin{bmatrix} 0 & 1 \\ 0 & 0 \end{bmatrix}, \begin{bmatrix} 0 & 0 \\ 1 & 0 \end{bmatrix}, \begin{bmatrix} 0 & 0 \\ 0 & 1 \end{bmatrix}\right\}$$

is an orthonormal basis of $\mathbb{R}^{2\times2}$. Also,

$$\left\{\begin{bmatrix} 3 & 0 \\ 0 & 0 \end{bmatrix}, \begin{bmatrix} 0 & 5 \\ 0 & 0 \end{bmatrix}\right\}$$

is not orthonormal, but it is an orthogonal basis of the subspace of $\mathbb{R}^{2\times2}$,

$$\left\{\begin{bmatrix} a & b \\ 0 & 0 \end{bmatrix} : a, b \in \mathbb{R}\right\}.$$

Consider the standard basis $\{\mathbf{e}_1, \mathbf{e}_2\}$ of $\mathbb{R}^2$. It is an orthonormal set when the inner product is the dot product. Rotate $\mathbf{e}_1$ and $\mathbf{e}_2$ through an angle of measure $\pi/4$ to obtain the set,

$$\mathscr{B} = \left\{\begin{bmatrix} \sqrt{2}/2 \\ \sqrt{2}/2 \end{bmatrix}, \begin{bmatrix} -\sqrt{2}/2 \\ \sqrt{2}/2 \end{bmatrix}\right\}.$$

This set is seen to be orthonormal. This is not surprising because a rotation is an isometry (Theorem 3.3.14). Because the vectors of $\mathscr{B}$ are nonzero and orthogonal to each other, every vector in $\mathbb{R}^2$ is a unique linear combination of the two vectors. The same thing can be done in $\mathbb{R}^3$. Rotating 2 orthogonal nonzero vectors in $\mathbb{R}^3$ also results in 2 orthogonal nonzero vectors, but, of course, 2 vectors in $\mathbb{R}^3$ will not form a basis. However, the orthogonal vectors will be linearly independent if none of the vectors equal zero, and this is true for any inner product space.

■ **Theorem 6.2.6**

Let V be an inner product space and take $\mathcal{S}$ be a subset of V. If $\mathcal{S}$ is orthogonal and $\mathbf{0} \notin \mathcal{S}$, then $\mathcal{S}$ is linearly independent.

Proof

Let $\mathbf{u}_1, \mathbf{u}_2, \dots, \mathbf{u}_k \in \mathcal{S}$, which are nonzero because $\mathbf{0} \notin \mathcal{S}$. Let $r_1, r_2, \dots, r_k \in \mathbb{R}$ such that

$$r_1\mathbf{u}_1 + r_2\mathbf{u}_2 + \cdots + r_k\mathbf{u}_k = \mathbf{0}.$$

Choose any $i = 1, 2, \dots, k$. Since $\langle \mathbf{u}_i, \mathbf{0} \rangle = 0$,

$$\langle \mathbf{u}_i, r_1\mathbf{u}_1 + r_2\mathbf{u}_2 + \cdots + r_k\mathbf{u}_k \rangle = 0.$$

Then,

$$\langle \mathbf{u}_i, r_1\mathbf{u}_1 \rangle + \cdots + \langle \mathbf{u}_i, r_i\mathbf{u}_i \rangle + \cdots + \langle \mathbf{u}_i, r_k\mathbf{u}_k \rangle = 0,$$

so

$$r_1 \langle \mathbf{u}_i, \mathbf{u}_1 \rangle + \cdots + r_i \langle \mathbf{u}_i, \mathbf{u}_i \rangle + \cdots + r_k \langle \mathbf{u}_i, \mathbf{u}_k \rangle = 0.$$

Because $\mathcal{S}$ is orthogonal, $\langle \mathbf{u}_i, \mathbf{u}_j \rangle = 0$ when $i \neq j$, so $r_i \langle \mathbf{u}_i, \mathbf{u}_i \rangle = 0$, but $\langle \mathbf{u}_i, \mathbf{u}_i \rangle \neq 0$ because $\mathbf{u}_i \neq \mathbf{0}$, which implies that $r_i = 0$ and $\mathcal{S}$ is linearly independent. ∎

■ Corollary 6.2.7

Let V be an inner product space. An orthogonal spanning set not containing the zero vector is a basis for V.

Orthogonal Complement

Consider the vector space $\mathbb{R}^5$. Let

$$W_1 = \mathsf{S}\left(\begin{bmatrix} 1 \\ 0 \\ 0 \\ 0 \\ 0 \end{bmatrix}, \begin{bmatrix} 0 \\ 1 \\ 0 \\ 0 \\ 0 \end{bmatrix}, \begin{bmatrix} 0 \\ 0 \\ 1 \\ 0 \\ 0 \end{bmatrix} \right) \quad \text{and} \quad W_2 = \mathsf{S}\left(\begin{bmatrix} 0 \\ 0 \\ 0 \\ 1 \\ 0 \end{bmatrix}, \begin{bmatrix} 0 \\ 0 \\ 0 \\ 0 \\ 1 \end{bmatrix} \right). \tag{6.12}$$

Observe that every vector in W_1 is orthogonal to every vector in W_2, and vice versa.

■ Definition 6.2.8

Let W be a subspace of an inner product space V. Define the **orthogonal complement** of W to be $W^\perp = \{ \mathbf{u} \in V : \langle \mathbf{u}, \mathbf{w} \rangle = 0 \text{ for all } \mathbf{w} \in W \}$.

Using (6.12) with Definition 6.2.8, it is evident that $W_1^\perp = W_2$ and $W_2^\perp = W_1$. This also suggests that the orthogonal complement is a subspace of its containing inner product space.

■ Theorem 6.2.9

Let V be an inner product space and W a subspace of V. The orthogonal complement of W is a subspace of V.

Proof

Because the zero vector is orthogonal to all vectors, $\mathbf{0} \in W^\perp$, so the orthogonal complement is nonempty. Let the vectors $\mathbf{u}, \mathbf{v}$ be elements of $W^\perp$ and r be a real number. Take $\mathbf{w} \in W$. Then, $\langle \mathbf{u}, \mathbf{w} \rangle = 0$ and $\langle \mathbf{v}, \mathbf{w} \rangle = 0$, so by Definition 6.1.1,

$$\langle \mathbf{u} + \mathbf{v}, \mathbf{w} \rangle = \langle \mathbf{u}, \mathbf{v} \rangle + \langle \mathbf{v}, \mathbf{w} \rangle = 0,$$

and

$$\langle r\mathbf{u}, \mathbf{w} \rangle = r \langle \mathbf{u}, \mathbf{w} \rangle = 0,$$

and this implies that $\mathbf{u} + \mathbf{v} \in W^\perp$ and $r\mathbf{u} \in W^\perp$. ∎

■ **Example 6.2.10**

Consider the plane P in $\mathbb{R}^3$ given by $3x + 4y - 2z = 0$. This plane is a subspace of $\mathbb{R}^3$ as noted in Example 5.2.16. Let L be the line containing $\mathbf{0}$ that is orthogonal to P. The parametric equations for L are

$$x = 3t,$$
$$y = 4t,$$
$$z = -2t.$$

It is evident that $P^\perp = L$ because the direction vector of L is a normal of P. That this is enough to prove the result is due to the next theorem.

■ **Theorem 6.2.11**

Let V be an inner product space with subspace W and $\mathbf{u} \in V$. Then, $\mathbf{u} \in W^\perp$ if and only if $\mathbf{u}$ is orthogonal to every vector in a spanning set of W.

Proof

Take S to be a spanning set of W such that $\mathbf{u}$ is orthogonal to all of its vectors. Let $\mathbf{v} \in W$ and find vectors $\mathbf{v}_1, \mathbf{v}_2, \dots, \mathbf{v}_k \in S$ and $r_1, r_2, \dots, r_k \in \mathbb{R}$ such that $\mathbf{v} = r_1 \mathbf{v}_1 + r_2 \mathbf{v}_2 + \cdots + r_k \mathbf{v}_k$. Then, $\mathbf{u} \in W^\perp$ because

$$\langle \mathbf{u}, \mathbf{v} \rangle = \langle \mathbf{u}, r_1 \mathbf{v}_1 + r_2 \mathbf{v}_2 + \cdots + r_k \mathbf{v}_k \rangle$$
$$= r_1 \langle \mathbf{u}, \mathbf{v}_1 \rangle + r_2 \langle \mathbf{u}, \mathbf{v}_2 \rangle + \cdots + r_k \langle \mathbf{u}, \mathbf{v}_k \rangle = 0.$$

The theorem follows because the converse is clear. ∎

■ **Example 6.2.12**

The plane of Example 6.2.10 is spanned by $\left\{ \begin{bmatrix} -4 \\ 3 \\ 0 \end{bmatrix}, \begin{bmatrix} 2 \\ 0 \\ 3 \end{bmatrix} \right\}$, and the line is

spanned by $\left\{ \begin{bmatrix} 3 \\ 4 \\ -2 \end{bmatrix} \right\}$. To illustrate Theorem 6.2.11, to prove that $P^\perp = L$, it is enough to compute $\begin{bmatrix} 3 \\ 4 \\ -2 \end{bmatrix} \cdot \begin{bmatrix} -4 \\ 3 \\ 0 \end{bmatrix} = 0$ and $\begin{bmatrix} 3 \\ 4 \\ -2 \end{bmatrix} \cdot \begin{bmatrix} 2 \\ 0 \\ 3 \end{bmatrix} = 0.$

■ **Example 6.2.13**

Let A be an $m \times n$ matrix. Take $\mathbf{u} \in [\mathrm{RS}\, A]^{\perp}$. This implies that $\mathbf{u} \cdot \mathbf{r}_i^T = 0$ for all $i = 1, 2, \ldots, n$ by Theorem 6.2.11. Using Theorem 3.1.17,

$$A\mathbf{u} = \begin{bmatrix} \mathbf{u} \cdot \mathbf{r}_1^T \\ \mathbf{u} \cdot \mathbf{r}_2^T \\ \vdots \\ \mathbf{u} \cdot \mathbf{r}_n^T \end{bmatrix} = \begin{bmatrix} 0 \\ 0 \\ \vdots \\ 0 \end{bmatrix}.$$

Therefore, $\mathbf{u} \in \mathrm{NS}\, A$. This proves that $(\mathrm{RS}\, A)^{\perp} \subseteq \mathrm{NS}\, A$. The previous steps reverse to prove that

$$(\mathrm{RS}\, A)^{\perp} = \mathrm{NS}\, A. \tag{6.13}$$

To make this concrete, take the matrix, $A = \begin{bmatrix} 1 & 2 & 3 \\ 4 & 5 & 6 \end{bmatrix}$. Its rows are linearly

independent, so $\left\{ \begin{bmatrix} 1 \\ 2 \\ 3 \end{bmatrix}, \begin{bmatrix} 4 \\ 5 \\ 6 \end{bmatrix} \right\}$ is a basis for $\mathrm{RS}\, A$. Moreover, since the reduced

row echelon form of A is $\begin{bmatrix} 1 & 0 & -1 \\ 0 & 1 & 2 \end{bmatrix}$, the set $\left\{ \begin{bmatrix} 1 \\ -2 \\ 1 \end{bmatrix} \right\}$ is a basis for $\mathrm{NS}\, A$.

Observe that

$$\begin{bmatrix} 1 \\ 2 \\ 3 \end{bmatrix} \cdot \begin{bmatrix} 1 \\ -2 \\ 1 \end{bmatrix} = 0 \quad \text{and} \quad \begin{bmatrix} 4 \\ 5 \\ 6 \end{bmatrix} \cdot \begin{bmatrix} 1 \\ -2 \\ 1 \end{bmatrix} = 0,$$

which illustrates (6.13). Furthermore, it should be noted that there is a similar relationship between the column space of A and the null space of A^T. Namely, as proved in Exercise 10,

$$[\mathrm{CS}\, A]^{\perp} = \mathrm{NS}\, A^T. \tag{6.14}$$

Direct Sum

There is another property to note regarding (6.12). It is their relationship to $\mathbb{R}^5$. Take any $\mathbf{u} \in \mathbb{R}^5$ with coordinates u_1, u_2, u_3, u_4, u_5. This vector is the sum of a vector from W_1 and a vector from W_2. In particular,

$$\begin{bmatrix} u_1 \\ u_2 \\ u_3 \\ u_4 \\ u_5 \end{bmatrix} = \begin{bmatrix} u_1 \\ u_2 \\ u_3 \\ 0 \\ 0 \end{bmatrix} + \begin{bmatrix} 0 \\ 0 \\ 0 \\ u_4 \\ u_5 \end{bmatrix},$$

where

$$\begin{bmatrix} u_1 \\ u_2 \\ u_3 \\ 0 \\ 0 \end{bmatrix} = u_1 \begin{bmatrix} 1 \\ 0 \\ 0 \\ 0 \\ 0 \end{bmatrix} + u_2 \begin{bmatrix} 0 \\ 1 \\ 0 \\ 0 \\ 0 \end{bmatrix} + u_3 \begin{bmatrix} 0 \\ 0 \\ 1 \\ 0 \\ 0 \end{bmatrix} \quad \text{and} \quad \begin{bmatrix} 0 \\ 0 \\ 0 \\ u_4 \\ u_5 \end{bmatrix} = u_4 \begin{bmatrix} 0 \\ 0 \\ 0 \\ 1 \\ 0 \end{bmatrix} + u_5 \begin{bmatrix} 0 \\ 0 \\ 0 \\ 0 \\ 1 \end{bmatrix}.$$

Furthermore, notice that the only vector common to W_1 and W_2 is $\mathbf{0}$. This leads to the next definition, which involves the notion of a sum of subspaces as given in Example 5.2.4.

■ **Definition 6.2.14**

Let $W_1, W_2, \ldots, W_k$ be subspaces of the vector space V. If for all $i = 1, 2, \ldots, k$,

$$W_i \cap \sum_{j \neq i} W_j = \{\mathbf{0}\}, \tag{6.15}$$

The subspace $W_1 + W_2 + \cdots + W_k$ is an **(internal) direct sum** of the **summands** $W_1, W_2, \ldots, W_k$. This direct sum is written as $W_1 \oplus W_2 \oplus \cdots \oplus W_k$.

When a direct sums has only two summands, (6.15) is equivalent to the condition, $W_1 \cap W_2 = \{\mathbf{0}\}$. For example, the work prior to the citation of Definition 6.2.14 shows that

$$\mathbb{R}^5 = \left\langle \begin{bmatrix} 1 \\ 0 \\ 0 \\ 0 \\ 0 \end{bmatrix}, \begin{bmatrix} 0 \\ 1 \\ 0 \\ 0 \\ 0 \end{bmatrix}, \begin{bmatrix} 0 \\ 0 \\ 1 \\ 0 \\ 0 \end{bmatrix} \right\rangle \oplus \left\langle \begin{bmatrix} 0 \\ 0 \\ 0 \\ 1 \\ 0 \end{bmatrix}, \begin{bmatrix} 0 \\ 0 \\ 0 \\ 0 \\ 1 \end{bmatrix} \right\rangle.$$

However, $\mathbb{R}^5$ can be written as a direct sum using up to five summands, such as

$$\mathbb{R}^5 = S(\mathbf{e}_1) \oplus S(\mathbf{e}_2) \oplus S(\mathbf{e}_3) \oplus S(\mathbf{e}_4) \oplus S(\mathbf{e}_5).$$

■ **Example 6.2.15**

The vector space $\mathbb{P}_3$ has $\mathcal{B} = \{1, 1 + x, x + x^2, x^2 + x^3\}$ as a basis (Exercise 4 of Section 5.4). Prove that

$$\mathbb{P}_3 = S(1, x + x^2) \oplus S(1 + x, x^2 + x^3).$$

First, assume that

$$r_1 + r_2(x + x^2) = s_1(1 + x) + s_2(x^2 + x^3),$$

so

$$r_1 + r_2(x + x^2) - s_1(1 + x) - s_2(x^2 + x^3) = 0.$$

Because $\mathcal{B}$ is a basis, $r_1 = r_2 = s_1 = s_2 = 0$, so

$$S(1, x + x^2) \cap S(1 + x, x^2 + x^3) = \{0 + 0x + 0x^2 + 0x^3\}.$$

Second, take a polynomial $p(x) = a_0 + a_1 x + a_2 x^2 + a_3 x^3$ with real coefficients. Writing

$$f(x) = (a_0 - a_1 + a_2 - a_3) + (a_2 - a_3)(x + x^2) \in S(1, x + x^2)$$

and

$$g(x) = (a_1 - a_2 + a_3)(1 + x) + a_3(x^2 + x^3) \in S(1 + x, x^2 + x^3),$$

it follows that $p(x) \in S(1, x + x^2) \oplus S(1 + x, x^2 + x^3)$ because

$$p(x) = f(x) + g(x). \tag{6.16}$$

It is important to realize that the representation of $p(x)$ in (6.16) is unique for the direct sum. Although the basis did play a role in this, the main reason for the uniqueness is that the only vector common to both summands is the zero vector. This holds for all direct sums within any vector space.

■ **Theorem 6.2.16**

For every vector $\mathbf{w}$ in $W_1 \oplus W_2 \oplus \cdots \oplus W_k$, where $W_1, W_2, \ldots, W_k$ are subspaces of a vector space V, there are unique $\mathbf{u}_1 \in W_1, \mathbf{u}_2 \in W_2, \ldots, \mathbf{u}_k \in W_k$ such that $\mathbf{w} = \mathbf{u}_1 + \mathbf{u}_2 + \cdots + \mathbf{u}_k$.

Proof

Since existence follows from the definition of the direct sum, prove uniqueness by taking $\mathbf{u}_i, \mathbf{v}_i \in W_i$ for all $i = 1, 2, \ldots, k$ such that

$$\mathbf{u}_1 + \mathbf{u}_2 + \cdots + \mathbf{u}_k = \mathbf{v}_1 + \mathbf{v}_2 + \cdots + \mathbf{v}_k.$$

After some subtractions,

$$\mathbf{u}_i - \mathbf{v}_i = \sum_{j \neq i} (\mathbf{v}_j - \mathbf{u}_j),$$

which implies that $\mathbf{u}_i = \mathbf{v}_i$ because

$$\mathbf{u}_i - \mathbf{v}_i \in W_i \cap \sum_{j \neq i} W_j = \{\mathbf{0}\}. \ \blacksquare$$

■ **Example 6.2.17**

Observe that

$$S\left(\begin{bmatrix} 1 & 0 \\ 0 & 2 \end{bmatrix}\right) \cap S\left(\begin{bmatrix} 1 & 0 \\ 0 & 0 \end{bmatrix}, \begin{bmatrix} 0 & 0 \\ 1 & 1 \end{bmatrix}\right) = \left\{\begin{bmatrix} 0 & 0 \\ 0 & 0 \end{bmatrix}\right\}.$$

Also,

$$A = \begin{bmatrix} 7 & 0 \\ -2 & 6 \end{bmatrix} = \begin{bmatrix} 4 & 0 \\ 0 & 8 \end{bmatrix} + \begin{bmatrix} 3 & 0 \\ -2 & -2 \end{bmatrix}, \tag{6.17}$$

with

$$\begin{bmatrix} 4 & 0 \\ 0 & 8 \end{bmatrix} \in S\left(\begin{bmatrix} 1 & 0 \\ 0 & 2 \end{bmatrix}\right)$$

and

$$\begin{bmatrix} 3 & 0 \\ -2 & -2 \end{bmatrix} \in S\left(\begin{bmatrix} 1 & 0 \\ 0 & 0 \end{bmatrix}, \begin{bmatrix} 0 & 0 \\ 1 & 1 \end{bmatrix}\right).$$

Therefore,

$$A \in S\left(\begin{bmatrix} 1 & 0 \\ 0 & 2 \end{bmatrix}\right) \oplus S\left(\begin{bmatrix} 1 & 0 \\ 0 & 0 \end{bmatrix}, \begin{bmatrix} 0 & 0 \\ 1 & 1 \end{bmatrix}\right),$$

and (6.17) is the unique expression of A as a sum of a matrix from the first summand with a matrix from the second summand.

That the direct sum of subspaces is a subspace follows from the fact that the sum of subspaces is a subspace as seen in Example 5.2.4. For this reason, every direct sum has a dimension. For example, using some familiar vector spaces, inspection will confirm that

$$\dim \mathbb{R}^5 = \dim S\left(\begin{bmatrix} 1 \\ 0 \\ 0 \\ 0 \\ 0 \end{bmatrix}, \begin{bmatrix} 0 \\ 1 \\ 0 \\ 0 \\ 0 \end{bmatrix}, \begin{bmatrix} 0 \\ 0 \\ 1 \\ 0 \\ 0 \end{bmatrix}\right) + \dim S\left(\begin{bmatrix} 0 \\ 0 \\ 0 \\ 1 \\ 0 \end{bmatrix}, \begin{bmatrix} 0 \\ 0 \\ 0 \\ 0 \\ 1 \end{bmatrix}\right),$$

and

$$\dim \mathbb{P}_3 = \dim S\left(1, x + x^2\right) + \dim S\left(1 + x, x^2 + x^3\right).$$

That this holds in general is the next result.

■ Theorem 6.2.18

For every direct sum of subspaces,

$$\dim W_1 \oplus W_2 \oplus \cdots \oplus W_n = \dim W_1 + \dim W_2 + \cdots + \dim W_n,$$

where $W_1, W_2, \dots, W_n$ are subspaces of the vector space V.

Proof

By using an induction argument, it suffices to prove the result for $n = 2$. Let $\mathcal{B} = \{\mathbf{u}_1, \mathbf{u}_2, \dots, \mathbf{u}_k\}$ be a basis for W_1 and $\mathcal{C} = \{\mathbf{v}_1, \mathbf{v}_2, \dots, \mathbf{v}_l\}$ be a basis for W_2. It is left to Exercise 12 to prove that $\mathcal{B} \cap \mathcal{C} = \varnothing$ and $\mathcal{B} \cup \mathcal{C}$ is a spanning set of $W_1 \oplus W_2$. To prove that $\mathcal{B} \cup \mathcal{C}$ is linearly independent, let $r_1, r_2, \dots, r_k, s_1, s_2, \dots, s_l \in \mathbb{R}$ such that

$$r_1\mathbf{u}_1 + r_2\mathbf{u}_2 + \cdots + r_k\mathbf{u}_k + s_1\mathbf{v}_1 + s_2\mathbf{v}_2 + \cdots + s_l\mathbf{v}_l = \mathbf{0}.$$

Write $\mathbf{w}_1 = r_1\mathbf{u}_1 + r_2\mathbf{u}_2 + \cdots + r_k\mathbf{u}_k$ and $\mathbf{w}_2 = s_1\mathbf{v}_1 + s_2\mathbf{v}_2 + \cdots + s_l\mathbf{v}_l$. This implies that $\mathbf{w}_1 = -\mathbf{w}_2$, so $\mathbf{w}_1, \mathbf{w}_2 \in W_1 \cap W_2$ by Theorem 5.2.8(b). Therefore,

$$r_1\mathbf{u}_1 + r_2\mathbf{u}_2 + \cdots + r_k\mathbf{u}_k = \mathbf{0} \quad \text{and} \quad s_1\mathbf{v}_1 + s_2\mathbf{v}_2 + \cdots + s_l\mathbf{v}_l = \mathbf{0}.$$

Because $\mathscr{B}$ is linearly independent, $r_i = 0$ for $i = 1, 2, \dots, k$, and because $\mathscr{C}$ is linearly independent, $s_i = 0$ for $i = 1, 2, \dots, l$. Therefore, $\mathscr{B} \cup \mathscr{C}$ is a basis for $W_1 \oplus W_2$, and its dimension is $k + l$. ∎

Gram–Schmidt Process

As $\{\mathbf{e}_1, \mathbf{e}_2, \dots, \mathbf{e}_n\}$ is an orthonormal basis of $\mathbb{R}^n$, the goal is to determine whether every inner product space has an orthonormal basis. Let V be an inner product space with basis $\mathscr{B} = \{\mathbf{u}_1, \mathbf{u}_2, \dots, \mathbf{u}_n\}$. From $\mathscr{B}$ will be constructed an orthogonal set that will be a basis for V. Start by defining

$$\mathbf{v}_1 = \mathbf{u}_1. \tag{6.18}$$

Next, project $\mathbf{u}_2$ onto $\mathbf{v}_1$, and define $\mathbf{v}_2$ to be the vector component of $\mathbf{u}_2$ orthogonal to $\mathbf{v}_1$. That is,

$$\mathbf{v}_2 = \mathbf{u}_2 - \text{proj}_{\mathbf{v}_1}\mathbf{u}_2 = \mathbf{u}_2 - \frac{\langle \mathbf{u}_2, \mathbf{v}_1 \rangle}{\langle \mathbf{v}_1, \mathbf{v}_1 \rangle}\mathbf{v}_1. \tag{6.19}$$

This is illustrated in Figure 6.1 using arrows to represent the abstract vectors. Since $\{\mathbf{v}_1, \mathbf{v}_2\}$ is an orthogonal set, it is a basis for a two-dimensional subspace of V (Theorem 6.2.6).

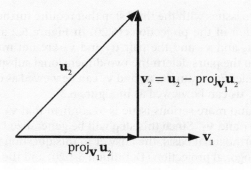

Figure 6.1 The projection of $\mathbf{u}_2$ onto $\mathbf{v}_1$.

This is when the algorithm gets interesting. The next vector must be orthogonal to $\mathsf{S}(\mathbf{v}_1, \mathbf{v}_2)$. This can be accomplished by finding a vector that is orthogonal to both $\mathbf{v}_1$ and $\mathbf{v}_2$ (Theorem 6.2.11). Do this by projecting $\mathbf{u}_3$ onto both $\mathbf{v}_1$ and $\mathbf{v}_2$ to find

$$\text{proj}_{\mathbf{v}_1}\mathbf{u}_3 = \frac{\langle \mathbf{u}_3, \mathbf{v}_1 \rangle}{\langle \mathbf{v}_1, \mathbf{v}_1 \rangle}\mathbf{v}_1 \quad \text{and} \quad \text{proj}_{\mathbf{v}_2}\mathbf{u}_3 = \frac{\langle \mathbf{u}_3, \mathbf{v}_2 \rangle}{\langle \mathbf{v}_2, \mathbf{v}_2 \rangle}\mathbf{v}_2. \tag{6.20}$$

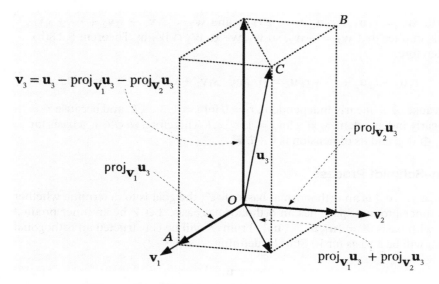

$$\mathbf{v}_3 = \mathbf{u}_3 - \text{proj}_{\mathbf{v}_1}\mathbf{u}_3 - \text{proj}_{\mathbf{v}_2}\mathbf{u}_3$$

Figure 6.2   Finding a vector orthogonal to $S(\mathbf{v}_1, \mathbf{v}_2)$.

As seen in Figure 6.2, the projection of $\mathbf{u}_3$ onto $S(\mathbf{v}_1, \mathbf{v}_2)$ is the sum of the two projections,

$$\text{proj}_{\mathbf{v}_1}\mathbf{u}_3 + \text{proj}_{\mathbf{v}_2}\mathbf{v}_3,$$

so the desired third basis vector that is orthogonal to $\mathbf{v}_1$ and $\mathbf{v}_2$ is

$$\mathbf{v}_3 = \mathbf{u}_3 - \text{proj}_{\mathbf{v}_1}\mathbf{u}_3 - \text{proj}_{\mathbf{v}_2}\mathbf{u}_3. \tag{6.21}$$

There are two issues with the third step that require further investigation. First, the representation of the projections (6.20) in Figure 6.2 are legitimate because both the pair $\mathbf{u}_3$ and $\mathbf{v}_1$ and the pair $\mathbf{u}_3$ and $\mathbf{v}_2$ are not multiples of each other (Exercise 19), so the pairs determine two-dimensional subspaces in which projections can be made. For example, $\mathbf{u}_3$ and $\mathbf{v}_1$ can be viewed as determining the plane $AOBC$, so $\text{proj}_{\mathbf{v}_1}\mathbf{u}_3$ can be viewed as in Figure 6.3.

The second and more serious issue is to confirm that $\mathbf{v}_3$ as defined in (6.21) is orthogonal to $\mathbf{v}_1$ and $\mathbf{v}_2$. Since this step will be generalized to find the remaining vectors of the orthogonal basis, the answer to this question requires a generalization of the orthogonal projection (Definition 6.1.20) and the theorem that follows.

■ **Definition 6.2.19**

Let V be an inner product space with subspace W. Let $\{\mathbf{v}_1, \mathbf{v}_2, \dots, \mathbf{v}_k\}$ be a basis for W. For all $\mathbf{u} \in V$, define the **orthogonal projection** (or simply the **projection**) of $\mathbf{u}$ onto W as

$$\text{proj}_W\mathbf{u} = \text{proj}_{\mathbf{v}_1}\mathbf{u} + \text{proj}_{\mathbf{v}_2}\mathbf{u} + \cdots + \text{proj}_{\mathbf{v}_k}\mathbf{u} = \sum_{i=1}^{k} \frac{\langle \mathbf{u}, \mathbf{v}_i \rangle}{\langle \mathbf{v}_i, \mathbf{v}_i \rangle}\mathbf{v}_i.$$

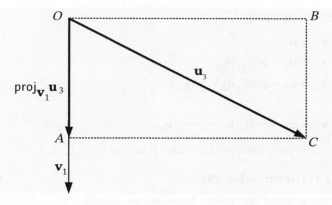

Figure 6.3   The projection of $\mathbf{u}_3$ onto $\mathbf{v}_1$.

The **vector component of u orthogonal to W** is

$$\mathbf{u} - \text{proj}_W \mathbf{u} = \mathbf{u} - \sum_{i=1}^{k} \frac{\langle \mathbf{u}, \mathbf{v}_i \rangle}{\langle \mathbf{v}_i, \mathbf{v}_i \rangle} \mathbf{v}_i.$$

■ Theorem 6.2.20

Let V be an inner production space with subspace W. If $\{\mathbf{v}_1, \mathbf{v}_2, \ldots, \mathbf{v}_k\}$ is an orthogonal basis for W and $\mathbf{u} \in V \setminus W$, then $\mathbf{u} - \text{proj}_W \mathbf{u}$ is orthogonal to $\mathbf{v}_i$ for all $i = 1, 2, \ldots, k$.

Proof

Let $\{\mathbf{v}_1, \mathbf{v}_2, \ldots, \mathbf{v}_k\}$ be an orthogonal set that spans W, which is a subspace of V. Suppose that $\mathbf{u}$ is a vector of V not in W. Let $j \in \mathbb{Z}$ such that $1 \leq j \leq k$. By Definition 6.1.1 and since $\langle \mathbf{v}_i, \mathbf{v}_j \rangle = 0$ if $i \neq j$,

$$\left\langle \mathbf{u} - \sum_{i=1}^{k} \frac{\langle \mathbf{u}, \mathbf{v}_i \rangle}{\langle \mathbf{v}_i, \mathbf{v}_i \rangle} \mathbf{v}_i, \mathbf{v}_j \right\rangle = \langle \mathbf{u}, \mathbf{v}_j \rangle - \left\langle \sum_{i=1}^{k} \frac{\langle \mathbf{u}, \mathbf{v}_i \rangle}{\langle \mathbf{v}_i, \mathbf{v}_i \rangle} \mathbf{v}_i, \mathbf{v}_j \right\rangle$$

$$= \langle \mathbf{u}, \mathbf{v}_j \rangle - \sum_{i=1}^{k} \frac{\langle \mathbf{u}, \mathbf{v}_i \rangle}{\langle \mathbf{v}_i, \mathbf{v}_i \rangle} \langle \mathbf{v}_i, \mathbf{v}_j \rangle$$

$$= \langle \mathbf{u}, \mathbf{v}_j \rangle - \frac{\langle \mathbf{u}, \mathbf{v}_j \rangle}{\langle \mathbf{v}_j, \mathbf{v}_j \rangle} \langle \mathbf{v}_j, \mathbf{v}_j \rangle = 0. \blacksquare$$

Suppose that $\mathbf{u}_3$ is a linear combination of $\mathbf{v}_1$ (6.18) and $\mathbf{v}_2$ (6.19). This would contradict the fact that $\mathcal{B}$ is a basis since $\mathbf{v}_1, \mathbf{v}_2 \in S(\mathbf{u}_1, \mathbf{u}_2)$. Thus, $\mathbf{u}_3 \notin S(\mathbf{v}_1, \mathbf{v}_2)$, so Theorem 6.2.20 implies that $\mathbf{v}_3$ (6.21) is orthogonal to $\mathbf{v}_1$ and $\mathbf{v}_2$. Repeat this construction, known as the **Gram–Schmidt process**, until all n vectors of the orthogonal basis for V are found:

$$\mathbf{v}_1 = \mathbf{u}_1,$$
$$\mathbf{v}_2 = \mathbf{u}_2 - \text{proj}_{\mathbf{v}_1} \mathbf{u}_2,$$
$$\mathbf{v}_3 = \mathbf{u}_3 - \text{proj}_{\mathbf{v}_1} \mathbf{u}_3 - \text{proj}_{\mathbf{v}_2} \mathbf{u}_3,$$
$$\vdots$$
$$\mathbf{v}_n = \mathbf{u}_n - \text{proj}_{\mathbf{v}_1} \mathbf{u}_n - \text{proj}_{\mathbf{v}_2} \mathbf{u}_n - \cdots - \text{proj}_{\mathbf{v}_{n-1}} \mathbf{u}_n.$$

This work implies the next theorem from which follow 2 corollaries.

■ Theorem 6.2.21 [Gram–Schmidt]

If V be an inner product space with basis $\{\mathbf{u}_1, \mathbf{u}_2, \dots, \mathbf{u}_n\}$, then V has an orthogonal basis $\{\mathbf{v}_1, \mathbf{v}_2, \dots, \mathbf{v}_n\}$, where $\mathbf{v}_1 = \mathbf{u}_1$, and if $1 < k \leq n$,

$$\mathbf{v}_k = \mathbf{u}_k - \sum_{i=1}^{k-1} \text{proj}_{\mathbf{v}_i} \mathbf{u}_k. \qquad (6.22)$$

■ Corollary 6.2.22

Every finite dimensional inner product space has an orthogonal basis.

■ Corollary 6.2.23

Every finite dimensional inner product space has an orthonormal basis.

■ Example 6.2.24

Viewing $\mathbb{R}^4$ as an inner product space with the dot product as its inner product, use the Gram–Schmidt process to find an orthonormal basis $\{\mathbf{v}_1, \mathbf{v}_2, \mathbf{v}_3\}$ of the subspace of $\mathbb{R}^4$ spanned by

$$\left\{ \begin{bmatrix} 1 \\ 0 \\ 1 \\ 2 \end{bmatrix}, \begin{bmatrix} 1 \\ -2 \\ 1 \\ 0 \end{bmatrix}, \begin{bmatrix} 0 \\ 1 \\ 2 \\ 1 \end{bmatrix} \right\}. \qquad (6.23)$$

First,

$$\mathbf{v}_1 = \begin{bmatrix} 1 \\ 0 \\ 1 \\ 2 \end{bmatrix}.$$

Second,

$$\mathbf{v}_2 = \begin{bmatrix} 1 \\ -2 \\ 1 \\ 0 \end{bmatrix} - \text{proj}_{\mathbf{v}_1} \begin{bmatrix} 1 \\ -2 \\ 1 \\ 0 \end{bmatrix} = \begin{bmatrix} 2/3 \\ -2 \\ 2/3 \\ -2/3 \end{bmatrix}.$$

Third,

$$\mathbf{v}_3 = \begin{bmatrix} 0 \\ 1 \\ 2 \\ 1 \end{bmatrix} - \text{proj}_{\mathbf{v}_1} \begin{bmatrix} 0 \\ 1 \\ 2 \\ 1 \end{bmatrix} - \text{proj}_{\mathbf{v}_2} \begin{bmatrix} 0 \\ 1 \\ 2 \\ 1 \end{bmatrix} = \begin{bmatrix} -1/2 \\ 1/2 \\ 3/2 \\ -1/2 \end{bmatrix}.$$

By Theorem 6.2.21, the set $\{\mathbf{v}_1, \mathbf{v}_2, \mathbf{v}_3\}$ is an orthogonal basis from which follows that

$$\left\{ \begin{bmatrix} \sqrt{6}/6 \\ 0 \\ \sqrt{6}/6 \\ \sqrt{6}/3 \end{bmatrix}, \begin{bmatrix} \sqrt{3}/6 \\ -\sqrt{3}/2 \\ \sqrt{3}/6 \\ -\sqrt{3}/6 \end{bmatrix}, \begin{bmatrix} -\sqrt{3}/6 \\ \sqrt{3}/6 \\ \sqrt{3}/2 \\ -\sqrt{3}/6 \end{bmatrix} \right\}$$

is an orthonormal basis.

■ Theorem 6.2.25

For every finite dimensional inner product space V, if W is a subspace of V, then $V = W \oplus W^\perp$.

Proof

First note that $W \cap W^\perp = \{\mathbf{0}\}$. To see this, take $\mathbf{u} \in W$ and $\mathbf{u} \in W^\perp$. This implies that $\langle \mathbf{u}, \mathbf{u} \rangle = 0$, but this is only possible if $\mathbf{u} = \mathbf{0}$. Next, let $\mathscr{C} = \{\mathbf{u}_1, \mathbf{u}_2, \dots, \mathbf{u}_k\}$ be an orthogonal basis of W by the Gram–Schmidt process (Theorem 6.2.21). Continue the Gram–Schmidt process to extend $\mathscr{C}$ to an orthogonal basis of V. Write this basis as

$$\mathscr{B} = \{\mathbf{u}_1, \mathbf{u}_2, \dots, \mathbf{u}_k, \mathbf{u}_{k+1}, \mathbf{u}_{k+2}, \dots, \mathbf{u}_n\},$$

where $n = \dim V$. Because $\mathscr{B}$ is orthogonal,

$$\langle r_1 \mathbf{u}_1 + r_2 \mathbf{u}_2 + \cdots + r_k \mathbf{u}_k, r_{k+1} \mathbf{u}_{k+1} + r_{k+2} \mathbf{u}_{k+2} + \dots, r_n \mathbf{u}_n \rangle = 0$$

for all $r_i \in \mathbb{R}$ with $i = 1, 2, \dots, n$. Therefore, $W^\perp = \mathrm{S}(\mathbf{u}_{k+1}, \mathbf{u}_{k+2}, \dots, \mathbf{u}_n)$, so conclude that $V = W \oplus W^\perp$. ■

■ Example 6.2.26

Extend the basis (6.23) of Example 6.2.24 to

$$\mathscr{B} = \left\{ \begin{bmatrix} 1 \\ 0 \\ 1 \\ 2 \end{bmatrix}, \begin{bmatrix} 1 \\ -2 \\ 1 \\ 0 \end{bmatrix}, \begin{bmatrix} 0 \\ 1 \\ 2 \\ 1 \end{bmatrix}, \begin{bmatrix} 1 \\ 0 \\ 0 \\ 0 \end{bmatrix} \right\}.$$

Apply the Gram–Schmidt process to $\mathscr{B}$ to obtain the orthogonal basis

$$\left\{ \begin{bmatrix} 1 \\ 0 \\ 1 \\ 2 \end{bmatrix}, \begin{bmatrix} 2/3 \\ -2 \\ 2/3 \\ -2/3 \end{bmatrix}, \begin{bmatrix} -1/2 \\ 1/2 \\ 3/2 \\ -/12 \end{bmatrix}, \begin{bmatrix} 2/3 \\ 1/3 \\ 0 \\ -1/3 \end{bmatrix} \right\}.$$

Letting

$$W = \mathsf{S}\left(\begin{bmatrix} 1 \\ 0 \\ 1 \\ 2 \end{bmatrix}, \begin{bmatrix} 2/3 \\ -2 \\ 2/3 \\ -2/3 \end{bmatrix}, \begin{bmatrix} -1/2 \\ 1/2 \\ 3/2 \\ -/12 \end{bmatrix} \right),$$

observe that

$$W^{\perp} = \mathsf{S}\left(\begin{bmatrix} 2/3 \\ 1/3 \\ 0 \\ -1/3 \end{bmatrix} \right),$$

so $\mathbb{R}^4 = W \oplus W^{\perp}$ by Theorem 6.2.25.

QR Factorization

Let A be an $m \times n$ matrix such that $\mathbf{c}_1, \mathbf{c}_2, \dots, \mathbf{c}_n \in \mathbb{R}^m$ are the columns of A and $m \geq n$. Assume that $\mathscr{B} = \{\mathbf{c}_1, \mathbf{c}_2, \dots, \mathbf{c}_n\}$ is linearly independent. Use the Gram–Schmidt process (Theorem 6.2.21) to find an orthonormal basis of the subspace of $\mathbb{R}^m$ spanned by $\mathscr{B}$. Call the vectors of this basis $\mathbf{v}_1, \mathbf{v}_2, \dots, \mathbf{v}_n$. By construction, for all $k = 1, 2, \dots, n$,

$$\mathbf{c}_k \in \mathsf{S}(\mathbf{v}_1, \mathbf{v}_2, \dots, \mathbf{v}_k). \tag{6.24}$$

To prove this, first observe that $\mathbf{c}_1 \in \mathsf{S}(\mathbf{v}_1)$ because $\mathbf{c}_1 = \mathbf{v}_1$. Next, let k be an integer such that $1 < k \leq n$. Then, by (6.22),

$$\mathbf{c}_k = \mathbf{v}_k + \sum_{i=1}^{k-1} \mathrm{proj}_{\mathbf{v}_i} \mathbf{c}_k.$$

Since $\mathrm{proj}_{\mathbf{v}_i} \mathbf{c}_k$ is a scalar multiple of $\mathbf{v}_i$ for each $i = 1, 2, \dots, k-1$,

$$\sum_{i=1}^{k-1} \mathrm{proj}_{\mathbf{v}_i} \mathbf{c}_k \in \mathsf{S}(\mathbf{v}_1, \mathbf{v}_2, \dots, \mathbf{v}_{k-1}),$$

proving (6.24). This means for $k = 1, 2, \dots, n$, there exists $r_{1k}, r_{2k}, \dots, r_{kk} \in \mathbb{R}$ such that

$$\mathbf{c}_k = r_{1k}\mathbf{v}_1 + r_{2k}\mathbf{v}_2 + \cdots + r_{kk}\mathbf{v}_k.$$

Define the $m \times n$ matrix,

$$Q = \begin{bmatrix} \mathbf{v}_1 & \mathbf{v}_2 & \cdots & \mathbf{v}_n \end{bmatrix},$$

and define $\mathbf{u}_k \in \mathbb{R}^n$ by

$$\mathbf{u}_k = \begin{bmatrix} r_{1k} \\ r_{2k} \\ \vdots \\ r_{kk} \\ 0 \\ \vdots \\ 0 \end{bmatrix}.$$

By Theorem 3.1.17,

$$Q\mathbf{u}_k = \mathbf{c}_k.$$

Define the $n \times n$ matrix R so that its columns are $\mathbf{u}_1, \mathbf{u}_2, \ldots, \mathbf{u}_n$. That is, R is the upper diagonal matrix,

$$R = \begin{bmatrix} r_{11} & r_{12} & r_{13} & \cdots & r_{1n} \\ 0 & r_{22} & r_{23} & \cdots & r_{2n} \\ 0 & 0 & r_{33} & \cdots & r_{3n} \\ \vdots & \vdots & \vdots & & \vdots \\ 0 & 0 & 0 & \cdots & r_{nn} \end{bmatrix}.$$

Therefore, by Definition 3.2.13,

$$QR = \begin{bmatrix} Q\mathbf{u}_1 & Q\mathbf{u}_2 & \cdots & Q\mathbf{u}_n \end{bmatrix} = \begin{bmatrix} \mathbf{c}_1 & \mathbf{c}_2 & \cdots & \mathbf{c}_n \end{bmatrix} = A.$$

This proves the next theorem.

■ Theorem 6.2.27 [QR Factorization]

If A is an $m \times n$ matrix with linearly independent columns, there exists an $m \times n$ matrix Q with columns that form an orthonormal subset of $\mathbb{R}^m$ and an $n \times n$ upper diagonal matrix R such that $A = QR$.

■ Example 6.2.28

Find the QR factorization of the matrix $A = \begin{bmatrix} 1 & 1 & 0 \\ 0 & -2 & 1 \\ 1 & 1 & 2 \\ 2 & 0 & 1 \end{bmatrix}$. The columns of Q are found by applying the Gram–Schmidt process to the columns of A as in Example 6.2.24, which means that

$$Q = \begin{bmatrix} \sqrt{6}/6 & \sqrt{3}/6 & -\sqrt{3}/6 \\ 0 & -\sqrt{3}/2 & \sqrt{3}/6 \\ \sqrt{6}/6 & \sqrt{3}/6 & \sqrt{3}/2 \\ \sqrt{6}/3 & -\sqrt{3}/6 & -\sqrt{3}/6 \end{bmatrix}.$$

Next, the columns of A must be written as linear combinations of the columns of Q. This is possible because the columns of Q form a basis for the spanning set of the columns of A. If the columns of Q are written as

$$
\mathbf{v}_1 = \begin{bmatrix} \sqrt{6}/6 \\ 0 \\ \sqrt{6}/6 \\ \sqrt{6}/3 \end{bmatrix}, \quad
\mathbf{v}_2 = \begin{bmatrix} \sqrt{3}/6 \\ -\sqrt{3}/2 \\ \sqrt{3}/6 \\ -\sqrt{3}/6 \end{bmatrix}, \quad
\mathbf{v}_3 = \begin{bmatrix} -\sqrt{3}/6 \\ \sqrt{3}/6 \\ \sqrt{3}/2 \\ -\sqrt{3}/6 \end{bmatrix},
$$

then by inspection it follows that

$$
\begin{bmatrix} 1 \\ 0 \\ 1 \\ 2 \end{bmatrix} = \sqrt{6}\,\mathbf{v}_1 \quad \text{and} \quad
\begin{bmatrix} 1 \\ -2 \\ 1 \\ 0 \end{bmatrix} = \frac{\sqrt{6}}{3}\mathbf{v}_1 + \frac{4\sqrt{3}}{3}\mathbf{v}_2,
$$

and by Example 3.5.24,

$$
\begin{bmatrix} 0 \\ 1 \\ 2 \\ 1 \end{bmatrix} = \frac{2\sqrt{6}}{3}\mathbf{v}_1 - \frac{\sqrt{3}}{3}\mathbf{v}_2 + \sqrt{3}\mathbf{v}_3.
$$

Therefore, it can be shown that $A = QR$, where

$$
R = \begin{bmatrix} \sqrt{6} & \sqrt{6}/3 & 2\sqrt{6}/3 \\ 0 & 4\sqrt{3}/3 & -\sqrt{3}/3 \\ 0 & 0 & \sqrt{3} \end{bmatrix}.
$$

Exercises

1. Determine whether each set is orthonormal using the given inner product.

   (a) $\left\{ \begin{bmatrix} 0 \\ 0 \\ 1 \end{bmatrix}, \begin{bmatrix} \sqrt{2}/2 \\ -\sqrt{2}/2 \\ 0 \end{bmatrix}, \begin{bmatrix} \sqrt{2}/2 \\ \sqrt{2}/2 \\ 0 \end{bmatrix} \right\}$, dot product.

   (b) $\left\{ \begin{bmatrix} 4 \\ 2 \end{bmatrix}, \begin{bmatrix} 6 \\ -3 \end{bmatrix} \right\}$, weighted inner product with weights $k_1 = 4$ and $k_2 = 2$.

   (c) $\{f, g\} \subset C[0, \pi]$, where $f(x) = \sin x$ and $g(x) = x^3$; integral inner product.

2. Let $\mathcal{S} = \{2 + x + 5x^2, -3 + 2x - x^2\}$. Let $\mathbb{P}_2$ be paired with the evaluation inner product with $r_1 = 0$, $r_2 = 1$, and $r_3 = 2$.

   (a) Show that $\mathcal{S}$ is an orthogonal set.

   (b) Normalize the polynomials of $\mathcal{S}$ to form an orthonormal set.

   (c) Find $p(x) \in \mathbb{P}_2$ so that $\mathcal{S} \cup \{p(x)\}$ is an orthogonal set.

3. Let V be a subspace of $\mathbb{R}^n$ spanned by an orthogonal set of n vectors. Prove that $V = \mathbb{R}^n$.

4. Let W be a subspace of the inner product space V. Prove that proj_W is a linear operator $V \to V$.

5. Let A be a square matrix and $\mathcal{S} = \{\mathbf{c}_1, \mathbf{c}_2, \dots, \mathbf{c}_n\}$ be the columns of A. Prove that if $\mathcal{S}$ is an orthogonal set with respect to the dot product, then A is invertible.

6. Explain why the orthogonal complement with respect to the dot product of the xy-plane in $\mathbb{R}^3$ is not the yz-plane.

7. View $\mathbb{R}^4$ as an inner product space using the dot product. Find the orthogonal complement of the plane,

$$\mathsf{S}\left(\begin{bmatrix} 1 \\ 2 \\ 3 \\ 4 \end{bmatrix}, \begin{bmatrix} 2 \\ 0 \\ 1 \\ 1 \end{bmatrix}\right).$$

8. Find the orthogonal complement of each subspace W of the inner product space V using the given inner product.

    (a) $V = \mathbb{R}^2$, $W = \mathsf{S}\left(\begin{bmatrix} 4 \\ -5 \end{bmatrix}\right)$, dot product.

    (b) $V = \mathbb{R}^3$, $W = \mathsf{S}\left(\begin{bmatrix} 2 \\ 0 \\ -3 \end{bmatrix}\right)$,

    weighted inner product with weights $k_1 = 3, k_2 = 4, k_3 = 2$.

    (c) $V = \mathbb{P}_4$, $W = \mathsf{S}(2 + x^3, x - 2x^4, 1 + x^2)$,

    evaluation inner product with $r_1 = 0, r_2 = 2$, and $r_3 = 1$.

    (d) $V = \mathsf{S}(\sin, \cos) \subset C[0, 1]$, $W = \mathsf{S}(\sin)$, integral inner product.

9. Let W be a subspace of an inner product space. Prove that $(W^\perp)^\perp = W$.

10. Prove (6.14) in Example 6.2.13.

11. Prove or show false: If W is a subspace of the inner product space V_1 and $T : V_1 \to V_2$ is an isomorphism, $T[W_1^\perp]$ is the orthogonal compliment of $T[W]$ in V_2.

12. Let W_1 and W_2 be subspaces of the vector space V so that $W_1 \cap W_2 = \{\mathbf{0}\}$. Prove.

    (a) There are no common vectors to the bases of W_1 and W_2.

    (b) The union of a basis of W_1 and a basis of W_2 is a basis of $W_1 \oplus W_2$.

13. Let V be a vector space such that $V = W_1 \oplus W_2$, where W_1 and W_2 are subspaces of V. Prove that if $\mathbf{u}_1 \in W_1$ and $\mathbf{u}_2 \in W_2$ such that $\mathbf{u}_1 + \mathbf{u}_2 = \mathbf{0}$, then $\mathbf{u}_1 = \mathbf{u}_2 = \mathbf{0}$.

14. Consider the isomorphism $T : V_1 \to V_2$. Let W_1 and W_2 be subspaces of V_1 such that $V_1 = W_1 \oplus W_2$. Prove that $V_2 = T[W_1] \oplus T[W_2]$.

15. Suppose that W_1 and W_2 are subspaces of the vector space V such that the intersection $W_1 \cap W_2 = \{\mathbf{0}\}$. Prove that $W_1 \oplus W_2$ is isomorphic to $W_1 \times W_2$, where the Cartesian product $W_1 \times W_2$ is a subspace of $V \times V$ as in Exercise 9 of Section 5.2.

16. Let A be an $m \times n$ matrix. Prove.

    (a) For all $\mathbf{u} \in \mathbb{R}^n$ there exists $\mathbf{v}_1 \in \operatorname{RS} A$ and $\mathbf{v}_2 \in \operatorname{NS} A$ such that $\mathbf{u} = \mathbf{v}_1 + \mathbf{v}_2$. Are the vectors $\mathbf{v}_1$ and $\mathbf{v}_2$ unique?

    (b) For all $\mathbf{b} \in \mathbb{R}^m$, if $A\mathbf{x} = \mathbf{b}$ is consistent, there exists $\mathbf{v} \in \operatorname{RS} A$ such that $A\mathbf{v} = \mathbf{b}$. Is the vector $\mathbf{v}$ unique?

17. For each inner product space V with the given inner product, use the Gram–Schmidt process on the basis $\mathscr{B}$ to find an orthonormal basis for V.

    (a) $V = \mathbb{R}^3, \mathscr{B} = \left\{ \begin{bmatrix} 6 \\ 1 \\ 5 \end{bmatrix}, \begin{bmatrix} 0 \\ 4 \\ 6 \end{bmatrix}, \begin{bmatrix} 1 \\ 1 \\ 3 \end{bmatrix} \right\}$, dot product.

    (b) $V = \operatorname{S}\left(\begin{bmatrix} 6 \\ 9 \\ 5 \\ 2 \end{bmatrix}, \begin{bmatrix} 4 \\ -1 \\ -3 \\ -2 \end{bmatrix} \right), \mathscr{B} = \left\{ \begin{bmatrix} 5 \\ 4 \\ 1 \\ 0 \end{bmatrix}, \begin{bmatrix} 1 \\ 5 \\ 4 \\ 2 \end{bmatrix} \right\},$

    weighted inner product with weights $k_1 = 1, k_2 = 2, k_3 = 3, k_4 = 4$.

    (c) $V = \mathbb{R}^{2\times 2}, \mathscr{B} = \left\{ \begin{bmatrix} 1 & 1 \\ 0 & 0 \end{bmatrix}, \begin{bmatrix} 0 & 1 \\ 0 & 1 \end{bmatrix}, \begin{bmatrix} 0 & 0 \\ 1 & 1 \end{bmatrix}, \begin{bmatrix} 1 & 0 \\ 1 & 0 \end{bmatrix} \right\},$

    Frobenius inner product.

    (d) $V = \mathbb{P}_2, \mathscr{B} = \{1, x, x^2\}$,

    evaluation inner product with $r_1 = 1$ and $r_2 = r_3 = 2$.

18. Find an orthonormal basis for the subspace $\operatorname{S}(\cos, \sin)$ of the inner product space $C[0, 1]$ using the integral inner product.

19. Prove that $\mathbf{u}_3$ and $\mathbf{v}_2$ in Figure 6.2 are not scalar multiples of each other.

20. Let $\mathscr{B}$ be an orthogonal basis for an inner product space V. Prove or show false: If the Gram–Schmidt process is applied to $\mathscr{B}$, the result will be $\mathscr{B}$.

21. Assume that V is an inner product space. Take two vectors $\mathbf{u}$ and $\mathbf{v}$ in V. Let $S = \{\operatorname{proj}_{\mathbf{v}} \mathbf{u}, \mathbf{u} - \operatorname{proj}_{\mathbf{v}} \mathbf{u}\}$.

    (a) Explain why S is linearly independent.

    (b) If $\dim V > 2$, find $\mathbf{w} \in V$ such that $S \cup \{\mathbf{w}\}$ is linearly independent.

22. Let Q be an $m \times n$ matrix such that its columns form an orthonormal set with respect to the dot product.

    (a) Prove that $Q^T Q = I_n$.

    (b) Let A be an $m \times n$ matrix. Explain how to use the QR factorization of A to solve $A\mathbf{x} = \mathbf{b}$ for any $\mathbf{b} \in \mathbb{R}^m$.

23. Use the QR factorization of each coefficient matrix to find the solution of the the corresponding system of linear equations.

(a) $\begin{bmatrix} 0 & 1 & 1 \\ 1 & 0 & 1 \\ 1 & 1 & 0 \end{bmatrix}$

(b) $\begin{bmatrix} 1 & -1 & 4 \\ 0 & 4 & 0 \\ 1 & 4 & -2 \\ 0 & -1 & 2 \end{bmatrix}$

24. Let A be an $m \times n$ matrix with QR factorization $A = QR$.

    (a) Prove that the columns of Q form an orthonormal basis for the column space of A.

    (b) Prove that the entries along the main diagonal of R can be taken to be positive.

25. Assume that the columns of the matrix A form an orthonormal set. Find the QR factorization of A.

CHAPTER 7

Matrix Theory

7.1 Eigenvectors and Eigenvalues

Consider the 2×2 diagonal matrix $D = \begin{bmatrix} 2 & 0 \\ 0 & 3 \end{bmatrix}$. This matrix operates on a vector $\mathbf{u} \in \mathbb{R}^2$ with coordinates a and b by doubling the first component and tripling the second component:

$$D\mathbf{u} = \begin{bmatrix} 2 & 0 \\ 0 & 3 \end{bmatrix} \begin{bmatrix} a \\ b \end{bmatrix} = \begin{bmatrix} 2a \\ 3b \end{bmatrix}. \tag{7.1}$$

If $\mathbf{u}$ is written as a linear combination of the standard vectors of $\mathbb{R}^2$, (7.1) becomes

$$D\mathbf{u} = D(a\mathbf{e}_1 + b\mathbf{e}_2) = 2a\mathbf{e}_1 + 3b\mathbf{e}_2$$

because D acts by scaling the coordinate vectors $\mathbf{e}_1$ and $\mathbf{e}_2$,

$$D\mathbf{e}_1 = 2\mathbf{e}_1 \quad \text{and} \quad D\mathbf{e}_2 = 3\mathbf{e}_2.$$

In general, for every positive integer k,

$$D^k\mathbf{e}_1 = 2^k\mathbf{e}_1 \quad \text{and} \quad D^k\mathbf{e}_2 = 3^k\mathbf{e}_2.$$

Linear Algebra, First Edition. Michael L. O'Leary.
© 2021 John Wiley & Sons, Inc. Published 2021 by John Wiley & Sons, Inc.
Companion Website: www.wiley.com/go/o'leary/linearalgebra

This implies that given any vector $\mathbf{u} \in \mathbb{R}^2$, to compute $D\mathbf{u}$, $D^2\mathbf{u}$, $D^3\mathbf{u}$, or any $D^k\mathbf{u}$, use the decomposition $\mathbf{u} = c_1\mathbf{e}_1 + c_2\mathbf{e}_2$ by computing

$$D^n\mathbf{u} = D^n(c_1\mathbf{e}_1 + c_2\mathbf{e}_2) = c_1D^n\mathbf{e}_1 + c_2D^n\mathbf{e}_2 = c_12^n\mathbf{e}_1 + c_23^n\mathbf{e}_2.$$

Indeed, for any polynomial $p(x)$, using Definition 3.2.25 it follows that

$$p(D)\mathbf{u} = c_1p(2)\mathbf{e}_1 + c_2p(3)\mathbf{e}_2. \tag{7.2}$$

As an example, let $p(x) = 4 + 7x + 9x^2$ and evaluate,

$$
\begin{aligned}
p(D)(c_1\mathbf{e}_1 + c_2\mathbf{e}_2) &= (4I_2 + 7D + 9D^2)(c_1\mathbf{e}_1 + c_2\mathbf{e}_2) \\
&= c_1(4I_2\mathbf{e}_1 + 7D\mathbf{e}_1 + 9D^2\mathbf{e}_1) + c_2(4I_2\mathbf{e}_2 + 7D\mathbf{e}_2 + 9D^2\mathbf{e}_2) \\
&= c_1(4\mathbf{e}_1 + 7(2)\mathbf{e}_1 + 9(2)^2\mathbf{e}_1) + c_2(4\mathbf{e}_2 + 7(3)\mathbf{e}_2 + 9(3)^2\mathbf{e}_2) \\
&= c_1p(2)\mathbf{e}_1 + c_2p(3)\mathbf{e}_2.
\end{aligned}
$$

The general proof of (7.2) is left to Exercise 17.

Of course, a nondiagonal matrix cannot be analyzed in such a straightforward manner. Consider, for example, the matrix $A = \begin{bmatrix} 0 & 1 \\ -2 & 3 \end{bmatrix}$, whose action on $\mathbf{e}_1$ and $\mathbf{e}_2$ is more complicated. Specifically,

$$A\mathbf{e}_1 = -2\mathbf{e}_2 \quad \text{and} \quad A\mathbf{e}_2 = \mathbf{e}_1 + 3\mathbf{e}_2.$$

For the diagonal matrix D, the standard vectors $\mathbf{e}_1$ and $\mathbf{e}_2$ in $\mathbb{R}^2$ are special directions in which D simply acts as scalar multiplication by 2 and 3, respectively. For the nondiagonal matrix A, the directions $\mathbf{e}_1$ and $\mathbf{e}_2$ do not have this property. Are there any directions in which the nondiagonal matrix A acts as scalar multiplication? Such a vector $\mathbf{u}$ would have the property that $A\mathbf{u}$ is equal to a scalar multiple of $\mathbf{u}$ of the form $\lambda\mathbf{u}$ for some scalar λ.

■ **Definition 7.1.1**

Let A be an $n \times n$ matrix. The nonzero vector $\mathbf{u} \in \mathbb{R}^n$ is an **eigenvector** of A if there exists $\lambda \in \mathbb{R}$ such that $A\mathbf{u} = \lambda\mathbf{u}$. Call λ the **eigenvalue** of A that **corresponds** to $\mathbf{u}$.

■ **Example 7.1.2**

Let $A = \begin{bmatrix} 2 & 6 \\ 5 & 3 \end{bmatrix}$. It is easy to check that $\lambda_1 = 8$ is an eigenvalue for A. Indeed,

$$\begin{bmatrix} 2 & 6 \\ 5 & 3 \end{bmatrix}\begin{bmatrix} 1 \\ 1 \end{bmatrix} = \begin{bmatrix} 8 \\ 8 \end{bmatrix} = 8\begin{bmatrix} 1 \\ 1 \end{bmatrix}$$

so that the vector $\mathbf{u}_1 = \begin{bmatrix} 1 \\ 1 \end{bmatrix}$ is an eigenvector of A corresponding to $\lambda_1 = 8$. Similarly, $\lambda_2 = -3$ is also an eigenvalue of A because

$$\begin{bmatrix} 2 & 6 \\ 5 & 3 \end{bmatrix} \begin{bmatrix} -6 \\ 5 \end{bmatrix} = \begin{bmatrix} 18 \\ -15 \end{bmatrix} = -3 \begin{bmatrix} -6 \\ 5 \end{bmatrix}$$

so that the vector $\mathbf{u}_2 = \begin{bmatrix} -6 \\ 5 \end{bmatrix}$ is an eigenvector of A corresponding to $\lambda_2 = -3$.

It should be emphasized what the eigenvalue and the eigenvectors reveal about a matrix. Assume that $\mathbf{u}$ is an eigenvalue of the square matrix A with corresponding eigenvalue λ. Notice that the normalization of $\mathbf{u}$ is also an eigenvector of A corresponding to λ because for any $r \in \mathbb{R}$,

$$A(r\mathbf{u}) = r(A\mathbf{u}) = r(\lambda\mathbf{u}) = \lambda(r\mathbf{u}). \tag{7.3}$$

For this reason if $\mathbf{u}$ is a unit vector, $\|A\mathbf{u}\| = \|\lambda\mathbf{u}\| = |\lambda| \|\mathbf{u}\| = |\lambda|$. This means that in the direction of $\mathbf{u}$, multiplication of $r\mathbf{u}$ by A results in a vector of length $|r\lambda|$. For example,

$$\|A\mathbf{u}\| = |\lambda|, \quad \|A(2\mathbf{u})\| = 2|\lambda|, \quad \|A(3\mathbf{u})\| = 3|\lambda|, \quad \dots,$$

illustrating that in the direction of $\mathbf{u}$, the eigenvalue λ acts like a type of slope.

Eigenspaces

A method to find all eigenvalues and eigenvectors of a given matrix will be presented shortly. At this point, however, it is important to note that the eigenvectors presented in Example 7.1.2 corresponding to λ_1 and λ_2 are not unique. In fact, any vector in

$$S(\mathbf{u}_1) = S\left(\begin{bmatrix} 1 \\ 1 \end{bmatrix}\right) \tag{7.4}$$

is an eigenvector of A corresponding to $\lambda_1 = 8$ by (7.3). Likewise, any vector in

$$S(\mathbf{u}_2) = S\left(\begin{bmatrix} -6 \\ 5 \end{bmatrix}\right) \tag{7.5}$$

is an eigenvector of A corresponding to $\lambda_2 = -3$. For this reason, it is advantageous to gather the eigenvectors that correspond to a particular eigenvalue into a set.

■ **Definition 7.1.3**

Let $\lambda \in \mathbb{R}$ be an eigenvalue of the $n \times n$ matrix A. The **eigenspace** of A corresponding to λ is

$$\text{ES}(A, \lambda) = \{\mathbf{u} \in \mathbb{R}^n : A\mathbf{u} = \lambda\mathbf{u}\} \cup \{\mathbf{0}\}.$$

For the matrix A of Example 7.1.2, (7.4) and (7.5) show that

$$\text{ES}(A, 8) = \text{S}\left(\begin{bmatrix} 1 \\ 1 \end{bmatrix}\right) \quad \text{and} \quad \text{ES}(A, -3) = \text{S}\left(\begin{bmatrix} -6 \\ 5 \end{bmatrix}\right).$$

Definition 7.1.1 requires that an eigenvector be nonzero, so the zero vector must be included as an element of $\text{ES}(A, \lambda)$ manually. This is done to ensure that each eigenspace is a vector space. It also enables a nice characterization.

■ **Theorem 7.1.4**

If $\lambda \in \mathbb{R}$ is an eigenvalue of the $n \times n$ matrix A, then $\text{ES}(A, \lambda) = \text{NS}(A - \lambda I_n)$.

Proof

A vector $\mathbf{u}$ is in $\text{ES}(A, \lambda)$ if and only if $A\mathbf{u} = \lambda \mathbf{u}$. This is equivalent to

$$A\mathbf{u} - \lambda\mathbf{u} = (A - \lambda I_n)\mathbf{u} = \mathbf{0}.$$

Hence, $\mathbf{u} \in \text{ES}(A, \lambda)$ if and only if $\mathbf{u} \in \text{NS}(A - \lambda I_n)$. ■

In Example 7.1.2 the eigenvalues and corresponding eigenvectors for a nondiagonal matrix were presented, but it was not indicated how these eigenvalues and eigenvectors can be found. By Definition 7.1.1, the goal is to find a scalar λ paired with a nonzero vector $\mathbf{u}$ such that λ and $\mathbf{u}$ satisfy

$$A\mathbf{u} = \lambda\mathbf{u}, \tag{7.6}$$

but (7.6) is equivalent to the condition,

$$(A - \lambda I_2)\mathbf{u} = \mathbf{0}. \tag{7.7}$$

Theorems 4.3.11 and 4.3.15 imply that a scalar λ is an eigenvalue of A if and only if (7.7) has a nontrivial solution, which is equivalent to the condition

$$\det(A - \lambda I_2) = 0, \tag{7.8}$$

the next theorem follows.

■ **Theorem 7.1.5**

A scalar λ is an eigenvalue of the $n \times n$ matrix A if and only if $\det(A - \lambda I_n) = 0$.

■ **Example 7.1.6**

For the matrix $A = \begin{bmatrix} 0 & 1 \\ -2 & 3 \end{bmatrix}$, condition (7.8) becomes

$$\det(A - \lambda I_2) = \det\left(\begin{bmatrix} 0 & 1 \\ -2 & 3 \end{bmatrix} - \begin{bmatrix} \lambda & 0 \\ 0 & \lambda \end{bmatrix}\right)$$

$$= \begin{vmatrix} 0 - \lambda & 1 \\ -2 & 3 - \lambda \end{vmatrix} = \lambda^2 - 3\lambda + 2 = 0.$$

This simplifies to $(\lambda - 2)(\lambda - 1) = 0$ so that the two eigenvalues for A are $\lambda_1 = 1$ and $\lambda_2 = 2$.

Characteristic Polynomial

The quantity $\det(A - xI_n)$ will in general be an nth degree polynomial with variable x when A is an $n \times n$ matrix. This polynomial, whose roots are the eigenvalues, is a property of A.

■ **Definition 7.1.7**

Let A be an $n \times n$ matrix.

   (a) The **characteristic polynomial** $p_A(x)$ of A is the polynomial

$$p_A(x) = \det(A - xI_n).$$

   (b) The **characteristic equation** of A is the equation

$$p_A(x) = 0.$$

   (c) The **spectrum** of A is the set

$$\sigma(A) = \{\lambda : p_A(\lambda) = 0\}.$$

   That is, the spectrum of A is the set of eigenvalues of A.

The above definitions immediately imply that the eigenvalues of a matrix are exactly the solutions of the characteristic equation of that same matrix. Once it has been determined that λ is an eigenvalue for A, the corresponding eigenspace can be discovered by finding the null space of $A - \lambda I_n$.

■ **Example 7.1.8**

Continuing with the case of the nondiagonal matrix $A = \begin{bmatrix} 0 & 1 \\ -2 & 3 \end{bmatrix}$, the characteristic polynomial was shown to be

$$p_A(x) = \det(A - xI_2) = x^2 - 3x + 2,$$

writing the polynomial with the more common decreasing exponents. Thus, the corresponding characteristic equation is

$$x^2 - 3x + 2 = 0,$$

the solutions of which are $x = 1$ and $x = 2$, the eigenvalues for A. These eigenvalues are associated with eigenvectors $\mathbf{u}_1$ and $\mathbf{u}_2$ with

$$A\mathbf{u}_1 = 1\mathbf{u}_1 \text{ and } A\mathbf{u}_2 = 2\mathbf{u}_2.$$

That is, A acts on $\mathbf{u}_1$ as scalar multiplication by 1 and A acts on $\mathbf{u}_2$ as scalar multiplication by 2.

To find the corresponding eigenvectors, it is useful to return again to condition (7.7). For $\lambda = 1$, (7.7) reduces to calculating the null space of $A - 1I_2$,

$$\mathsf{ES}(A, 1) = \mathsf{NS}\,(A - 1I_2) = \{\mathbf{x} : (A - 1I_2)\mathbf{x} = 0\},$$

and similarly for $\lambda = 2$, the null space is

$$\mathsf{ES}(A, 2) = \mathsf{NS}\,(A - 2I_2) = \{\mathbf{x} : (A - 2I_2)\mathbf{x} = 0\}.$$

A quick calculation shows that $\mathsf{ES}(A, 1) = \mathsf{S}\left(\begin{bmatrix} 1 \\ 1 \end{bmatrix}\right)$ and $\mathsf{ES}(A, 2) = \mathsf{S}\left(\begin{bmatrix} 1 \\ 2 \end{bmatrix}\right)$.

■ **Example 7.1.9**

Consider again the matrix $A = \begin{bmatrix} 0 & 1 \\ -2 & 3 \end{bmatrix}$. Knowing the eigenspaces of the matrix A simplifies many matrix computations involving A. For example, to compute $A^n\mathbf{u}$ for some vector $\mathbf{u}$, it is convenient to first decompose $\mathbf{u}$ into a linear combination of the normalized eigenvectors of A,

$$\mathbf{u}_1 = \begin{bmatrix} \sqrt{2}/2 \\ \sqrt{2}/2 \end{bmatrix} \quad \text{and} \quad \mathbf{u}_2 = \begin{bmatrix} \sqrt{5}/5 \\ 2\sqrt{5}/5 \end{bmatrix}.$$

Then, to compute $A^{10}\mathbf{u}$ where $\mathbf{u} = \begin{bmatrix} 2 \\ 3 \end{bmatrix}$ it is helpful to write $\mathbf{u}$ as a linear combination of the eigenvectors of A,

$$\mathbf{u} = \sqrt{2}\mathbf{u}_1 + \sqrt{5}\mathbf{u}_2,$$

so that

$$\begin{aligned}
A^{10}\mathbf{u} &= A^{10}\left(\sqrt{2}\mathbf{u}_1 + \sqrt{5}\mathbf{u}_2\right) \\
&= \sqrt{2}A^{10}\mathbf{u}_1 + \sqrt{5}A^{10}\mathbf{u}_2 \\
&= \sqrt{2}(1^{10})\mathbf{u}_1 + \sqrt{5}(2^{10})\mathbf{u}_2 \\
&= \sqrt{2}\mathbf{u}_1 + 1024\sqrt{5}\mathbf{u}_2 \\
&= \begin{bmatrix} 1 \\ 1 \end{bmatrix} + \begin{bmatrix} 1024 \\ 2048 \end{bmatrix} = \begin{bmatrix} 1025 \\ 2049 \end{bmatrix}.
\end{aligned}$$

Example 7.1.9 used in an essential way the fact that the eigenvectors of the matrix A form a basis for $\mathbb{R}^2$. In particular, the two eigenvectors for the matrix A, one corresponding to $\lambda = 1$ and the other to $\lambda = 2$, are linearly independent. This is always the case in general.

■ **Theorem 7.1.10**

Eigenvectors corresponding to distinct eigenvalues are linearly independent.

Proof

Let A be an $n \times n$ matrix and suppose that $\lambda_1, \lambda_2, \dots, \lambda_k \in \sigma(A)$ are distinct eigenvalues with corresponding eigenvectors $\mathbf{u}_1, \mathbf{u}_2, \dots, \mathbf{u}_k$. This means that

$$A\mathbf{u}_1 = \lambda_1\mathbf{u}_1, \quad A\mathbf{u}_2 = \lambda_2\mathbf{u}_2, \quad \dots, \quad A\mathbf{u}_k = \lambda_k\mathbf{u}_k.$$

Suppose further that $r_1\mathbf{u}_1 + r_2\mathbf{u}_2 + \dots + r_k\mathbf{u}_k = \mathbf{0}$ for some $r_1, r_2, \dots r_k \in \mathbb{R}$. To show that $\{\mathbf{u}_1, \mathbf{u}_2, \dots, \mathbf{u}_k\}$ is linearly independent, it is required to show that $r_1 = r_2 = \dots = r_k = 0$. Towards this end, for $j = 1, 2, \dots, k$ consider the polynomial

$$p_j(x) = \prod_{i \neq j}(x - \lambda_i)$$

formed by multiplying all factors of the form $x - \lambda_i$ skipping only $x - \lambda_j$. For each $i \neq j$, since $A - \lambda_i I$ is a factor of $p_j(A)$, there exists a polynomial $q(x)$ such that

$$p_j(A)\mathbf{u}_i = q(A)(A - \lambda_i I_n)\mathbf{u}_i = q(A)\mathbf{0} = \mathbf{0}.$$

On the other hand, $p_j(A)\mathbf{u}_j \neq \mathbf{0}$ since

$$p_j(A)\mathbf{u}_j = \Big[\prod_{i \neq j}(A - \lambda_i I_n)\Big]\mathbf{u}_j = \Big[\prod_{i \neq j}(\lambda_j - \lambda_i)\Big]\mathbf{u}_j \neq \mathbf{0}.$$

Multiplying $p_j(A)$ by each side of $r_1\mathbf{u}_1 + r_2\mathbf{u}_2 + \dots + r_k\mathbf{u}_k = \mathbf{0}$ yields

$$\begin{aligned}
\mathbf{0} &= p_j(A)(r_1\mathbf{u}_1 + r_2\mathbf{u}_2 + \dots + r_k\mathbf{u}_k) \\
&= r_1 p_j(A)\mathbf{u}_1 + r_2 p_j(A)\mathbf{u}_2 + \dots + r_k p_j(A)\mathbf{u}_k \\
&= r_j p_j(A)\mathbf{u}_j \\
&= r_j \prod_{i \neq j}(\lambda_j - \lambda_i)\mathbf{u}_i,
\end{aligned}$$

and since $\prod_{i \neq j}(\lambda_j - \lambda_i) \neq 0$, it follows that $r_j = 0$. ■

Note that the converse of Theorem 7.1.10 is not true. Eigenvectors corresponding to one and the same eigenvalue can be linearly independent. Indeed, the identity matrix, $\begin{bmatrix} 1 & 0 \\ 0 & 1 \end{bmatrix}$, provides a trivial example. The linearly independent vectors $\mathbf{e}_1$ and $\mathbf{e}_2$ both correspond to the eigenvalue $\lambda = 1$.

■ **Example 7.1.11**

Let $A = \begin{bmatrix} 1 & 1 & 1 \\ 0 & 1 & 2 \\ 2 & 1 & 0 \end{bmatrix}$. The characteristic polynomial for A is

$$p_A(x) = \det(A - xI_3) = \begin{vmatrix} 1-x & 1 & 1 \\ 0 & 1-x & 2 \\ 2 & 1 & -x \end{vmatrix} = -x^3 + 2x^2 + 3x.$$

The corresponding characteristic equation is $-x^3 + 2x^2 + 3x = 0$, which is simplified as

$$x(x - 3)(x + 1) = 0$$

so that the spectrum of A is $\sigma(A) = \{0, 3, -1\}$ because A has eigenvalues

$$\lambda_1 = 0, \quad \lambda_2 = 3, \quad \lambda_3 = -1.$$

The eigenspaces are the null spaces NS (A) for $\lambda_1 = 0$, NS $(A - 3I_3)$ for $\lambda_2 = 3$, and NS $(A + 1I_3)$ for $\lambda_3 = -1$. Because eigenvectors corresponding to distinct eigenvalues are linearly independent (Theorem 7.1.10), and because $\mathbb{R}^3$ has dimension 3, each of these three eigenspaces must be one-dimensional. Indeed, a quick calculation shows

$$\text{ES}(A, 0) = S\left(\begin{bmatrix} 1 \\ -2 \\ 1 \end{bmatrix}\right), \quad \text{ES}(A, 3) = S\left(\begin{bmatrix} 1 \\ 1 \\ 1 \end{bmatrix}\right), \quad \text{ES}(A, -1) = S\left(\begin{bmatrix} 0 \\ 1 \\ -1 \end{bmatrix}\right).$$

As expected, the eigenvectors $\begin{bmatrix} 1 \\ -2 \\ 1 \end{bmatrix}, \begin{bmatrix} 1 \\ 1 \\ 1 \end{bmatrix}$, and $\begin{bmatrix} 0 \\ 1 \\ -1 \end{bmatrix}$ corresponding to the eigenvalues $\lambda = -1, \lambda = 0$, and $\lambda = 3$, respectively, form a linearly independent set.

In Example 7.1.11, it was fortunate that the matrix A had a set of eigenvectors which formed a basis for $\mathbb{R}^3$. When this is the case, any vector can be written as a linear combination of the eigenvectors of the matrix and the operation of the matrix is particularly easy to describe. This, however, is not always the case as illustrated in the following example.

■ **Example 7.1.12**

Consider the 2×2 matrix $B = \begin{bmatrix} 1 & 1 \\ 0 & 1 \end{bmatrix}$. The characteristic polynomial of B is computed as

$$\det(B - xI_2) = \begin{vmatrix} 1-x & 1 \\ 0 & 1-x \end{vmatrix} = (x - 1)^2, \tag{7.9}$$

so B has only one eigenvalue, $\lambda = 1$, with eigenspace

$$\text{ES}(B, 1) = \text{NS}\left(\begin{bmatrix} 0 & 1 \\ 0 & 0 \end{bmatrix}\right) = \text{S}\left(\begin{bmatrix} 1 \\ 0 \end{bmatrix}\right).$$

Hence, the eigenvectors of B do not span $\mathbb{R}^2$.

Although straightforward, Example 7.1.12 contains two points worthwhile to mention. First, the eigenvalue of the triangular matrix B is found on its main diagonal. This is not a coincidence. A glance at (7.9) and a recollection of how to find the determinant of a triangular matrix using Corollary 4.3.7(b) should confirm this. The proof of the generalization of this fact is left to Exercise 18, but the result is stated here for convenience.

■ **Theorem 7.1.13**

The eigenvalues of a triangular matrix are exactly those entries of its main diagonal, including repeats.

The second point is that the eigenvalue $\lambda = 1$ that appeared in Example 7.1.12 is a root of multiplicity 2 in the characteristic polynomial for B (7.9). The eigenvectors for the matrix B in Example 7.1.12 did not form a basis for $\mathbb{R}^2$ because the eigenspace $\text{ES}(B, 1)$ has dimension 1. Both of these facts are important.

■ **Definition 7.1.14**

Let the real number λ be an eigenvalue of the matrix A.

    (a) The **algebraic multiplicity** of λ is the multiplicity of λ as a root of the characteristic polynomial of A.

    (b) The **geometric multiplicity** of λ equals $\dim \text{ES}(A, \lambda)$.

■ **Example 7.1.15**

Both eigenvalues of Example 7.1.8 have algebraic and geometric multiplicities equal to 1, while the eigenvalue of Example 7.1.12 has an algebraic multiplicity of 2 and a geometric multiplicity of 1.

Because eigenvalues are roots of polynomials in $\mathbb{P}$, it is possible for an eigenvalue to be a nonreal complex number. For example, consider the matrix $A = \begin{bmatrix} 0 & -1 \\ 1 & 0 \end{bmatrix}$. The characteristic polynomial for A is $\det(A - xI_2) = x^2 + 1$, which has complex conjugate roots $\lambda_1 = i$ and $\lambda_2 = -i$ with corresponding eigenvectors $\begin{bmatrix} 1 \\ -i \end{bmatrix}$ and $\begin{bmatrix} 1 \\ i \end{bmatrix}$, respectively. The eigenspaces for these eigenvalues are

$$\text{ES}(A, i) = \text{NS}(A - iI_2) = \text{S}\left(\begin{bmatrix} 1 \\ -i \end{bmatrix}\right) \quad \text{and} \quad \text{ES}(A, -i) = \text{NS}(A + iI_2) = \text{S}\left(\begin{bmatrix} 1 \\ i \end{bmatrix}\right).$$

It should be noted, however, that these sets do not form vector spaces with scalars from $\mathbb{R}$. It should be clear that addition with vectors with nonreal coordinates is not a binary operation $\mathbb{R}^n \times \mathbb{R}^n \to \mathbb{R}^n$ and scalar multiplication with the same type of vectors is not a scalar operation $\mathbb{R} \times \mathbb{R}^n \to \mathbb{R}^n$. The solution to this is to modify the definition of a vector space to include complex number scalars and view sets such as $ES(A, i)$ as subspaces of

$$\mathbb{C}^n = \left\{ \begin{bmatrix} z_1 \\ z_2 \\ \vdots \\ z_n \end{bmatrix} : z_i \in \mathbb{C} \text{ for } i = 1, 2, \dots, n \right\},$$

which has the same vector space properties as $\mathbb{R}^n$ with vector addition and scalar multiplication defined as extensions of the operations on $\mathbb{R}^n$. This being said, such vector spaces will not be a subject for this book. Nonetheless, the possibility of nonreal roots to polynomials will require subsequent results to be written carefully.

Cayley–Hamilton Theorem

The characteristic polynomial of $A = \begin{bmatrix} 0 & 1 \\ -2 & 3 \end{bmatrix}$ is $p_A(x) = x^2 - 3x + 2$ (Example 7.1.8). This polynomial is used to find the eigenvalues of A. Also,

$$p_A(A) = \begin{bmatrix} 0 & 1 \\ -2 & 3 \end{bmatrix}^2 - 3 \begin{bmatrix} 0 & 1 \\ -2 & 3 \end{bmatrix} + 2 \begin{bmatrix} 1 & 0 \\ 0 & 1 \end{bmatrix} = \begin{bmatrix} 0 & 0 \\ 0 & 0 \end{bmatrix}.$$

Such an equation holds true for every square matrix. The result is credited to Arthur Cayley and William Rowan Hamilton.

■ **Theorem 7.1.16 [Cayley–Hamilton]**

$p_A(A) = 0$ for every square matrix A.

Proof

By Theorem 4.4.5, using the classical adjoint gives

$$(A - xI_n) \operatorname{adj}(A - xI_n) = [\det(A - xI_n)]I_n = p_A(x)I_n.$$

Because each entry of $\operatorname{adj}(A - xI_n)$ is calculated using the determinant of an $(n-1) \times (n-1)$ matrix as noted in Definition 4.4.4, these entries have degree at most $n - 1$. Therefore, by combining like terms, there exist $n \times n$ matrices B_i for $i = 1, 2, \dots, n - 1$ such that $\operatorname{adj}(A - xI_n) = \sum_{i=0}^{n-1} x^i B_i$. Because

$$(A - xI_n) \sum_{i=0}^{n-1} x^i B_i = \sum_{i=0}^{n-1} x^i A B_i - \sum_{i=0}^{n-1} x^{i+1} B_i = \sum_{i=0}^{n-1} (x^i A - x^{i+1} I_n) B_i,$$

it follows that

$$p_A(x)I_n = AB_0 + \sum_{i=1}^{n-1} x^i(AB_i - B_{i-1}) - x^n B_{n-1}. \tag{7.10}$$

Write $p_A(x) = c_0 + c_1 x + \cdots + c_{n-1} x^{n-1} + c_n x^n$, so

$$p_A(x)I_n = c_0 I_n + c_1 x I_n + \cdots + c_{n-1} x^{n-1} I_n + c_n x^n I_n. \tag{7.11}$$

Comparing the matrices on the powers of x in (7.10) and (7.11) implies that

$$
\begin{aligned}
AB_0 &= c_0 I_n, \\
AB_1 - B_0 &= c_1 I_n, \\
AB_2 - B_1 &= c_2 I_n, \\
&\vdots \\
AB_{n-1} - B_{n-2} &= c_{n-1} I_n, \\
-B_{n-1} &= c_n I_n.
\end{aligned}
\tag{7.12}
$$

Multiplying each of the equations of (7.12) by increasing powers of A gives

$$
\begin{aligned}
AB_0 &= c_0 I_n, \\
A^2 B_1 - AB_0 &= c_1 A, \\
A^3 B_2 - A^2 B_1 &= c_2 A^2, \\
&\vdots \\
A^n B_{n-1} - A^{n-1} B_{n-2} &= c_{n-1} A^{n-1}, \\
-A^n B_{n-1} &= c_n A^n.
\end{aligned}
\tag{7.13}
$$

Summing the matrices on the left side of the equations of (7.13) yields the zero matrix, and summing the matrices on the right side of those same equations gives $p_A(A)$. ∎

Exercises

1. Verify that the real number λ is an eigenvalue for the given matrix A with corresponding eigenvector $\mathbf{u}$.

(a) $A = \begin{bmatrix} 1 & 2 \\ 3 & 6 \end{bmatrix}$, $\lambda = 0$, $\mathbf{u} = \begin{bmatrix} -2 \\ 1 \end{bmatrix}$

(b) $A = \begin{bmatrix} 1 & 1 & 0 \\ 0 & 1 & 1 \\ 0 & 0 & 1 \end{bmatrix}$, $\lambda = 1$, $\mathbf{u} = \begin{bmatrix} 1 \\ 0 \\ 0 \end{bmatrix}$

(c) $A = \begin{bmatrix} 1 & 2 & 3 \\ 4 & 5 & 6 \\ 7 & 8 & 9 \end{bmatrix}$, $\lambda = 0$, $\mathbf{u} = \begin{bmatrix} 1 \\ -2 \\ 1 \end{bmatrix}$

(d) $A = \begin{bmatrix} 1 & 2 & 1 \\ 0 & 1 & 1 \\ 2 & 0 & 1 \end{bmatrix}$, $\lambda = 3$, $\mathbf{u} = \begin{bmatrix} 2 \\ 1 \\ 2 \end{bmatrix}$

2. Find all eigenvalues and corresponding eigenvectors for each matrix and identify its characteristic polynomial.

(a) $\begin{bmatrix} 5 & 0 \\ 2 & 5 \end{bmatrix}$

(b) $\begin{bmatrix} 4 & 3 \\ 0 & -1 \end{bmatrix}$

(c) $\begin{bmatrix} 3 & -12 \\ 4 & -11 \end{bmatrix}$

(d) $\begin{bmatrix} 2 & 2 \\ 3 & -2 \end{bmatrix}$

(e) $\begin{bmatrix} 2 & 1 & 1 \\ 0 & -2 & 0 \\ 1 & -4 & 2 \end{bmatrix}$

(f) $\begin{bmatrix} 5 & 0 & 0 \\ 4 & 5 & 0 \\ 3 & 9 & 5 \end{bmatrix}$

(g) $\begin{bmatrix} 1 & 0 & 0 \\ 5 & 8 & 7 \\ 9 & 0 & 2 \end{bmatrix}$

(h) $\begin{bmatrix} 2 & 4 & 9 \\ 0 & 5 & 7 \\ 0 & 0 & 2 \end{bmatrix}$

(i) $\begin{bmatrix} 7 & 0 & 8 \\ 3 & 3 & 7 \\ -4 & 0 & -5 \end{bmatrix}$

(j) $\begin{bmatrix} 3 & 12 & 0 \\ 0 & -3 & 0 \\ 5 & 7 & 2 \end{bmatrix}$

3. Find all eigenvalues and corresponding eigenvectors for each matrix and identify its characteristic polynomial.

(a) $\begin{bmatrix} 3 & 4 & 0 & 4 \\ 0 & -3 & 8 & 2 \\ 0 & 0 & 7 & 1 \\ 0 & 0 & 0 & -7 \end{bmatrix}$

(b) $\begin{bmatrix} 5 & 0 & 0 & 0 & 0 \\ 0 & -2 & 5 & 4 & 6 \\ 7 & 0 & 1 & 2 & 8 \\ 3 & 0 & 0 & -2 & 0 \\ 7 & 0 & 0 & 0 & 0 \end{bmatrix}$

(c) $\begin{bmatrix} 2 & 0 & 0 & 0 \\ 4 & -2 & 5 & 9 \\ 7 & 0 & 1 & 2 \\ 0 & 0 & 0 & -2 \end{bmatrix}$

(d) $\begin{bmatrix} 7 & 0 & 0 & 0 & 0 \\ 5 & 7 & 0 & 0 & 0 \\ 6 & 1 & 7 & 0 & 0 \\ 9 & 2 & 0 & 3 & 0 \\ 1 & 3 & 5 & 4 & 3 \end{bmatrix}$

4. Find all eigenvalues and corresponding eigenvectors of the standard matrix for a reflection in $\mathbb{R}^2$ (Definition 3.3.6) and a reflection in $\mathbb{R}^3$ (Definition 3.3.6).

5. Find the eigenvalues and corresponding eigenvectors of the standard matrix for a rotation in $\mathbb{R}^2$ (3.25) and a rotation in $\mathbb{R}^3$ (3.26–3.28).

6. Find all eigenvalues and corresponding eigenvectors of an identity matrix and a zero matrix.

7. Prove that an $n \times n$ matrix has at most n eigenvalues.

8. Let A be a 2×2 matrix. Prove that $p_A(x) = x^2 - (\text{tr } A)x + \det A$.

9. Use the method of Example 7.1.9 to evaluate.

(a) $\begin{bmatrix} 3 & 1 \\ 0 & 2 \end{bmatrix}^6 \begin{bmatrix} 2 \\ 4 \end{bmatrix}$

(b) $\begin{bmatrix} 3 & -4 & 3 \\ 0 & 2 & 0 \\ 0 & 1 & -2 \end{bmatrix}^{10} \begin{bmatrix} 1 \\ 2 \\ 3 \end{bmatrix}$

10. Find all complex eigenvalues and corresponding eigenvalues.

(a) $\begin{bmatrix} 1 & 3 \\ -1 & 2 \end{bmatrix}$ (b) $\begin{bmatrix} 1 & 2 & 1 \\ 0 & 1 & 1 \\ 2 & 0 & 1 \end{bmatrix}$ (c) $\begin{bmatrix} 1 & 2 & 1 & 0 \\ 0 & 1 & 1 & 0 \\ 2 & 0 & 1 & 1 \\ 0 & 0 & 0 & 1 \end{bmatrix}$

11. Given that $p_A(x)$ is the characteristic polynomial of A, find the characteristic polynomial of $-A$. Prove the result.

12. Find a 4×4 matrix A with the given characteristic polynomials.

    (a) $p_A(x) = x^4 - 2x^2 + 1$
    (b) $p_A(x) = x^4 + x^2 + 1$

13. Find the eigenspace for each eigenvalue for the matrices from Exercise 2. In addition, give the algebraic multiplicity and geometric multiplicity of each eigenvalue.

14. Let A be a 2×2 matrix with only real eigenvalues.

    (a) Find A so that it has only 1 eigenvalue.
    (b) Find A so that has 2 eigenvalues.
    (c) When does A have only 1 eigenvalue? Prove the result.

15. Find 3×3 matrices A_1, A_2, A_3 and $\lambda \in \mathbb{R}$ such that each matrix has only λ as an eigenvalue but the geometric multiplicity of A_i is i.

    (a) What is the characteristic polynomial of each matrix?
    (b) How does the algebraic multiplicity relate to the geometric multiplicity?

16. Let A be a 4×4 matrix. Let $\lambda_1, \lambda_2 \in \mathbb{R}$ enumerate its eigenvalues.

    (a) Explain why the geometric multiplicity of λ_1 is at least 1 and at most 3.
    (b) Find all possible characteristic polynomials of A for each of the possible geometric multiplicities of λ_1.
    (c) How would this problem change if A was a 5×5 matrix with $\lambda_1, \lambda_2, \lambda_3 \in \mathbb{R}$ as its only eigenvalues?

17. Prove (7.2).

18. Prove Theorem 7.1.13.

19. Prove that A is invertible if and only if $\lambda = 0$ is not an eigenvalue of A.

20. Let λ be an eigenvalue of the invertible matrix A. Prove that $1/\lambda$ is an eigenvalue of A^{-1}.

21. Demonstrate that λ is an eigenvalue of A if and only if λ is an eigenvalue of A^T.

22. Use two methods to prove that if $\mathbf{u}$ is an eigenvector of A, then $r\mathbf{u}$ is an eigenvector of A for all nonzero $r \in \mathbb{R}$.

23. Let A have eigenvalue λ with corresponding eigenvector $\mathbf{u}$. Prove that λ^2 is an eigenvalue of A^2 with corresponding eigenvector $\mathbf{u}$.

24. Is a linear combination of eigenvectors guaranteed to be an eigenvector? Explain.

25. Suppose that A is a square matrix such that $A^2 = 0$. Prove that $\lambda = 0$ is the only eigenvalue of A.

26. The linear operator $T : \mathbb{R}^n \to \mathbb{R}^n$ has **eigenvalue** λ with corresponding **eigenvector u** means that $T\mathbf{u} = \lambda\mathbf{u}$. Find the eigenvalues and corresponding eigenvectors of the given linear transformations.

(a) $T : \mathbb{R}^2 \to \mathbb{R}^2,\ T\begin{bmatrix} x \\ y \end{bmatrix} = \begin{bmatrix} 5x \\ 3y \end{bmatrix}$

(b) $T : \mathbb{R}^2 \to \mathbb{R}^2,\ T\begin{bmatrix} x \\ y \end{bmatrix} = \begin{bmatrix} 4x + 2y \\ -5x - 3y \end{bmatrix}$

(c) $T : \mathbb{R}^3 \to \mathbb{R}^3,\ T\begin{bmatrix} x \\ y \\ z \end{bmatrix} = \begin{bmatrix} -2x + 2y - z \\ 2x + y - 2z \\ -3x - 6y \end{bmatrix}$

(d) $T : \mathbb{R}^3 \to \mathbb{R}^3,\ T\begin{bmatrix} x \\ y \\ z \end{bmatrix} = \begin{bmatrix} 5x + 2y + 5z \\ 3y + z \\ 7z \end{bmatrix}$

27. Verify the result of the Cayley–Hamilton Theorem (7.1.16) using the matrices and characteristic polynomials in Exercise 2.

28. Explain why the following is not a proof of the Cayley–Hamilton Theorem: $p_A(A) = \det(A - AI_n) = \det(A - A) = \det 0 = 0$.

7.2 Minimal Polynomial

Let A be an $n \times n$ matrix. The set $\{I_n, A, A^2, \ldots, A^{n^2}\}$ is linearly dependent because $\dim \mathbb{R}^{n \times n} = n^2$ [Theorem 5.4.27(a)]. This implies that there exists a positive integer $k \le n^2$ such that $\{I_n, A, A^2, \ldots, A^{k-1}\}$ is linearly independent but $\{I_n, A, A^2, \ldots, A^k\}$ is linearly dependent. Therefore, there exists unique $a_0, a_1, \ldots, a_k \in \mathbb{R}$ where

$$A^k = a_{k-1}A^{k-1} + \cdots + a_2 A^2 + a_1 A + a_0 I_n. \tag{7.14}$$

Define the **minimal polynomial** of A to be the monic polynomial

$$m_A(x) = x^k - a_{k-1}x^{k-1} - \cdots - a_2 x^x - a_1 x - a_0. \tag{7.15}$$

Notice that the degree of $m_A(x)$ is at least 1, and

$$m_A(A) = 0. \tag{7.16}$$

■ **Example 7.2.1**

Find the minimal polynomial of $A = \begin{bmatrix} 3 & 0 \\ 0 & 5 \end{bmatrix}$. Because A is diagonal,

$$A^n = \begin{bmatrix} 3^n & 0 \\ 0 & 5^n \end{bmatrix},$$

so $\{I_2, A\}$ is linearly independent, but A^2 is a linear combination of I_2 and A. Specifically, $A^2 = 8A - 15I_2$ This implies that $m_A(x) = x^2 - 8x + 15$.

■ **Example 7.2.2**

Find the minimal polynomial of $B = \begin{bmatrix} 1 & 0 & 1 \\ 0 & 1 & 0 \\ 0 & 1 & 2 \end{bmatrix}$. Notice that

$$B^2 = \begin{bmatrix} 1 & 1 & 3 \\ 0 & 1 & 0 \\ 0 & 3 & 4 \end{bmatrix} \quad \text{and} \quad B^3 = \begin{bmatrix} 1 & 4 & 7 \\ 0 & 1 & 0 \\ 0 & 7 & 8 \end{bmatrix}.$$

Hence, $\{I_3, B, B^2\}$ is linearly independent, but $B^3 = 4B^2 - 5B + 2I_3$, so the minimal polynomial of B is $m_B(x) = x^3 - 4x^2 + 5x - 2$.

The proof of the key property of minimal polynomials requires some facts about polynomials learned in precalculus. Let $f(x), g(x) \in \mathbb{P}$. Recall, as with Definition 1.3.1, $g(x)$ **divides** $f(x)$ means that $g(x) \neq 0$ and there exists $q(x) \in \mathbb{P}$ such that $f(x) = g(x)q(x)$. For example, $x + 2$ divides $x^2 + 3x + 2$ because $x^2 + 3x + 2 = (x + 2)(x + 1)$. However, $x + 2$ does not divide $x^2 + 3x + 3$. Instead, when $x^2 + 3x + 3$ is divided by $x + 2$ there is a remainder of 1, and $x^2 + 3x + 3$ can be written as $(x + 2)(x + 1) + 1$. The theorem that guarantees that this can be done for any two polynomials in $\mathbb{P}$ is called the **Division Algorithm**. It states that given $f(x)$ and $g(x)$, there exists $q(x), r(x) \in \mathbb{P}$ such that

$$f(x) = g(x)q(x) + r(x)$$

and the degree of $r(x)$ [written as $\deg r(x)$] is less than the degree of $g(x)$ [written as $\deg g(x)$]. Notice, of course, that $g(x)$ divides $f(x)$ if $r(x) = 0$.

■ **Theorem 7.2.3**

Let A be an $n \times n$ matrix. If $f(x) \in \mathbb{P}$ such that $f(A) = 0$, then $m_A(x)$ divides $f(x)$.

Proof

Take $f(x) \in \mathbb{P}$ and assume that $f(A) = 0$. By the Division Algorithm there exists polynomials $q(x)$ and $r(x)$ such that the degree of $r(x)$ is less than the degree of $m_A(x)$, and

$$f(x) = m_A(x)q(x) + r(x).$$

Because $m_A(A) = 0$,

$$r(A) = f(A) - m_A(A)r(A) = 0. \qquad (7.17)$$

Write $r(x) = r_0 + r_1 x + \cdots + r_{k-1} x^{k-1}$. Because $\{I_n, A, A^2, \dots, A^{k-1}\}$ is linearly independent, (7.17) implies that $r_0 = r_1 = \cdots = r_{k-1} = 0$, proving that the remainder $r(x)$ is the zero polynomial. Hence, $m_A(x)$ divides $f(x)$. ∎

Let A be a square matrix with characteristic polynomial $p_A(x)$. The Cayley–Hamilton Theorem (7.1.16) states that $p_A(A) = 0$, so $m_A(x)$ divides $p_A(x)$ by Theorem 7.2.3. Now, suppose that $f(x)$ divides $p_A(x)$ and $f(A) = 0$. Theorem 7.2.3 implies that $m_A(x)$ also divides $f(x)$, and this means that the degree of $f(x)$ is at least the degree of $m_A(x)$. Putting these two quick points together leads to the corollary that explains how the minimal polynomial is minimal.

■ **Corollary 7.2.4**

For every square matrix A, the minimal polynomial $m_A(x)$ is the monic polynomial of least degree that divides the characteristic polynomial of A with the property $m_A(A) = 0$.

Let $A = \begin{bmatrix} 3 & 0 \\ 0 & 5 \end{bmatrix}$ from Example 7.2.1. Recall that the Cayley–Hamilton Theorem (7.1.16) states that $p_A(A) = 0$. Therefore, because the characteristic polynomial of A has degree 2 and the minimal polynomial must divide $p_A(x)$ by Theorem 7.2.3, the degree of $m_A(x)$ is at most 2. Hence, $m_A(x) = x^2 - 8x + 15$.

Next, consider $B = \begin{bmatrix} 1 & 1 & 0 \\ 0 & 1 & 0 \\ 0 & 0 & 6 \end{bmatrix}$. The eigenvalues of B are $\lambda = 1$ and $\lambda = 6$ (Theorem 7.1.13), with 1 having algebraic multiplicity 2. This means that the characteristic polynomial of B is $p_B(x) = (x - 1)^2(x - 6)$, and because the minimal polynomial divides the characteristic polynomial, it follows that the factorization of $m_B(x)$ into powers of prime polynomials of $\mathbb{P}$, the so-called **prime power decomposition**, contains only positive powers of $x - 1$ and powers of $x - 6$. Hence, $m_B(x) = (x - 1)(x - 6)$ or $m_B(x) = (x - 1)^2(x - 6)$. The reason this is always the case is the next result.

■ **Theorem 7.2.5**

For every $n \times n$ matrix A, the real number λ is a root of $p_A(x)$ if and only if λ is a root of $m_A(x)$.

Proof

Let λ be a root of $p_A(x)$. This means that there exists a nonzero vector $\mathbf{u} \in \mathbb{R}^n$ such that $A\mathbf{u} = \lambda \mathbf{u}$. Write $m_A(x) = a_k x^k + \cdots + a_2 x^2 + a_1 x + a_0$. Then,

$$\mathbf{0} = m_A(A)\mathbf{u} = (a_k x^k + a_{k-1} x^{k-1} + \cdots + a_1 x + a_0)\mathbf{u}$$
$$= a_k(A^k \mathbf{u}) + a_{k-1}(A^{k-1} \mathbf{u}) + \cdots + a_1(A\mathbf{u}) + a_0 \mathbf{u}$$

$$= a_k(\lambda^k \mathbf{u}) + a_{k-1}(\lambda^{k-1}\mathbf{u}) + \cdots + a_1(\lambda\mathbf{u}) + a_0\mathbf{u}$$

$$= (a_k\lambda^k + a_{k-1}\lambda^{k-1} + \cdots + a_1\lambda + a_0 I_n)\mathbf{u} = m_A(\lambda)\mathbf{u}.$$

Therefore, $m_A(\lambda) = 0$ because $\mathbf{u} \neq \mathbf{0}$. The converse holds because $m_A(x)$ divides $p_A(x)$. ∎

What does this mean for a square matrix A? There exists scalars $\lambda_1, \lambda_2, \dots, \lambda_k$ such that the characteristic polynomial of A can be written as

$$p_A(x) = r(x - \lambda_1)^{a_1}(x - \lambda_2)^{a_2} \cdots (x - \lambda_k)^{a_k}.$$

Since $m_A(x)$ is a monic polynomial that divides $p_A(x)$ and shares its roots, it must be the case that

$$m_A(x) = (x - \lambda_1)^{b_1}(x - \lambda_2)^{b_2} \cdots (x - \lambda_k)^{b_k}, \tag{7.18}$$

where $0 < b_i \leq a_i$ for all $i = 1, 2, \dots, k$, and the scalars b_i taken together are minimal so that $m_A(A) = 0$.

Invariant Subspaces

The $n \times n$ matrix A acts on the vectors from its eigenspace $\mathsf{ES}(A, \lambda)$ as the operation of scalar multiplication by λ. That is to say, if $\mathbf{u} \in \mathsf{ES}(A, \lambda)$, then $A\mathbf{u} = \lambda\mathbf{u}$. Consequently, when A operates on a vector from any of its eigenspaces, it always returns a vector from that same eigenspace. This is confirmed by calculating

$$A(A\mathbf{u}) = A(\lambda\mathbf{u}) = \lambda(A\mathbf{u}).$$

This notion is generalized next.

■ **Definition 7.2.6**

A subspace W of $\mathbb{R}^n$ is an **A-invariant subspace** for an $n \times n$ matrix A if $A\mathbf{u} \in W$ for all $\mathbf{u} \in W$.

From Example 7.1.8, $\begin{bmatrix} 1 \\ 2 \end{bmatrix}$ is an eigenvector for $A = \begin{bmatrix} 0 & 1 \\ -2 & 3 \end{bmatrix}$ with corresponding eigenvalue $\lambda = 2$. Then,

$$\begin{bmatrix} 0 & 1 \\ -2 & 3 \end{bmatrix} \begin{bmatrix} 1 \\ 2 \end{bmatrix} = \begin{bmatrix} 2 \\ 4 \end{bmatrix} \in \mathsf{ES}(A, 2),$$

confirming that the eigenspace is A-invariant. This has been proved directly, but because $\mathsf{ES}(A, 2) = \mathsf{NS}[f(A)]$ where $f(x) = x - 2$, the result can also be proved by an appeal to the next theorem.

■ **Theorem 7.2.7**

If A is a square matrix and $f(x) \in \mathbb{P}$, then $\mathsf{NS}[f(A)]$ is A-invariant.

Proof

Let $\mathbf{u} \in \text{NS}[f(A)]$. This means that $f(A)\mathbf{u} = \mathbf{0}$. Then, $A\mathbf{u} \in \text{NS}[f(A)]$ because Theorem 3.2.26(a) implies

$$f(A)(A\mathbf{u}) = [f(A)A]\mathbf{u} = [Af(A)]\mathbf{u} = A[f(A)\mathbf{u}] = A\mathbf{0} = \mathbf{0}. \blacksquare$$

Because $\text{ES}(A, \lambda) = \text{NS}[f(A)]$ where $f(x) = x - \lambda$, the corollary immediately follows.

■ Corollary 7.2.8

An eigenspace of the square matrix A is A-invariant.

Note that although every eigenspace is an invariant subspace of $\mathbb{R}^n$, the converse is false. For example, the trivial subspace $\{\mathbf{0}\}$ and the entire vector space are always invariant subspaces but are not always eigenspaces.

■ Example 7.2.9

Let $f(x) = x^2 + (2 + \sqrt{5})x$ and $A = \begin{bmatrix} -4 & 1 \\ 1 & 0 \end{bmatrix}$. Then,

$$f(A) = \begin{bmatrix} -4 & 1 \\ 1 & 0 \end{bmatrix}^2 + (2 + \sqrt{5}) \begin{bmatrix} -4 & 1 \\ 1 & 0 \end{bmatrix} = \begin{bmatrix} 9 - 4\sqrt{5} & -2 + \sqrt{5} \\ -2 + \sqrt{5} & 1 \end{bmatrix}.$$

Since the reduced row echelon form of $f(A)$ is $\begin{bmatrix} 1 & 2 + \sqrt{5} \\ 0 & 0 \end{bmatrix}$,

$$\text{NS}[f(A)] = \text{S}\left(\begin{bmatrix} 2 + \sqrt{5} \\ -1 \end{bmatrix} \right).$$

This null space is A-invariant by Theorem 7.2.7. As a check, compute

$$\begin{bmatrix} -4 & 1 \\ 1 & 0 \end{bmatrix} \begin{bmatrix} 2 + \sqrt{5} \\ -1 \end{bmatrix} = \begin{bmatrix} -9 - 4\sqrt{5} \\ 2 + \sqrt{5} \end{bmatrix} = (-2 - \sqrt{5}) \begin{bmatrix} 2 + \sqrt{5} \\ -1 \end{bmatrix}.$$

■ Example 7.2.10

Let $A = \begin{bmatrix} 2 & 0 & 4 \\ 0 & 2 & 7 \\ 0 & 0 & 3 \end{bmatrix}$. The eigenvalues of A are 2 and 3, and its characteristic polynomial is $p_A(x) = (x - 2)^2(x - 3)$. It can be shown that the eigenspace for $\lambda = 2$ is

$$\text{ES}(A, 2) = \text{S}\left(\begin{bmatrix} 1 \\ 0 \\ 0 \end{bmatrix}, \begin{bmatrix} 0 \\ 1 \\ 0 \end{bmatrix} \right).$$

To confirm Corollary 7.2.8, compute for all $r_1, r_2 \in \mathbb{R}$,

$$A(r_1 \mathbf{e}_1 + r_2 \mathbf{e}_2) = r_1 \begin{bmatrix} 2 \\ 0 \\ 0 \end{bmatrix} + r_2 \begin{bmatrix} 0 \\ 2 \\ 0 \end{bmatrix} \in \text{ES}(A, 2).$$

Generalized Eigenvectors

Consider the matrix,

$$A = \begin{bmatrix} 1 & 0 & 1 \\ 0 & 1 & 0 \\ 0 & 1 & 2 \end{bmatrix}. \tag{7.19}$$

The eigenvalues of A are 1 and 2 because

$$\det(A - \lambda I_3) = \begin{vmatrix} 1-\lambda & 0 & 1 \\ 0 & 1-\lambda & 0 \\ 0 & 1 & 2-\lambda \end{vmatrix} = (1-\lambda)^2(2-\lambda),$$

and the eigenspaces of A are

$$\text{ES}(A, 1) = \text{NS}(A - I_3) = \text{S}\left(\begin{bmatrix} 1 \\ 0 \\ 0 \end{bmatrix}\right) \quad \text{and} \quad \text{ES}(A, 2) = \text{NS}(A - 2I_3) = \text{S}\left(\begin{bmatrix} 1 \\ 0 \\ 1 \end{bmatrix}\right).$$

It is natural at this point to try to generalize the notions of eigenvalue and eigenvector by taking higher powers of $A - \lambda I_3$. For this reason, try computing

$$\det(A - \lambda I_3)^2 = [\det(A - \lambda I_3)]^2 = (1-\lambda)^4(2-\lambda)^2,$$

so Theorem 4.3.19 implies that generalizing eigenvalues will not lead to new values of λ. Furthermore,

$$\text{NS}(A - 2I_3)^2 = \text{S}\left(\begin{bmatrix} 1 \\ 0 \\ 1 \end{bmatrix}\right), \tag{7.20}$$

so again, no new vectors are obtained. However, because

$$(A - I_3)^2 = \begin{bmatrix} 0 & 1 & 1 \\ 0 & 0 & 0 \\ 0 & 1 & 1 \end{bmatrix},$$

it is evident that

$$\text{NS}(A - 2I_3) \subset \text{NS}(A - I_3)^2 = \text{S}\left(\begin{bmatrix} 1 \\ 0 \\ 0 \end{bmatrix}, \begin{bmatrix} 0 \\ -1 \\ 1 \end{bmatrix}\right). \tag{7.21}$$

Whether this path will prove fruitful is unclear at the moment, but at least there was a result, so some terminology is in order.

■ **Definition 7.2.11**

Let A be an $n \times n$ matrix with eigenvalue $\lambda \in \mathbb{R}$. The vector $\mathbf{u} \in \mathbb{R}^n$ is a **generalized eigenvector** of A corresponding to λ if there exists a positive integer k such that

$$(A - \lambda I_n)^k \mathbf{u} = 0.$$

The **generalized eigenspace** of A with respect to λ is

$$G(A, \lambda) = \bigcup_{i=0}^{\infty} \mathrm{NS}\,(A - \lambda I_n)^i.$$

Continuing work with the matrix A of (7.19), $\begin{bmatrix} 0 \\ -1 \\ 1 \end{bmatrix}$ is a generalized eigenvector of A corresponding to $\lambda = 1$ because of (7.21), but it is not an eigenvector of A. To find all generalized eigenspaces of A, continue to take powers of the appropriate matrices. However, (7.20) suggests that the null spaces of further powers of $A - 2I_3$ will equal $\mathrm{NS}\,(A - 2I_3) = \mathrm{NS}\,(A - 2I_3)^2$, and

$$\mathrm{NS}\,(A - I_3)^3 = \mathrm{S}\left(\begin{bmatrix} 1 \\ 0 \\ 0 \end{bmatrix}, \begin{bmatrix} 0 \\ -1 \\ 1 \end{bmatrix} \right)$$

suggests that the null spaces of further powers of $A - I_3$ will equal $\mathrm{NS}\,(A - I_3)^2$. This is evidence for

$$G(A, 1) = \mathrm{NS}\,(A - I_3) \cup \mathrm{NS}\,(A - I_3)^2 \cup \mathrm{NS}\,(A - I_3)^3 \cup \cdots = \mathrm{NS}\,(A - I_3)$$

and

$$G(A, 2) = \mathrm{NS}\,(A - 2I_3) \cup \mathrm{NS}\,(A - 2I_3)^2 \cup \mathrm{NS}\,(A - 2I_3)^3 \cup \cdots = \mathrm{NS}\,(A - 2I_3)^2.$$

This example is generalized with the next theorem.

■ **Theorem 7.2.12**

Let A be an $n \times n$ matrix with eigenvalue $\lambda \in \mathbb{R}$. There exists an integer $m \leq n$ such that

$$\mathrm{NS}\,(A - \lambda I_n) \subset \mathrm{NS}\,(A - \lambda I_n)^2 \subset \cdots \subset \mathrm{NS}\,(A - \lambda I_n)^m$$

and $\mathrm{NS}\,(A - \lambda I_n)^m = \mathrm{NS}\,(A - \lambda I_n)^k$ for all $k > m$.

Proof

Let $\mathbf{v} \in \mathbb{R}^n$ such that $(A - \lambda I_n)^i \mathbf{v} = \mathbf{0}$ for some positive integer i. Because

$$(A - \lambda I_n)^{i+1} \mathbf{v} = (A - \lambda I_n)(A - \lambda I_n)^i \mathbf{v} = \mathbf{0},$$

it follows that

$$\text{NS}(A - \lambda I_n)^i \subseteq \text{NS}(A - \lambda I_n)^{i+1}. \tag{7.22}$$

Let m be minimal so that

$$\text{NS}(A - \lambda I_n)^m = \text{NS}(A - \lambda I_n)^{m+1}.$$

This implies that there is a set of vectors $\mathscr{B} = \{\mathbf{u}_1, \mathbf{u}_2, \ldots, \mathbf{u}_m\} \subset \mathbb{R}^n$ such that

$$\mathbf{u}_1 \in \text{NS}(A - \lambda I_n),$$
$$\mathbf{u}_i \in \text{NS}(A - \lambda I_n)^i \setminus \text{NS}(A - \lambda I_n)^{i-1} \text{ if } 1 < i \le m.$$

Hence, (7.22) is a proper subset, and Exercise 13 of Section 5.3 implies that $\mathscr{B}$ is linearly independent, so $m \le n$ by Theorem 5.4.22(a). Also, the chain stabilizes at m. Prove this by induction on $i > 0$ by examining the null space of $(A - \lambda I_n)^{m+i}$. The minimality of m is the basis case of $i = 1$, so assume that

$$(A - \lambda I_n)^{m+i+1}\mathbf{v} = \mathbf{0}.$$

Then,

$$(A - \lambda I_n)^{m+i}(A - \lambda I_n)\mathbf{v} = \mathbf{0},$$

so by induction,

$$(A - \lambda I_n)\mathbf{v} \in \text{NS}(A - \lambda I_n)^{m+i} = \text{NS}(A - \lambda I_n)^m.$$

Therefore, $(A - \lambda I_n)^m(A - \lambda I_n)\mathbf{v} = \mathbf{0}$, which implies that

$$\mathbf{v} \in \text{NS}(A - \lambda I_n)^{m+1} = \text{NS}(A - \lambda I_n)^m. \blacksquare$$

Because the chain of null spaces in Theorem 7.2.12 stabilizes at a power not greater than n (this power is called the **index** of λ), the general eigenspace of A with respect to λ can be determined by simply finding the null space of $(A - \lambda I_n)^n$. This follows immediately from Theorem 7.2.12. The second corollary follows directly form Theorem 7.2.7.

■ **Corollary 7.2.13**

$G(A, \lambda) = \text{NS}(A - \lambda I_n)^n$ for every $n \times n$ matrix A with eigenvalue $\lambda \in \mathbb{R}$.

■ **Corollary 7.2.14**

A generalized eigenspace of a square matrix A is A-invariant.

Primary Decomposition Theorem

The minimal polynomial of a square matrix can be used to decompose $\mathbb{R}^n$ as a direct sum of its generalized eigenspaces. What is required to get this to work is a set of vectors formed by multiplying every vector in a subspace of $\mathbb{R}^n$ by a fixed square matrix. Since a matrix represents a linear transformation, the resulting set is basically the image of a set under a linear transformation as in Definition 1.4.32.

■ Definition 7.2.15

Define $A[W] = \{A\mathbf{u} : \mathbf{u} \in W\}$, where W is a subspace of $\mathbb{R}^n$ and A is an $n \times n$ matrix.

To illustrate Definition 7.2.15, consider the subspace of $\mathbb{R}^3$ given by

$$W = \left\{ \begin{bmatrix} a+b \\ a \\ b \end{bmatrix} : a, b \in \mathbb{R} \right\}.$$

Let $A = \begin{bmatrix} 1 & 1 & 0 \\ 0 & 1 & 1 \\ 1 & 0 & 1 \end{bmatrix}$ and multiply to find $A[W] = \left\{ \begin{bmatrix} 2a+b \\ a+b \\ a+2b \end{bmatrix} : a, b \in \mathbb{R} \right\}.$

There is another requirement for the decomposition. Recall that an $n \times n$ matrix A is an idempotent if $A^2 = A$. This was discussed in Example 3.2.23. Notice that if A is an idempotent, $I_n - A$ is also an idempotent. This is because

$$(I_n - A)^2 = I_n - 2A + A^2 = I_n - 2A + A = I_n - A.$$

Therefore, if an idempotent is found, there will always be another. Also, notice that

$$A(I_n - A) = A - A^2 = A - A = 0.$$

A pair of idempotents of the form A and $I_n - A$ satisfies the premises of the next lemma.

■ Lemma 7.2.16

Let A_i be nonzero $n \times n$ matrices for $i = 1, 2, \ldots, k$ such that

- $A_1 + A_2 + \cdots + A_k = I_n$,
- $A_i A_j = 0$ if $i \neq j$.

Then, A_i is idempotent for all $i = 1, 2, \ldots, k$, and

$$\mathbb{R}^n = A_1[\mathbb{R}^n] \oplus A_2[\mathbb{R}^n] \oplus \cdots \oplus A_k[\mathbb{R}^n].$$

Proof

The matrix A_i is an idempotent because the hypotheses imply

$$A_i = A_i I_n = A_i(A_1 + A_2 + \cdots + A_k) = A_i^2 + \sum_{i \neq j} A_i A_j = A_i^2.$$

To prove that $\mathbb{R}^n$ is the indicated direct sum, take $\mathbf{u} \in \mathbb{R}^n$. Then,

$$\mathbf{u} = I_n \mathbf{u} = (A_1 + A_2 + \cdots + A_k)\mathbf{u} = A_1 \mathbf{u} + A_2 \mathbf{u} + \cdots + A_k \mathbf{u}.$$

Lastly, to show that the sum is direct, take $\mathbf{v}_i \in \mathbb{R}^n$ with $i = 1, 2, \dots, k$ such that

$$A_i\mathbf{v}_i = \sum_{i \neq j} A_j\mathbf{v}_j.$$

Using the second hypothesis and the fact that A_i is an idempotent,

$$A_i\mathbf{v}_i = A_i^2\mathbf{v}_i = \sum_{i \neq j} A_i(A_j\mathbf{v}_i) = \sum_{i \neq j} (A_iA_j)\mathbf{v}_i = \mathbf{0}.$$

Therefore, $A_i[\mathbb{R}^n] \cap \sum_{i \neq j} A_j[\mathbb{R}^n] = \{\mathbf{0}\}.$ ■

■ **Example 7.2.17**

Take $A_1 = \begin{bmatrix} 1 & 0 \\ 1 & 0 \end{bmatrix}$ and $A_2 = \begin{bmatrix} 0 & 0 \\ -1 & 1 \end{bmatrix}$. Notice that

$$\begin{bmatrix} 1 & 0 \\ 1 & 0 \end{bmatrix} + \begin{bmatrix} 0 & 0 \\ -1 & 1 \end{bmatrix} = \begin{bmatrix} 1 & 0 \\ 0 & 1 \end{bmatrix} \text{ and } \begin{bmatrix} 1 & 0 \\ 1 & 0 \end{bmatrix}\begin{bmatrix} 0 & 0 \\ -1 & 1 \end{bmatrix} = \begin{bmatrix} 0 & 0 \\ 0 & 0 \end{bmatrix}.$$

Also, A_1 and A_2 are idempotents. To confirm that

$$\mathbb{R}^2 = A_1[\mathbb{R}^2] \oplus A_2[\mathbb{R}^2],$$

first observe that $A_1\mathbb{R}^2 = \mathsf{S}\left(\begin{bmatrix} 1 \\ 1 \end{bmatrix}\right)$ and $A_2\mathbb{R}^2 = \mathsf{S}\left(\begin{bmatrix} 0 \\ 1 \end{bmatrix}\right)$. Then,

$$A_1[\mathbb{R}^2] \cap A_2[\mathbb{R}^2] = \left\{\begin{bmatrix} 0 \\ 0 \end{bmatrix}\right\},$$

and for all $a, b \in \mathbb{R}$, $\begin{bmatrix} a \\ b \end{bmatrix} = a\begin{bmatrix} 1 \\ 1 \end{bmatrix} + (b - a)\begin{bmatrix} 0 \\ 1 \end{bmatrix}.$

As suggested by the polynomial version of the Division Algorithm, polynomials behave very much like integers. So much so, much of the terminology that is applied to the integers is also applied to polynomials. The term *divides* is one of those. Here are some more. Let $f_1(x), f_2(x), \dots, f_k(x) \in \mathbb{P}$. A **greatest common divisor** $g(x)$ of $f_1(x), f_2(x), \dots, f_k(x)$ is a polynomial in $\mathbb{P}$ such that $g(x)$ divides $f_i(x)$ for all $i = 1, 2, \dots, k$, and if $h(x) \in \mathbb{P}$ also divides each $f_i(x)$, then $h(x)$ divides $g(x)$. For example, a greatest common divisor of $(x - 1)^3(x - 2)^5$ and $(x - 1)^2(x - 2)^7$ is $(x - 1)^2(x - 2)^5$. Notice that greatest common divisors are not unique. That $a(x - 1)^2(x - 2)^5$ is also a greatest common divisor of $(x - 1)^3(x - 2)^5$ and $(x - 1)^2(x - 2)^7$ for all $a \in \mathbb{R}$ shows this. Once the greatest common divisor is in hand, an equation can be found that connects it to the polynomials that it divides. The result is due to Étienne Bézout.

■ **Lemma 7.2.18 [Bézout's Identity]**

Let $f_1(x), f_2(x), \ldots, f_k(x) \in \mathbb{P}$ have degree of at least 1 and greatest common divisor $g(x) \in \mathbb{P}$. There exists $a_1(x), a_2(x), \ldots, a_k(x) \in \mathbb{P}$ such that

$$a_1(x)f_1(x) + a_2(x)f_2(x) + \cdots + a_k(x)f_k(x) = g(x).$$

Proof

If $k = 1$, the result is trivially true. For the $k = 2$ case, define

$$S = \{u_1(x)f_1(x) + u_2(x)f_2(x) \, : \, u_1(x), u_2(x) \in \mathbb{P}\} \setminus \{0\}.$$

There exists $m(x) \in S$ of minimal degree (Exercise 5). Write

$$m(x) = a_1(x)f_1(x) + a_2(x)f_2(x)$$

for some $a_1(x), a_2(x) \in \mathbb{P}$. By the Division Algorithm, there are polynomials $q(x)$ and $r(x)$ so that $0 \leq \deg r(x) < \deg m(x)$ and $f_1(x) = q(x)m(x) + r(x)$. Therefore,

$$\begin{aligned} r(x) = f_1(x) - q(x)m(x) &= f_1(x) - q(x)[a_1(x)f_1(x) + a_2(x)f_2(x)] \\ &= [1 - q(x)a_1(x)]f_1(x) - q(x)a_2(x)f_2(x). \end{aligned}$$

Suppose that $r(x) \neq 0$. Then, $r(x) \in S$, which is impossible because $m(x)$ is defined to have minimal degree. Hence, $r(x) = 0$, and $m(x)$ divides $f_1(x)$. Similarly, $m(x)$ divides $f_2(x)$, so $m(x)$ divides $g(x)$. Furthermore, because $g(x)$ is a common divisor of $f_1(x)$ and $f_2(x)$, it follows that $g(x)$ divides $m(x)$ by Exercise 6, so $g(x) = rm(x)$ for some $r \in \mathbb{R}$, proving the $k = 2$ case. Complete the proof by induction (Exercise 7). ■

All of the machinery for the promised decomposition theorem for $\mathbb{R}^n$ is in place except for one. Polynomials $f_1(x), f_2(x), \ldots, f_k(x) \in \mathbb{P}$ are **relatively prime** if 1 is one of their greatest common divisors. For example, $(x - 1)^3(x - 2)^5$ and $(x - 1)^2(x - 2)^7$ are not relatively prime, but $(x - 1)^3(x - 2)^5$ and $(x - 3)^8(x - 4)^6$ are relatively prime. Using Bézout's Identity with relatively prime polynomials often proves to be a useful technique.

■ **Theorem 7.2.19 [Primary Decomposition Theorem]**

Let A be an $n \times n$ matrix. Let $m_A(x)$ be the minimal polynomial of A with prime power decomposition $m_A(x) = [p_1(x)]^{e_1}[p_2(x)]^{e_2} \cdots [p_k(x)]^{e_k}$. Then,

$$\mathbb{R}^n = \mathsf{NS}\,[p_1(A)]^{e_1} \oplus \mathsf{NS}\,[p_2(A)]^{e_2} \oplus \cdots \oplus \mathsf{NS}\,[p_k(A)]^{e_k}.$$

Proof

Define $f_i(x) = \prod_{j \neq i}[p_j(x)]^{e_j}$ for all $i = 1, 2, \ldots, k$. This implies that the poly-nomials $f_1(x), f_2(x), \ldots, f_k(x)$ are relatively prime polynomials in $\mathbb{P}$, so by Bézout's Identity (7.2.18), there exists $a_1(x), a_2(x), \ldots, a_k(x) \in \mathbb{P}$ such that

$$a_1(x)f_1(x) + a_2(x)f_2(x) + \cdots + a_k(x)f_k(x) = 1.$$

Applying Theorem 3.2.26 gives

$$a_1(A)f_1(A) + a_2(A)f_2(A) + \cdots + a_k(A)f_k(A) = I_n. \tag{7.23}$$

Further, suppose that $i \neq j$. This implies that $m_A(x)$ divides $f_i(x)f_j(x)$, so $f_i(x)f_j(x) = m_A(x)g(x)$ for some $g(x) \in \mathbb{P}$, and

$$\begin{aligned}[a_i(A)f_i(A)][a_j(A)f_j(A)] &= a_i(A)a_j(A)f_i(A)f_j(A) \\ &= a_i(A)a_j(A)m_A(A)g(A) = 0.\end{aligned} \tag{7.24}$$

(7.23) and (7.24) satisfy the hypotheses of Lemma 7.2.16. Hence,

$$\mathbb{R}^n = a_1(A)f_1(A)[\mathbb{R}^n] \oplus a_2(A)f_2(A)[\mathbb{R}^n] \oplus \cdots \oplus a_k(A)f_k(A)[\mathbb{R}^n].$$

It now suffices to prove that each of these summands is the appropriate null space. To accomplish this, first take $\mathbf{u} = a_i(A)f_i(A)\mathbf{v}$ with $\mathbf{v} \in \mathbb{R}^n$. Then,

$$[p_i(A)]^{e_i}\mathbf{u} = [p_i(A)]^{e_i}a_i(A)f_i(A)\mathbf{v} = a_i(A)m_A(A)\mathbf{v} = \mathbf{0}.$$

Conversely, suppose that $[p_i(A)]^{e_i}\mathbf{u} = \mathbf{0}$. By (7.23),

$$\mathbf{u} = a_1(A)f_1(A)\mathbf{u} + \cdots + a_i(A)f_i(A)\mathbf{u} + \cdots + a_k(A)f_k(A)\mathbf{u}.$$

When $j \neq i$, it follows that $[p_i(x)]^{e_i}$ divides $f_j(x)$, so $a_j(A)f_j(A)\mathbf{u} = \mathbf{0}$. Therefore, $\mathbf{u} = a_i(A)f_i(A)\mathbf{u}$, and it is proved that for all $i = 1, 2, \ldots, k$,

$$[a_i(A)f_i(A)]\mathbb{R}^n = \text{NS}\,[p_i(A)]^{e_i}. \blacksquare$$

Suppose that the real number λ is the only eigenvalue for the 3×3 matrix A, so its characteristic polynomial is $(x - \lambda)^3$. Assume that an examination of the general eigenspaces of A reveals that they stabilize at $m = 3$. That is,

$$\text{NS}\,(A - \lambda I_3) \subset \text{NS}\,(A - \lambda I_3)^2 \subset \text{NS}\,(A - \lambda I_3)^3 = G(A, \lambda).$$

As in the proof of Theorem 7.2.12, take

$$\begin{aligned}\mathbf{u}_1 &\in \text{NS}\,(A - \lambda I_3), \\ \mathbf{u}_2 &\in \text{NS}\,(A - \lambda I_3)^2 \setminus \text{NS}\,(A - \lambda I_3), \\ \mathbf{u}_3 &\in \text{NS}\,(A - \lambda I_3)^3 \setminus \text{NS}\,(A - \lambda I_3)^2,\end{aligned}$$

so $\{\mathbf{u}_1, \mathbf{u}_2, \mathbf{u}_3\}$ is a basis for $\mathbb{R}^3$. Take $\mathbf{v} \in \mathbb{R}^3 \setminus \{\mathbf{0}\}$ and write $\mathbf{v} = r_1\mathbf{u}_1 + r_2\mathbf{u}_2 + r_3\mathbf{u}_3$ for some $r_1, r_2, r_3 \in \mathbb{R}$. Let $f(x) = (x - \lambda)^3$. Then,

$$f(A)\mathbf{v} = (A - \lambda I_3)^3(r_1\mathbf{u}_1 + r_2\mathbf{u}_2 + r_3\mathbf{u}_3)$$
$$= r_1(A - \lambda I_3)^3\mathbf{u}_1 + r_2(A - \lambda I_3)^3\mathbf{u}_2 + r_3(A - \lambda I_3)^3\mathbf{u}_3 = \mathbf{0}$$

because $(A - \lambda I_3)^i\mathbf{u}_i = \mathbf{0}$ for $i = 1, 2, 3$. Therefore, $f(A) = 0$. Furthermore, $(A - \lambda I_3)^2 \neq 0$ because $\mathbf{u}_3 \notin \text{NS}(A - \lambda I_3)^2$, which implies that

$$(A - \lambda I_3)^2(\mathbf{u}_1 + \mathbf{u}_2 + \mathbf{u}_3) = \mathbf{u}_3 \neq \mathbf{0}.$$

All of this means that $f(x)$ is the minimal polynomial of A. This argument can be generalized (Exercise 28) to prove the connection between the minimal polynomial and the general eigenspaces.

■ Lemma 7.2.20

Let $\lambda_1, \lambda_2, \ldots, \lambda_k \in \mathbb{R}$ be the distinct eigenvalues of the $n \times n$ matrix A with indices $m_1, m_2, \ldots, m_k$, respectively. Then,

$$m_A(x) = (x - \lambda_1)^{m_1}(x - \lambda_2)^{m_2} \cdots (x - \lambda_k)^{m_k}.$$

Because $\text{NS}(A - \lambda I_n)^m = \text{NS}\, p(A)$ when $p(x) = (x - \lambda)^m$, Lemma 7.2.20 pairs with the Primary Decomposition Theorem (7.2.19) to immediately give the next theorem.

■ Theorem 7.2.21

Let A be an $n \times n$ matrix. If $\lambda_1, \lambda_2, \ldots, \lambda_k \in \mathbb{R}$ are the distinct eigenvalues of A, then

$$\mathbb{R}^n = \text{G}(A, \lambda_1) \oplus \text{G}(A, \lambda_2) \oplus \cdots \oplus \text{G}(A, \lambda_k).$$

■ Example 7.2.22

The minimal polynomial of $A = \begin{bmatrix} 1 & 0 & 1 \\ 0 & 1 & 0 \\ 0 & 1 & 2 \end{bmatrix}$ from Example 7.2.2 has prime power decomposition,

$$x^3 - 4x^2 + 5x - 2 = (x - 1)^2(x - 2).$$

Letting $p_1(x) = x - 1$ and $p_2(x) = x - 2$,

$$[p_1(A)]^2 = \begin{bmatrix} 0 & 0 & 1 \\ 0 & 0 & 0 \\ 0 & 1 & 1 \end{bmatrix}^2 = \begin{bmatrix} 0 & 1 & 1 \\ 0 & 0 & 0 \\ 0 & 1 & 1 \end{bmatrix} \quad \text{and} \quad p_2(A) = \begin{bmatrix} -1 & 0 & 1 \\ 0 & -1 & 0 \\ 0 & 1 & 0 \end{bmatrix}.$$

Because

$$\text{NS}\,[p_1(A)]^2 = \text{S}\left(\begin{bmatrix} 1 \\ 0 \\ 0 \end{bmatrix}, \begin{bmatrix} 0 \\ -1 \\ 1 \end{bmatrix}\right) \quad \text{and} \quad \text{NS}\,[p_2(A)] = \text{S}\left(\begin{bmatrix} 1 \\ 0 \\ 1 \end{bmatrix}\right),$$

the Primary Decomposition Theorem (7.2.19) implies that

$$\mathbb{R}^3 = S\left(\begin{bmatrix} 1 \\ 0 \\ 0 \end{bmatrix}, \begin{bmatrix} 0 \\ -1 \\ 1 \end{bmatrix}\right) \oplus S\left(\begin{bmatrix} 1 \\ 0 \\ 1 \end{bmatrix}\right).$$

■ **Example 7.2.23**

Let $A = \begin{bmatrix} 1 & 0 & 1 \\ 0 & 1 & 0 \\ 0 & 0 & 2 \end{bmatrix}$, a slight modification of the matrix from Example 7.2.22.

Observe that $p_A(x)$ is still $(x - 1)^2(x - 2)$, but $(A - I_3) = (A - I_3)^2$, so

$$NS(A - I_3) = G(A, 1) = S\left(\begin{bmatrix} 1 \\ 0 \\ 0 \end{bmatrix}, \begin{bmatrix} 0 \\ 1 \\ 0 \end{bmatrix}\right).$$

Also,

$$NS(A - 2I_3) = G(A, 2) = S\left(\begin{bmatrix} -1 \\ 0 \\ 1 \end{bmatrix}\right).$$

Therefore, $m_A(x) = (x - 1)(x - 2)$ by Lemma 7.2.20, and by Theorem 7.2.21,

$$\mathbb{R}^n = G(A, \lambda_1) \oplus G(A, \lambda_2) = S\left(\begin{bmatrix} 1 \\ 0 \\ 0 \end{bmatrix}, \begin{bmatrix} 0 \\ 1 \\ 0 \end{bmatrix}\right) \oplus S\left(\begin{bmatrix} -1 \\ 0 \\ 1 \end{bmatrix}\right).$$

Exercises

1. Find a greatest common divisor of the polynomials $(x - 1)^3(x + 2)^3(x - 3)^7$, $(x - 1)^2(x + 2)^4(x - 3)^6$, and $(x - 1)^5(x + 2)^4(x - 3)^5$, and then find another.

2. Show that $x^2 + 3x + 1$ divides $2x^4 + 5x^3 + 2x^2 + 8x + 3$.

3. Verify the Division Algorithm by finding polynomials $q(x)$ and $r(x)$ such that $3x^5 - x^4 + 6x^2 - x + 3 = (x^2 + 3)q(x) + r(x)$ with $\deg r(x) < 2$.

4. Find the prime power decomposition of $x^5 - 8x^4 + 14x^3 + 20x^2 - 39x - 36$.

5. If S is a nonempty set of positive integers, there exists $m \in S$ such that $m \leq n$ for all $n \in S$. This property is called the **Well Ordering Principle**, and the number m is the **minimum** of the set S. Prove the Well Ordering Principle using mathematical induction.

6. Let $f(x), g_1(x), g_2(x) \in \mathbb{P}$. Prove that if $f(x)$ divides $g_1(x)$ and $g_2(x)$, then $f(x)$ divides $u_1(x)g_1(x) + u_2(x)g_2(x)$ for all $u_1(x), u_2(x) \in \mathbb{P}$.

7. Finish the proof of Bézout's Identity (Theorem 7.2.18).

8. Show that $\begin{bmatrix} 1 & 0 \\ 0 & 1 \end{bmatrix}$ and $\begin{bmatrix} 1 & 1 \\ 0 & 1 \end{bmatrix}$ have the same characteristic polynomial but different minimal polynomials.

9. Find the minimal polynomial of each matrix, and use it to find the eigenvalues of each matrix.

(a) $\begin{bmatrix} 4 & 2 \\ 0 & 4 \end{bmatrix}$

(c) $\begin{bmatrix} 7 & 0 & 0 \\ 1 & 7 & 0 \\ 6 & 4 & 2 \end{bmatrix}$

(e) $\begin{bmatrix} 1 & 0 & 0 \\ 5 & 8 & 7 \\ 9 & 0 & 2 \end{bmatrix}$

(b) $\begin{bmatrix} 3 & 4 \\ 12 & 11 \end{bmatrix}$

(d) $\begin{bmatrix} 2 & 1 & 1 \\ 0 & -2 & 0 \\ 1 & -4 & 2 \end{bmatrix}$

(f) $\begin{bmatrix} 3 & 2 & 0 \\ 0 & 3 & 8 \\ 0 & 0 & 3 \end{bmatrix}$

10. Given that the characteristic polynomial of A is $p_A(x) = (x-3)^2(x+2)^3(x-1)$, what are the possible minimal polynomials of A?

11. Let A be an idempotent matrix. Find all possible minimal polynomials of A. What does this imply about the possible eigenvalues of A?

12. Suppose that A is an $n \times n$ matrix such that $A^2 = I_n$, a so-called **involutory matrix**. Find all possible minimal polynomials of A. What does this imply about the possible eigenvalues of A?

13. Prove that A is invertible if and only if $m_A(0) \neq 0$.

14. Let $f(x) \in \mathbb{P}$. Prove that if λ is an eigenvalue of A, then $f(\lambda)$ is an eigenvalue of $f(A)$.

15. Assume that A is a square matrix. Prove that $p_A(x) = m_A(x)$ if and only if the geometric multiplicity of every eigenvalue of A is 1.

16. Let A and B be square matrices. Prove that the minimal polynomial of $\begin{bmatrix} A & 0 \\ 0 & B \end{bmatrix}$ is $m_A(x)m_B(x)$. See Exercise 16 of Section 4.2.

17. Prove that each subspace W of $\mathbb{R}^n$ is A-invariant.

(a) $W = S\left(\begin{bmatrix} -1 \\ -5 \end{bmatrix}\right)$, $A = \begin{bmatrix} 2 & 1 \\ 10 & 5 \end{bmatrix}$

(b) $W = \left\{ \begin{bmatrix} 6x + 10y \\ 2x + 8y \\ x + 4y \end{bmatrix} : x, y \in \mathbb{R} \right\}$, $A = \begin{bmatrix} 6 & 5 & 1 \\ 2 & 4 & -2 \\ 1 & 2 & -1 \end{bmatrix}$

(c) $W = S\left(\begin{bmatrix} -1 \\ 1 \\ 1 \end{bmatrix}\right)$, $A = \begin{bmatrix} 6 & 5 & 1 \\ 2 & 4 & -2 \\ 1 & 2 & -1 \end{bmatrix}$

18. Prove that the null space of a square matrix A is A-invariant.

19. Prove that the column space of a square matrix A is A-invariant.

20. Let A be an $n \times n$ matrix and W_1 and W_2 be A-invariant subspaces of $\mathbb{R}^n$. Prove that both $W_1 \cap W_2$ and $W_1 + W_2$ are A-invariant.

21. Let A be a square matrix. Prove directly that $\mathbf{u}$ is an eigenvector of A if and only if $S(\mathbf{u})$ is A-invariant.

22. Let A be an $n \times n$ matrix and W be a subspace of the inner product space $\mathbb{R}^n$ with the dot product. Demonstrate that if W is A-invariant, $W^\perp$ is A^T-invariant. (*Hint:* Exercise 10 from Section 6.1.)

23. Show that each vector $\mathbf{u}$ is a generalized eigenvector of the matrix A.

    (a) $\mathbf{u} = \begin{bmatrix} 1 \\ 0 \end{bmatrix}$, $A = \begin{bmatrix} 2 & -1 \\ 1 & 4 \end{bmatrix}$

    (b) $\mathbf{u} = \begin{bmatrix} 2 \\ 3 \\ 0 \end{bmatrix}$, $A = \begin{bmatrix} 4 & 3 & 0 \\ 0 & 4 & 2 \\ 0 & 0 & 2 \end{bmatrix}$

24. Find the generalized eigenspace of the matrix A with respect to eigenvalue $\lambda = 1$ and give the index of $\lambda = 1$.

    (a) $A = \begin{bmatrix} 1 & 1 \\ 0 & 1 \end{bmatrix}$

    (b) $A = \begin{bmatrix} 1 & 0 \\ 0 & 1 \end{bmatrix}$

25. Find the generalized eigenspace of the matrix A with respect to eigenvalue $\lambda = 4$ and give the index of $\lambda = 4$.

    (a) $A = \begin{bmatrix} 4 & 1 & 0 \\ 0 & 4 & 1 \\ 0 & 0 & 4 \end{bmatrix}$

    (b) $A = \begin{bmatrix} 4 & 0 & 0 \\ 0 & 4 & 1 \\ 0 & 0 & 4 \end{bmatrix}$

    (c) $A = \begin{bmatrix} 4 & 0 & 0 \\ 0 & 4 & 0 \\ 0 & 0 & 4 \end{bmatrix}$

26. Given the $n \times n$ matrix A and subspace W of $\mathbb{R}^n$, find $A[W]$.

    (a) $A = \begin{bmatrix} 2 & 5 \\ 1 & 0 \end{bmatrix}$, $W = \left\{ \begin{bmatrix} 2x + 4y \\ x - y \end{bmatrix} : x, y \in \mathbb{R} \right\}$

    (b) $A = \begin{bmatrix} 3 & 1 \\ 6 & 2 \end{bmatrix}$, $W = \left\{ \begin{bmatrix} x + y \\ 3x + 3y \end{bmatrix} : x, y \in \mathbb{R} \right\}$

    (c) $A = \begin{bmatrix} 3 & 0 & 0 \\ 0 & 5 & 0 \\ 0 & 0 & -1 \end{bmatrix}$, $W = \left\{ \begin{bmatrix} 3x + 2y \\ 4x \\ 6y \end{bmatrix} : x, y \in \mathbb{R} \right\}$

27. Let $T : \mathbb{R}^n \to \mathbb{R}^n$ be a linear transformation. Prove that $T[W] = [T][W]$ for every subspace W of $\mathbb{R}^n$.

28. Prove Lemma 7.2.20.

29. Write $\mathbb{R}^n$ as the direct sum of generalized eigenspaces of the given matrices.

    (a) $\begin{bmatrix} 5 & 8 & 7 \\ 0 & 5 & 0 \\ 0 & -4 & 5 \end{bmatrix}$

    (b) $\begin{bmatrix} 3 & 0 & 0 \\ 1 & 3 & 0 \\ 2 & 0 & 3 \end{bmatrix}$

    (c) $\begin{bmatrix} 4 & 1 & 2 \\ 0 & 6 & 0 \\ 0 & 4 & 6 \end{bmatrix}$

30. Find a basis of $\mathbb{R}^n$ consisting of generalized eigenvectors of the given matrices.

    (a) $\begin{bmatrix} 5 & 1 & 4 & 7 \\ 0 & 5 & 0 & 0 \\ 0 & 4 & 1 & 0 \\ 0 & 8 & 0 & 1 \end{bmatrix}$

    (b) $\begin{bmatrix} 0 & 1 & 1 & 1 \\ 1 & 0 & 1 & 1 \\ 1 & 1 & 0 & 1 \\ 1 & 1 & 1 & 0 \end{bmatrix}$

    (c) $\begin{bmatrix} 4 & 1 & 0 & 0 \\ 0 & 4 & 1 & 0 \\ 0 & 0 & 4 & 1 \\ 0 & 0 & 0 & 4 \end{bmatrix}$

31. Let $T : \mathbb{R}^n \to \mathbb{R}^n$ be a linear operator. Take $S \subseteq \mathbb{R}^n$. Define the **restriction** of T to S to be the function $T \upharpoonright S : S \to \mathbb{R}^n$ such that $(T \upharpoonright S)\mathbf{u} = T\mathbf{u}$ for all $\mathbf{u} \in S$. Prove that $T \upharpoonright G([T], \lambda) : G([T], \lambda) \to G([T], \lambda)$ is a linear transformation.

32. Let V be an n-dimensional vector space and $T : V \to V$ be a linear operator. Take $\lambda \in \mathbb{R}$ to be an eigenvalue of T (Exercise 26 of Section 7.1). Find subspaces W_1 and W_2 of V such that W_1 is isomorphic to $G([T], \lambda)$ and $V = W_1 \oplus W_2$.

7.3   Similar Matrices

Consider the 2×2 matrix $A = \begin{bmatrix} 1 & 1 \\ 2 & 0 \end{bmatrix}$. It is the standard matrix of the linear trans-

formation $T : \mathbb{R}^2 \to \mathbb{R}^2$ defined by $T\mathbf{x} = A\mathbf{x}$ for all $\mathbf{x} \in \mathbb{R}^2$. Let $\mathcal{B} = \left\{ \begin{bmatrix} 1 \\ 0 \end{bmatrix}, \begin{bmatrix} 1 \\ 1 \end{bmatrix} \right\}$

be an ordered basis and perform a change of basis to rewrite T in terms of the new basis. Referencing (5.42), the standard matrix $\mathcal{B} \to \mathcal{B}$ for T is

$$[T]_{\mathcal{B}\mathcal{B}} = [T_{\mathcal{B}}][T][T_{\mathcal{B}}^{-1}],$$

where by Theorem 5.6.21,

$$[T_{\mathcal{B}}^{-1}] = \begin{bmatrix} 1 & 1 \\ 0 & 1 \end{bmatrix}.$$

Since $[T_{\mathcal{B}}^{-1}]$ is easy to find because its columns are simply the vectors of $\mathcal{B}$, it is traditional to replace $[T_{\mathcal{B}}^{-1}]$ with P and write

$$P = \begin{bmatrix} 1 & 1 \\ 0 & 1 \end{bmatrix}.$$

It is worth emphasizing that

$$P\mathbf{e}_1 = \begin{bmatrix} 1 \\ 0 \end{bmatrix} \quad \text{and} \quad P\mathbf{e}_2 = \begin{bmatrix} 1 \\ 1 \end{bmatrix}.$$

This means, following Example 5.6.24, that P is a change in basis matrix because P is the standard matrix of the identity transformation on $\mathbb{R}^2$ composed with $T_{\mathcal{B}}^{-1}$ Moreover, it follows that

$$P^{-1} = [T_{\mathcal{B}}] = \begin{bmatrix} 1 & -1 \\ 0 & 1 \end{bmatrix},$$

so P^{-1} is the standard matrix for the coordinate map that maps the new basis $\mathcal{B}$ back to the standard basis,

$$P^{-1} \begin{bmatrix} 1 \\ 0 \end{bmatrix} = \mathbf{e}_1 \quad \text{and} \quad P^{-1} \begin{bmatrix} 1 \\ 1 \end{bmatrix} = \mathbf{e}_2.$$

Therefore, because $[T] = A$,

$$B = P^{-1}AP = \begin{bmatrix} -1 & 0 \\ 2 & 2 \end{bmatrix}$$

is the standard matrix $\mathcal{B} \rightarrow \mathcal{B}$ of the linear transformation with standard matrix A. The matrices A and $B = P^{-1}AP$ describe the same linear transformation, only written using different bases. Since matrices are linear transformations, it is reasonable to slightly modify the diagram of Figure 5.6 and observe that the following diagram commutes:

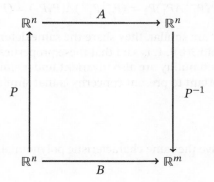

The relationship just described between the matrices A and B will play an important role throughout the rest of this book.

■ Definition 7.3.1

If A and B are $n \times n$ matrices, A is **similar** to B means that there exists an invertible matrix P such that $P^{-1}AP = B$.

The matrices $B = \begin{bmatrix} 1 & 1 \\ 2 & 0 \end{bmatrix}$ and $A = \begin{bmatrix} -1 & 0 \\ 2 & 2 \end{bmatrix}$ were shown to be similar using $P = \begin{bmatrix} 1 & 1 \\ 0 & 1 \end{bmatrix}$. Also, notice that

$$A = PBP^{-1} \quad \text{and} \quad AP = PB,$$

which can always be written when A and B are similar.

Since similar matrices represent the same linear transformation up to a change of basis, this means that similar matrices are equal in some sense as with Theorem 5.6.6.

■ Theorem 7.3.2

Let A, B, and C be $n \times n$ matrices.

(a) A is similar to A.

(b) If A is similar to B, then B is similar to A.

(c) If A is similar to B and B is similar to C, then A is similar to C.

Proof

It is obvious that every matrix is similar to itself. Simply take P to be the identity matrix. If A is similar to B, say by $B = P^{-1}AP$, then B is similar to A since $A = (P^{-1})^{-1}BP^{-1}$. Finally, suppose that A is similar to B and B is similar to C so that $B = P_1^{-1}AP_1$ and $C = P_2^{-1}BP_2$. Then, A is similar to C because

$$C = P_2^{-1}BP_2 = P_2^{-1}(P_1^{-1}AP_1)P_2 = (P_2^{-1}P_1^{-1})A(P_1P_2) = (P_1P_2)^{-1}A(P_1P_2). \blacksquare$$

If matrices A and B are similar, they share the same determinant and the same trace [Exercises 3(a) and 3(b)]. It is said that these properties are **invariant** under similarity. The rank and nullity are also invariant under similarity [Exercises 3(c) and 3(d)]. More important to present concerns is that similar matrices will share eigenvalues.

■ **Theorem 7.3.3**

Similar matrices have the same characteristic polynomial.

Proof

Let A and B be $n \times n$ matrices. Suppose that A is similar to B so that there is an invertible matrix P with $A = PBP^{-1}$. By Theorem 4.3.19, it follows that

$$
\begin{aligned}
p_A(x) &= \det(A - xI_n) \\
&= \det(PBP^{-1} - xPI_nP^{-1}) \\
&= \det[P(B - xI_n)P^{-1}] \\
&= \det P \det(B - xI_n) \det P^{-1} \\
&= \det P \det P^{-1} \det(B - xI_n) \\
&= \det(B - xI_n) = p_B(x). \blacksquare
\end{aligned}
$$

The next result is a corollary to Theorem 7.3.3 because the eigenvalues of a matrix are exactly the roots of the characteristic polynomial. However, the given proof does not rely on Theorem 7.3.3 and is instructive.

■ **Corollary 7.3.4**

Similar matrices have the same eigenvalues.

Proof

Suppose that the $n \times n$ matrices A and B are similar so that $A = PBP^{-1}$ for some invertible matrix P. Let $\lambda \in \sigma(A)$ with $A\mathbf{u} = \lambda\mathbf{u}$ for some eigenvector $\mathbf{u} \in \mathbb{R}^n$. Therefore, $PBP^{-1}\mathbf{u} = \lambda\mathbf{u}$. Applying P^{-1} to both sides yields

$BP^{-1}\mathbf{u} = \lambda P^{-1}\mathbf{u}$ so that $P^{-1}\mathbf{u}$ is an eigenvector for B with corresponding eigenvalue λ. This shows that $\sigma(A) \subseteq \sigma(B)$. The proof that $\sigma(B) \subseteq \sigma(A)$ follows by reversing the roles of A and B. ■

Schur's Lemma

It is a known fact that a given matrix A is the standard matrix for a linear transformation. It is desirable that A is a simple matrix so that working with it is relatively easy, such as when performing calculations, but there is no reason to expect that A will be nice. This is where similar matrices come into play. The goal is to find a basis so that when the standard matrix of the coordinate map P^{-1} is applied, the resulting matrix $P^{-1}AP$ is as simple as possible. Since both A and $P^{-1}AP$ represent the same linear transformation, work with the nicer $P^{-1}AP$.

There are many forms of a matrix that are considered nice. The first is due to Issai Schur and uses a particular type of coordinate map. Matrices A and B are **orthogonally similar** if there exists an orthogonal P such that $B = P^{-1}AP$. The significance of this is that the coordinate maps given by P and its inverse are isometries (Corollary 4.4.19), so that the functions that change basis do so without distorting segments and angles.

■ Lemma 7.3.5 [Schur]

Let A be an $n \times n$ matrix with eigenvalues $\lambda_1, \lambda_2, \dots, \lambda_n \in \mathbb{R}$. Then, A is orthogonally similar to an upper triangular matrix with the eigenvalues of A as the entries of its main diagonal.

Proof

Proceed by induction on k. The basis case holds because a 1×1 matrix is upper triangular. Suppose that A is an $(k+1) \times (k+1)$ matrix and let λ_1 have corresponding eigenvector $\mathbf{u}$ with $\|\mathbf{u}\| = 1$. Let $\{\mathbf{u}, \mathbf{v}_1, \dots, \mathbf{v}_{k-1}\}$ be an orthonormal basis of $\mathbb{R}^n$, using the Gram–Schmidt process (Theorem 6.2.21) to find it if necessary. Define the orthogonal matrix,

$$P_1 = \begin{bmatrix} \mathbf{u} & \mathbf{v}_1 & \cdots & \mathbf{v}_{k-1} \end{bmatrix},$$

and then let $B = P_1^T A P_1$. That is,

$$P_1 B = A P_1 \tag{7.25}$$

and

$$B = P_1^T \begin{bmatrix} A\mathbf{u} & A\mathbf{v}_1 & \cdots & A\mathbf{v}_{n-1} \end{bmatrix} = P_1^T \begin{bmatrix} \lambda_1\mathbf{u} & A\mathbf{v}_1 & \cdots & A\mathbf{v}_{n-1} \end{bmatrix}.$$

This implies that there exists a $k \times k$ matrix C such that B can be viewed as the block matrix. Using $*$ to represent an appropriately sized block of random entries,

$$B = \begin{bmatrix} \lambda_1 & * \\ 0 & C \end{bmatrix}.$$

Before the induction hypothesis is applied to C, note that the eigenvalues of C are $\lambda_2, \lambda_3, \dots, \lambda_{k+1}$. This follows because

$$\det(B - \lambda I_{k+1}) = (\lambda_1 - \lambda)\det(C - \lambda I_k)$$

and

$$
\begin{aligned}
\det(B - I\lambda) &= \det(P_1^T A P_1 - \lambda I_{k+1}) \\
&= \det(P_1^T A P - P^T(\lambda I_{k+1})P_1) \\
&= \det(P_1^T(A - \lambda I_{k+1})P_1) \\
&= \det(A - \lambda I_{k+1}) = (\lambda_1 - \lambda)(\lambda_2 - \lambda) \cdots (\lambda_{k+1} - \lambda).
\end{aligned}
$$

Thus, by induction there exists a $k \times k$ orthogonal P_2 and $k \times k$ matrix U such that $U = P_2^T C P_2$ is upper triangular with diagonal entries of $\lambda_2, \lambda_3, \dots, \lambda_{k+1}$. It should be emphasized that

$$P_2 U = C P_2. \tag{7.26}$$

To finish the proof, let

$$P = P_1 \begin{bmatrix} 1 & 0 \\ 0 & P_2 \end{bmatrix},$$

which is orthogonal because it is the product of orthogonal matrices (Exercise 16 of Section 4.4), so by (7.25) and (7.26),

$$
\begin{aligned}
AP = AP_1 \begin{bmatrix} 1 & 0 \\ 0 & P_2 \end{bmatrix} \\
= P_1 B \begin{bmatrix} 1 & 0 \\ 0 & P_2 \end{bmatrix} \\
= P_1 \begin{bmatrix} B\mathbf{e}_1 & B \begin{bmatrix} 0 \\ P_2 \end{bmatrix} \end{bmatrix} \\
= P_1 \begin{bmatrix} \lambda_1 \mathbf{e}_1 & \begin{bmatrix} \lambda_1 & * \\ 0 & C \end{bmatrix}\begin{bmatrix} 0 \\ P_2 \end{bmatrix} \end{bmatrix} \\
= P_1 \begin{bmatrix} \lambda_1 \mathbf{e}_1 & \begin{bmatrix} * \\ C P_2 \end{bmatrix} \end{bmatrix} \\
= P_1 \begin{bmatrix} \lambda_1 \mathbf{e}_1 & \begin{bmatrix} * \\ P_2 U \end{bmatrix} \end{bmatrix} \\
= P_1 \begin{bmatrix} 1 & 0 \\ 0 & P_2 \end{bmatrix}\begin{bmatrix} \lambda_1 & * \\ 0 & U \end{bmatrix} = P \begin{bmatrix} \lambda_1 & * \\ 0 & U \end{bmatrix}, \tag{7.27}
\end{aligned}
$$

where the last matrix of (7.27) is upper triangular. ∎

■ **Example 7.3.6**

The matrix of Example 7.1.11, $A = \begin{bmatrix} 1 & 1 & 1 \\ 0 & 1 & 2 \\ 2 & 1 & 0 \end{bmatrix}$, has eigenvalues -1, 0, and

3 with corresponding eigenvectors, $\begin{bmatrix} 0 \\ 1 \\ -1 \end{bmatrix}$, $\begin{bmatrix} 1 \\ -2 \\ 1 \end{bmatrix}$, and $\begin{bmatrix} 1 \\ 1 \\ 1 \end{bmatrix}$, respectively. Use

the procedure of the proof of Schur's Lemma (7.3.5) to find an upper triangular matrix to which A is orthogonally similar. First, the matrix P_1 is orthogonal, where

$$P_1 = \begin{bmatrix} 0 & 1 & 0 \\ \sqrt{2}/2 & 0 & \sqrt{2}/2 \\ -\sqrt{2}/2 & 0 & \sqrt{2}/2 \end{bmatrix}.$$

Because

$$P_1^{-1} = P_1^T = \begin{bmatrix} 0 & \sqrt{2}/2 & -\sqrt{2}/2 \\ 1 & 0 & 0 \\ 0 & \sqrt{2}/2 & \sqrt{2}/2 \end{bmatrix},$$

it follows that

$$P_1^T A P_1 = \begin{bmatrix} -1 & -\sqrt{2} & 1 \\ 0 & 1 & \sqrt{2} \\ 0 & \sqrt{2} & 2 \end{bmatrix}.$$

Let $C = \begin{bmatrix} 1 & \sqrt{2} \\ \sqrt{2} & 2 \end{bmatrix}$, so $P_1^T A P_1 = \begin{bmatrix} -1 & * \\ 0 & C \end{bmatrix}$. Because

$$\begin{vmatrix} 1-\lambda & \sqrt{2} \\ \sqrt{2} & 2-\lambda \end{vmatrix} = (1-\lambda)(2-\lambda) - 2 = \lambda(\lambda - 3),$$

the eigenvalues of C are $\lambda = 0$ and $\lambda = 3$, which are the unused eigenvalues of A. It can be shown that $\begin{bmatrix} -\sqrt{6}/3 \\ \sqrt{3}/3 \end{bmatrix}$ is an eigenvector corresponding to $\lambda = 0$ of C. Therefore,

$$P_2 = \begin{bmatrix} -\sqrt{6}/3 & \sqrt{3}/3 \\ \sqrt{3}/3 & \sqrt{6}/3 \end{bmatrix}$$

is a desired matrix for C. Specifically, because

$$P_2^T = \begin{bmatrix} -\sqrt{6}/3 & \sqrt{3}/3 \\ \sqrt{3}/3 & \sqrt{6}/3 \end{bmatrix},$$

it follows that

$$P_2^T C P_2 = \begin{bmatrix} 0 & 0 \\ 0 & 3 \end{bmatrix},$$

which is upper triangular with the eigenvalues of C along the main diagonal. This work implies that a desired change of basis matrix is

$$P = P_1 \begin{bmatrix} 1 & 0 & 0 \\ 0 & -\sqrt{6}/3 & \sqrt{3}/3 \\ 0 & \sqrt{3}/3 & \sqrt{6}/3 \end{bmatrix} = \begin{bmatrix} 0 & -\sqrt{6}/3 & \sqrt{3}/3 \\ \sqrt{2}/2 & \sqrt{6}/6 & \sqrt{3}/3 \\ -\sqrt{2}/2 & \sqrt{6}/6 & \sqrt{3}/3 \end{bmatrix},$$

so the final result is the upper triangular matrix,

$$U = P^T A P = \begin{bmatrix} -1 & \sqrt{3} & 0 \\ 0 & 0 & 0 \\ 0 & 0 & 3 \end{bmatrix}, \tag{7.28}$$

which has the eigenvalues of A along the main diagonal. Notice that the equation in (7.28) can be rewritten as $A = PUP^T$. This decomposition of A is called a **Schur factorization**.

Block Diagonal Form

Consider the matrix

$$B = \begin{bmatrix} 1 & 2 & 0 & 0 & 0 \\ 3 & 1 & 0 & 0 & 0 \\ 0 & 0 & 3 & 2 & 1 \\ 0 & 0 & 1 & 0 & 4 \\ 0 & 0 & 2 & 3 & 1 \end{bmatrix}. \tag{7.29}$$

Under the right condition, this matrix is similar to a 5×5 matrix with the same entries except that the 0 block in the upper right corner can be replaced with any 2×3 matrix. The next lemma gives the condition.

■ **Lemma 7.3.7**

Let A, B, and C be matrices such that the size of A is $n \times n$, the size of B is $n \times m$, and the size of C is $m \times m$. If A and C have no common eigenvalues,

$$\begin{bmatrix} A & B \\ 0 & C \end{bmatrix} \text{ is similar to } \begin{bmatrix} A & 0 \\ 0 & C \end{bmatrix}.$$

Proof

Suppose that $\sigma(A) \cap \sigma(C) = \varnothing$. Let X be an arbitrary $n \times m$ matrix and define the $(n + m) \times (n + m)$ matrix,

$$P = \begin{bmatrix} I_n & X \\ 0 & I_m \end{bmatrix} \text{ so that } P^{-1} = \begin{bmatrix} I_n & -X \\ 0 & I_m \end{bmatrix}.$$

Notice that the zero matrix in the lower left corner is $m \times n$. Now, find the matrix PAP^{-1} by computing

$$\begin{bmatrix} I_n & X \\ 0 & I_m \end{bmatrix}\begin{bmatrix} A & 0 \\ 0 & C \end{bmatrix}\begin{bmatrix} I_n & -X \\ 0 & I_m \end{bmatrix} = \begin{bmatrix} A & XC \\ 0 & C \end{bmatrix}\begin{bmatrix} I_n & -X \\ 0 & I_m \end{bmatrix} = \begin{bmatrix} A & XC - AX \\ 0 & C \end{bmatrix}.$$

To prove the theorem it is required to find X so that $B = XC - AX$. Define the linear transformation $T : \mathbb{R}^{n \times m} \rightarrow \mathbb{R}^{n \times m}$ by $TX = XC - AX$. To show that there exists the required X, it is sufficient to demonstrate that T is onto. However, in order to accomplish this, by Corollary 5.5.4 it is enough to show that T is one-to-one, so assume that $XC - AX = 0$. Let $p_A(x)$ be the characteristic polynomial of A. By Exercise 16,

$$0 = p_A(A)X = Xp_A(C). \tag{7.30}$$

Let $\lambda_1, \lambda_2, \ldots, \lambda_k$ be the eigenvalues of A. This implies that

$$p_A(C) = (C - \lambda_1 I_m)(C - \lambda_2 I_m) \cdots (C - \lambda_k I_m).$$

Because A and C do not share eigenvalues, for every $i = 1, 2, \ldots, k$,

$$\det(C - \lambda_i I_m) \neq 0,$$

so $p_A(C)$ is invertible. Hence, (7.30) implies that $X = 0$, from which follows that T is one-to-one by Theorem 5.2.20. ∎

Lemma 7.3.7 implies that the matrix

$$A = \begin{bmatrix} 1 & 2 & 3 & 4 & 5 \\ 3 & 1 & 6 & 7 & 8 \\ 0 & 0 & 3 & 2 & 1 \\ 0 & 0 & 1 & 0 & 4 \\ 0 & 0 & 2 & 3 & 1 \end{bmatrix}$$

is similar to matrix B of (7.29) because the upper left block does not have any eigenvalues in common with the lower right block. It is expected that mathematical induction would enable similarity results using matrices with any number of diagonal blocks. This is where Schur's Lemma (7.3.5) comes into play.

■ Theorem 7.3.8

Let A be a square matrix and $\lambda_1, \lambda_2, \ldots, \lambda_k \in \mathbb{R}$ enumerate the distinct eigenvalues of A. Then, A is similar to a block diagonal matrix of the form

$$\begin{bmatrix} B_1 & 0 & 0 & \cdots & 0 \\ 0 & B_2 & 0 & \cdots & 0 \\ 0 & 0 & B_3 & \cdots & 0 \\ \vdots & \vdots & & \ddots & \vdots \\ 0 & 0 & 0 & \cdots & B_k \end{bmatrix},$$

where for all $i = 1, 2, \ldots, k$, the submatrix B_i is upper triangular with each entry along its main diagonal equal to λ_i.

Proof

By following the proof Schur's Lemma in a particular order, A is similar to an upper triangular matrix B such that

$$B = \begin{bmatrix} \lambda_1 & * & * & * & * & * \\ 0 & \lambda_1 & * & * & * & * \\ & & \ddots & & & \\ 0 & 0 & \lambda_1 & * & * & * \\ & & & \ddots & & \\ 0 & 0 & 0 & \lambda_k & * & * \\ 0 & 0 & 0 & 0 & \lambda_k & * \\ & & & & & \ddots \\ 0 & 0 & 0 & 0 & 0 & \lambda_k \end{bmatrix}.$$

Proceed by induction on k, the number of eigenvalues of A. If $k = 1$, the theorem holds immediately, so assume that $k > 1$. Write

$$B = \begin{bmatrix} A & * \\ 0 & C \end{bmatrix},$$

where A is upper triangular with each entry on its main diagonal equal to λ_1 and C is upper triangular with the remaining eigenvalues on its main diagonal. Let n be the algebraic multiplicity of λ_1 so that the size of A is $n \times n$. By Lemma 7.3.7, there exists an invertible P_1 such that

$$P_1^{-1}BQ_1 = \begin{bmatrix} A & 0 \\ 0 & C \end{bmatrix}.$$

By induction, there exists an invertible matrix P_2 such that $P_2^{-1}CP_2$ is a block diagonal matrix as described in the conclusion of the theorem. Therefore,

$$\begin{bmatrix} I_n & 0 \\ 0 & P_2^{-1} \end{bmatrix} P_1^{-1}BP_1 \begin{bmatrix} I_n & 0 \\ 0 & P_2 \end{bmatrix}$$

has the desired form. ■

■ **Example 7.3.9**

Continuing Example 7.2.22 where $A = \begin{bmatrix} 1 & 0 & 1 \\ 0 & 1 & 0 \\ 0 & 1 & 2 \end{bmatrix}$ and

$$\mathbb{R}^3 = \mathsf{S}\left(\begin{bmatrix} 1 \\ 0 \\ 0 \end{bmatrix}, \begin{bmatrix} 0 \\ -1 \\ 1 \end{bmatrix} \right) \oplus \mathsf{S}\left(\begin{bmatrix} 1 \\ 0 \\ 1 \end{bmatrix} \right).$$

Take the found bases of the summands of $\mathbb{R}^3$ and define

$$P = \begin{bmatrix} 1 & 0 & 1 \\ 0 & -1 & 0 \\ 0 & 1 & 1 \end{bmatrix}.$$

View P as a linear transformation $\mathbb{R}^3 \to \mathbb{R}^3$ such that

$$P\mathbf{e}_1 = \begin{bmatrix} 1 \\ 0 \\ 0 \end{bmatrix}, \quad P\mathbf{e}_2 = \begin{bmatrix} 0 \\ -1 \\ 1 \end{bmatrix}, \quad P\mathbf{e}_3 = \begin{bmatrix} 1 \\ 0 \\ 1 \end{bmatrix}.$$

Then,

$$P^{-1} = \begin{bmatrix} 1 & -1 & -1 \\ 0 & -1 & 0 \\ 0 & 1 & 1 \end{bmatrix},$$

and this results in the block matrix,

$$P^{-1}AP = \begin{bmatrix} 1 & 1 & 0 \\ 0 & 1 & 0 \\ 0 & 0 & 2 \end{bmatrix}.$$

Notice that the blocks are square upper triangular matrices. To illustrate why $P^{-1}AP$ looks as it does, trace the path of $\mathbf{e}_2$ under the action of $P^{-1}AP$ using the diagram:

The multiplication of $P\mathbf{e}_1$ by A gives $A\begin{bmatrix} 0 \\ -1 \\ 1 \end{bmatrix} = \begin{bmatrix} 1 \\ -1 \\ 1 \end{bmatrix}$. Write this vector using

the basis $\left\{ \begin{bmatrix} 1 \\ 0 \\ 0 \end{bmatrix}, \begin{bmatrix} 0 \\ -1 \\ 1 \end{bmatrix}, \begin{bmatrix} 1 \\ 0 \\ 1 \end{bmatrix} \right\}$,

$$\begin{bmatrix} 1 \\ -1 \\ 1 \end{bmatrix} = \begin{bmatrix} 1 \\ 0 \\ 0 \end{bmatrix} + \begin{bmatrix} 0 \\ -1 \\ 1 \end{bmatrix} + 0\begin{bmatrix} 1 \\ 0 \\ 1 \end{bmatrix}.$$

Because $\left\{ \begin{bmatrix} 1 \\ 0 \\ 1 \end{bmatrix} \right\}$ is the selected basis for the second summand, the scalar on

$\begin{bmatrix} 1 \\ 0 \\ 1 \end{bmatrix}$ must be zero because $\begin{bmatrix} 1 \\ -1 \\ 1 \end{bmatrix}$ is a vector of the first summand. This was

guaranteed because the summands are A-invariant (Theorem 7.2.7). Therefore, the second column of $P^{-1}AP$ is

$$P^{-1} \begin{bmatrix} 1 \\ -1 \\ 1 \end{bmatrix} = P^{-1} \left(\begin{bmatrix} 1 \\ 0 \\ 0 \end{bmatrix} + \begin{bmatrix} 0 \\ -1 \\ 1 \end{bmatrix} + 0 \begin{bmatrix} 1 \\ 0 \\ 1 \end{bmatrix} \right) = \begin{bmatrix} 1 \\ 1 \\ 0 \end{bmatrix}.$$

Nilpotent Matrices

Consider the following matrix sequences:

$$\begin{bmatrix} 0 & 1 \\ 0 & 0 \end{bmatrix}, \begin{bmatrix} 0 & 1 \\ 0 & 0 \end{bmatrix}^2 = \begin{bmatrix} 0 & 0 \\ 0 & 0 \end{bmatrix}$$

and

$$\begin{bmatrix} 0 & 1 & 0 \\ 0 & 0 & 1 \\ 0 & 0 & 0 \end{bmatrix}, \begin{bmatrix} 0 & 1 & 0 \\ 0 & 0 & 1 \\ 0 & 0 & 0 \end{bmatrix}^2 = \begin{bmatrix} 0 & 0 & 1 \\ 0 & 0 & 0 \\ 0 & 0 & 0 \end{bmatrix}, \begin{bmatrix} 0 & 1 & 0 \\ 0 & 0 & 1 \\ 0 & 0 & 0 \end{bmatrix}^3 = \begin{bmatrix} 0 & 0 & 0 \\ 0 & 0 & 0 \\ 0 & 0 & 0 \end{bmatrix}.$$

As sequences of the null spaces of Theorem 7.2.12 stabilize resulting in generalized eigenspaces that play a role in representing the vector space $\mathbb{R}^n$, these matrices also stabilize at a certain power and play a role in representing elements of $\mathbb{R}^{n \times n}$.

■ **Definition 7.3.10**

A square matrix A is **nilpotent** if there is a positive integer k so that $A^k = 0$, where the minimum value of k is the **index** of A.

Under the right conditions, a square matrix is similar to the sum of a diagonal matrix and a nilpotent matrix.

■ **Definition 7.3.11**

Let $\lambda \in \mathbb{R}$. Define the matrix $N_k = [n_{ij}]$ to be the $k \times k$ matrix such that for all $i, j = 1, 2, \ldots, k$,

$$n_{ij} = \begin{cases} 1 & \text{if } j = i + 1, \\ 0 & \text{otherwise.} \end{cases}$$

That is, each entry of the **superdiagonal** of N_k is 1, so

$$N_k = \begin{bmatrix} 0 & 1 & 0 & \cdots & 0 & 0 \\ 0 & 0 & 1 & \cdots & 0 & 0 \\ 0 & 0 & 0 & \cdots & 0 & 0 \\ \vdots & \vdots & \vdots & & \vdots & \vdots \\ 0 & 0 & 0 & \cdots & 1 & 0 \\ 0 & 0 & 0 & \cdots & 0 & 1 \\ 0 & 0 & 0 & \cdots & 0 & 0 \end{bmatrix}.$$

Notice that N_k is a nilpotent matrix (Exercise 17).

■ **Lemma 7.3.12**

Let A be an $n \times n$ upper triangular matrix with only one eigenvalue λ. If $\lambda \in \mathbb{R}$, there exists positive integers $k_1, k_2, \ldots, k_m$ with $k_1 + k_2 + \cdots + k_m = n$ such that $A - \lambda I_n$ is similar to the block matrix

$$\begin{bmatrix} N_{k_1} & 0 & \cdots & 0 \\ 0 & N_{k_2} & \cdots & 0 \\ \vdots & & \ddots & \vdots \\ 0 & 0 & \cdots & N_{k_m} \end{bmatrix}.$$

Proof

Let $\lambda \in \mathbb{R}$. Prove this by induction on the number of blocks. Let $A \in \mathbb{R}^{n \times n}$ with $n > 1$. Consider the chain of null spaces,

$$\mathrm{NS}(A - \lambda I_n) \subset \mathrm{NS}(A - \lambda I_n)^2 \subset \cdots \subset \mathrm{NS}(A - \lambda I_n)^i \subset \cdots$$

By Theorem 7.2.12 the sequence stabilizes at some integer $k_m \leq n$, so there exists $\mathbf{u} \in \mathbb{R}^n$ such that

$$\mathbf{u} \in \mathrm{NS}(A - \lambda I_n)^{k_m} \setminus \mathrm{NS}(A - \lambda I_n)^{k_m - 1}. \tag{7.31}$$

Because

$$(A - \lambda I_n)^{i-1}\mathbf{u} \in \mathrm{NS}(A - \lambda I_n)^{k_m - i + 1} \setminus \mathrm{NS}(A - \lambda I_n)^{k_m - i},$$

the columns of the $n \times k_m$ matrix,

$$P_1 = [(A - \lambda I_n)^{k_m - 1}\mathbf{u} \ \ (A - \lambda I_n)^{k_m - 2}\mathbf{u} \ \ \cdots \ \ (A - \lambda I_n)\mathbf{u} \ \ \mathbf{u}], \tag{7.32}$$

form a linearly independent set. It should be noted at this time that

$$(A - \lambda I_n)P_1 = [\mathbf{0} \ \ (A - \lambda I_n)^{k_m - 1}\mathbf{u} \ \ \cdots \ \ (A - \lambda I_n)^2\mathbf{u} \ \ (A - \lambda I_n)\mathbf{u}],$$

so using the notation of Definition 7.3.11,

$$(A - \lambda I_n)P_1 = P_1 \begin{bmatrix} 0 & 1 & 0 & \cdots & 0 & 0 \\ 0 & 0 & 1 & \cdots & 0 & 0 \\ 0 & 0 & 0 & \cdots & 0 & 0 \\ \vdots & \vdots & \vdots & & \vdots & \vdots \\ 0 & 0 & 0 & \cdots & 0 & 1 \\ 0 & 0 & 0 & \cdots & 0 & 0 \end{bmatrix} = P_1 N_{k_m}.$$

If $k_m = n$, then P_1 is invertible by Theorem 5.5.14, and $P_1^{-1}(A - \lambda I_n)P_1$ is in the desired form. Otherwise, let B be the top left $(n - k_m) \times (n - k_m)$ upper triangular block of A. Because λ is the only eigenvalue of B, by induction there is an invertible $(n - k_m) \times (n - k_m)$ matrix Q such that

$$Q^{-1}(B - \lambda I_{n-k_m})Q = \begin{bmatrix} N_{k_1} & 0 & \cdots & 0 \\ 0 & N_{k_2} & \cdots & 0 \\ \vdots & & \ddots & \vdots \\ 0 & 0 & \cdots & N_{k_{m-1}} \end{bmatrix} = N.$$

Letting 0 represent the $k_m \times (n - k_m)$ zero matrix, define the $n \times (n - k_m)$ matrix, $P_2 = \begin{bmatrix} Q \\ 0 \end{bmatrix}$. Then, writing A as the block matrix, $A = \begin{bmatrix} B & D \\ 0 & C \end{bmatrix}$, where C is an $k_m \times k_m$ upper triangular matrix, D is an $(n - k_m) \times k_m$ matrix, and 0 is the $k_m \times (n - k_m)$ zero matrix, it follows that

$$(A - \lambda I_n)P_2 = \begin{bmatrix} B - \lambda I_{n-k_m} & D \\ 0 & C - \lambda I_{k_m} \end{bmatrix}\begin{bmatrix} Q \\ 0 \end{bmatrix}$$

$$= \begin{bmatrix} (B - \lambda I_{n-k_m})Q \\ 0 \end{bmatrix} = \begin{bmatrix} QN \\ 0 \end{bmatrix} = P_2N.$$

To finish, let $P = [P_2 \ P_1]$. Because the columns of P are linearly independent, P is invertible, and

$$(A - \lambda I_n)P = (A - \lambda I_n)[P_2 \ P_1]$$
$$= [(A - \lambda I_n)P_2 \ (A - \lambda I_n)P_1]$$
$$= [P_2N \ P_1N_{k_m}].$$

This implies that $P^{-1}(A - \lambda I_n)P$ is of the desired form because

$$[P_2N \ P_1N_{k_m}] = [P_2 \ P_1]\begin{bmatrix} N & 0 \\ 0 & N_{k_m} \end{bmatrix}. \blacksquare$$

■ **Example 7.3.13**

The matrix $A = \begin{bmatrix} 2 & 1 & 0 \\ 0 & 2 & 4 \\ 0 & 0 & 2 \end{bmatrix}$ has only $\lambda = 2$ as an eigenvalue. This implies that the characteristic polynomial of A is $p_A(x) = (x - 2)^3$. Because

$$(A - 2I_3) = \begin{bmatrix} 0 & 1 & 0 \\ 0 & 0 & 4 \\ 0 & 0 & 0 \end{bmatrix}, \ (A - 2I_3)^2 = \begin{bmatrix} 0 & 0 & 4 \\ 0 & 0 & 0 \\ 0 & 0 & 0 \end{bmatrix}, \ (A - 2I_3)^3 = \begin{bmatrix} 0 & 0 & 0 \\ 0 & 0 & 0 \\ 0 & 0 & 0 \end{bmatrix},$$

it follows that

$$\mathsf{NS}(A - 2I_3) = \mathsf{S}\left(\begin{bmatrix} 1 \\ 0 \\ 0 \end{bmatrix}\right), \ \mathsf{NS}(A - 2I_3)^2 = \mathsf{S}\left(\begin{bmatrix} 1 \\ 0 \\ 0 \end{bmatrix}, \begin{bmatrix} 0 \\ 1 \\ 0 \end{bmatrix}\right), \ \mathsf{NS}(A - 2I_3)^3 = \mathbb{R}^3.$$

As in (7.31), make an easy choice for $\mathbf{u} \in$ NS$(A - 2I_3)^3 \setminus$ NS$(A - 2I_3)^2$ like

$$\mathbf{u} = \begin{bmatrix} 0 \\ 0 \\ 1 \end{bmatrix}, \quad (A - 2I_3)\mathbf{u} = \begin{bmatrix} 0 \\ 4 \\ 0 \end{bmatrix}, \quad (A - 2I_3)^2\mathbf{u} = \begin{bmatrix} 4 \\ 0 \\ 0 \end{bmatrix},$$

so as in (7.32), define

$$P = \begin{bmatrix} 4 & 0 & 0 \\ 0 & 4 & 0 \\ 0 & 0 & 1 \end{bmatrix}.$$

Finally, because

$$P^{-1} = \begin{bmatrix} 1/4 & 0 & 0 \\ 0 & 1/4 & 0 \\ 0 & 0 & 1 \end{bmatrix},$$

conclude that $A - 2I_3$ is similar to

$$P^{-1}(A - 2I_3)P = \begin{bmatrix} 0 & 1 & 0 \\ 0 & 0 & 1 \\ 0 & 0 & 0 \end{bmatrix}.$$

Jordan Canonical Form

Of course, not every square matrix is upper triangular. Nonetheless, under the right conditions, a square matrix is similar to a matrix in a very nice form, one that is nicer than the upper triangular matrix given by Schur's Lemma (7.3.5) and nicer than the block diagonal matrix of Theorem 7.3.8. This next form requires an initial definition.

■ **Definition 7.3.14**

Let $D_{\lambda,k}$ be the $k \times k$ diagonal matrix such that each entry on its main diagonal is λ. The $k \times k$ matrix $J_{\lambda,k}$ is a **Jordan block** when $J_{\lambda,k} = D_{\lambda,k} + N_k$.

The Jordan block $J_{\lambda,k}$ has λ as its only eigenvalue, so its characteristic polynomial is $(x - \lambda)^k$. This means that Theorem 7.3.3 and Corollary 7.3.4 imply that any matrix similar to $J_{\lambda,k}$ will have characteristic polynomial $(x - \lambda)^k$ and λ as its only eigenvalue.

■ **Definition 7.3.15**

An $n \times n$ matrix J is in **Jordan canonical form** if there are $\lambda_1, \lambda_2, \ldots, \lambda_m \in \mathbb{R}$ and $k_1, k_2, \ldots, k_m \in \mathbb{Z}^+$ with $k_1 + k_2 + \cdots + k_m = n$ such that J is the block diagonal matrix

$$J = \begin{bmatrix} J_{\lambda_1,k_1} & 0 & \cdots & 0 \\ 0 & J_{\lambda_2,k_2} & \cdots & 0 \\ \vdots & \vdots & & \vdots \\ 0 & 0 & & J_{\lambda_m,k_m} \end{bmatrix}.$$

■ **Lemma 7.3.16**

Let A be an $n \times n$ upper triangular matrix. If A has only one eigenvalue $\lambda \in \mathbb{R}$, then A is similar to a matrix in Jordan canonical form.

Proof

Assume that $\lambda \in \mathbb{R}$ is the only eigenvalue of A. By Lemma 7.3.12 there exists an invertible matrix P such that

$$P^{-1}(A - \lambda I_n)P = \begin{bmatrix} N_{k_1} & 0 & \cdots & 0 \\ 0 & N_{k_2} & \cdots & 0 \\ \vdots & \vdots & \ddots & \vdots \\ 0 & 0 & \cdots & N_{k_m} \end{bmatrix}$$

for some $k_1, k_2, \ldots, k_m \in \mathbb{Z}^+$. Because $P^{-1}(A - \lambda I_n)P = P^{-1}AP - \lambda I_n$, it follows that $P^{-1}AP$ is in Jordan canonical form since

$$P^{-1}AP = \lambda I_n + P^{-1}(A - \lambda I_n)P. \blacksquare$$

■ **Example 7.3.17**

Continuing Example 7.3.13 where $A = \begin{bmatrix} 2 & 1 & 0 \\ 0 & 2 & 4 \\ 0 & 0 & 2 \end{bmatrix}$ and it was found that

$$\begin{bmatrix} 1/4 & 0 & 0 \\ 0 & 1/4 & 0 \\ 0 & 0 & 1 \end{bmatrix} \begin{bmatrix} 0 & 1 & 0 \\ 0 & 0 & 4 \\ 0 & 0 & 0 \end{bmatrix} \begin{bmatrix} 4 & 0 & 0 \\ 0 & 4 & 0 \\ 0 & 0 & 1 \end{bmatrix} = \begin{bmatrix} 0 & 1 & 0 \\ 0 & 0 & 1 \\ 0 & 0 & 0 \end{bmatrix}.$$

The proof of Lemma 7.3.16 shows that

$$\begin{bmatrix} 1/4 & 0 & 0 \\ 0 & 1/4 & 0 \\ 0 & 0 & 1 \end{bmatrix} \begin{bmatrix} 2 & 1 & 0 \\ 0 & 2 & 4 \\ 0 & 0 & 2 \end{bmatrix} \begin{bmatrix} 4 & 0 & 0 \\ 0 & 4 & 0 \\ 0 & 0 & 1 \end{bmatrix} = \begin{bmatrix} 2 & 0 & 0 \\ 0 & 2 & 0 \\ 0 & 0 & 2 \end{bmatrix} + \begin{bmatrix} 0 & 1 & 0 \\ 0 & 0 & 1 \\ 0 & 0 & 0 \end{bmatrix}.$$

That is, A is similar to $J_{2,3}$.

■ **Example 7.3.18**

Consider the upper triangular matrix

$$A = \begin{bmatrix} 2 & 3 & 0 & 0 \\ 0 & 2 & 0 & 4 \\ 0 & 0 & 2 & 0 \\ 0 & 0 & 0 & 2 \end{bmatrix}.$$

Its only eigenvalue is $\lambda = 2$, but it is not similar to a Jordan block. Some calculation will reveal that

$$\text{ES}(A, 2) = \text{NS}(A - 2I_4) = \text{S}\left(\begin{bmatrix} 1 \\ 0 \\ 0 \\ 0 \end{bmatrix}, \begin{bmatrix} 0 \\ 0 \\ 1 \\ 0 \end{bmatrix} \right),$$

$$NS(A - 2I_4)^2 = S\left(\begin{bmatrix} 1 \\ 0 \\ 0 \\ 0 \end{bmatrix}, \begin{bmatrix} 0 \\ 1 \\ 0 \\ 0 \end{bmatrix}, \begin{bmatrix} 0 \\ 0 \\ 1 \\ 0 \end{bmatrix}\right),$$

and

$$NS(A - 2I_4)^3 = \mathbb{R}^4.$$

Choose

$$\mathbf{u} = \begin{bmatrix} 0 \\ 0 \\ 0 \\ 1 \end{bmatrix} \in NS(A - 2I_4)^3 \setminus NS(A - 2I_4)^2.$$

Then,

$$(A - 2I_4)\mathbf{u} = \begin{bmatrix} 0 \\ 4 \\ 0 \\ 0 \end{bmatrix} \quad \text{and} \quad (A - 2I_4)^2\mathbf{u} = \begin{bmatrix} 12 \\ 0 \\ 0 \\ 0 \end{bmatrix}.$$

Lastly, take $\mathbf{v} \notin S\left((A - 2I_4)^2\mathbf{u}, (A - 2I_4)\mathbf{u}, \mathbf{u}\right)$, say

$$\mathbf{v} = \begin{bmatrix} 0 \\ 0 \\ 1 \\ 0 \end{bmatrix},$$

and define

$$P = \left[(A - 2I_4)^2\mathbf{u} \ (A - 2I_4)\mathbf{u} \ \mathbf{u} \ \mathbf{v}\right] = \begin{bmatrix} 12 & 0 & 0 & 0 \\ 0 & 4 & 0 & 0 \\ 0 & 0 & 0 & 1 \\ 0 & 0 & 1 & 0 \end{bmatrix}.$$

Then,

$$P^{-1}AP = \begin{bmatrix} 2 & 1 & 0 & 0 \\ 0 & 2 & 1 & 0 \\ 0 & 0 & 2 & 0 \\ 0 & 0 & 0 & 2 \end{bmatrix} = \begin{bmatrix} J_{2,3} & 0 \\ 0 & J_{2,1} \end{bmatrix}.$$

This representation is not unique because if, instead,

$$Q = \left[\mathbf{v} \ (A - 2I_4)^2\mathbf{u} \ (A - 2I_4)\mathbf{u} \ \mathbf{u}\right] = \begin{bmatrix} 0 & 12 & 0 & 0 \\ 0 & 0 & 4 & 0 \\ 1 & 0 & 0 & 0 \\ 0 & 0 & 0 & 1 \end{bmatrix},$$

then

$$Q^{-1}AQ = \begin{bmatrix} 2 & 0 & 0 & 0 \\ 0 & 2 & 1 & 0 \\ 0 & 0 & 2 & 1 \\ 0 & 0 & 0 & 2 \end{bmatrix} = \begin{bmatrix} J_{2,1} & 0 \\ 0 & J_{2,3} \end{bmatrix}.$$

■ **Theorem 7.3.19**

Every $n \times n$ matrix A is similar to a matrix in Jordan canonical form if the eigenvalues of A are elements of $\mathbb{R}$.

Proof

There exists an invertible matrix Q such that $Q^{-1}AQ$ is a block diagonal matrix as described in Theorem 7.3.8. Let $\lambda_1, \lambda_2, \dots, \lambda_k \in \mathbb{R}$ be the distinct eigenvalues of A and assume that $B_1, B_2, \dots, B_k$ are the upper triangular blocks on the diagonal of $Q^{-1}AQ$ corresponding to these eigenvalues. Write

$$Q^{-1}AQ = \begin{bmatrix} B_1 & 0 & \cdots & 0 \\ 0 & B_2 & \cdots & 0 \\ \vdots & & \ddots & \vdots \\ 0 & 0 & \cdots & B_k \end{bmatrix}.$$

By Lemma 7.3.16, there exists invertible $P_1, P_2, \dots, P_k$ so that $P_i^{-1}B_iP_i$ is in Jordan canonical form for all $i = 1, 2, \dots, k$. Define the block matrix

$$P = \begin{bmatrix} P_1 & 0 & \cdots & 0 \\ 0 & P_2 & \cdots & 0 \\ \vdots & & \ddots & \vdots \\ 0 & 0 & \cdots & P_k \end{bmatrix}.$$

This implies that

$$P^{-1} = \begin{bmatrix} P_1^{-1} & 0 & \cdots & 0 \\ 0 & P_2^{-1} & \cdots & 0 \\ \vdots & & \ddots & \vdots \\ 0 & 0 & \cdots & P_k^{-1} \end{bmatrix}.$$

Therefore,

$$P^{-1}Q^{-1}AQP = \begin{bmatrix} P_1^{-1}B_1P_1 & 0 & \cdots & 0 \\ 0 & P_2^{-1}B_2P_2 & \cdots & 0 \\ \vdots & & \ddots & \vdots \\ 0 & 0 & \cdots & P_k^{-1}B_kP_k \end{bmatrix},$$

which is in Jordan canonical form. ■

■ **Corollary 7.3.20**

If A is an $n \times n$ matrix with only real eigenvalues, A is similar to the sum of a diagonal matrix and a nilpotent matrix.

Because the proof of Theorem 7.3.19 relies on Lemma 7.3.16, which, in turn, relies on the block diagonal matrix given by Theorem 7.3.8, which, in turn, relies on the upper triangular matrix of Schur's Lemma (7.3.5), it can appear as a daunting task to find the Jordan canonical form of a matrix. Fortunately, the method found in the proof of Lemma 7.3.12 can be used in the general case.

■ **Example 7.3.21**

Take the matrix,

$$A = \begin{bmatrix} 3 & 82 & -3 & 82 & -48 \\ 0 & -23 & 1 & -26 & 16 \\ 2 & 16 & 3 & 10 & 0 \\ 0 & 22 & -1 & 25 & -16 \\ 0 & 8 & 0 & 8 & -1 \end{bmatrix}.$$

This matrix has 2 eigenvalues, $\lambda = 3$ (algebraic multiplicity 3) and $\lambda = -1$ (algebraic multiplicity 2). Since the eigenvalues are real, A is similar to a matrix in Jordan canonical form. To find the matrix P, first examine powers of $(A - 3I_5)$ to find that

$$\text{NS}(A - 3I_5) = \text{S}\left(\begin{bmatrix} 3 \\ -1 \\ 0 \\ 1 \\ 0 \end{bmatrix}\right), \quad \text{NS}(A - 3I_5)^2 = \text{S}\left(\begin{bmatrix} 3 \\ -1 \\ 0 \\ 1 \\ 0 \end{bmatrix}, \begin{bmatrix} 0 \\ 0 \\ 1 \\ 0 \\ 0 \end{bmatrix}\right),$$

and

$$\text{NS}(A - 3I_5)^3 = \text{NS}(A - 3I_5)^4 = \text{S}\left(\begin{bmatrix} 1 \\ 0 \\ 0 \\ 0 \\ 0 \end{bmatrix}, \begin{bmatrix} 0 \\ 0 \\ 1 \\ 0 \\ 0 \end{bmatrix}, \begin{bmatrix} 0 \\ -1 \\ 0 \\ 1 \\ 0 \end{bmatrix}\right).$$

Pick $\mathbf{u} \in \text{NS}(A - 3I_5)^3 \setminus \text{NS}(A - 3I_5)^2$ and compute two more vectors for P:

$$\mathbf{u} = \begin{bmatrix} 1 \\ 0 \\ 0 \\ 0 \\ 0 \end{bmatrix}, \quad (A - 3I_5)\mathbf{u} = \begin{bmatrix} 0 \\ 0 \\ 2 \\ 0 \\ 0 \end{bmatrix}, \quad (A - 3I_5)^2\mathbf{u} = \begin{bmatrix} -6 \\ 2 \\ 0 \\ -2 \\ 0 \end{bmatrix}.$$

Next, examine powers of $(A + I_5)$ to find that

$$\text{NS}(A + I_5) = \text{S}\left(\begin{bmatrix} 12 \\ -4 \\ 0 \\ 4 \\ 1 \end{bmatrix}\right), \quad \text{NS}(A + I_5)^2 = \text{NS}(A + I_5)^3 = \text{S}\left(\begin{bmatrix} 1 \\ -1/2 \\ -1 \\ 1 \\ 0 \end{bmatrix}, \begin{bmatrix} 8 \\ -2 \\ 4 \\ 0 \\ 1 \end{bmatrix}\right).$$

Take $\mathbf{v} \in \text{NS}(A + I_5)^2 \setminus \text{NS}(A + I_5)$ and compute another vector for P:

$$\mathbf{v} = \begin{bmatrix} 2 \\ -1 \\ -2 \\ 2 \\ 0 \end{bmatrix} \quad \text{and} \quad (A + I_5)\mathbf{v} = \begin{bmatrix} 96 \\ -32 \\ 0 \\ 32 \\ 8 \end{bmatrix}.$$

Finally, set

$$P = \left[(A - 3I_5)^2\mathbf{u} \ \ (A - 3I_5)\mathbf{u} \ \ \mathbf{u} \ \ (A + I_5)\mathbf{v} \ \ \mathbf{v} \right] = \begin{bmatrix} -6 & 0 & 1 & 96 & 2 \\ 2 & 0 & 0 & -32 & -1 \\ 0 & 2 & 0 & 0 & -2 \\ -2 & 0 & 0 & 32 & 2 \\ 0 & 0 & 0 & 8 & 0 \end{bmatrix},$$

and then calculate

$$P^{-1}AP = \begin{bmatrix} 3 & 1 & 0 & 0 & 0 \\ 0 & 3 & 1 & 0 & 0 \\ 0 & 0 & 3 & 0 & 0 \\ 0 & 0 & 0 & -1 & 1 \\ 0 & 0 & 0 & 0 & -1 \end{bmatrix}.$$

Exercises

1. Let A be an $n \times n$ matrix. Prove that if A is similar to I_n, then $A = I_n$, and if A is similar to the $n \times n$ zero matrix, then A is a zero matrix.

2. Determine whether the given pairs of matrices are similar. Prove each result.

   (a) $\begin{bmatrix} 1 & 2 \\ 0 & 3 \end{bmatrix}, \begin{bmatrix} 1 & 4 \\ 0 & 3 \end{bmatrix}$

   (b) $\begin{bmatrix} 1 & 1 & 0 \\ 0 & 1 & 1 \\ 0 & 0 & 2 \end{bmatrix}, \begin{bmatrix} 1 & 1 & 0 \\ 0 & 2 & 1 \\ 0 & 0 & 2 \end{bmatrix}$

   (c) $\begin{bmatrix} 1 & 4 \\ 0 & 3 \end{bmatrix}, \begin{bmatrix} 1 & 4 \\ 1 & 3 \end{bmatrix}$

   (d) $\begin{bmatrix} 4 & 0 & 0 \\ 0 & 3 & 0 \\ 0 & 0 & 5 \end{bmatrix}, \begin{bmatrix} 3 & 0 & 0 \\ 0 & 5 & 0 \\ 0 & 0 & 4 \end{bmatrix}$

3. Let A be similar to B. Prove.

   (a) $\det A = \det B$

   (b) $\operatorname{tr} A = \operatorname{tr} B$

   (c) $\operatorname{rank} A = \operatorname{rank} B$

   (d) $\operatorname{null} A = \operatorname{null} B$

4. Let A be an idempotent matrix that is similar to matrix B. Show that B is idempotent.

5. Let A be a square matrix. Prove that A is similar to A^T.

6. Assume that A is similar to B. Show that A^T is similar to B^T.

7. Let A be an invertible matrix that is similar to B. Prove that B is invertible and A^{-1} is similar to B^{-1}.

8. Prove that if A is similar to B, then A^2 is similar to B^2.

9. Prove that the converse of Corollary 7.3.4 is false, namely, that matrices can have the same eigenvalues but not be similar.

10. Suppose that A is similar to B. Prove that $\dim \operatorname{ES}(\lambda, A) = \dim \operatorname{ES}(\lambda, B)$.

11. Let $T : V \to V$ be a linear operator. Let $\mathcal{B}_1$ and $\mathcal{B}_2$ be bases for V. Prove that $[T]_{\mathcal{B}_1 \mathcal{B}_1}$ is similar to $[T]_{\mathcal{B}_2 \mathcal{B}_2}$.

12. Given the $n \times n$ matrix A, find the upper triangular matrix U given by Schur's Lemma (7.3.5) that has the eigenvalues of A on its main diagonal and is similar to A.

   (a) $A = \begin{bmatrix} 1 & 4 \\ 2 & 3 \end{bmatrix}$

   (c) $A = \begin{bmatrix} 2 & 0 & 0 \\ 1 & 1 & 1 \\ 1 & 0 & 3 \end{bmatrix}$

   (b) $A = \begin{bmatrix} 4 & -5 & -1 \\ 1 & 0 & -1 \\ 1 & -3 & 2 \end{bmatrix}$

   (d) $A = \begin{bmatrix} -3 & -2 & 3 \\ 3 & 4 & -9 \\ 1 & 2 & -5 \end{bmatrix}$

13. Reprove Schur's Lemma but replace upper triangular with lower triangular.

14. Let A be an $n \times n$ matrix, P be an orthogonal matrix, and U be an upper triangular such that $A = PUP^T$. That is, suppose that A has a Schur factorization. Prove that, counting multiplicities, A has n real eigenvalues.

15. For each matrix A, find the block diagonal matrix similar to A that is guaranteed to exist by Theorem 7.3.8.

   (a) $A = \begin{bmatrix} 4 & -5 & 7 \\ 0 & 2 & 0 \\ 0 & -3 & 4 \end{bmatrix}$

   (b) $A = \begin{bmatrix} 2 & 0 & 3 & 0 \\ 7 & 4 & 5 & -4 \\ 0 & 0 & 2 & 0 \\ -1 & 0 & 2 & 4 \end{bmatrix}$

16. Prove (7.30) in the proof of Theorem 7.3.7.

17. Prove that N_k is nilpotent for all positive integers k.

18. Let A be a nilpotent matrix with index k. Prove that the minimal polynomial of A is $m_A(x) = x^k$. What is the characteristic polynomial of A?

19. Prove that if each entry on the main diagonal of a matrix is 0, the matrix is nilpotent. Show that the converse is false.

20. Prove that both the determinant and the trace of a nilpotent matrix is 0.

21. Prove or show false: Every noninvertible matrix is the product of nilpotent matrices.

22. Let N be an $n \times n$ nilpotent matrix. Prove that $\det(I_n + N) = 1$.

23. Prove that A is nilpotent if and only if $\lambda = 0$ is the only eigenvalue of A.

24. Given a matrix J in Jordan canonical form, find $p_J(x)$ and $m_J(x)$.

   (a) $J = \begin{bmatrix} 5 & 0 & 0 & 0 \\ 0 & 3 & 1 & 0 \\ 0 & 0 & 3 & 1 \\ 0 & 0 & 0 & 3 \end{bmatrix}$

   (b) $J = \begin{bmatrix} 5 & 0 & 0 & 0 \\ 0 & 3 & 0 & 0 \\ 0 & 0 & 3 & 1 \\ 0 & 0 & 0 & 3 \end{bmatrix}$

(c) $J = \begin{bmatrix} 5 & 1 & 0 & 0 \\ 0 & 5 & 0 & 0 \\ 0 & 0 & 3 & 1 \\ 0 & 0 & 0 & 3 \end{bmatrix}$

(d) $J = \begin{bmatrix} 5 & 0 & 0 & 0 \\ 0 & 5 & 0 & 0 \\ 0 & 0 & 3 & 1 \\ 0 & 0 & 0 & 3 \end{bmatrix}$

25. Use Exercise 24 to draw a conclusion about the relationship between the characteristic and minimal polynomials of a matrix and its Jordan canonical form.

26. Given the size of a matrix A along with its characteristic and minimal polynomials, find all possible matrices J in Jordan canonical form to which A could be similar.

   (a) 4×4, $p_A(x) = (x - 3)^2(x - 1)^2$, $m_A(x) = (x - 3)(x - 1)^2$
   (b) 4×4, $p_A(x) = (x - 3)^2(x - 1)^2$, $m_A(x) = (x - 3)(x - 1)$
   (c) 5×5, $p_A(x) = (x - 3)^2(x - 1)^2$, $m_A(x) = (x - 3)^2(x - 1)^2$
   (d) 5×5, $p_A(x) = (x - 3)(x - 1)^4$, $m_A(x) = (x - 3)(x - 1)^2$
   (e) 6×6, $p_A(x) = (x - 3)^4(x - 1)^2$, $m_A(x) = (x - 3)^2(x - 1)^2$
   (f) 6×6, $p_A(x) = (x - 3)(x - 1)^5$, $m_A(x) = (x - 3)(x - 1)^2$

27. For each matrix A, if possible, find the matrix in Jordan canonical form to which A is similar.

   (a) $A = \begin{bmatrix} 5 & 8 & 7 \\ 0 & 5 & 0 \\ 0 & -4 & 5 \end{bmatrix}$

   (c) $A = \begin{bmatrix} 3 & 0 & 0 \\ 1 & 3 & 0 \\ 2 & 0 & 3 \end{bmatrix}$

   (b) $A = \begin{bmatrix} 5 & 1 & 4 & 7 \\ 0 & 5 & 0 & 0 \\ 0 & 4 & 1 & 0 \\ 0 & 8 & 0 & 1 \end{bmatrix}$

   (d) $A = \begin{bmatrix} 4 & 0 & 0 & 0 & 0 \\ 5 & 4 & 0 & 0 & 0 \\ 0 & 2 & 4 & 0 & 0 \\ 5 & 0 & 5 & 3 & 0 \\ 0 & 2 & 0 & 2 & 3 \end{bmatrix}$

28. Let $A = D + N$ be an $n \times n$ matrix in Jordan canonical form, where D is a diagonal matrix and N is nilpotent. Let B be an $n \times n$ matrix. Prove that $AB = BA$ if and only if $DB = BD$ and $NB = BN$.

7.4   Diagonalization

The Jordan canonical form (Definition 7.3.15) appears as the culmination of a sequence of forms to which matrices are similar under the right conditions. The question is whether there is a next step? The likely candidate is to remove the 1s from the superdiagonal of the Jordan canonical form and try to find matrices that are similar to diagonal matrices.

■ **Definition 7.4.1**

A matrix A is **diagonalizable** if it is similar to a diagonal matrix with the eigenvalues of A as the entries of its main diagonal.

Theorem 7.3.3 and Corollary 7.3.4 declare that similar matrices share their characteristic polynomials and, thus, their eigenvalues. However, the converses of these two results are false. For example, the matrices,

$$J = \begin{bmatrix} 1 & 1 \\ 0 & 1 \end{bmatrix} \quad \text{and} \quad I_2 = \begin{bmatrix} 1 & 0 \\ 0 & 1 \end{bmatrix},$$

both have characteristic polynomial $p(x) = (x - 1)^2$ and both have a single eigenvalue $\lambda = 1$. However, the matrix I is diagonal and the matrix J is not similar to any diagonal matrix as the next theorem will show.

■ Theorem 7.4.2

A square matrix A is diagonalizable if and only if it has a set of eigenvectors that are a basis of $\mathbb{R}^n$.

Proof

Suppose that the $n \times n$ matrix A is diagonalizable so that $D = P^{-1}AP$, where D is a diagonal matrix. This can be rewritten as $AP = PD$. Because D is diagonal, there are constants $\lambda_1, \lambda_2, \ldots, \lambda_n \in \mathbb{R}$ such that for $i = 1, 2, \ldots, n$,

$$D\mathbf{e}_i = \lambda_i \mathbf{e}_i.$$

It follows that each $P\mathbf{e}_k$ is an eigenvector for A with corresponding eigenvalue λ_k. Indeed,

$$A(P\mathbf{e}_k) = PD\mathbf{e}_k = P(\lambda_k \mathbf{e}_k) = \lambda_k(P\mathbf{e}_k). \tag{7.33}$$

Exercise 14 of Section 5.4 implies that $\{P\mathbf{e}_1, P\mathbf{e}_2, \ldots, P\mathbf{e}_n\}$ is linearly independent since P is invertible and thus is a basis of $\mathbb{R}^n$.

To prove the converse, let $\{\mathbf{u}_1, \mathbf{u}_2, \ldots, \mathbf{u}_n\}$ be a basis of $\mathbb{R}^n$ consisting entirely of eigenvectors of A. Define $P = [\mathbf{u}_1 \ \mathbf{u}_2 \ \cdots \ \mathbf{u}_n]$, which is invertible. It follows that $P^{-1}AP$ is a diagonal matrix. To see this, observe that for all $i = 1, 2, \ldots, n$, letting λ_i be the corresponding eigenvalue for $\mathbf{u}_i$,

$$P^{-1}AP\mathbf{e}_i = P^{-1}(A\mathbf{u}_i) = P^{-1}(\lambda_i \mathbf{u}_i) = \lambda_i P^{-1}\mathbf{u}_i = \lambda_i \mathbf{e}_i,$$

so $P^{-1}AP$ is the diagonal matrix,

$$P^{-1}AP = \begin{bmatrix} \lambda_1 & 0 & \cdots & 0 \\ 0 & \lambda_2 & \cdots & 0 \\ \vdots & \vdots & & \vdots \\ 0 & 0 & \cdots & \lambda_n \end{bmatrix}. ■$$

■ Example 7.4.3

Consider again the matrix $J = \begin{bmatrix} 1 & 1 \\ 0 & 1 \end{bmatrix}$. This matrix cannot be diagonalizable because its eigenvectors are multiples of the vector $\mathbf{e}_1$ since $\mathsf{ES}(J, 1) = \mathsf{S}(\mathbf{e}_1)$. This implies that J is not similar to any diagonal matrix because it does not have a basis of eigenvectors.

It is possible for an $n \times n$ matrix to be diagonalizable when it has fewer than n eigenvalues. Consider the matrix $A = \begin{bmatrix} 1 & 0 & 4 \\ 0 & 1 & 0 \\ 0 & 0 & 2 \end{bmatrix}$. Its eigenvalues are $\lambda = 1$, with corresponding eigenvectors $\begin{bmatrix} 1 \\ 0 \\ 0 \end{bmatrix}$ and $\begin{bmatrix} 0 \\ 1 \\ 0 \end{bmatrix}$, and $\lambda = 2$, with corresponding eigenvector $\begin{bmatrix} 4 \\ 0 \\ 1 \end{bmatrix}$. These vectors form a basis of $\mathbb{R}^3$, so A is diagonalizable by Theorem 7.4.2.

Specifically, let $P = \begin{bmatrix} 1 & 0 & 4 \\ 0 & 1 & 0 \\ 0 & 0 & 1 \end{bmatrix}$, so $PAP^{-1} = \begin{bmatrix} 1 & 0 & 0 \\ 0 & 1 & 0 \\ 0 & 0 & 2 \end{bmatrix}$. This being said, it is the case that the converse is true.

■ Theorem 7.4.4

If an $n \times n$ matrix A has n distinct eigenvalues, A is diagonalizable.

Proof

Suppose that A has a set of n distinct eigenvalues with corresponding eigenvectors $\mathbf{u}_1, \mathbf{u}_2, \ldots, \mathbf{u}_n$. Theorem 7.1.10 implies that $\mathcal{B} = \{\mathbf{u}_1, \mathbf{u}_2, \ldots, \mathbf{u}_n\}$ is linearly independent, so $\mathcal{B}$ is a basis for $\mathbb{R}^n$ by Theorem 5.4.19. Thus, A is similar to a diagonal matrix by Theorem 7.4.2. ■

Given an $n \times n$ matrix A, the process of diagonalization is straightforward. To diagonalize a matrix:

1. Find a basis $\{\mathbf{u}_1, \mathbf{u}_2, \ldots, \mathbf{u}_n\}$ consisting entirely of eigenvectors of A. If no such basis exists, A is not diagonalizable.

2. Form the invertible matrix $P = [\mathbf{u}_1 \ \mathbf{u}_2 \ \cdots \ \mathbf{u}_n]$ with columns that are the eigenvector basis found in the first step.

3. The diagonalization of A is $D = P^{-1}AP$. Moreover, due to (7.33) in the proof of Theorem 7.4.2, if λ_i is the eigenvalue that corresponds to $\mathbf{u}_i$, the entries along the main diagonal of D will be $\lambda_1, \lambda_2, \ldots, \lambda_n$ in this order but not necessarily distinct.

■ Example 7.4.5

To diagonalize the matrix $A = \begin{bmatrix} 1 & 1 \\ 2 & 0 \end{bmatrix}$, first observe that the eigenvectors,

$$\mathbf{u}_1 = \begin{bmatrix} -1 \\ 2 \end{bmatrix} \text{ for } \lambda = -1 \quad \text{and} \quad \mathbf{u}_2 = \begin{bmatrix} 1 \\ 1 \end{bmatrix} \text{ for } \lambda = 2,$$

form a basis for $\mathbb{R}^2$. The similarity matrix P is formed by using these eigenvectors as the columns, so

$$P = [\mathbf{u}_1 \ \mathbf{u}_2] = \begin{bmatrix} -1 & 1 \\ 2 & 1 \end{bmatrix}, \quad \text{which has inverse} \quad P^{-1} = \begin{bmatrix} -1/3 & 1/3 \\ 2/3 & 1/3 \end{bmatrix}.$$

It is easily checked that $P^{-1}AP = \begin{bmatrix} -1 & 0 \\ 0 & 2 \end{bmatrix}$.

Because an $n \times n$ matrix can have fewer than n eigenvalues and still be diagonalizable, there is not a reasonable diagonalization test involving the characteristic polynomial. There is one, however, that involves the minimal polynomial.

■ **Theorem 7.4.6**

Let A be an $n \times n$ matrix. Then, A is diagonalizable if and only if the minimal polynomial of A has real roots, each of multiplicity 1.

Proof

Suppose that A is diagonalizable. Theorem 7.4.2 implies that there exists a basis $\mathcal{B} = \{\mathbf{u}_1, \mathbf{u}_2, \dots, \mathbf{u}_n\}$ of $\mathbb{R}^n$ consisting entirely of eigenvectors. Let $\lambda_1, \lambda_2, \dots, \lambda_k$ be the distinct eigenvalues of A. For each $i = 1, 2, \dots, n$, there exists eigenvalue λ_{k_i} that corresponds to $\mathbf{u}_i$. Therefore, $A\mathbf{u}_i = \lambda_{k_i}\mathbf{u}_i$, so

$$(A - \lambda_{k_i}I_n)\mathbf{u}_i = \mathbf{0}. \tag{7.34}$$

Write $m(x) = (x - \lambda_1)(x - \lambda_2) \cdots (x - \lambda_k)$. Then, by (7.34),

$$m(A)\mathbf{u}_i = (A - \lambda_1 I_n)(A - \lambda_2 I_n) \cdots (A - \lambda_k I_n)\mathbf{u}_i = \mathbf{0}.$$

Thus, $m(A) = 0$, so by Theorem 7.2.3, $m_A(x)$ divides $m(x)$. However, since the degree of $m(x)$ is k and $m(x)$ has k distinct roots, it must be the case that $m_A(x) = m(x)$.

To prove the converse, assume that

$$m_A(x) = (x - \lambda_1)(x - \lambda_2) \cdots (x - \lambda_k)$$

with $\lambda_1, \lambda_2, \dots, \lambda_k$ being distinct scalars. By the Primary Decomposition Theorem (7.2.19),

$$\mathbb{R}^n = \text{NS}(A - \lambda_1 I_n) \oplus \text{NS}(A - \lambda_2 I_n) \oplus \cdots \oplus \text{NS}(A - \lambda_k I_n).$$

For each $i = 1, 2, \dots, k$, let $\mathcal{B}_i$ be a basis of $\text{NS}(A - \lambda_i I)$. As in the proof of Theorem 6.2.18, $\mathcal{B} = \mathcal{B}_1 \cup \mathcal{B}_2 \cup \cdots \cup \mathcal{B}_n$ is a basis for $\mathbb{R}^n$. Each vector in $\mathcal{B}$ is an eigenvector of A because if $\mathbf{u}$ is a vector in $\text{NS}(A - \lambda_i I)$ for some i, it follows that $(A - \lambda_i I)\mathbf{u} = \mathbf{0}$, so $A\mathbf{u} = \lambda_i\mathbf{u}$. Hence, A is diagonalizable by Theorem 7.4.2. ■

■ **Example 7.4.7**

Example 7.2.2 showed that the minimal polynomial of $B = \begin{bmatrix} 1 & 0 & 1 \\ 0 & 1 & 0 \\ 0 & 1 & 2 \end{bmatrix}$ is

$m_B(x) = (x-1)^2(x-2)$. Conclude by Theorem 7.4.6 that B is not diagonalizable. This can be confirmed by checking its eigenvectors. Some calculations will show that only multiples of $\mathbf{e}_1$ are eigenvectors corresponding to $\lambda = 1$ and only multiples of $\mathbf{e}_1 + \mathbf{e}_3$ are eigenvectors corresponding to $\lambda = 2$, so any set of linearly independent eigenvectors of B has only two vectors, which by Theorem 7.4.2 implies that B is not diagonalizable.

Orthogonal Diagonalization

Consider the symmetric matrix $A = \begin{bmatrix} 1 & 0 & 0 \\ 0 & 1 & 2 \\ 0 & 2 & -2 \end{bmatrix}$. It has eigenvalues $\lambda = 1, \lambda = 2$,

and $\lambda = -3$ with corresponding eigenvectors $\begin{bmatrix} 1 \\ 0 \\ 0 \end{bmatrix}, \begin{bmatrix} 0 \\ 2 \\ 1 \end{bmatrix}$, and $\begin{bmatrix} 0 \\ -1 \\ 2 \end{bmatrix}$, respectively.

Since these three eigenvectors form a basis for $\mathbb{R}^3$, the matrix A is diagonalizable by Theorem 7.4.2. In addition, it should be noted that any two of these vectors are orthogonal, and if normalized, the vectors form an orthonormal set. Therefore,

$$P = \begin{bmatrix} 1 & 0 & 0 \\ 0 & 2\sqrt{5}/5 & -\sqrt{5}/5 \\ 0 & \sqrt{5}/5 & 2\sqrt{5}/5 \end{bmatrix}$$

is an orthogonal matrix with inverse

$$P^{-1} = P^T = \begin{bmatrix} 1 & 0 & 0 \\ 0 & 2\sqrt{5}/5 & \sqrt{5}/5 \\ 0 & -\sqrt{5}/5 & 2\sqrt{5}/5 \end{bmatrix},$$

such that

$$P^{-1}AP = \begin{bmatrix} 1 & 0 & 0 \\ 0 & 2 & 0 \\ 0 & 0 & -3 \end{bmatrix}$$

showing that A is orthogonally similar to a diagonal matrix. Examining this example reveals two characteristics of symmetric matrices. Here is the first.

■ **Theorem 7.4.8**

The eigenvectors corresponding to distinct eigenvalues of a symmetric matrix are orthogonal with respect to the dot product.

Proof

Suppose that A is a symmetric matrix and that λ_1 and λ_2 are distinct eigenvalues of A with corresponding eigenvectors $\mathbf{u}_1$ and $\mathbf{u}_2$, respectively. Then, by Theorem 2.2.6(c) and Corollary 4.4.9,

$$\lambda_1(\mathbf{u}_1 \cdot \mathbf{u}_2) = \lambda_1\mathbf{u}_1 \cdot \mathbf{u}_2 = A\mathbf{u}_1 \cdot \mathbf{u}_2 = \mathbf{u}_1 \cdot A\mathbf{u}_2 = \mathbf{u}_1 \cdot \lambda_2\mathbf{u}_2 = \lambda_2(\mathbf{u}_1 \cdot \mathbf{u}_2).$$

It follows that $(\lambda_1 - \lambda_2)(\mathbf{u}_1 \cdot \mathbf{u}_2) = \mathbf{0}$. Because λ_1 and λ_2 are distinct, it is the case that $(\lambda_1 - \lambda_2) \neq 0$. Hence, $\mathbf{u}_1 \cdot \mathbf{u}_2 = 0$. ■

Shur's Lemma (7.3.5) introduced the concept of matrices being orthogonally similar. Combining this with diagonalization leads to the second characteristic of symmetric matrices.

■ **Definition 7.4.9**

The square matrix A is **orthogonally diagonalizable** if there exists an orthogonal matrix P such that $P^{-1}AP$ is a diagonal matrix with the eigenvalues of A as the entries of its main diagonal.

■ **Theorem 7.4.10**

If A is an $n \times n$ symmetric matrix with n distinct eigenvalues, A is orthogonally diagonalizable.

Proof

Suppose that the n distinct eigenvalues of A are $\lambda_1, \lambda_2, \ldots, \lambda_n$ have corresponding eigenvectors $\mathbf{u}_1, \mathbf{u}_2, \ldots, \mathbf{u}_n$. Suppose further that these eigenvectors have length 1. This is always possible since if $\mathbf{u}$ is an eigenvector for A with eigenvalue λ, the normalization of $\mathbf{u}$ is a unit eigenvector corresponding to λ. By Theorem 7.4.8, the eigenvectors $\{\mathbf{u}_1, \mathbf{u}_2, \ldots, \mathbf{u}_n\}$ form an orthonormal basis for $\mathbb{R}^n$. Hence, if $P = [\mathbf{u}_1 \ \mathbf{u}_2 \ \cdots \ \mathbf{u}_n]$, then P is an orthogonal matrix by Theorem 4.4.15. Lastly, using the Kronecker delta (Example 5.4.5),

$$\mathbf{u}_i \cdot \mathbf{u}_j = \delta_{ij},$$

so $P^{-1}AP$ is the desired diagonal matrix because

$$\begin{aligned}
P^{-1}AP = P^TAP &= P^TA[\mathbf{u}_1 \ \mathbf{u}_2 \ \cdots \ \mathbf{u}_n] \\
&= P^T[A\mathbf{u}_1 \ A\mathbf{u}_2 \ \cdots \ A\mathbf{u}_n] \\
&= P^T[\lambda_1\mathbf{u}_1 \ \lambda_2\mathbf{u}_2 \ \cdots \ \lambda_n\mathbf{u}_n] \\
&= [\mathbf{u}_1 \ \mathbf{u}_2 \ \cdots \ \mathbf{u}_n]^T[\lambda_1\mathbf{u}_1 \ \lambda_2\mathbf{u}_2 \ \cdots \ \lambda_n\mathbf{u}_n] \\
&= \begin{bmatrix} \lambda_1 & 0 & \ldots & 0 \\ 0 & \lambda_2 & \ldots & 0 \\ \ldots & & & \\ 0 & 0 & \ldots & \lambda_n \end{bmatrix}. \blacksquare
\end{aligned}$$

Consider the symmetric matrix $A = \begin{bmatrix} 2 & 1 & 1 \\ 1 & 2 & 1 \\ 1 & 1 & 2 \end{bmatrix}$. It has eigenvalues $\lambda = 1$ (algebraic multiplicity 2) and $\lambda = 4$. Its eigenvectors are $\begin{bmatrix} -1 \\ 1 \\ 0 \end{bmatrix}$ and $\begin{bmatrix} -1 \\ 0 \\ 1 \end{bmatrix}$, which correspond to $\lambda = 1$, and $\begin{bmatrix} 1 \\ 1 \\ 1 \end{bmatrix}$, which corresponds to $\lambda = 4$. The problem here is that A has too few eigenvalues, so Theorem 7.4.10 does not apply. There are enough eigenvectors to form the matrix P, but the resulting matrix will not be orthogonal. There is, however, a fix for this. Use the Gram–Schmidt process (Theorem 6.2.21) to find an orthonormal basis for $ES(A, 1)$, the result being

$$\left\{ \begin{bmatrix} -\sqrt{2}/2 \\ \sqrt{2}/2 \\ 0 \end{bmatrix}, \begin{bmatrix} -\sqrt{6}/6 \\ -\sqrt{6}/6 \\ \sqrt{6}/3 \end{bmatrix} \right\}.$$

Adding the normalization of the eigenvector corresponding to $\lambda = 4$ gives the orthogonal matrix,

$$P = \begin{bmatrix} -\sqrt{2}/2 & -\sqrt{6}/6 & \sqrt{3}/3 \\ \sqrt{2}/2 & -\sqrt{6}/6 & \sqrt{3}/3 \\ 0 & \sqrt{6}/3 & \sqrt{3}/3 \end{bmatrix}, \quad \text{so} \quad P^T AP = \begin{bmatrix} 1 & 0 & 0 \\ 0 & 1 & 0 \\ 0 & 0 & 4 \end{bmatrix}.$$

This means that the result of Theorem 7.4.10 can be greatly improved upon.

■ Theorem 7.4.11

A square matrix A is orthogonally diagonalizable if and only if A is symmetric.

Proof

First, suppose that there exists a orthogonal matrix P such that $P^{-1}AP = D$ for some diagonal matrix D. It follows that A is symmetric because

$$A^T = (PDP^{-1})^T = (P^{-1})^T D^T P^T = (P^T)^T D^T P^T = PDP^{-1} = A. \quad (7.35)$$

To prove the converse, let A be symmetric. Schur's Lemma (7.3.5) implies that there exists an orthogonal matrix P such that $U = P^{-1}AP$ is upper triangular with the eigenvalues of A along the main diagonal. Therefore, using a sequence of equations similar to (7.35), it is seen that $U^T = U$, which implies that U is the desired diagonal matrix. ■

Simultaneous Diagonalization

Let A and B be $n \times n$ matrices. If A and B are diagonalizable, is it possible to use the same matrix P for both matrices? That is, can P be found so that $P^{-1}AP$ is diagonal with the eigenvalues of A along its main diagonal and $P^{-1}BP$ is diagonal with the eigenvalues of B on its main diagonal? If such a matrix P can be found, A and B are **simultaneously diagonalizable**. This can be done under the right condition, and this condition is simple as the following results show.

◼ Lemma 7.4.12

Let A and B be $n \times n$ matrices such that $AB = BA$. If $\mathbf{u} \in ES(A, \lambda)$, then $B\mathbf{u} \in ES(A, \lambda)$.

Proof

Take $\mathbf{u} \in \mathbb{R}^n$ such that $A\mathbf{u} = \lambda\mathbf{u}$. Then, $BA\mathbf{u} = \lambda B\mathbf{u}$, so $AB\mathbf{u} = \lambda B\mathbf{u}$. Hence, $B\mathbf{u}$ is an eigenvector of A with corresponding eigenvalue λ. ◼

◼ Theorem 7.4.13

The $n \times n$ matrices A and B are simultaneously diagonalizable if and only if A and B are diagonalizable and $AB = BA$.

Proof

Let P be an invertible matrix such that $P^{-1}AP$ and $P^{-1}BP$ are diagonal matrices. Because diagonal matrices commute,

$$AB = P(P^{-1}AP)(P^{-1}BP)P^{-1} = P(P^{-1}BP)(P^{-1}AP)P^{-1} = BA.$$

To prove the converse, assume that A and B commute with each other and are diagonalizable. There exists an invertible matrix P so that

$$P^{-1}AP = \begin{bmatrix} D_{\lambda_1} & 0 & \cdots & 0 \\ 0 & D_{\lambda_2} & & 0 \\ \vdots & & \ddots & \vdots \\ 0 & 0 & \cdots & D_{\lambda_k} \end{bmatrix},$$

where $\lambda_1, \lambda_2, \ldots, \lambda_k$ are the eigenvalues of A and D_{λ_i} is the $n_i \times n_i$ diagonal matrix in which each entry on its main diagonal equals λ_i. Let $\mathbf{u}_1, \mathbf{u}_2, \ldots, \mathbf{u}_{n_1}$ form a basis for $ES(A, \lambda_1)$. It is legal to assume that $\mathbf{u}_1, \mathbf{u}_2, \ldots, \mathbf{u}_{n_1}$ are the first n_1 columns of P. Take an integer j such that $1 \le j \le n_1$ and note that $B\mathbf{u}_j \in ES(A, \lambda_1)$ by Lemma 7.4.12 because $AB = BA$. Then, there exists scalars $r_1, r_2, \ldots, r_{n_1}$ such that $B\mathbf{u}_j = r_1\mathbf{u}_1 + r_2\mathbf{u}_2 + \cdots + r_{n_1}\mathbf{u}_{n_1}$. Therefore,

$$\begin{aligned} P^{-1}BP\mathbf{e}_j &= P^{-1}B\mathbf{u}_j \\ &= P^{-1}(r_1\mathbf{u}_1 + r_2\mathbf{u}_2 + \cdots + r_{n_1}\mathbf{u}_{n_1}) \\ &= r_1P^{-1}\mathbf{u}_1 + r_2P^{-1}\mathbf{u}_2 + \cdots + r_{n_1}P^{-1}\mathbf{u}_{n_1} \\ &= r_1\mathbf{e}_1 + r_2\mathbf{e}_2 + \cdots + r_{n_1}\mathbf{e}_{n_1}, \end{aligned}$$

which means that the upper left $n_1 \times n_1$ block of $P^{-1}BP$ possibly has nonzero entries. Call this block B_1. Repeat the argument for the other diagonal blocks of $P^{-1}AP$ to find for each $i = 2, 3, \ldots, k$, an $n_i \times n_i$ block B_i such that

$$P^{-1}BP = \begin{bmatrix} B_1 & 0 & \cdots & 0 \\ 0 & B_2 & & 0 \\ \vdots & & \ddots & \vdots \\ 0 & 0 & \cdots & B_k \end{bmatrix}.$$

Because B is diagonalizable, $P^{-1}BP$ is a diagonalizable (Exercise 6), and this implies that each B_i is diagonalizable. By Theorem 7.4.2 it follows that B_i has a basis $\mathcal{B}_i$ of $\mathbb{R}^{n_i}$ consisting of eigenvectors of B_i. Since D_{λ_i} is an $n_i \times n_i$ diagonal matrix with only λ_i on its main diagonal, the vectors of $\mathcal{B}_i$ are also eigenvectors of D_{λ_i}. Let Q_i be the $n_i \times n_i$ matrix with the vectors of $\mathcal{B}_i$ as its columns, so Q_i diagonalizes both B_i and D_i. Define the invertible block matrix

$$Q = \begin{bmatrix} Q_1 & 0 & \cdots & 0 \\ 0 & Q_2 & & 0 \\ \vdots & & \ddots & \vdots \\ 0 & 0 & \cdots & Q_k \end{bmatrix}.$$

Finally, compute to find that both

$$Q^{-1}P^{-1}APQ = \begin{bmatrix} Q_1^{-1}D_{\lambda_1}Q_1 & 0 & \cdots & 0 \\ 0 & Q_2^{-1}D_{\lambda_2}Q_2 & & 0 \\ \vdots & & \ddots & \vdots \\ 0 & 0 & \cdots & Q_k^{-1}D_{\lambda_k}Q_k \end{bmatrix}$$

and

$$Q^{-1}P^{-1}BPQ = \begin{bmatrix} Q_1^{-1}B_1Q_1 & 0 & \cdots & 0 \\ 0 & Q_2^{-1}B_2Q_2 & & 0 \\ \vdots & & \ddots & \vdots \\ 0 & 0 & \cdots & Q_k^{-1}B_kQ_k \end{bmatrix}$$

are diagonal matrices, so the invertible PQ diagonalizes both A and B. ■

■ **Example 7.4.14**

Let $A = \begin{bmatrix} 0 & 0 & 2 & 0 \\ 0 & 0 & 0 & 2 \\ 2 & 0 & 0 & 0 \\ 0 & 2 & 0 & 0 \end{bmatrix}$ and $B = \begin{bmatrix} 0 & 5 & 0 & 0 \\ 5 & 0 & 0 & 0 \\ 0 & 0 & 0 & 5 \\ 0 & 0 & 5 & 0 \end{bmatrix}$. Observe that

$$AB = BA = \begin{bmatrix} 0 & 0 & 0 & 10 \\ 0 & 0 & 10 & 0 \\ 0 & 10 & 0 & 0 \\ 10 & 0 & 0 & 0 \end{bmatrix}.$$

Also, A has eigenvalues $\lambda = -2$ and $\lambda = 2$ with set of linearly independent eigenvectors,

$$\left\{ \begin{bmatrix} 1 \\ 0 \\ 1 \\ 0 \end{bmatrix}, \begin{bmatrix} 0 \\ 1 \\ 0 \\ 1 \end{bmatrix}, \begin{bmatrix} -1 \\ 0 \\ 1 \\ 0 \end{bmatrix}, \begin{bmatrix} 0 \\ -1 \\ 0 \\ 1 \end{bmatrix} \right\}.$$

The matrix B has eigenvalues $\lambda = -5$ and $\lambda = 5$ with set of linearly independent eigenvectors,

$$\left\{ \begin{bmatrix} 1 \\ 1 \\ 0 \\ 0 \end{bmatrix}, \begin{bmatrix} 0 \\ 0 \\ 1 \\ 1 \end{bmatrix}, \begin{bmatrix} -1 \\ 1 \\ 0 \\ 0 \end{bmatrix}, \begin{bmatrix} 0 \\ 0 \\ -1 \\ 1 \end{bmatrix} \right\}.$$

Therefore, since these sets are bases for $\mathbb{R}^4$, both A and B are (orthogonally) diagonalizable by Theorem 7.4.2. Hence, Theorem 7.4.13 implies that A and B are simultaneously diagonalizable. Let

$$P = \begin{bmatrix} 1 & 0 & -1 & 0 \\ 0 & 1 & 0 & -1 \\ 1 & 0 & 1 & 0 \\ 0 & 1 & 0 & 1 \end{bmatrix}, \quad \text{so } P^{-1} = \begin{bmatrix} 1/2 & 0 & 1/2 & 0 \\ 0 & 1/2 & 0 & 1/2 \\ -1/2 & 0 & 1/2 & 0 \\ 0 & -1/2 & 0 & 1/2 \end{bmatrix}.$$

Next, compute

$$P^{-1}BP = \begin{bmatrix} 0 & 5 & 0 & 0 \\ 5 & 0 & 0 & 0 \\ 0 & 0 & 0 & 5 \\ 0 & 0 & 5 & 0 \end{bmatrix}.$$

The matrix $B_1 = \begin{bmatrix} 0 & 5 \\ 5 & 0 \end{bmatrix}$ is diagonalizable using $Q_1 = \begin{bmatrix} -1 & 1 \\ 1 & 1 \end{bmatrix}$. Because the eigenvalues of B_1 are $\lambda = -5$ and $\lambda = 5$,

$$Q_1^{-1}B_1Q_1 = \begin{bmatrix} -5 & 0 \\ 0 & 5 \end{bmatrix}.$$

Finally, set

$$Q = \begin{bmatrix} -1 & 1 & 0 & 0 \\ 1 & 1 & 0 & 0 \\ 0 & 0 & -1 & 1 \\ 0 & 0 & 1 & 1 \end{bmatrix}$$

and compute

$$PQ = \begin{bmatrix} -1 & 1 & 1 & -1 \\ 1 & 1 & -1 & -1 \\ -1 & 1 & -1 & 1 \\ 1 & 1 & 1 & 1 \end{bmatrix}, \quad \text{so } (PQ)^{-1} = \begin{bmatrix} -1/4 & 1/4 & -1/4 & 1/4 \\ 1/4 & 1/4 & 1/4 & 1/4 \\ 1/4 & -1/4 & -1/4 & 1/4 \\ -1/4 & -1/4 & 1/4 & 1/4 \end{bmatrix}.$$

That PQ is the desired invertible matrix is confirmed by

$$(PQ)^{-1}A(PQ) = \begin{bmatrix} 2 & 0 & 0 & 0 \\ 0 & 2 & 0 & 0 \\ 0 & 0 & -2 & 0 \\ 0 & 0 & 0 & -2 \end{bmatrix}$$

and

$$(PQ)^{-1}B(PQ) = \begin{bmatrix} -5 & 0 & 0 & 0 \\ 0 & 5 & 0 & 0 \\ 0 & 0 & -5 & 0 \\ 0 & 0 & 0 & 5 \end{bmatrix}.$$

Quadratic Forms

Although linear algebra is the study of linear functions, there are instances when the subject can be applied to nonlinear functions to obtain important results. Consider the equations,

$$x^2 + 4y^2 = 16 \tag{7.36}$$

and

$$x^2 + 20xy + 4y^2 = 9. \tag{7.37}$$

The graphs of these equations are conic sections. The graph of (7.36) is immediately recognized as an ellipse, and its graph is easily found because the equation can be written as

$$\frac{x^2}{16} + \frac{y^2}{4} = 1.$$

It is also the case that (7.37) is an ellipse. This should also be apparent, but because of the xy-term, its graph is not readily available.

Linear algebra can help the investigation of (7.37) by finding a basis so that the equation has no xy-term when the equation is written in terms of that basis. To accomplish this, start by writing the second degree expressions from (7.36) and (7.37) in matrix form:

$$\begin{bmatrix} x & y \end{bmatrix} \begin{bmatrix} 1 & 0 \\ 0 & 4 \end{bmatrix} \begin{bmatrix} x \\ y \end{bmatrix} = \begin{bmatrix} x & 4y \end{bmatrix} \begin{bmatrix} x \\ y \end{bmatrix} = x^2 + 4y^2,$$

and

$$\begin{bmatrix} x & y \end{bmatrix} \begin{bmatrix} 1 & 10 \\ 10 & 4 \end{bmatrix} \begin{bmatrix} x \\ y \end{bmatrix} = \begin{bmatrix} x + 10y & 10x + 4y \end{bmatrix} \begin{bmatrix} x \\ y \end{bmatrix} = x^2 + 20xy + 4y^2.$$

This means that the equations of (7.36) and (7.37) can be written as

$$\begin{bmatrix} x & y \end{bmatrix} \begin{bmatrix} 1 & 0 \\ 0 & 4 \end{bmatrix} \begin{bmatrix} x \\ y \end{bmatrix} = 16 \quad \text{and} \quad \begin{bmatrix} x & y \end{bmatrix} \begin{bmatrix} 1 & 10 \\ 10 & 4 \end{bmatrix} \begin{bmatrix} x \\ y \end{bmatrix} = 9.$$

Generalizing this work leads to the next definition.

■ **Definition 7.4.15**

A function $f : \mathbb{R}^n \to \mathbb{R}$ is a **quadratic form** if there exists an $n \times n$ symmetric matrix A such that $f(\mathbf{x}) = \mathbf{x}^T A \mathbf{x}$ for all $\mathbf{x} \in \mathbb{R}^n$.

Consider again the equation in (7.37). Choose $P = \begin{bmatrix} 4 & 0 \\ 2 & 3 \end{bmatrix}$ and write

$$P \begin{bmatrix} u \\ v \end{bmatrix} = \begin{bmatrix} 4u \\ 2u + 3v \end{bmatrix} = \begin{bmatrix} x \\ y \end{bmatrix}. \tag{7.38}$$

Making the substitution into (7.37),

$$(4u)^2 + 20(4u)(2u + 3v) + 4(2u + 3v)^2 = 9$$

gives

$$192u^2 + 288uv + 36v^2 = 9. \tag{7.39}$$

This means that (7.39) represents the same set of values as (7.37) assuming the relationship between $\begin{bmatrix} u \\ v \end{bmatrix}$ and $\begin{bmatrix} x \\ y \end{bmatrix}$ given by the **change of variable** (7.38). The goal that will enable a better understanding of a quadratic equation with a middle term is to find a change of variable so that the middle term will disappear when the change is applied.

■ **Theorem 7.4.16 [Principal Axis Theorem]**

For every $n \times n$ symmetric matrix A with eigenvalues $\lambda_1, \lambda_2, \dots, \lambda_n$, there exists an orthogonal matrix P and diagonal matrix D such that $P^{-1}AP = D$, and with the change of variable $\mathbf{x} = P\mathbf{y}$, where $\mathbf{y}$ has coordinates $y_1, y_2, \dots, y_n$,

$$\mathbf{x}^T A \mathbf{x} = \mathbf{y}^T D \mathbf{y} = \lambda_1 y_1^2 + \lambda_2 y_2^2 + \cdots + \lambda_n y_n^2.$$

Proof

Since A is symmetric, Theorem 7.4.11 implies that there exists an orthogonal matrix P such that $D = P^{-1}AP = P^T AP$ is a diagonal matrix. Let $\mathbf{x} = P\mathbf{y}$ be a change of variable. Then, writing $A = PDP^T$ it follows that

$$\begin{aligned} \mathbf{x}^T A \mathbf{x} &= \mathbf{x}^T (PDP^T)\mathbf{x} \\ &= (\mathbf{x}^T P)D(P^T \mathbf{x}) \\ &= ([P\mathbf{y}]^T P)D(P^T P\mathbf{y}) \\ &= (\mathbf{y}^T P^T P)D(P^T P\mathbf{y}) \\ &= \mathbf{y}^T D \mathbf{y} = \lambda_1 y_1^2 + \lambda_2 y_2^2 + \cdots + \lambda_n y_n^2. \ \blacksquare \end{aligned}$$

■ **Example 7.4.17**

The polynomial $7x^2 + 4xy + 4y^2$ is identified by the quadratic form

$$\begin{bmatrix} x & y \end{bmatrix} A \begin{bmatrix} x \\ y \end{bmatrix}, \quad \text{where} \quad A = \begin{bmatrix} 7 & 2 \\ 2 & 4 \end{bmatrix}.$$

The symmetric matrix A has eigenvalues $\lambda_1 = 8$ and $\lambda_2 = 3$ with corresponding eigenvectors

$$\begin{bmatrix} 2 \\ 1 \end{bmatrix} \quad \text{and} \quad \begin{bmatrix} -1/2 \\ 1 \end{bmatrix}.$$

To form the orthogonal change of basis matrix P, normalize the two eigenvectors and enter the results as columns in P. In other words,

$$P = \begin{bmatrix} 2\sqrt{5}/5 & -\sqrt{5}/5 \\ \sqrt{5}/5 & 2\sqrt{5}/5 \end{bmatrix},$$

which implies that

$$P^{-1}AP = \begin{bmatrix} 8 & 0 \\ 0 & 3 \end{bmatrix}.$$

Therefore, letting $\mathbf{y} = \begin{bmatrix} u \\ v \end{bmatrix}$, the change of variable $\mathbf{x} = P\mathbf{y}$ gives

$$\begin{bmatrix} x & y \end{bmatrix} \begin{bmatrix} 7 & 2 \\ 2 & 4 \end{bmatrix} \begin{bmatrix} x \\ y \end{bmatrix} = \begin{bmatrix} u & v \end{bmatrix} \begin{bmatrix} 8 & 0 \\ 0 & 3 \end{bmatrix} \begin{bmatrix} u \\ v \end{bmatrix} = 8u^2 + 3v^2.$$

As an example, this work implies that with respect to the standard basis, the equation

$$7x^2 + 4xy + 4y^2 = 16, \tag{7.40}$$

describes the same conic section as the equation

$$8u^2 + 3v^2 = 16 \tag{7.41}$$

with respect to the basis

$$\mathcal{B} = \left\{ \begin{bmatrix} 2\sqrt{5}/5 \\ \sqrt{5}/5 \end{bmatrix}, \begin{bmatrix} -\sqrt{5}/5 \\ 2\sqrt{5}/5 \end{bmatrix} \right\}.$$

To see how this works, consider $\mathbf{x} = \begin{bmatrix} 0 \\ 2 \end{bmatrix}$. This vector satisfies (7.40). Because

$$P^{-1} = P^T = \begin{bmatrix} 2\sqrt{5}/5 & \sqrt{5}/5 \\ -\sqrt{5}/5 & 2\sqrt{5}/5 \end{bmatrix},$$

it follows that

$$\mathbf{y} = P^{-1}\mathbf{x} = \begin{bmatrix} 2\sqrt{5}/5 \\ 4\sqrt{5}/5 \end{bmatrix}. \tag{7.42}$$

Three facts should now be emphasized.

- The vectors in the new basis are computed by $\mathbf{b}_1 = P\mathbf{e}_1$ and $\mathbf{b}_2 = P\mathbf{e}_2$, and because P is orthogonal, $\mathbf{b}_1$ is orthogonal to $\mathbf{b}_2$ by Corollary 4.4.19.

- Because $P^{-1} = [T_{\mathscr{B}}]$ is used to find the $\mathscr{B}$-coordinate form of a vector, (7.42) implies that the representation of $\mathbf{x}$ with respect to the new basis is $\mathbf{x}_{\mathscr{B}} = \mathbf{y}$. This is confirmed by

$$\begin{bmatrix} 0 \\ 2 \end{bmatrix} = \frac{2\sqrt{5}}{5}\begin{bmatrix} 2\sqrt{5}/5 \\ \sqrt{5}/5 \end{bmatrix} + \frac{4\sqrt{5}}{5}\begin{bmatrix} -\sqrt{5}/5 \\ 2\sqrt{5}/5 \end{bmatrix}.$$

- The Principal Axis Theorem (7.4.16) guarantees that $\mathbf{y}$ satisfies (7.41). This is confirmed by

$$8\left(\frac{2\sqrt{5}}{5}\right)^2 + 3\left(\frac{4\sqrt{5}}{5}\right)^2 = \frac{32}{5} + \frac{48}{5} = 16.$$

The example is illustrated in Figure 7.1.

Because the matrix P guaranteed to exist by the Principal Axis Theorem (7.4.16) is an orthogonal matrix, $\det P = \pm 1$ by Theorem 4.4.16. In particular, this means that P represents a rotation when $\det P = 1$ or a reflection when $\det P = -1$ (Example 4.4.17). Simply check $\det P$ to see which one it is, but most of the time it will

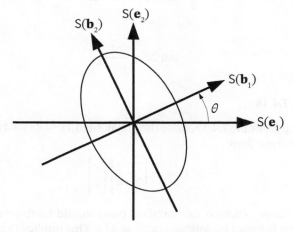

Figure 7.1 The ellipse $7x^2 + 4xy + 4y^2 = 16$ with respect to two bases.

be a rotation. If P is a rotation, there exists θ such that

$$P = \begin{bmatrix} \cos\theta & -\sin\theta \\ \sin\theta & \cos\theta \end{bmatrix}.$$

The measure θ represents the angle formed by e_1 and b_1 and also by e_2 and b_2 as in Figure 7.1. To find the value of θ, write the symmetric matrix A as

$$A = \begin{bmatrix} a & c \\ c & b \end{bmatrix}$$

and compute

$$P^T AP = \begin{bmatrix} a\cos\theta + c\sin\theta & c\cos\theta + b\sin\theta \\ -a\sin\theta + c\cos\theta & -c\sin\theta + b\cos\theta \end{bmatrix} \begin{bmatrix} \cos\theta & -\sin\theta \\ \sin\theta & \cos\theta \end{bmatrix}.$$

Because $P^T AP$ is a diagonal matrix,

$$(a\cos\theta + c\sin\theta)(-\sin\theta) + (c\cos\theta + b\sin\theta)\cos\theta = 0,$$

from which is obtained

$$(b - a)\sin\theta\cos\theta + c(\cos^2\theta - \sin^2\theta) = 0.$$

An appeal to double angle identities gives

$$\frac{b - a}{2}\sin 2\theta + c\cos 2\theta = 0.$$

Hence,

$$\tan 2\theta = \frac{2c}{a - b}.$$

■ **Example 7.4.18**

Continuing Example 7.4.17, since the polynomial $7x^2 + 4xy + 4y^2$ is identified by the quadratic form

$$\begin{bmatrix} x & y \end{bmatrix} \begin{bmatrix} 7 & 2 \\ 2 & 4 \end{bmatrix} \begin{bmatrix} x \\ y \end{bmatrix},$$

the angle through which the standard basis should be rotated to eliminate the xy-term is found by solving $\tan 2\theta = 4/3$. This implies that $\theta \approx 26.57°$.

Exercises

1. Determine whether each matrix A is diagonalizable. If so, find a diagonal matrix to which A is similar. If not, explain why.

(a) $\begin{bmatrix} 5 & 0 \\ 2 & 5 \end{bmatrix}$

(b) $\begin{bmatrix} 3 & -12 \\ 4 & -11 \end{bmatrix}$

(c) $\begin{bmatrix} 2 & 1 & 1 \\ 0 & -2 & 0 \\ 1 & -4 & 2 \end{bmatrix}$

(d) $\begin{bmatrix} 7 & 1 & 3 \\ 0 & 7 & 5 \\ 0 & 0 & 7 \end{bmatrix}$

(e) $\begin{bmatrix} 0 & 0 & 8 \\ 0 & 8 & 0 \\ 8 & 0 & 0 \end{bmatrix}$

(f) $\begin{bmatrix} 2 & 0 & 0 & 0 \\ 4 & -2 & 5 & 9 \\ 7 & 0 & 1 & 2 \\ 0 & 0 & 0 & -2 \end{bmatrix}$

(g) $\begin{bmatrix} 3 & 2 & 0 & 0 \\ 1 & 1 & 0 & 0 \\ 0 & 0 & 2 & 4 \\ 0 & 0 & 3 & 2 \end{bmatrix}$

(h) $\begin{bmatrix} 5 & 0 & 5 & 0 & 5 \\ 0 & 2 & 0 & 2 & 0 \\ 5 & 0 & 5 & 0 & 5 \\ 0 & 2 & 0 & 2 & 0 \\ 5 & 0 & 5 & 0 & 5 \end{bmatrix}$

2. Given the characteristic or minimal polynomial for some 3×3 matrix A, determine whether A is diagonalizable. If it cannot be determined, explain why.

(a) $p_A(x) = (x - 1)^3$

(b) $m_A(x) = (x - 3)(x - 7)(x + 1)$

(c) $p_A(x) = (x + 4)(x + 9)(x + 2)$

(d) $m_A(x) = (x + 2)(x + 4)^2$

3. Let A be a 4×4 matrix with exactly two eigenvalues λ_1 and λ_2. Suppose that $\dim ES(A, \lambda_1) = 1$ and $\dim ES(A, \lambda_2) = 3$. Is A diagonalizable? Explain.

4. Let A be a 3×3 matrix with three distinct eigenvalues. Is A diagonalizable? Explain.

5. Let A be a 7×7 matrix with exactly three eigenvalues, $\lambda_1, \lambda_2,$ and λ_3. Suppose that $\dim ES(A, \lambda_1) = 2$, the algebraic multiplicity of λ_2 is 1, and the geometric multiplicity of λ_3 is 3. Is A diagonalizable? Explain.

6. Prove that if A is diagonalizable and A is similar to B, then B is diagonalizable.

7. Let A be a square matrix. Prove or show false.

(a) If A is invertible, A is diagonalizable.

(b) If A is diagonalizable, A is invertible.

8. Let A be diagonalizable. Prove or show false.

(a) If A is also invertible, A^{-1} is diagonalizable.

(b) A^k is diagonalizable for all positive integers k.

(c) A^T is diagonalizable.

9. Find diagonalizable matrices A and B of the same size so that the given matrices are not diagonalizable.

(a) $A + B$

(b) AB

10. Are the 2×2 and 3×3 rotation and reflection matrices diagonalizable? Explain.

11. Assume that N is a nilpotent matrix that is also diagonalizable. Prove that N is a zero matrix.

12. Prove that if A is a diagonalizable $n \times n$ matrix with only 1 eigenvalue, $A = rI_n$ for some $r \in \mathbb{R}$.

13. Let A and B be diagonal matrices of the same size. Prove that A is similar to B if and only if the entries on the main diagonal of A are the entries on the main diagonal of B, possibly in a different order.

14. Let A be an $n \times n$ diagonalizable matrix with eigenvalues $\lambda_1, \lambda_2, \ldots, \lambda_n \in \mathbb{R}$, with possible repeats. Prove.

    (a) $\det A = \lambda_1 \lambda_2 \cdots \lambda_n$. (b) $\operatorname{tr} A = \lambda_1 + \lambda_2 + \cdots + \lambda_n$.

15. Determine whether each matrix A is orthogonally diagonalizable. If so, find a diagonal matrix to which A is similar. If not, explain why.

    (a) $\begin{bmatrix} 1 & 1 & 0 \\ 1 & 2 & 0 \\ 0 & 0 & 3 \end{bmatrix}$ (b) $\begin{bmatrix} 1 & 1 & 0 & 2 \\ 1 & 1 & 0 & 0 \\ 0 & 0 & 2 & 0 \\ 2 & 0 & 0 & 1 \end{bmatrix}$ (c) $\begin{bmatrix} 0 & 1 & 0 & 1 \\ 4 & 2 & 3 & 0 \\ 0 & 0 & 2 & 1 \\ 1 & 0 & 4 & 0 \end{bmatrix}$

16. Let A be orthogonally diagonalizable. Prove or show false.

    (a) If A is also invertible, A^{-1} is orthogonally diagonalizable.

    (b) A^k is orthogonally diagonalizable for all positive integers k.

    (c) A^T is orthogonally diagonalizable.

17. If the given matrices A and B are simultaneously diagonalizable, find an invertible matrix P such that $P^{-1}AP$ and $= P^{-1}BP$ are diagonal matrices.

    (a) $A = \begin{bmatrix} 4 & 2 \\ 2 & 4 \end{bmatrix}$, $B = \begin{bmatrix} 1 & 3 \\ 3 & 1 \end{bmatrix}$

    (b) $A = \begin{bmatrix} 0 & 1 & 0 \\ 1 & 0 & 1 \\ 0 & 1 & 0 \end{bmatrix}$, $B = \begin{bmatrix} 1 & 0 & 1 \\ 0 & 0 & 0 \\ 1 & 0 & 1 \end{bmatrix}$

    (c) $A = \begin{bmatrix} 0 & 0 & 1 & 0 \\ 0 & 0 & 0 & 1 \\ 1 & 0 & 0 & 0 \\ 0 & 1 & 0 & 0 \end{bmatrix}$, $B = \begin{bmatrix} 0 & 6 & 0 & 0 \\ 6 & 0 & 0 & 0 \\ 0 & 0 & 0 & 6 \\ 0 & 0 & 6 & 0 \end{bmatrix}$

18. Find the quadratic form $\mathbf{x}^T A \mathbf{x}$ that represents each polynomial.

    (a) $5x^2 + 6xy + 9y^2$ (b) $x^2 - 8xy - 3y^2$ (c) $9x^2 - 2y^2$

19. The **standard form** equation of an ellipse centered at the origin with horizontal major axis is $x^2/a^2 + y^2/b^2 = 1$ $(a > b > 0)$. If $b > a > 0$, the major axis is vertical. The **standard form** equation of a hyperbola centered at the origin with horizontal transverse axis is $x^2/a^2 - y^2/b^2 = 1$. The x and y variables exchange position if the transverse axis is vertical. For each equation, find the following:

    - P such that the change in variable $P \begin{bmatrix} u \\ v \end{bmatrix} = \begin{bmatrix} x \\ y \end{bmatrix}$ results in an equation without a middle term.

- The standard form equation with respect to the basis $\mathscr{B} = \{P^{-1}\mathbf{e}_1, P^{-1}\mathbf{e}_2\}$.
- The measure θ of the angle formed by $\mathbf{e}_1$ and $P^{-1}\mathbf{e}_1$.
- $\mathbf{x}_{\mathscr{B}}$, where the coordinates of $\mathbf{x}$ satisfy the original equation.

(a) $x^2 - 4xy + y^2 = 4, \ \mathbf{x} = \begin{bmatrix} 2 \\ 0 \end{bmatrix}$

(b) $3x^2 + 4xy + 6y^2 = 16, \ \mathbf{x} = \begin{bmatrix} 0 \\ 2\sqrt{6}/3 \end{bmatrix}$

(c) $2x^2 + 6xy - 6y^2 = 25, \ \mathbf{x} = \begin{bmatrix} 5\sqrt{2}/2 \\ 0 \end{bmatrix}$

(d) $2x^2 - 4xy - y^2 = 9, \ \mathbf{x} = \begin{bmatrix} 2 \\ \sqrt{15} - 4 \end{bmatrix}$

20. Suppose that $\det P = -1$ for the orthogonal matrix P given by the Principal Axis Theorem (7.4.16). What does this say about the original quadratic form?

21. Let A and B be symmetric matrices of the same size. Prove that if A and B have only positive eigenvalues, $A + B$ has only positive eigenvalues. Can the same be said about negative eigenvalues?

22. The quadratic form $\mathbf{x}^T A \mathbf{x}$ is **positive definite** means that $\mathbf{x}^T A \mathbf{x} > 0$ for all $\mathbf{x} \neq \mathbf{0}$. The quadratic form $\mathbf{x}^T A \mathbf{x}$ is **negative definite** means that $\mathbf{x}^T A \mathbf{x} < 0$ for all $\mathbf{x} \neq \mathbf{0}$. Prove.

(a) $\mathbf{x}^T A \mathbf{x}$ is positive definite if and only if all eigenvalues of A are positive.

(b) $\mathbf{x}^T A \mathbf{x}$ is negative definite if and only if all eigenvalues of A are negative.

23. Let $f(x, y) = [x \ y] A \begin{bmatrix} x \\ y \end{bmatrix}$ be a quadratic form. What does the graph of f look like if the quadratic form is positive definite? What does it look like if it is negative definite?

Further Reading

Andrews-Larson, Christine. 2015. "Roots of Linear Algebra: An Historical Exploration of Linear Systems." *PRIMUS* 25 (6): 506–528.

Curtis, Charles W. 1984. *Linear Algebra: An Introductory Approach*. Undergraduate Texts in Mathematics. New York: Springer-Verlag.

Enderton, Herbert B. 1977. *Elements of Set Theory*. San Diego, CA: Academic Press.

Fearnley-Sander, Desmond. 1979. "Hermann Grassmann and the Creation of Linear Algebra." *The American Mathematical Monthly* 10: 809–817.

Fraleigh, John B. 1999. *A First Course in Abstract Algebra*. 7th ed. Reading, MA: Addison-Wesley.

Halmos, Paul R. 1960. *Naive Set Theory*. New York: Springer-Verlag.

Jacobson, Nathan. 1985. *Basic Algebra I*. 2nd ed. New York: W. H. Freeman.

Lax, Peter D. 2007. *Linear Algebra*. 2nd ed. Pure and Applied Mathematics. Hoboken, NJ: John Wiley & Sons.

Meyer, Carl. 2000. *Matrix Analysis and Applied Linear Algebra*. Philadelphia, PA: SIAM.

Steele, J. Michael. 2004. *The Cauchy-Schwarz Master Class*. Cambridge, UK: Cambridge University Press.

Linear Algebra, First Edition. Michael L. O'Leary
© 2021 John Wiley & Sons, Inc. Published 2021 by John Wiley & Sons, Inc
Companion Website: www.wiley.com/go/o'leary/linearalgebra

Index

Printed and bound by CPI Group (UK) Ltd, Croydon, CR0 4YY

16/04/2025